Scientific Research and Methodology

This textbook is designed for teaching quantitative research in the scientific, health and engineering disciplines at first-year undergraduate level, with an emphasis on statistics. It covers the research process, including asking research questions, research design, data collection, summarising data, analysis and communication. Many real journal articles are used throughout the text as examples that demonstrate the use of the techniques.

Students are introduced to statistics as a method for answering questions. Descriptive research questions lead to analysis of single proportions and means. Repeated-measures research questions are answered using paired quantitative data. Relational research questions compare proportions, odds and means in different groups. Correlational research questions are studied using correlation and regression techniques.

Statistical topics include numerical summary methods (such as means, odds ratios and identification of outliers), graphing (such as histograms, case-profile plots and scatterplots), confidence intervals and hypothesis testing. Emphasis is placed on understanding and concepts; while calculations are shown in simple situations, they are deferred to software when the computations become tedious and disruptive to understanding.

Almost every dataset used is a real dataset and is available online or in an associated R package SRMData. Software output is often used when calculations become onerous. The output is sufficiently generic that the book can be used in conjunction with any statistical software.

Peter K. Dunn is Associate Professor of Biostatistics in the School of Science, Technology and Engineering at the University of the Sunshine Coast. He has published in the areas of generalized linear models, Tweedie distributions and statistical education, and has authored numerous R packages. He is an award-winning statistical educator who has been teaching statistics for over 30 years.

Scientific Research and Methodology

An Introduction to Quantitative Research and Statistics

Peter K. Dunn

CRC Press
Taylor & Francis Group
Boca Raton London New York

CRC Press is an imprint of the
Taylor & Francis Group, an **informa** business
A CHAPMAN & HALL BOOK

Designed cover image: Shutterstock_2478201895

First edition published 2026
by CRC Press
2385 NW Executive Center Drive, Suite 320, Boca Raton FL 33431

and by CRC Press
4 Park Square, Milton Park, Abingdon, Oxon, OX14 4RN

CRC Press is an imprint of Taylor & Francis Group, LLC

ISBN: 978-1-032-49672-6 (hbk)
ISBN: 978-1-032-49673-3 (pbk)
ISBN: 978-1-003-39493-8 (ebk)

DOI: 10.1201/9781003394938

Typeset in Latin Modern font
by KnowledgeWorks Global Ltd.

Publisher's note: This book has been prepared from camera-ready copy provided by the authors.

To a random sample of seven people,
drawn from the scores of people who have contributed to the production of this book
and whose contributions I greatly appreciate.

Contents

Preface

This book introduces quantitative research in the scientific and health disciplines, with an emphasis on introductory statistics. Unlike many introductory statistics textbooks, this textbook gives context to the statistics by first covering the basics of the research design process; it connects the research question with the means to answer that question. I believe this is crucial to understanding the need and purpose of using statistics. The research process is broken into six steps, which provide the framework for the content.

The book is designed for teaching at first-year undergraduate level, with examples mostly drawn from science, health and engineering. Many real journal articles are used throughout the text in examples, to demonstrate the use of the techniques. Almost every dataset used in this book is real and available in the **R** package **SRMData** (see App. A).

The main focus of the book is the analysis of data, with an emphasis on understanding the underlying concepts rather than a focus on using mathematics. Software output is often used to help when calculations become onerous. The output is from jamovi [The jamovi Project, 2022], but is sufficiently generic that no knowledge of jamovi is necessary to use this book, and this book can be read without relying on any specific statistical software. (jamovi, however, is *free* to download and use.)

The following call-outs are used in this book:

> These chunks introduce the objectives for the chapters of the book.

> ⚠ These chunks highlight common mistakes or warnings, about a particular concept or about using a formula.

> ⓘ These chunks offer helpful information.

> 💻 These chunks refer to information about using software or a calculator.

> 🎤 These chunks indicate how certain symbols and terms are pronounced.

> ✔ These end-of-chapter chunks provide answers to the end-of-chapter *Quick review questions.*

This book was made using **R** [R Core Team, 2018] with the **bookdown** package [Xie, 2016], using **Markdown** syntax and **knitr** [Xie, 2015] and numerous other **R** packages. All of this software is *free* and open source. Other resources used include:

- various icons from **iconmonstr** (freely available).
- the images of the cards (e.g., in Sect. 25.1), which are in the public domain and available from https://code.google.com/archive/p/vector-playing-cards/.

Earlier drafts of this textbook have been used to teach thousands of students, and the book has been used by many fantastic teaching assistants. I thank all of them for their feedback. Special thanks to Dr Amanda Shaker (La Trobe University), who reported numerous issues in earlier editions (and often provided corrections).

Learning Outcomes

In this book, you will learn to:

- develop quantitative research questions and testable hypotheses.
- design quantitative studies to answer simple quantitative research questions.
- select and produce appropriate graphical, numerical and statistical analyses.
- select, apply and interpret the results of the correct statistical technique to analyse data.
- comprehend, apply and communicate in the language of research and statistics.
- demonstrate professional integrity in planning, interpreting and reporting the results of quantitative studies.

1

Research: an introduction

In this chapter, you will learn to:

- identify quantitative and qualitative research.
- identify the steps in the quantitative research process.

1.1 Introduction: how we know what we know

Scientists once believed that all life regularly and commonly arose spontaneously from non-living matter. *Recipes* even existed; for example, van Helmont [1671] gave this recipe for making a mouse [Pasteur, 1922]:

> If a soiled shirt is placed in the opening of a vessel containing grains of wheat, the reaction of the leaven in the shirt with fumes from the wheat will, after approximately twenty-one days, transform the wheat into mice.

This was called 'spontaneous generation' (or 'abiogenesis'). This theory is clearly incorrect, so how did the idea emerge? How was it disproven? Through *observation* and *research*.

Spontaneous generation was consistent with *observations*: following the above recipe *did* produce mice. However, the hypothesis ('possible explanation') of spontaneous generation was rejected when later evidence, in better-designed research studies, contradicted the hypothesis. A new hypothesis was proposed to explain the appearance of the mice, which was tested against the evidence, and so on. Briefly, this is the *evidence-based, scientific process*.

As a more recent example, the dangers of smoking were still being debated into the 1990s:

> . . . a causal role for smoking [has] not been proved beyond reasonable doubt.
>
> — Eysenck [1991], p. 429

All scientific knowledge emerges in a similar way: observations lead to questions and hypotheses, which are tested against *evidence*. If the evidence *contradicts* the hypothesis, the hypothesis is rejected. If the evidence is *consistent* with the hypothesis, the hypothesis is *temporarily* accepted (until any contradictory evidence emerges, if ever).

Hypotheses not contradicted by large amounts of evidence, over a long time, are sometimes called *laws* or *theories* (such as the 'Law of conservation of energy'). Theories and laws can be disproven if contradictory evidence emerges. Knowledge in all scientific disciplines is accumulated using a similar evidence-based process.

1.2 Evidence-based research

Every discipline changes, develops, improves, and adapts—usually through *research*. Your discipline is not the same as it was ten years ago; it will change in the next ten years. Scientists, engineers and health practitioners need to know how to understand and adapt to this change.

Remaining current in your discipline requires understanding research, even if you will not be researching yourself. You still need to know the language, tools, concepts and ideas of research, and you need to be able to critique research. Research is the foundation of science.

Science seeks *evidence-based answers*: reaching conclusions based on *evidence*, rather than hunches, feelings, intuition, hopes, or tradition. The *evidence* comes from analysing *data*.

Definition 1.1 (Data). *Data* refers to information (observations or measurements), such as numbers, labels, recordings, videos, text, etc.

A *dataset* refers to an organised and structured collection of data.

Research involves asking research questions, designing studies to collect data, analysing data, and accurately reporting the results. This book covers all these components.

1.3 Quantitative, qualitative, and mixed-methods research

Research can be broadly classified as *qualitative* or *quantitative*. These are different yet complementary approaches to answering research questions (Table 1.1). Both methods have advantages and disadvantages, and can be used together (called *mixed-methods* research). The decision to use qualitative, quantitative or mixed-methods approaches depends on the purpose of the research.

TABLE 1.1: Concisely comparing qualitative and quantitative research.

Qualitative	Aspect	Quantitative
Feelings, opinions	**What**	Measured or observed data
Suggest hypotheses, explore, depth	**Why**	Make objective conclusions
Non-random samples	**Studied**	Representative samples; random samples
Very detailed; for specific groups	**Conclusions**	General
Words, audio, video, pictures, ...	**Data**	Numbers, measurements, counts, ...
Usually small samples are studied	**Size**	Often large samples are studied
Often time-consuming	**Time**	Usually more efficient
Rarely generalisable	**Applicability**	Often generalisable
Thematic analysis, content analysis, etc.	**Analysis**	Numerical summaries, test hypotheses, etc.
Interviews, focus groups	**Examples**	Experiments, closed surveys, lab. studies

Briefly, *qualitative research* leads to a deeper understanding, usually for a very specific group. Meanings, motivations, opinions or themes often emerge from qualitative research. In contrast, *quantitative research* summarises and analyses data usually from large groups, using *numerical* methods, such as averages and percentages. In quantitative research, typically information about a large group of interest (a *population*) is found from a subset of the population (a *sample*).

Definition 1.2 (Quantitative research). *Quantitative research* summarises and analyses data using numerical methods, such as averages, proportions and percentages.

⚠ This book is about *quantitative* research.

Example 1.1 (Types of research). Oliveira et al. [2020] used mixed-methods research to study the adoption of electric taxis in Nottingham (UK).

In the *quantitative* component of the study, taxis were tracked for over 32 000 km and 9 764 h. The researchers determined where taxis often stopped (as potential charging locations). The Trent Street taxi rank (near the main train station) was the most-used stop with an average wait time of 20 mins; Milton St (17 mins) and Wheeler gate (10 mins) also recorded high average stop-times. Numerical information about speeds was also obtained.

In the *qualitative* component of the study, nine taxis drivers participated in interviews and focus groups. This allowed the researchers to (p. 6)

> ... explore themes and motivations in a way that was not possible with the initial quantitative analysis.

Participants' responses were classified as *barriers* (safety; costs; speeds) or *facilitators* (opportunities to charge regularly; convenient locations) to the proposed charging locations.

Example 1.2 (Quantitative research). During 1988/1989, an unusually high number of *Legionella longbeachae* infections were observed in South Australians. The researchers [O'Connor et al., 2007] wanted to identify the source to prevent further infections.

The researchers noticed that many of those infected were gardeners who had recently handled potting mix, so they hypothesised that the infection was associated with using potting mix. They designed a study to test this hypothesis, then collected data from 100 people (25 *with* the infection, and 75 people of similar age and sex *without* the infection).

The researchers classified and summarised the data, then analysed the data to reach an evidence-based conclusion: potting mix was partially, but not solely, responsible for the infections. The researchers communicated their recommendations to reduce the risks of people contracting the infection in the future.

1.4 The steps in research

The research process (for both qualitative and quantitative research) ideally follows the process summarised in Fig. 1.1, but this is not always possible or practical. The process is not always linear: researchers may jump from step to step as necessary, and research often leads to new research questions (RQs) so that the process restarts. Each step is important.

- *Asking* the RQ (Chap. 2). Research begins with a research question to answer.
- *Designing* the study (Chaps. 3 to 8). Evidence-based research uses data to answer the RQ. A study is designed to obtain that data: determining who or what to study; finding

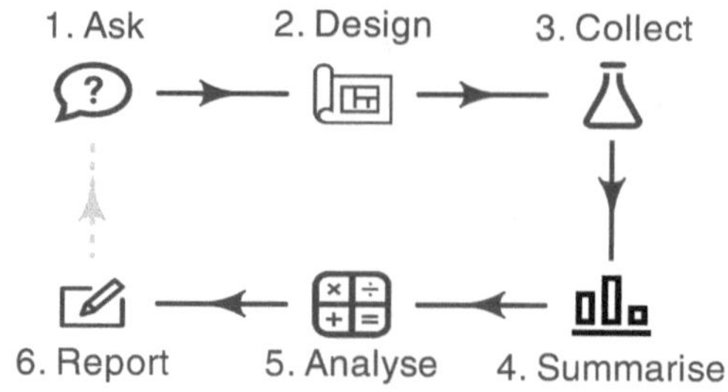

FIGURE 1.1: The six basic steps in research.

those to study; deciding what information to obtain; and ensuring data are obtained ethically.

- *Collecting* the data (Chap. 9). The data collection process must be ethical, reproducible and clearly documented.
- *Classifying* and *summarising* the data (Chaps. 10 to 17). Before analysis, the data must be classified and summarised to inform the analysis. (A computer is useful.)
- *Analysing* the data (Chaps. 18 to 34). Analysis refers to using the data to find an answer to the research question. (A computer is useful.)
- *Reporting* the results (Chaps. 35 and 36). Communicating the results appropriately, accurately and ethically is important, including identifying any limitations of the research.

1.5 Using computers in research

Statistical software (such as Python, jamovi, R, SAS, SPSS, Stata, etc.) is useful for summarising and analysing data. Statistical software:

- is designed for working with large datasets.
- encourages reproducible research (Sect. 5.3).
- allows high-precision formatting and graphics.
- is powerful; with some programming skills, almost anything is possible.
- is specifically designed for analysing and working with data.

> This book sometimes shows output from jamovi [The jamovi Project, 2022], but using jamovi is *not* essential for understanding this book.

Using spreadsheets for storing and analysing data requires care. Expensive and dangerous errors have been made due to using spreadsheets [AlTarawneh and Thorne, 2016]. Some challenges with using spreadsheets include that:

- spreadsheets may *change data* (e.g., reformatting entries as dates) when not appropriate [Ziemann et al., 2016].
- spreadsheets may include *formulas with errors* that are difficult to locate and hence fix [Panko, 2015, London and Slagter, 2021].
- spreadsheets *do not leave a record* of how the data were analysed or prepared. Keeping a record of the analysis, preparation of variables, and other operations with the data is good scientific practice (*reproducible research*; see Sect. 5.3) [Simons and Holmes, 2019].
- spreadsheets often produce poor graphs [Su, 2008].

Problems with spreadsheets, as with any software, are often due to human error, but *spreadsheets make errors hard to find and hence hard to fix.* Spreadsheets are useful for data collection and basic data manipulation, but are not designed for scientific analysis. Be careful using spreadsheets for research and analysis.

1.6 Exercises

Answers to odd-numbered exercises are given at the end of the book.

Exercise 1.1. Consider this RQ: 'For three different junctional tourniquets, which is the quickest, on average, to apply?'

1. What data would be likely be needed to answer this research question?
2. Is this RQ likely to be answered using a *quantitative* or *qualitative* research study? Explain.

Exercise 1.2. Consider this RQ: 'Why do people dump rubbish in mangroves?'

1. What data would be likely be needed to answer this research question?
2. Is this RQ likely to be answered using a *quantitative* or *qualitative* research study? Explain.

Exercise 1.3. Consider this RQ: 'What percentage of Egyptians experience side effects from a specific medication?'

1. What data would be likely be needed to answer this research question?
2. Is this RQ likely to be answered using a *quantitative* or *qualitative* research study? Explain.

Exercise 1.4. Consider this RQ: 'What is the average number of rooftop solar panels on domestic homes in a certain city?'

1. What data would be likely be needed to answer this research question?
2. Is this RQ likely to be answered using a *quantitative* or *qualitative* research study? Explain.

Exercise 1.5. Frost and Murtagh [2023] conducted a study to better understand the views of adults regarding planting greenery in front gardens. To do so, they conducted (p. 80):

> ... five online focus groups with 20 participants aged 20–64 in England... [then] audio recorded each focus group, transcribed it verbatim and analysed transcripts using thematic analysis.

What type of research study is this: qualitative, quantitative or mixed-methods?

Exercise 1.6. Häußermann et al. [2023] studied Germany's transition to green energy, and (p. 1):

> ... investigated social acceptance of green hydrogen at an early stage in its implementation, before wider rollout.

To do so, they used (p. 1):

> ... semi-structured interviews ($n = 24$) and two participatory workshops ($n = 51$) in a selected region in central Germany serve alongside a representative survey ($n = 2\,054$)...

What type of research study is this: qualitative, quantitative or mixed-methods?

Part I

Asking research questions

2

Research questions

In this chapter, you will learn to:

- identify and write quantitative research questions.
- identify the variables implied by a quantitative research question.
- identify and distinguish observational and experimental studies.
- identify and distinguish the units of analysis and units of observations in a study.
- write operational and conceptual definitions.

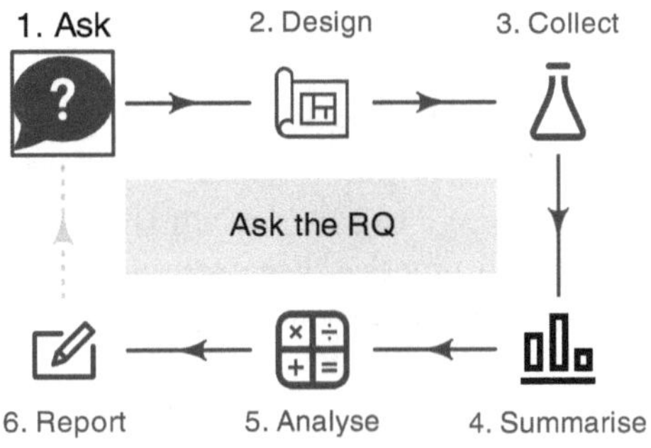

2.1 Introduction

The research question (RQ) directs all other components of the research. Since quantitative research summarises and analyses data using numerical methods (like averages or percentages), the RQ must be *written* carefully so it can be *answered* effectively. Four different types of RQs are studied:

- descriptive RQs (Sect. 2.2).
- relational RQs (Sect. 2.3).
- repeated-measures RQs (Sect. 2.4).
- correlational RQs (Sect. 2.6).

Since the RQ directs all other components of the research, writing RQs should be the first step of any research study. Specifically, RQs should be asked before data are collected.

⚠ RQs should be written *before data are collected.*

2.2 Descriptive RQs

All RQs identify a large group of interest to be studied (called a *population*), and study something *about* that population (called the *outcome*).

The population is any broad group of interest; for example:

- all German males between 18 and 35 years of age.

- all bamboo flooring materials manufactured in China.
- all elderly females with glaucoma in Canada.
- all *Pinguicula grandiflora* growing in Europe.

Definition 2.1 (Population). A *population* is a group of *individuals* from which the total set of observations of interest *could* be made, and to which the results will generalise.

Populations comprise many *individuals* (or *cases*). If the individuals are people, individuals may also be called *subjects*.

⚠ The words *population*, *individuals* and *cases* do *not* just refer to people, though they may be commonly used that way in general conversation.

Data are rarely taken from all the individuals in the population: *all* individuals are rarely accessible in practice. For example, testing a new drug cannot possibly study *all* people who might use the drug (some may not even be born yet). In contrast, a *sample* is a *subset* of the population from which data are obtained (Chap. 6). Countless samples are possible from any given population, but only one is studied.

Definition 2.2 (Sample). A *sample* is a subset of individuals from the population. The data are collected from the sample.

⚠ The *population* in an RQ is *not* just those studied; it is the whole group to which results could generalise.

Example 2.1 (Samples). A study of American college women [Woolf et al., 2009] compared iron status in highly-active and sedentary women.

The study compared 28 active and 28 sedentary American college women, from which data were collected. The *population* was *all* active and sedentary American college women. The group of 56 subjects was the *sample*.

Descriptive RQs study something *about* the identified population, called the *outcome*. Because the RQ concerns a large group (the population), the outcome numerically describes a *group* of individuals (not single individuals). The outcome is, for example, an *average* or *proportion* summarising a group of individuals.

Definition 2.3 (Outcome). The *outcome* in an RQ is the result, output, consequence or effect of interest in a study, numerically summarised for a group of individuals.

The outcome of interest in a population may be (for example) the

- *average* amount of wear after 1 000 h of use.
- *proportion* of people whose pupils dilate.
- *average* weight loss after three weeks on a diet.
- *percentage* of seedlings that die.

⚠ The *outcome* in an RQ summarises a *population*; it does not describe the *individuals* in the population.

Descriptive RQs can now be introduced.

Definition 2.4 (Descriptive RQ). *Descriptive RQs* have a population and an outcome.

Some RQs ask about the *value* of some population quantity (such as: what is the average internal body temperature?); these are called *estimation* RQs. Some RQs require *making a decision* about the population (such as: is the average internal body temperature the same for females and males?); these are called *decision-making* RQs. Descriptive RQs have one of these forms, depending on what information is sought (Sect. 2.8):

- *estimation* RQs: Among {*the population*}, what is {*the outcome*}?
- *decision-making* RQs: Among {*the population*}, is {*the outcome*} equal to {*a given value*}?

⚠ These templates are *not* 'recipes', but guidelines.

Answering *estimation* descriptive RQs is studied in Chaps. 22 and 23. Answering *decision-making* descriptive RQs is studied in Chaps. 26 and 27.

Example 2.2 (Descriptive RQ). Mackowiak et al. [1992] studied men and women aged 18 to 40; this is the *population*. The *outcome* of interest in this population is the *average body temperature*. The sample comprised 148 'healthy men and women' aged 18 to 40. One descriptive RQ was:

> What is the average body temperature?

This is an *estimation* RQ. They also studied a *decision-making* descriptive RQ (where 98.6°F (or 37.0°C) is a commonly-accepted value for the internal body temperature):

> Is the average body temperature really 98.6°F (37.0°C)?

2.3 Relational RQs

Studying relationships usually is more interesting than simply describing a population. *Relational RQs* compare the outcome for groups of different individuals in the population, or compare two different sub-populations. These comparisons are called *between-individuals* comparisons, as they compare the outcome *between* (or among) groups of *different* individuals. Examples include:

- comparing the average amount of wear in floorboards *between* two different groups: standard wooden floorboards, and bamboo floorboards.
- comparing the average heart rate *across* three groups of people: those not receiving the drug, those receiving a weekly dose, and those receiving a daily dose of the drug.

Definition 2.5 (Comparison (between individuals)). The *between-individuals comparison* in an RQ identifies the small number of groups of different individuals for which the outcome is compared.

Example 2.3 (Between-individuals comparison). Williams et al. [2022] compared the average weight of female and male Leadbeater's possums. 'Sex of the possum' is the *between-individuals* comparison; average weight is the outcome.

Relational RQs can now be introduced.

Definition 2.6 (Relational RQ). *Relational RQs* have a population, outcome, and a *between*-individuals comparison.

Relational RQs have one of these forms, depending on what information is sought:

- *estimation* RQ: Among {*the population*}, what is the difference in {*the outcome*} for {*the groups being compared*}?
- *decision-making* RQ: Among {*the population*}, is {*the outcome*} the same for {*the groups being compared*}?

Example 2.4 (Relational RQ). Consider this RQ (based on Estévez-Báez et al. [2019]):

> Among Cubans between 13 and 20 years of age, is the average heart rate the same for females and males?

The *population* is 'Cubans between 13 and 20 years of age', the *outcome* is '*average* heart rate', and the *between-individuals comparison* is between two separate groups: 'between females and males'. This is a *relational RQ*.

This is a *decision-making RQ*, since it asks if the average heart rate is the same for females and males. An *estimation*-type relational RQ would ask about the *size* of difference in the average heart rate between females and males.

2.4 Repeated-measures RQs

Rather than comparing the outcome for groups of different individuals, *repeated-measures RQs* compare the outcome multiple times within the *same* individuals.

These comparisons are called *within-individuals* comparisons, as they compare the outcome *within the same individuals*, not across groups of *different* individuals. The multiple measurements may be different points in time (e.g., the height of the same trees at one, two and five years after planting), but do not have to be time points.

Examples include:

- comparing the average strength of hind legs of horses to the forelegs of the same horses.
- comparing the average thickness of the cornea in left eyes and right eyes of the same individuals.
- comparing the average amount of wear in many individual floorboards after one, five and ten years of use.

Definition 2.7 (Within-individuals comparison). The *within-individuals comparison* in the RQ identifies the small number of different, distinct situations for which the outcome is compared for each individual.

Example 2.5 (Between- and within-individual comparisons). Consider comparing the strength of the dominant and non-dominant legs of professional football players.

A *between*-individuals comparison would compare the average strengths of the dominant and non-dominant legs *between different* groups of footballers: one group would have their dominant-leg strength measured, and the other would have their non-dominant-leg strength measured. This is a *between*-individuals comparison.

In contrast, the strengths of the dominant and non-dominant legs could be recorded on the *same* individuals. This study examines *within*-individuals changes: the average differences between the strengths of the dominant and non-dominant legs *within* the same individuals. In this study, *no between-individuals comparison* exists: different groups are not being compared.

Studies may use *both* within- and between-individuals comparisons (see Sect. 30.8). For instance, a study may examine the *change* in individuals' heart rate (the *within*-individuals comparison), for two drugs given to different groups (the *between*-groups comparison).

Repeated-measures RQs can now be introduced.

Definition 2.8 (Repeated-measures RQ). *Repeated-measures RQs* have a population, outcome and a *within*-individuals comparison.

Repeated-measures RQs have one of these forms, depending on what information is sought:

- *estimation* RQ: Among {*the population*}, what is the change in {*the outcome*} for {*the alternatives being compared within individuals*}?
- *decision-making* RQ: Among {*the population*}, is {*there a change in the outcome*} for {*the alternatives being compared within individuals*}?

Example 2.6 (Repeated-measure RQ). Rowland et al. [2017] compared the temperature in the *same* tree hollows in summer and winter:

> For tree hollows in the Strathbogie Ranges, Australia, what is the average temperature difference between summer and winter?

The comparison is *within individuals*, as the temperature is measured for the *same* tree hollows at the two times. This is a repeated-measures, estimation-type RQ.

Repeated-measures RQs with only two within-individual comparisons are often called *paired.*

Example 2.7 (Paired repeated-measures study). Levitsky et al. [2004] compared the weights of the same university students at the beginning of university, and then after 12 weeks. The comparison is *within* individuals, and the study is a *repeated-measures* study. Since each student has a *pair* of weight measurements, this is a *paired* study.

2.5 Variables

RQs are about *populations*. However, the data to answer an RQ come from *individuals* in that population. The aspects or characteristics that can *vary* called *variables*.

Definition 2.9 (Variable). A *variable* is a single aspect or characteristic, associated with the individuals, whose values can vary.

Example 2.8 (Variables). Examples of variables include: the duration of cold symptoms; sex; tree girth; response to a survey question (Yes, Maybe, No); city of birth; hair colour.

Some variables change from one individual to another individual, such as sex and height. These are called *between*-individuals variables. In repeated-measures studies, some variables of interest change over repeated measurements from the same individuals; these are called *within*-individuals variables.

Definition 2.10 (Between- and within-individuals variables). *Between*-individuals variables vary from one individual to another individual. *Within*-individuals variables vary from one recording or measurement to another *within* the same individuals.

⚠ A between-individuals variable is a single aspect that can vary from *individual to individual.* While *your* city of birth does not change, 'city of birth' is a variable because it varies from *individual* to *individual.*

Example 2.9 (Within-individuals variables). Rowland et al. [2017] compared the temperature in the *same* tree hollows in summer and winter (Example 2.6). The comparison is *within individuals*: the temperature is measured for the *same* tree hollows (the *individuals*) at two different times.

'Season' is a within-individuals variable, as each tree hollow is studied for two different seasons. 'Temperature' is also a within-individuals variable, as it is measured twice for each tree hollow.

Example 2.10 (Between-individuals comparison). Williams et al. [2022] compared the average weight of female and male Leadbeater's possums (Example 2.3).

'Sex of the possum' is a *between-individuals* variable; it can vary from possum to possum. 'Weight' is also a *between-individuals* variable; it can vary from possum to possum.

Example 2.11 (Variables). 'Duration of cold symptoms' is a between-individuals *variable*: its value can vary from individual to individual. The '*average* duration of cold symptoms' is the *outcome*, a numerical summary of many individuals' cold durations.

While many variables can be recorded, two essential variables are (Table 2.1):

- the *response variable*, which records information to determine the outcome.
- the *explanatory variable*, which records information to determine the comparison.

Usually, one variable can be considered as perhaps influencing the value of the other variable. This variable is called the *explanatory variable* (which may *explain* changes in the other variable). The other is the *response variable* (whose values *respond* to changes in the explanatory variable). To be able to influence the response variable, the explanatory variable must occur before (or at the same time) as the response variable.

TABLE 2.1: The relationship between the population and the individuals.

Population		**Individuals**
Outcome	→	Response variable
Comparison	→	Explanatory variable

The value of the *response* variable may change in *response* to the value of the explanatory variable. The value of the *explanatory* variable may *explain* changes in the value of the response variable.

Definition 2.11 (Explanatory variable). An *explanatory variable* may (partially) explain or be associated with changes in another variable of interest (the response variable).

Definition 2.12 (Response variable). A *response variable* records the result, output, consequence or effect of interest from changes in another variable (the explanatory variable).

⚠ The *response variable* is sometimes called the *dependent variable*, and the *explanatory variable* is sometimes called the *independent variable*. We avoid these terms, since the words 'dependent' and 'independent' have many meanings in research.

The RQ cannot be answered without data for the response and explanatory variables. The *outcome* is a numerical summary of the values of the response variable (Table 2.2) recorded from many individuals. The values of the explanatory variable distinguish between the values of the *comparison* for the individuals (Tables 2.3 and 2.4) being made.

TABLE 2.2: Outcomes and corresponding response variable.

Outcome describing the population		**Response variable in individuals**
Average diastolic blood pressure	→	Diastolic blood pressure of *individuals*
Percentage of seedlings that sprout	→	Whether an *individual* seedling sprouts
Proportion owning iPad	→	Whether an *individual* owns an iPad
Average cold duration	→	Cold duration for *individuals*

TABLE 2.3: *Between-individuals* comparisons and the corresponding *between-individuals* explanatory variable.

Comparison being made		Explanatory variable in individuals
Between jarrah, beech, bamboo boards	→	Type of floorboard in *different* individual homes
Between 3 kg/ha, 4 kg/ha fertiliser rates	→	Application rate in *different* individual paddocks
Between people in 20s, 30s and 40s	→	Age group for each *different* individual person

TABLE 2.4: *Within-individuals* comparison and corresponding *within-individuals* explanatory variable.

Comparison being made		Explanatory variable in individuals
Before and after receiving a drug	→	When measured on *each* individual person
Between left and right arms	→	Which arm in *each* individual person is used
Between forelegs and hind legs	→	Which legs are measured on *each* individual horse

Example 2.12 (Variables). Consider a study of the ground surface temperature of public playgrounds in Boston in summer.

The *population* comprises all public playgrounds in Boston; each public playground is an *individual*. The *outcome* is the *average* ground surface temperature in summer over many playgrounds; the *response variable* is the ground surface temperature for *individual* ground surfaces in summer.

The between-individuals *comparison* is between the four types of ground surfaces (rubber, soil, sand, mulch). The *explanatory variable* is the type of surface for individual playgrounds.

2.6 Correlational RQs

Correlational RQs are not concerned with summarising outcomes in comparison *groups*. Instead, correlational RQs explore relationships between two variables measured or observed on or about the individuals.

Definition 2.13 (Correlational RQ). *Correlational RQs* explore the relationship between two variables.

Correlational RQs have one of these forms, depending on what information is sought:

- *estimation* RQ: Among {*the population*}, how strong is the relationship between {*the response variable*} and {*the explanatory variable*}?
- *decision-making* RQ: Among {*the population*}, is {*the response variable*} related to {*the explanatory variable*}?

Examples include studying the relationship between:

- the height of plants (response variable) and the number of hours of sunlight per day (explanatory variable).
- heart rate (response variable) and the number of grams of caffeine consumed that day (explanatory variable).

Usually, one variable can be considered as the explanatory variable, and the other as the response variable (Sect. 2.5). To be able to influence the response variable, the explanatory variable must occur before (or at the same time) as the response variable. Explanatory and response variables may be either within- or between-individuals variables.

Example 2.13 (Correlational RQ). Consider studying marathon runners. An RQ exploring the relationship between the individuals' water intake on the day before the race and the individuals' race times would be a correlational RQ. The water intake on the day before the race *may* influence the race time.

The water intake on the day before the race is the explanatory variable, and the race time is the response variable.

Example 2.14 (Correlational RQ). The Wollemi pine was discovered by science in 1994. Offord and Zimmer [2023] studied the growth of these rare plants.

One correlational RQ concerned the relationship between the diameter of trees at breast height (DBH; response variable), and the pH of the soil (explanatory variable). The two variables are the DBH and pH, both recorded for many trees.

Also studied was the relationship between the DBH for each tree at various times after the planting date (a repeated-measure RQ). Each tree has the DBH measured over time, for many time points. Time is the *within*-individuals comparison.

In some situations, the variables are neither response nor explanatory variable; the interest is just in the association between the two variables.

Example 2.15 (Correlation RQ). González-Acosta et al. [2024] recorded the length and weight of 14 040 fish for 39 demersal fish species. The study has two variables (fish length; fish weight), but identifying a response variable and explanatory variable is meaningless. The estimation-type correlational RQ is:

> Among demersal fish, how strong is the relationship between length and weight?

2.7 Interventions

Sometimes, the explanatory variable naturally occurs without manipulation by the researchers (e.g., the height of people; the sex of oxen; the pH of forest soil). Sometimes, however, the explanatory variable is manipulated by researchers (e.g., the dose of fertiliser applied; the dose of drug given); this is called an *intervention.*

Definition 2.14 (Intervention). An *intervention* is present when *researchers* can manipulate (or impose) the values of the *explanatory variable* on the individuals to determine the impact on the response variable.

When an intervention is present, the values of the explanatory variable are *manipulated* by the researchers, and are called *treatments.* When an intervention is *not* present, the values of the explanatory variable are *not* manipulated by the researchers, and are called *conditions.* The *analysis* is the same whether an intervention is used or not, but the *interpretation* of the results depend on whether an intervention is used (Sect. 4.5).

Definition 2.15 (Treatments). The *treatments* are the values of the explanatory variable that the researchers can manipulate and impose upon the individuals.

Definition 2.16 (Condition). The *conditions* are the values of the explanatory variable that those in the study have or experience, but are not manipulated or imposed by the researchers.

An intervention is present when the researchers:

- explicitly give a dose of a new drug to patients.
- explicitly apply wear-testing loads to two different flooring materials.
- explicitly expose people to different stimuli.
- explicitly apply different doses of fertiliser.

Example 2.16 (Intervention). Bird et al. [2008] *supplied* one group of participants with a diet using refined flour, and *supplied* another group of participants with a diet using a new flour variety. 'Type of diet' is the (between-individuals) explanatory variable. Since the researchers manipulate which subjects ate which flour, this study has an intervention. 'Type of diet' is the treatment.

Example 2.17 (No intervention). To compare the average blood pressure in female and male Scots, blood pressure was measured using a blood pressure machine (a sphygmomanometer). The researchers interact with the participants to measure blood pressure, but there is *no* intervention. Using the sphygmomanometer is just a way to measure blood pressure, to *obtain* the data.

The *comparison* is between females and males (the conditions), which cannot be manipulated or imposed on the individuals by the researchers; *there is no intervention.*

Often, one of the comparison groups is the *control group*. The *control group* is a comparison group *not* receiving the treatment being studied, or *not* having the condition being studied, but *as similar as possible* to the other individuals in all other ways. The control group is like a benchmark for detecting changes in the outcome due to the treatment or condition of interest (Sect. 7.5). Sometimes the control group is given a *placebo*: a non-effective treatment that appears to be the real treatment.

Definition 2.17 (Control). A *control* is an individual without the treatment or condition of interest, but as similar as possible in *every other way* to other individuals. A *control group* is a group of controls.

Definition 2.18 (Placebo). A *placebo* is a treatment with no intended effect or active ingredient, but appears to be the real treatment.

Example 2.18 (Control group). To test the effectiveness of a new medication, patients report to a doctor to receive injections of the new drug. Patients assigned to the *control group* do not receive the drug. The controls should also report to a doctor and receive an injection (like those receiving the drug); the injection, however, would contain no active ingredients (a placebo).

Together, the **P**opulation, **O**utcome, **C**omparison and **I**ntervention form the POCI acronym (sometimes written as PICO) to aid remembering the elements of RQs. The POCI acronym is not helpful for correlational RQs.

Example 2.19 (POCI). Woolf et al. [2009] measured iron status in highly-active and sedentary American college women.

The *outcome* is the 'average iron status'. The between-individuals *comparison* is between highly-active and sedentary women. For this comparison to be an intervention, the *researchers* would need to tell each individual woman to be highly active or sedentary. This seems unlikely, so the study does not have an intervention.

2.8 Estimation and decision-making RQs

As noted earlier, RQs can be written with one of two purposes. *Estimation RQs* ask how precisely an unknown *value* in the *population* is estimated by the *sample*. Estimation RQs are answered using *confidence intervals*, which are discussed in Chaps. 22 to 23, Chaps. 29 to 31, plus Sects. 33.2.2 and 33.4.3.

Decision-making RQs require a decision to be made about the unknown values in the population. They are answered using *hypothesis tests*, and discussed in Chaps. 26 to 27, Chaps. 29 to 31, plus Sects. 33.2.2 and 33.4.3.

Example 2.20 (Decision-making RQs). Thane et al. [2004] studied 'British young people aged 4–18' and asked numerous RQs. One *decision-making* relational RQ was:

> In British young people aged 4–18, is the average daily zinc intake the same for boys and girls?

Decision-making RQ have two possible answers. For the example above, the average zinc intake either *is* the same for boys and girls, or *is not* the same for boys and girls, in the *population* (Fig. 2.1). These two options are *hypotheses*: potential answers to the RQ. However, answers are rarely clear in practice, since only one of the countless possible samples from the population is studied. Instead, researchers decide *how strongly* the sample evidence supports a particular hypothesis about the *population*.

Evidence may *support* or *contradict* a hypothesis; evidence rarely *proves* a hypothesis (at least, without any other support, such as theoretical support). Ultimately, after collecting data from a *sample*, a decision must be made about which explanation about the *population* is more consistent with the data collected.

Decision-making RQs can be asked in different ways. For the zinc-intake study above (Fig. 2.1), the RQ could ask (about the population):

- is the average zinc intake *the same* for boys and girls?
- is the average zinc intake *different* for boys and girls?
- is the average zinc intake *lower* for boys, compared to girls?
- is the average zinc intake *higher* for boys, compared to girls?

The first two are *two-tailed RQs* (and are essentially asking the same question but in different ways): the average zinc intake could be higher for girls or higher for boys. We are just

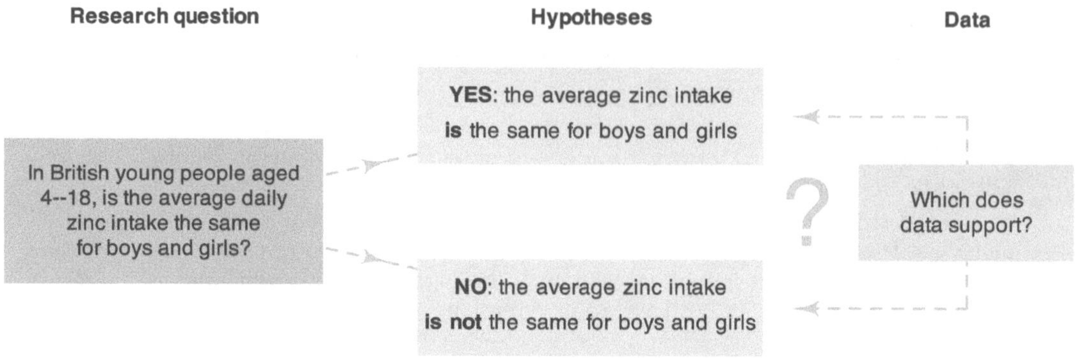

FIGURE 2.1: Two possible answers to the RQ (two hypotheses) about zinc intake in children.

interested in whether any difference is present; that is, two options are being considered. The last two are *one-tailed RQs*, since they ask specifically about a difference in just one direction: boys lower than girls, or boys higher than girls.

Most RQs are two-tailed, unless a good reason exists to ask a one-tailed RQ *before* the data are collected (e.g., a drug has been developed specifically to *reduce* blood pressure). RQs should be formed before the data are collected.

⚠ In general, RQs should be two-tailed RQs, unless a justifiable reason exists for asking a one-tailed question *before data are collected.*

2.9 Units of observation and analysis

Units of observation and *units of analysis* are different yet similar concepts that must be distinguished to properly identify a population.

Consider this descriptive RQ:

> In English 20-something men, what is the average thickness of head-hair strands?

To answer this question, the thickness of individual hair strands needs to be measured. The 'things' from or about which measurements are taken are called *units of observation.*

> **Definition 2.19** (Unit of observation). The *unit of observation* is the entity that is observed, from or about which measurements are taken and data collected.

For this RQ, the unit of observation is the hair strand: the thickness measurements are taken from the hair strands. Suppose the thickness of 100 hair strands is recorded. These 100 hair strands could be obtained in many different ways. Two options are to:

- take 100 hair strands, all from the same man.
- take one hair strand from each of 100 different men.

While each approach gives 100 measurements, these two approaches are *very* different. Only one man is represented in the first scenario, so every hair strand is likely to be similar.

However, 100 different men are represented in the second. The difference is related to the concept of *unit of analysis*.

The purpose of the study is to make conclusions about 'men': the RQ is asking about 'men'. Each different man provides a separate, independent measurement of hair strand thickness. The 'man' is the unit of analysis; each man provides a unique example of a hair strand.

The first scenario above has one unit of analysis (which provided all 100 units of observation). The second scenario has 100 units of analysis (each providing one unit of observation).

Identifying units of analysis takes care. The units of analysis:

- can be single units of observation, or *collections* of units of observations (as in the hair-strand example).
- are usually determined by the RQ: what is being compared or studied?
- must be distinct, and separate to, each other (or nearly so).

Definition 2.20 (Unit of analysis)**.** The *unit of analysis* is the smallest collection of units of observations (and perhaps the units of observations themselves) about which conclusions are made; the smallest distinct elements of the population for which information is analysed.

ⓘ Sometimes the *units of analysis* and *units of observation* are the same.

In the hair-strand study, all the hair strands from the same man have essentially 'lived their life together': they are all washed together with the same shampoo, exposed to the same amount of sunlight and exercise, share the same genetics, etc. However, different men potentially use different shampoo, exercise differently, have different genetics, and so on. The hair of different men tends to exhibit distinct characteristics. Each man is a collection of units of observations (hair strands).

Definition 2.21 (Sample size)**.** The sample size n is the number of units of analysis.

Example 2.21 (Units of analysis, observation)**.** To compare the average amount of fibre in wholemeal and white bread, researchers take ten slices from one loaf of wholemeal bread, and ten slices from one loaf of white bread. The amount of fibre (in grams) in each slice is determined. The units of *observation* are the 'slices': the type of bread (explanatory variable) and the amount of fibre (response variable) are observed on individual slices.

The unit of *analysis* is the 'loaf' (a collection of slices), because the RQ is comparing types of *bread*, and the slices for each type of bread are all from the same loaf. Slices from the same loaf share the same baker and bakery; they were made with the same ingredients, in the same oven, baked at the same temperature, etc. The sample size is $n = 2$ loaves.

Example 2.22 (Units of analysis, observation)**.** The *Spectrum* website reported a study where researchers examined '10 neurons from each of the 16 mice' (November 2022). The researchers treated each neuron as an independent observation, so $n = 16 \times 10 = 160$.

However, neurons in the brain of the same animal are *not* independent observations. The unit of analysis is the mouse; the unit of observation is the neuron. The actual sample size was $n = 16$; each unit of analysis has 10 units of observation. A total of 160 neurons from 16 mice is very different to a study of 160 neurons from 160 genetically-different mice.

The units of observation and units of analysis *may* be the same, and often are the same. However, they are sometimes different, and identifying these situations is *crucial*. Importantly, studies compare units of analysis, not units of observation.

Example 2.23 (Units of analysis, observation). Suppose researchers record the diastolic blood pressure (DBP) from 15 patients aged under 40 years of age, and 15 different patients aged 40 or older. The DBP is measured on every patients' right arm, so there are 15 observations for the 'Under 40' group, and 15 observations for the '40 and over' group.

Provided the patients are not closely related, the patients are independent of each other. (If all 15 observations were all from the same family, for example, this would not be true.) The 'patient' is the unit of analysis *and* the unit of observation.

Later, the researchers decide to take measurements from the left *and* right arms of every patient. Thus, there are now 30 observations for the 'Under 40' group, and 30 observations for the '40 and over' group. However, the left and right arm measurements for each person are likely to be very similar. The 'patient' is the unit of analysis, and each patient provides two observations (one from each arm).

In both cases, the sample size is $n = 30$: both have 30 units of analysis.

Example 2.24 (Units of analysis). A study compared two physical activity (PA) programs. Each of 44 children in the study, chosen from schools across the region, was allocated to one of two PA programs (with parental agreement). The children's fitness was measured for every student at the end of the six-month study.

The *units of observation* are the students: fitness measurements are taken from each student. The *units of analysis* are also the students: students using the different programs are being compared. In addition, the PA program was *allocated* to each student individually, and each student has their own family routines and activities, etc. and lives separate, distinct lives. Each unit of analysis (student) has one unit of observation.

The study has 44 units of analysis, each with one unit of observation.

Example 2.25 (Units of analysis). Consider comparing the percentage of females and males wearing hats at a specific beach.

People in *groups* at the beach will probably not be independent: people in groups tend to behave similarly. For example, a couple will often (but not always) *both* be wearing or not wearing hats; friends often behave in similar ways.

Hence, the researchers may decide to use data from individual people, and not groups ('person' is the unit of analysis *and* unit of observation). Alternatively, the researchers may decide to use people *groups* as the *unit of analysis* (some will be groups of one), and record data from just *one* person in any group (e.g., the person closest to the researchers when the group is noticed).

2.10 Definitions

Research studies usually include terms that must be carefully and precisely defined, so that others know *exactly* what words and terms mean, without ambiguity. Two types of definitions can be given when necessary.

Definition 2.22 (Conceptual definition). A *conceptual definition* articulates precisely *what* words or phrases mean in a study.

Definition 2.23 (Operational definition). An *operational definition* articulates exactly *how* something will be identified, measured, observed or assessed.

In many cases, a clear *operational definition* is needed to describe how data will be collected to ensure repeatability and consistent data collection, by removing any ambiguity about how data are obtained.

Example 2.26 (Operational and conceptual definitions). Consider a study examining stress in students. A *conceptual definition* would describe *what is meant* by 'stress' (in contrast to, say, 'anxiety').

An *operational definition* would describe *how* 'stress' is *measured*, since stress cannot be measured directly (like height, for example). 'Stress' could be *measured* using a questionnaire or measuring physical characteristics, for instance. Other ways of measuring stress are also possible, and all have advantages and disadvantages.

Sometimes the definitions themselves are not important; a clear definition is simply needed. To avoid confusion, commonly-accepted definitions should be used unless good reasons exist for using a different definition. When a commonly-accepted definition does not exist, the definition being used should be very clearly articulated, and the reason given if necessary.

Example 2.27 (Operational and conceptual definitions). A research article [Gillet et al., 2018] entitled 'Shoulder range of motion and strength in young competitive tennis players with and without history of shoulder problems' provided these necessary conceptual definitions (among others):

- 'young': 8–15 years of age.
- 'competitive tennis players': the best players in their age category in France, and members of a French tennis centre of excellence.

An operational definition was provided for 'Shoulder strength': as measured using a hand-held dynamometer.

Example 2.28 (Operational and conceptual definitions). Consider a study requiring water temperature to be measured.

An *operational definition* would explain *how* the temperature is measured: the thermometer type, how the thermometer was positioned, how long was it left in the water; and so on.

A *conceptual* definition would describe the scientific definition of temperature, and would not be needed (as 'temperature' is a well-understood term).

2.11 Example: writing a RQ

Suppose you notice some people taking echinacea (a herb) after they get a common cold. You may wonder: does taking echinacea help in any way with a cold? You may ask:

Is it better to take echinacea when you have a cold?

This RQ is clearly poor, but is a starting point. This RQ can be refined by clarifying the POCI elements. For example, what *population* is of interest? Many options exist: all residents of your country, or just adults in a specific part of your country. Some of these may not be practical (i.e., when a sample cannot easily be obtained that represents the population).

What *outcome* could be used to determine echinacea's effectiveness? Options include the *average* cold duration, or the *percentage* of people who take days off work due to the cold.

The initial RQ is also vague: better than *what*? The outcome could be *compared* between groups (between those taking echinacea and the controls (those who do not)). A within-individuals comparison seems unsuitable for this RQ.

The study could also have *intervention* or not, which has implications for how the study is conducted and how the results are interpreted. If the study *did not have an intervention*, the subjects would decide for themselves how to treat their cold. If the study *did have an intervention*, the use of echinacea would be imposed by the researchers.

Many terms need defining, too. What is meant by 'echinacea' (fresh? tablet form? as a tea?); 'cold' (self-diagnosed? diagnosed by a doctor?), and so on.

Based on the above, this RQ could be considered (based on Barrett et al. [2010]):

Among Australian teenagers with a common cold, is the average duration of cold symptoms shorter for teens given a daily dose of echinacea, compared to teens taking no echinacea?

2.12 Preparing software

Statistical software packages are used to store data for subsequent analyses. Datasets that *do not* contain any within-individuals variables are organised so that:

- each *row* represents one unit of analysis.
- each *column* represents one between-individuals variable.

An additional column of identifying information may also appear, such as the person's name, or concrete batch number.

In statistical software, the variable *names* are not placed in a row (say, in Row 1, above the data itself), which might happen when using a spreadsheet. The *names* of the variables are the names of the columns.

Example 2.29 (Preparing statistical software)**.** In Sect. 2.11, an RQ was asked about whether using echinacea reduced the duration of the common cold.

For this RQ, the two between-individuals *variables* are 'Duration of cold symptoms' (response variable), and 'Type of treatment' (explanatory variable). The person is the unit of analysis, so the number of *rows* in the data worksheet is the sample size. The data worksheet needs at least two columns (Fig. 2.2):

- one for duration of each individual's cold symptoms.
- one for whether the individual received a dose of echinacea or received no medication.

An additional column may record the name or ID of each individual, and more columns may record other within-individuals variables (such as age and height of the individuals).

	Name	Duration	Treatment	Age
1	Mary	6	Echinacea	13
2	Rupesh	4	None	17
3	Samuel	5	None	18
4	Vidush	5	Echinacea	13
5	Jessica	4	Echinacea	14
6	Cooper-Jay	3	Echinacea	15
7	Bishal	6	Echinacea	17
8	Keenan	6	Echinacea	19

FIGURE 2.2: Software prepared for data with no within-individuals variable. Each row represents an individual; each column represents a between-individuals variable.

Datasets that *do* contain within-individuals variables can be organised in *wide* or *long* format. Some analyses are easier using wide format, and some using long format.

In *wide* format:

- each *row* represents one unit of analysis.
- each between-individuals variable is represented in a column.
- each within-individuals variable is represented in *multiple* columns, one for each measurement of that variable on the individuals.

In *long* format:

- each unit of analysis is represented by *multiple* rows.
- each between-individuals variable is represented in a column, and the data repeated in each row corresponding to that unit of analysis.
- each within-individuals variables is represented by one column.

Example 2.30 (Long and wide data formats)**.** Example 2.7 discussed a study where the weights of university students were recorded in both Weeks 1 and 12.

In *wide* format, each *row* represents one individual (Fig. 2.3, left panel). In *long* format, each individual is represented by multiple rows (Fig. 2.3, right panel).

2.13 Chapter summary

In this chapter, you have learnt to write *research questions* for quantitative analysis. All research questions (RQs) study a *population* (P). Descriptive RQs study some *outcome* (O) in the population. Relational RQs *compare* the outcome between different groups of individuals (a between-individuals comparison). Repeated-measures RQs compare the *same* outcome when measured on the same individuals multiple times (a within-individuals comparison). Some RQs also have an *intervention* (I): when the values of the comparison can

	Student	WtWk1	WtWk12	Age
1	Michael	77.0	75.6	19
2	Erin	49.5	50.0	22
3	Pravash	60.3	61.2	18
4	Tenzin	51.8	53.6	21
5	Maya	67.5	69.8	22
6	Donald	46.8	47.7	20
7	Takoda	63.9	66.6	21
8	Vraj	54.0	55.8	19

	Student	When	Weight	Age
1	Michael	1	77.0	19
2	Michael	12	75.6	19
3	Erin	1	49.5	22
4	Erin	12	50.0	22
5	Pravash	1	60.3	18
6	Pravash	12	61.2	18
7	Tenzin	1	51.8	21
8	Tenzin	12	53.6	21

FIGURE 2.3: Software prepared for data with a within-individuals variable; the same data is shown in both panels. Left: in *wide* format, with one individual per row. Right: in *long* format, with multiple rows per individual. Both include a column of identifying information.

be manipulated by the researchers. Correlational RQs ask about the relationship between variables. RQs may be *decision-making* RQs (one- or two-tailed) or *estimation* RQs.

Data comes from a sample of *individuals* in the population. The *outcome* is a numerical summary of the values of the response variable from many individuals. Similarly, the data concerning the comparison comes from measuring or observing the values of the *explanatory* variables from individuals.

The *who* or *what* that observations are made from are called the *units of observation.* The smallest independent collections of units of observations (that is, independent examples of the population) are called the *units of analysis.*

2.14 Quick review questions

Consider this RQ:

> In elite female netball players, do players in defence positions have the same average number of knee injuries (per player, per season) compared to players in attacking positions?

Are the following statements *true* or *false*?

1. The *comparison* is 'between knee injuries and other types of injuries'.
2. The *comparison* in this RQ is a *between*-individuals comparison.
3. The *outcome* is 'the average number of knee injuries per player, per season'.
4. The *response variable* is 'the average number of knee injuries per season'.
5. The *unit of analysis* is 'the number of knee injuries'.
6. The *unit of observation* is 'the elite netball player'.
7. This RQ is a descriptive RQ.
8. This RQ is an estimation-type RQ.

2.15 Exercises

Answers to odd-numbered exercises are given at the end of the book.

Exercise 2.1. For the following *response* variables, what are the corresponding *outcomes*?

1. Whether a vehicle crashes or not.
2. The height people can jump.
3. The number of tomatoes per plant.

Exercise 2.2. For the following *response* variables, what are the corresponding *outcomes*?

1. Whether a person owns a car.
2. The time it takes for seedlings to sprout.
3. The amount of caffeine in cola drinks.

Exercise 2.3. For the following *comparisons*, what are the corresponding *explanatory* variables?

1. Between vegans and non-vegans.
2. Between caffeinated and decaffeinated coffee.
3. Between taking zero, one or two 7 mg iron tablets per day.

Exercise 2.4. For the following *comparisons*, what are the corresponding *explanatory* variables?

1. Between frozen vegetables and fresh vegetables.
2. Between 91-octane, 95-octane, and ethanol-blended car fuel.
3. Between large cities and small cities.

Exercise 2.5. For the following studies, determine whether the study is likely to use a *between*-individuals comparison or a *within*-individuals comparison. In each case, identify the outcome.

1. A study to determine if a higher percentage of people at a particular city park wear hats in summer compared to winter.
2. A study to determine if the average yield of a specific variety of tomato plants is the same when three different fertilisers are applied.

Exercise 2.6. For the following studies, determine whether the study is likely to use a *between*-individuals comparison or a *within*-individuals comparison. In each case, identify the outcome.

1. A study to determine if the average balance time on right legs is the same as on left legs.
2. A study to determine if average cholesterol levels are the same when measured on the same people before and after a diet change.

Exercise 2.7. A study of Phu Quoc Ridgeback dogs (*Canis familiaris*) explored the relationship between body length and body height [Quan et al., 2017].

1. What type of RQ would be asked about the dogs?
2. What are the response and explanatory variables?

Exercise 2.8. Pinet et al. [2022] recorded typing speed and age for 1 301 students.

1. What type of RQ could be asked in this study?
2. What are the response and explanatory variables?

Exercise 2.9. Consider this RQ:

> Among Danish university students, is the average resting diastolic blood pressure the same for students who regularly drive to university and those who regularly ride bicycles to university?

1. For this RQ, identify the population, outcome, and comparison (if any).
2. For this RQ, is there an intervention? Explain.
3. What *type* of question is used (descriptive; relational; repeated measures; correlational)?
4. What is the *purpose* of the RQ: estimation or decision-making?
5. What *operational* and *conceptual definitions* would be needed?
6. What information *must* be collected from each individual to answer the RQ (i.e., the variables)?
7. Identify the units of analysis and the units of observation.

Exercise 2.10. Checkley et al. [2002] (p. 210) conducted:

> a 4-year (1995–1998) field study in a Peruvian peri-urban community... to examine the relation between diarrhea and nutritional status in 230 children < 3 years of age

For this study:

1. identify P, O, C and I (where relevant).
2. infer the primary research question.
3. what *type* of question is used (descriptive; relational; repeated measures; correlational)?
4. what is the *purpose* of the RQ: estimation or decision-making?
5. what *operational definitions* would be needed?
6. what are the *response* and *explanatory* variables?
7. what are the units of observation and units of analysis?

Exercise 2.11. Consider this RQ: 'Is the average walking speed the same when texting and talking on a mobile phone?'

1. What *type* of question is used (descriptive; relational; repeated measures; correlational)?
2. Is this RQ one- or two-tailed?
3. Is there an intervention? Explain.
4. What is the *explanatory* variable?
5. What is the *response* variable?
6. What is the *outcome*?
7. What are the units of observation and units of analysis?

Exercise 2.12. Consider this RQ, with an intervention:

> For Japanese adults with a common cold, do people who take vitamin C tablets daily have, on average, a shorter cold duration than people who do not take any vitamin C tablets?

1. Identify the population, comparison and outcome.
2. What is the response variable?
3. What is the explanatory variable?
4. What type of RQ is this: estimation or decision-making?
5. Is the RQ one-tailed or two-tailed?

Exercise 2.13. Animals in an experiment are divided into pens (three animals per pen), and feed is allocated to each pen [Sterndale et al., 2017]. Animals in different pens receive different feed; animals in the same pen receive the same feed. The weight gain of each animal is recorded.

1. What is the *unit of observation*? Why?
2. What is the *unit of analysis*? Why?
3. Identify the between-individuals comparison.

Exercise 2.14. A research study was comparing the average length of Blue Gum eucalypt leaves in two areas of Queensland. A student takes 40 leaves from each of ten trees in Area A, and 40 leaves from each of ten trees in Area B.

Are the following statements *true* or *false*?

1. The unit of analysis is the individual leaf.
2. The unit of observation is the individual leaf.
3. The unit of analysis is the tree.

What is the size of the sample in the study?

Exercise 2.15. Consider this actual student RQ from the university where I work.

> Among 10 Australian adults, does the time taken to read a passage of text change when different fonts are used?

Critique the RQ, and write a better RQ (if necessary).

Exercise 2.16. Consider this actual student RQ from the university where I work.

> Of students that study at (a University), do males have a larger lung capacity than females?

Critique the RQ, and write a better RQ (if necessary).

Exercise 2.17. Prinz and Murray [2023] examined the strength needed to pull out nose-hairs. Fifty nose-hairs were pulled from one author's nose, and 50 nose hairs pulled from the other author's nose, and the average pull-out strengths for each man compared.

1. What are the units of analysis and units of observation?
2. What is the sample size in this study?

Exercise 2.18. Huang et al. [2020] placed different people into one of three different virtual-reality (VR) environments: trees, grass or concrete. Stress levels were measured using 'skin conductance level' (SCL) for each individual, before and after exposure to the VR environment.

1. Identify the *between*-individuals comparisons.
2. Identify the *within*-individuals comparisons.
3. Is their definition for SCL (p. 2) *conceptual* or *operational*?

> SCLs are an unbiased measure of sympathetic activity via the electric impulses on the skin's surface and sweat glands, which are innervated only by the sympathetic nervous system...

Exercise 2.19. Consider this two-tailed RQ (based on Tudor-Locke et al. [2015]):

> For American adults, is the average number of recorded steps per day the same when recorded using both a waist accelerometer, and a wrist accelerometer?

1. Identify the population and the individuals.
2. Identify the outcome.
3. Identify the response and explanatory variables.
4. Determine if the comparison is *between*- or *within*-individuals.

Exercise 2.20. Studies can incorporate many types of RQs. For example, Thane et al. [2004] studied 'British young people aged 4–18' and answered numerous RQs, including:

- what is the average zinc intake of the children?
- does the average zinc intake meet recommended dietary guidelines?
- what is the strength of the association between plasma zinc and retinol concentrations?
- is the average zinc intake the same for boys and girls?

For each RQ, classify these RQs as descriptive, relational, repeated-measures, or correlational RQs. Then, classify them as estimation or decision-making RQs. Does the study have an invention?

Exercise 2.21. Stern et al. [2021] studied the relationship between daily sodium excretion and whether people had been diagnosed with diabetes or not, in Israeli adults. The study also explored the strength of the relationship between the daily sodium excretion and the systolic blood pressure.

Classify the two RQs as descriptive, relational, repeated-measures, or correlational RQs. Then, classify them as estimation or decision-making RQs. Does the study have an invention?

Exercise 2.22. Ghasemi and Pirzadeh [2019] studied the incidence of musculoskeletal disorders in Iranian bus drivers. They introduced a program that aimed to provide relief for the drivers. Each bus driver was evaluated both before and after the intervention.

Classify the RQ as descriptive, relational, repeated-measures, or correlational RQs. Then, classify the RQ as estimation or decision-making RQs. Does the study have an invention?

Exercise 2.23. To determine the average length of the legs of emus, 27 emus from various zoos were studied. For each emu, the length of the left and right leg were recorded, resulting in 54 measurements.

What is the sample size for this study? Explain.

Exercise 2.24. A study compared the percentage of females and males that wear closed-in shoes to the supermarket. For each person they observed, the type of shoe on each person's left and right foot (as either closed-in; not closed-in) was recorded. This approach resulted in 310 observations.

What is the sample size for this study? Explain.

Exercise 2.25. A study compares the wear on two brands of car tyres. Four tyres of Brand A are allocated to each of Cars 1–5, and four tyres of Brand B are allocated to each of Cars 6–10. After 12 months, the amount of wear is recorded on each tyre, and the two brands compared.

What are the units of analysis, the units of observation and the sample size?

Exercise 2.26. Parsons et al. [2018] discuss a scenario where six subjects with colorectal cancer underwent therapy. Another six similar subjects did not receive the therapy. The size of all the subjects' removed lymph nodes were then measured. Each subject's specimen (p. 6):

> was divided into two sub-samples after collection [...] processed and analysed at two occasions, by different members of the laboratory team [...] Three slices per sub-sample were collected for each subject.

How many units of analysis and the units of observation are present?

Exercise 2.27. Bamboo is a fast-growing, strong grass often used for green building practices. A small research study explored the hardness of bamboo when used as flooring material.

The *Janka hardness*[1] of bamboo flooring provided by Bamboo Flooring Australia Pty Ltd was measured by the Queensland Department of Primary Industries [Gerber, 2004]. Five floorboards were taken, and two hardness measurements were taken on *each* board (units not given, but probably kilonewtons; Table 2.5).

1. What is the unit of analysis: the test, the board, each measurement, kilonewtons, or something else? Explain your answer.
2. How many units of analysis are there?
3. How many units of observation are there?
4. Comment on the amount of variation *between* the boards compared to the amount of variation *within* boards.
5. Suppose the measurements were taken from 10 *different* places on the *same* board (rather than from five different boards). How many units of analysis are there now? Explain your answer.

TABLE 2.5: Two Janka hardness measurements from five different bamboo boards.

Board 1	Board 2	Board 3	Board 4	Board 5
10.5	8.0	11.5	10.3	10.2
7.5	8.0	11.2	9.9	9.3

Exercise 2.28. Critique the following research questions, outlining how and why they can be improved (if at all).

1. Among domestic water tanks used in south-east Queensland, are lead concentrations in water in concrete tanks higher than in poly tanks?
2. Are lower-limb amputees more likely to die?
3. Is the amount of salt the same for home brand as for non-home brand beans?
4. Among zoo animals, is the weight of adult elephants greater than juvenile kangaroos (joeys)?
5. Is the average reaction time related to gender?

What terms might need defining for each RQ?

Answers to *Quick review* questions: 1. False. **2.** True. **3.** True. **4.** False. **5.** False. **6.** True. **7.** False. **8.** False.

[1]The force required to embed an 11.28 mm steel ball into wood to half the diameter of the ball.

Part II

Research design

3

Overview of research design

So far, you have learnt to ask an RQ. **In this chapter**, you will learn why research design is important, by learning to:

- identify reasons why the value of the response variable varies.
- identify and distinguish extraneous, confounding and lurking variables.
- understand how chance impacts the values of the response variable.
- explain external and internal validity.

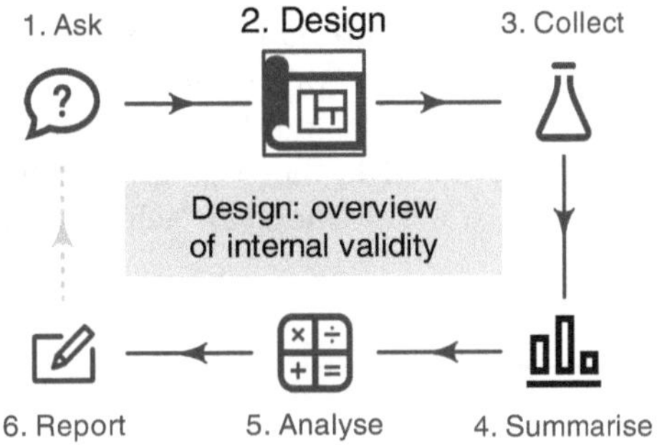

3.1 Introduction: internal and external validity

An RQ asks about a *population*. However, studying every member of a population is generally impossible due to cost, time, ethics, logistics and/or practicality. A subset of the population (a *sample*) is studied, comprising some *individuals* from the population. *Countless* different samples are possible.

⚠ One challenge of research is learning about a population from studying just one of the countless possible samples.

Being able to generalise about the population of interest from studying a sample is called *external validity*. Chapter 6 discusses how to select a suitable sample to study to enhance external validity.

Definition 3.1 (External validity). *External validity* refers to the ability to generalise the results to the rest of the population, beyond just those in the sample studied.

Apart from being external valid, well-designed research studies should be *internally valid*. An internally valid study allows the researchers to focus on the relationship of interest between the response and explanatory variables, by eliminating, or accounting for, other sources of variation in the values of the response variable. These other sources are discussed in the rest of this chapter.

Definition 3.2 (Internal validity). *Internal validity* refers to the extent to which a cause-and-effect relationship can be established in a study.

A study with *high* internal validity shows that the changes in the response variable can be (at least partially) attributed to changes in the explanatory variables; other explanations have been ruled out.

⚠ One challenge of research is learning about the relationship between the response and explanatory variables, when the value of the response variable can also be influenced by other factors.

Studies with *low* internal validity leave open other possibilities, apart from changes in the value of the explanatory variable, to explain changes in the value of the response variable. Ideally, all studies should be designed to be *internally valid* as far as possible. Internal validity is studied in more detail in Chap. 7. Different research studies (Chap. 4) differ in the extent to which they can achieve internal validity.

3.2 Variation in the values of the response variable

In any study, the values of the response variable vary from individual to individual. Many reasons explain *why* these values vary.

Example 3.1 (Study design). Consider this RQ:

> For students in a large university course, is the average typing speed (in words per minute) the same for those aged under 25 ('younger') and 25 or over ('older')?

The typing speed (the *response variable*) of the many individuals will vary: every student in the study recording exactly the same typing speed is highly unlikely. The variation in the values of students' typing speeds (Fig. 3.1) may be due to:

- *the explanatory variable* (Sect. 3.3). The values of the explanatory variable may influence the values of the response variable. Of course, they may not either; the purpose of the study is to find if, or to what extent, this is true. In this example, the *explanatory variable* is the age group of the student, which may impact typing speed.
- *other variables*, called *extraneous variables* (Sect. 3.4). Other variables (apart from the explanatory variable) may influence the response variable (perhaps more than the explanatory variable), such as 'sex of the person', or 'whether the person wears glasses'. The impact of these variables can be accommodated if the study is designed appropriately.
- *chance* (or *randomness*, or *natural variation*) (Sect. 3.5). The same person doing the same thing repeatedly under the same conditions will not record exactly the same typing speed every attempt. This is unavoidable, but good research design can minimise the size of this variation.

Designing studies to maximise internal validity requires identifying important extraneous variables, and eliminating (as far as possible) anything that obscures the relationship between the response and explanatory variables.

Example 3.2 (Design). In the typing-speed study, suppose younger students were *always* asked to use their dominant hand, and older students *always* asked to use their non-dominant hand. Younger students would probably have a faster average typing speed, simply because they use their dominant hands (not due to age). This research design would produce a study with poor internal validity.

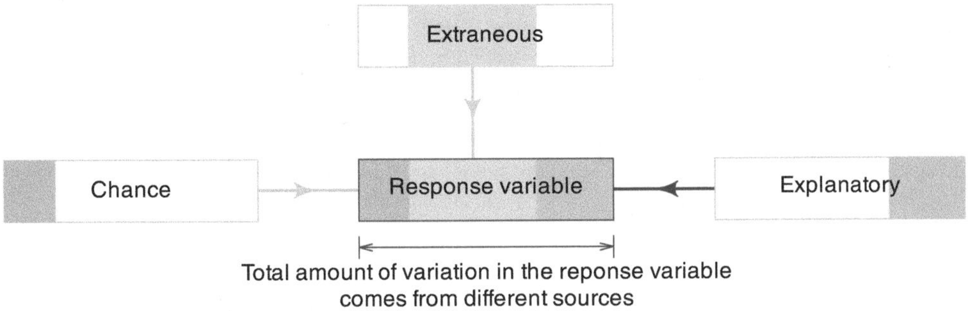

FIGURE 3.1: Other factors can influence the values of the response variable, besides the explanatory variable.

Definition 3.3 (Research design). *Research design* refers to the decisions made by the researchers to maximise *external validity* and *internal validity*.

Internal validity is one of the most important properties of scientific studies, and is relevant for reasoning about evidence more generally. Designing studies to maximise internal validity is the focus of Chap. 7.

ⓘ Data collection is often tedious, time-consuming and expensive: you usually get one chance to collect data. In contrast, data (once collected) can be analysed as many times as necessary. Design the study properly the first time!

3.3 Variation due to changes in the explanatory variable

Changes in the values of the explanatory variable may, or may not, be associated with changes in the values of the response variable. If nothing else influenced the values of the response variable, life would be easy: any change of a given size in the value of the explanatory variable would *always* result in a change of the same size in the value of the response variable.

Example 3.3 (Explanatory variable). In the typing-speed study (Example 3.1), the explanatory variable is the age group. If nothing else influenced typing speed, every younger student would record the same typing speed every time, and every older student would record the same typing speed every time. This is clearly unreasonable.

3.4 Variation due to changes in the extraneous variables

Other variables (besides the explanatory variable) almost certainly exist which are associated with changes in the value of the response variable. These are called *extraneous variables.*

Definition 3.4 (Extraneous variable). An *extraneous variable* is any variable associated with the response variable, but is not the explanatory variable.

Example 3.4 (Extraneous variables). In the typing-speed study (Example 3.1), potential extraneous variables may include age, the presence or absence of certain medical conditions, the level of familiarity with computers, whether the person wears glasses, etc.

The impact of some extraneous variables on the response variable can be reduced by fixing the values of the variable. These variables are called *control variables.*

Definition 3.5 (Control variables). *Control (or controlled) variables* are extraneous variables whose values are fixed for the study.

A *control variable* is different from a *control group* (Def. 2.17).

Example 3.5 (Control variables). In the typing-speed study (Example 3.1), typing speeds would vary greatly if students used different types of keyboards; for example, if some students used mechanical keyboards, and some used on-screen keyboards (e.g., on a tablet). The impact of age is easier to detect if all students use the *same* keyboards, as this would reduce the variation in the typing speeds.

'Type of keyboard' is a *control variable.*

If *many* other variables are fixed in value (i.e., are control variables), the relationship between the explanatory and response variables is far easier to detect and measure. However, using too many control variables may limit the population, and hence the generalisability of the results. In the typing-speed study, for example, restricting the study to left-handed males who do not wear glasses would restrict the results to a very narrow group of people.

All extraneous variables are, by definition, related to the response variable. They may or may not also be associated with the explanatory variable. Extraneous variable *also* related to the explanatory variable are called *confounding variables* (or *confounders*); see Fig. 3.2 (left panel). A confounding variable can obscure the true relationship between the response and explanatory variables.

Definition 3.6 (Confounding variable). A *confounding variable* (or a *confounder*) is an extraneous variable associated with the response *and* explanatory variables.

Definition 3.7 (Confounding). *Confounding* is when a third variable influences the observed relationship between the response and explanatory variable.

⚠ Confounding variables are *associated* with both the response and explanatory variables. This does not necessarily mean the value of the confounding variable *causes* changes in the values of the response or explanatory variables.

Example 3.6 (Confounding variables and associations). Consider a study comparing the proportion of females and males wearing sunglasses while walking in a local park. To determine if the variable 'whether it is raining' is an *extraneous* variable, we ask:

1. Is the wearing of sunglasses (the response variable) more or less likely if it is raining?

The absence of rain may influence people to be more likely to wear sunglasses. Hence, 'whether it is raining' is very likely an extraneous variable.

To determine if it is a *confounding* variable, we also ask:

2. Is one sex (the explanatory variable) more likely to be walking in the park depending on whether it is raining?

We do *not* ask if the presence of rain *changes* the sex of the person; we ask if the presence of rain is *associated* with different proportions of males and females walking in the presence of rain. It *may* be the case (for example) that males are more likely to walk in the rain than females, so 'whether it is raining' *may* be an extraneous variable (but it is not obvious).

A relationship between the response and explanatory variables may be apparent, but only because *both* of these variables are associated with the confounding variable (Fig. 3.2). No relationship actually exists between the response and explanatory variables.

Example 3.7 (Confounding variables). People who carry cigarette lighters are more likely to get lung cancer. The reason this relationship exists, however, is because of a *confounding variable*. 'Whether the person is a smoker' is probably associated with *both* the response and explanatory variables:

- smokers are more likely to carry a cigarette lighter (the explanatory variable) than non-smokers.
- smokers are more likely to develop lung cancer (the response variable) than non-smokers.

No relationship actually exists between carrying a cigarette lighter are getting lung cancer.

Managing confounding is *very* important, as ignoring confounding can completely change the observed relationship between the response and explanatory variables (see Sect. 15.7) and hence can compromise internal validity. Managing confounding is discussed in Sect. 7.2.

If the values of potential confounding variables are recorded, their impact can be managed. However, sometimes the values of the confounding variables are not recorded (perhaps due to poor design); then, they are called *lurking variables* (Fig. 3.2, left panel). Lurking variables can lead to wrong conclusions.

Definition 3.8 (Lurking variable). A *lurking variable* is an extraneous variable associated with the response *and* explanatory variables (that is, is a *confounding* variable), but whose values *are not* recorded in the study data.

Example 3.8 (Lurking variables). Joiner [1981; Wilson Jr, 1952] wanted to determine if the time in the production mould influenced the strength of plastic parts (p. 55–56):

FIGURE 3.2: Confounding variables are extraneous variables associated with the response and explanatory variables. Left: If the confounding variable is not measured (and so a lurking variable is present), an apparent association does exist between the response and explanatory variables. Usually, confounding is not as extreme as shown in this diagram, and the confounding variable may slightly change the relationship between response and explanatory variables. Right: In extreme confounding situations, as shown here, no real association between exists between the response and explanatory variables; the association is explained by a confounding variable.

> Hot plastic was introduced in the mold, pressed for 10 s, and removed. Another batch was then introduced into the same mold, pressed for 20 s, and so on, the time increasing with each batch.

Greater time in the mould (explanatory variable) was found to be associated with greater plastic strength (response variable). However, mould temperature was later found to be a *lurking variable*, since it was associated with *both* the response and explanatory variables:

- higher mould temperatures (the lurking variable) were associated with greater strength (the response variable).
- higher mould temperatures (the lurking variable) were experienced by later batches with longer mould times (the explanatory variable), since the mould was hotter for the later batches.

The cause of the greater strength was *not* the time in the mould; it was the higher temperature experienced by the later moulds (Fig. 3.3).

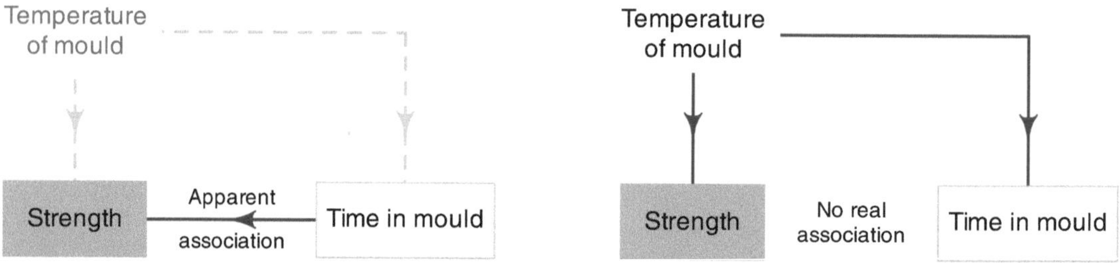

FIGURE 3.3: An example of a lurking variable. Left: the relationship as originally understood. Right: the relationship after the lurking variable was eventually exposed.

To clarify the language (Table 3.1):

- extraneous variables are, by definition, always associated with the response variable. If they are not recorded, and so the researchers are unaware of them, they become part of unexplained chance.
- extraneous variables are called *confounding variables* if they are also related to the explanatory variable.
- confounding variables are called *lurking variables*, if they are not recorded.

These terms are not always used consistently by all researchers, but the *ideas* are important nonetheless.

TABLE 3.1: The relationships between extraneous, confounding and lurking variables. Entries in *italics* indicate different types of extraneous variables.

	Related to response only	**Related to response and explanatory**
Measured/observed	*Extraneous*	*Extraneous (confounding)*
Not measured/observed	Chance	*Extraneous (lurking)*

To avoid lurking variables, researchers generally collect lots of information that may be relevant about the *individuals in the study* (such as age and sex if the study involves people) and *circumstances of the individuals in the study* (such as the temperature at the time of data collection), in case they are confounding variables.

Example 3.9 (Low internal validity). Larson et al. [2021] reviewed numerous studies that used double-fortified salt to manage iodine and iron deficiencies. They concluded that the internal validity of studies was 'generally weak' (p. 265) due, in part, to 'unaccounted confounders' (i.e., lurking variables).

3.5 Variation due to natural variation (chance)

Chance variation (or natural variation) refers to variation that cannot otherwise be explained: even repeating a study exactly the same way every time on the same individuals will not always produce the same values of the response variable.

The influence of the explanatory variable is hard to detect if the amount of chance variation contributing to the response variable overwhelms the amount of variation produced by changes in the value of the explanatory variable. *Minimising the amount of the chance variation* requires using good design principles, and measuring as many other extraneous variables that may explain variation in the response variable as reasonable.

Chance can impact the values of the response variable in different ways:

- each *individual* can produce different values of the response variable each time the response variable is measured (*within*-individuals *variation*).
- each individual in the study can produce different values of the response variable compared to *other* individuals (*between*-individuals *variation*).

Different strategies are needed to understand each of these sources of variation:

- to estimate the amount of variation *within* individuals, multiple observations are needed from each unit of analysis (each individual).
- to estimate the amount of variation *between* individuals, multiple units of analysis (individuals) are needed.

Example 3.10 (Three ways to sample). Consider the typing-speed study (Example 3.1) again, and these three sampling approaches:

- taking 30 observations from one younger student would tell us about variation in that student's typing speed, but very little about variation in younger students' typing speeds more generally.
- taking one observation from 30 different younger students would tell us about variation in younger students' typing speeds in general. We only have one measurement from each student, but since we might expect that the same person to produce similar (not identical) typing speeds, this should not be a big problem.
- taking three observations from each of 10 different younger students would tell us about variation in younger students' typing speeds in general, and a little about the variation in each students' typing speeds too.

3.6 Chapter summary

Research questions are about populations, but samples are studied in practice. Studies that use a sample that represents the population of interest are called *externally valid.*

In a research study, the main interest is usually the relationship between a *response variable* and *explanatory variable.* Well-designed studies that allow the researchers to focus on this relationship have good *internal validity.* Such studies eliminate, or account for, other explanations for the variation in the values of the response variable.

The values of the response variable can be influenced by more than just the explanatory variable, such as *extraneous variables* (other variables not of primary interest), and *chance.*

Some extraneous variables are also related to the explanatory variable, and are called *confounding variables* (and are *lurking variables* if not recorded). If the research design makes it difficult to separate the relationship between the response and explanatory variable from other possible causes, the study has poor *internal validity.*

3.7 Quick review questions

Martnes and Bere [2023] compared the average time to complete a journey when (p. 1)

> ... riding an electric-assisted bicycle with cargo (30 kg) and without cargo...

They recorded the age, height, weight, and resting metabolic rate of all subjects who completed the 4.5 km ride. Each subject was allocated to ride both with *and* without cargo.

Are the following statements *true* or *false*?

1. The explanatory variable is 'the age of the subjects'.
2. 'The height of the subjects' is a lurking variable.
3. The explanatory variable is 'whether the bicycle is ridden with or without cargo'.
4. 'Weight' is an extraneous variable.
5. The response variable is 'the time to complete the journey'.
6. 'Age' is a possible confounding variable.
7. 'Resting metabolic rate' is a possible confounding variable.

3.8 Exercises

Answers to odd-numbered exercises are given at the end of the book.

Exercise 3.1. The *Giant Mine* in Yellowknife, Canada, ceased operation in 1999 after 50 years, during which 237 000 tonnes of arsenic trioxide was released. Houben et al. [2016] examined the arsenic concentration in 25 lakes within a 25 km radius of the mine 11 years after the mine closed, to determine if the arsenic concentration was related to the distance of the lake from the mine. They also recorded:

- the type of bedrock (volcanic; sedimentary; grandiorite).
- the ecology type (lowland; upland).
- the elevation of the lake (in metres).
- the lake area (in hectares).
- the catchment area (in hectares).

Use this information to answer the following.

1. What is the *response* variable?
2. What is the *explanatory* variable?
3. Is the variable 'Catchment area' likely to be a *lurking* variable?
4. Is the variable 'Type of bedrock' likely to be a *confounding* variable?
5. What is the *best* description of the variable 'Ecology type': response, explanatory, confounding, or lurking variable?

Exercise 3.2. A study examined the relationship between diet quality and depression in Australian adolescents [Jacka et al., 2010]. The researchers used a sample of 7 114 adolescents aged 10–14 years old, and also measured information about (p. 435):

> ... age, gender, socioeconomic status, parental education, parental work status, family conflict, poor family management, dieting behaviours, body mass index, physical activity, and smoking...

1. Identify the response and explanatory variables.
2. Which of the other listed variable reasonably could be considered *extraneous variables*, *confounding variables* and *lurking variables*?

Exercise 3.3. A newspaper article [Anonymous, 2012] reported that 'Women who drank green tea at least three times a week were 14 per cent less likely to develop a cancer of the digestive system'. However, the final paragraph of the article notes that:

> Nobody can say whether green tea itself is the reason, since green tea lovers are often more health-conscious in general.

Identify the explanatory and response variables, and explain the quotation using language introduced in this chapter.

Exercise 3.4. A study recorded the lung capacity (using Forced Expiratory Volume, or FEV, in litres) of children aged 3 to 19 [Tager et al., 1979, Kahn, 2005], and also recorded whether not the children were smokers. One finding was that children who smoke have a *larger* average FEV (i.e., larger average lung capacity) than children who do *not* smoke, in general.

Name a confounding variable that may explain this surprising finding. Would it be likely that this variable is a *lurking* variable?

Exercise 3.5. Consider a study to determine if the percentage of children who consume Ready-To-Eat-Cereals (RTEC) for breakfast is the same for children aged between 5 and 10, as for children aged between 11 and 15. The researchers also measured the age of the child, the number of siblings living with the child, and the sex of the child.

1. Which of these variables are extraneous variables?
    - The sex of the child.
    - Whether the child consumes RTEC.
    - The age group of the child.
    - The age of the child.
    - The number of siblings living with the child.
2. Is the variable 'the sex of the child' a lurking variable?
3. Is it reasonable to consider the weight of the child as a lurking variable?

Exercise 3.6. Which of the following types of variables are special types of extraneous variables? (a) Lurking variables; (b) explanatory variables; (c) confounding variables.

Exercise 3.7. A study of New Zealanders found that people wearing hearing aids were more likely to have grey hair than people *not* wearing hearing aids. What confounding variable is likely to be present?

Exercise 3.8. Researchers are studying a new (but expensive) insecticide that is claimed to be more effective for use in apple orchards than other (cheaper) insecticides. A study found that apple orchards where the farmers chose to use the new insecticide had a similar number of insects per tree than orchards where the farmers chose *not* to use the new insecticide, What confounding variable is likely to be present?

Exercise 3.9. An agricultural study recorded the wheat yield for 18 organic farms and 29 conventional farms. Farms across North Dakota and Kansas (USA) were used for the study, and the yield (in tonnes per hectare) was recorded from each farm. The organic farms were generally smaller than the non-organic farms, and located in areas with better soil quality.

Which of these are likely to be confounding variables (if any)? Which may be useful control variables (if any)? Explain your reasoning.

1. Crop yield.
2. Soil quality.
3. Climate.
4. The colour of the farmer's main tractor.
5. The size of the farm.
6. The hours of sunlight per day over the growing season.

Exercise 3.10. A study of school teachers found a relationship between the average number of children plus grandchildren for the teacher, and having high blood pressure.

Which of these is likely to be a confounding variable? Which may be useful control variables? Explain.

1. Age of the teacher.
2. Sex of the teacher.
3. The colour of the teacher's car.
4. Whether the teacher is a smoker.
5. Whether the teacher is very health conscious.
6. Whether the teacher has high blood pressure.
7. Whether the teacher teaches a health subject.

Answers to *Quick review* questions: 1. False. **2.** False. **3.** True. **4.** True. **5.** True. **6.** False. **7.** False (confounding variables are potentially related to the response *and* explanatory variables).

4

Types of research studies

You have learnt how to ask an RQ and understand the main principles of research design. **In this chapter**, you will learn to:

- identify and describe the types of quantitative research studies.
- compare and distinguish experimental and observational studies.
- describe and identify the directionality in observational studies.
- describe and identify true experimental and quasi-experimental studies.

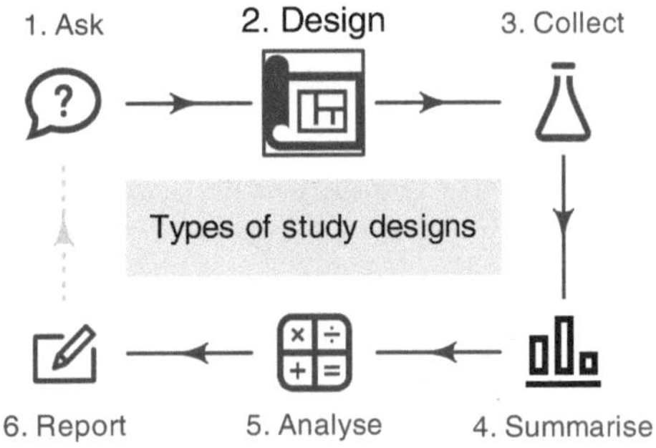

4.1 Introduction

Chapter 2 introduced four types of research questions: descriptive, relational, repeated-measures and correlational. This chapter discusses the types of research studies needed to answer these RQs, while Chaps. 5 to 9 discuss the details of designing these studies and collecting the data.

Different types of studies can be used to collect the data needed to answer RQs:

- *descriptive* studies (Sect. 4.2) answer descriptive RQs.
- *observational* studies (Sect. 4.3) answer RQs with an explanatory variable, but *no intervention*.
- *experimental* studies (Sect. 4.4) answer RQs with an explanatory variable and *an intervention*.

Observational and experimental studies are sometimes collectively called *analytical studies*.

4.2 Descriptive studies

Definition 4.1 (Descriptive study). *Descriptive studies* answer descriptive RQs.

Descriptive studies are not explicitly studied further, as the relevant ideas are present in the discussion of observational and experimental studies.

Example 4.1 (Descriptive study). Lee et al. [2020] studied the percentage of people in Hong Kong wearing face masks in various situations. A descriptive RQ is being asked: the population is 'residents of Hong Kong', and the outcome is (for example) 'the percentage who wear face masks when taking care of family members with fever'. Answering this RQ requires a *descriptive study*.

4.3 Observational studies

Observational studies are used for RQs with no intervention. They are commonly-used, and sometimes are the only type of research design possible. Observational studies do not have an intervention, and hence have *conditions* (Def. 2.16) rather than *treatments*.

Definition 4.2 (Observational study). *Observational studies* study relationships *without* an intervention.

Example 4.2 (Between-individuals observational study). Consider again this one-tailed, decision-making RQ (based on the ideas in Sect. 2.11):

> Among Australian teenagers with a common cold, is the average duration of cold symptoms *shorter* for teens taking a daily dose of echinacea compared to teens taking no medication?

This RQ has a *between-individuals* comparison, so is a relational RQ. If the researchers *do not* impose the taking of echinacea (that is, the individuals make this decision themselves), the study is observational. The two *conditions* are 'taking echinacea', and 'not taking echinacea' (Fig. 4.1).

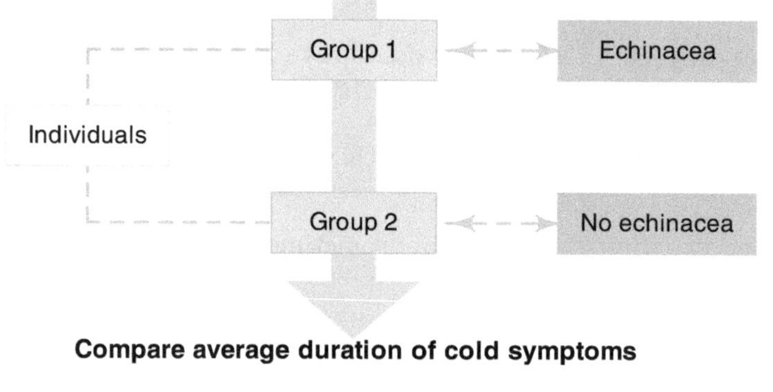

FIGURE 4.1: Observational studies with a relational RQ. The dashed lines indicate steps not under the control of the researchers.

Example 4.3 (Within-individuals observational study). Levitsky et al. [2004] recorded the weights of university students at the beginning of university, and then after 12 weeks from the same students. The comparison is *within* individuals; this is a *repeated-measures* (paired) RQ. Since the researchers do not impose anything on the students, there is *no intervention* (Fig. 4.2).

The *outcome* is the average weight. The *response variable* is the weight of individuals. The *within-individuals comparison* is the week of the university semester (1 and 12).

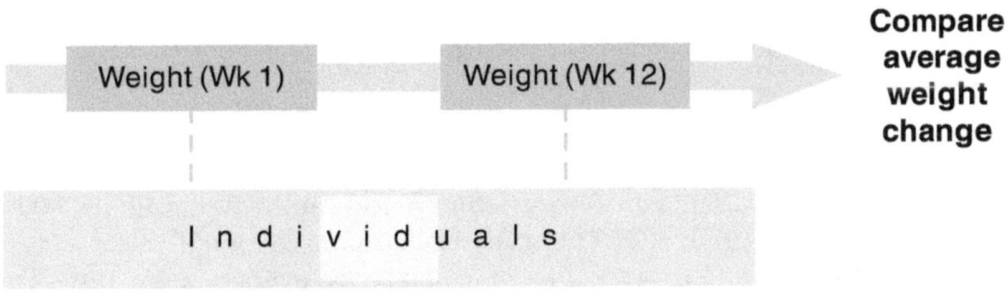

FIGURE 4.2: Observational studies with a repeated-measures RQ. The dashed lines indicate steps not under the control of the researchers.

Example 4.4 (Correlational observational study)**.** Poovaragavan et al. [2023] explored the relationship between time since death, and the concentration of sodium in synovial (knee) fluid. This is a correlational RQ as groups are not being compared. The time since death is the explanatory variable, and the concentration of sodium in synovial fluid is the response variable. The researchers do not impose the time since death, so there is *no intervention* (Fig. 4.3).

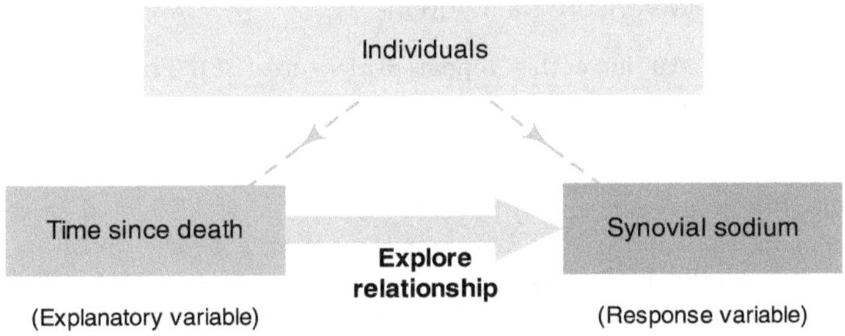

FIGURE 4.3: Observational studies with a correlational RQ. The dashed lines indicate steps not under the control of the researchers.

4.4 Experimental studies

Experimental studies, or *experiments*, are used for RQs with an intervention, and are commonly-used. Well-designed experimental studies can establish a *cause-and-effect relationship* between the response and explanatory variables. However, using experimental studies is not always possible. In general, well-designed experimental studies are more likely to be internally valid than observational studies. Experimental studies have an intervention, and hence *treatments* (Def. 2.15).

Definition 4.3 (Experiment)**.** *Experimental studies* (or *experiments*) study relationships *with* an intervention.

⚠ In an *experimental study*, the unit of analysis (Def. 2.20) is the smallest collection of units of observations that can be randomly allocated to separate treatments.

Example 4.5 (Within-individuals experimental study)**.** Consider this RQ:

> For obese men over 60 years-of-age, what is the average increase in heart rate after walking 400 m?

This RQ uses a *within-individuals comparison* (before and after walking 400 m) so is a repeated-measures (and paired) RQ. The study has an intervention if researchers impose the 400 m walk on the subjects (Fig. 4.4). The *outcome* is the average heart rate. The *response variable* is the heart rate for each individual man.

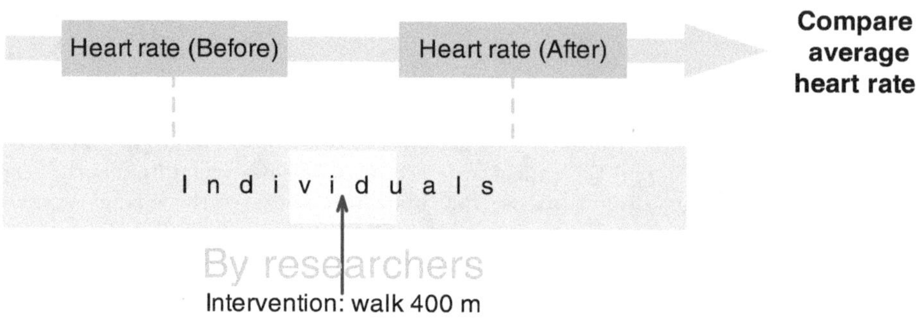

FIGURE 4.4: Experimental studies with a repeated-measures RQ. The dashed lines indicate steps not under the control of the researchers.

Example 4.6 (Correlational experimental study)**.** Xu et al. [2023] studied leaf-drip irrigation, exploring the relationship between the water pressure and flow rate. This is a correlational RQ, where the hydraulic pressure time is the explanatory variable, and the flow rate is the response variable. The researchers imposed nine different values for water pressure, so there is an intervention (Fig. 4.5).

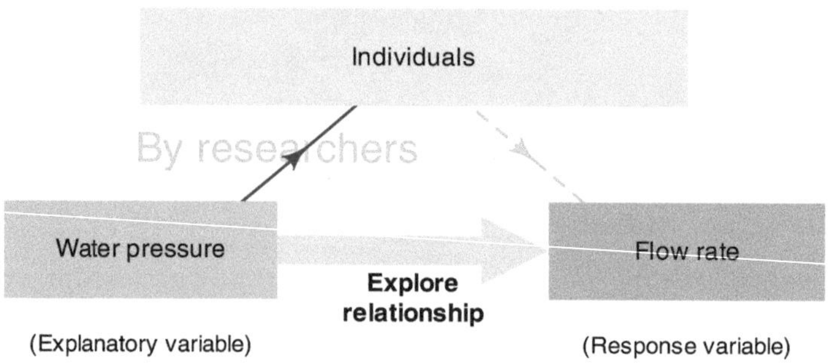

FIGURE 4.5: Experimental studies with a correlational RQ. The dashed lines indicate steps not under the control of the researchers.

Between-individuals experimental studies can be either *true experiments* (Sect. 4.4.1) or *quasi-experiments* (Sect. 4.4.2); see Table 4.1.

TABLE 4.1: Comparing analytical designs with a between-individuals comparison.

Study type	Individuals allocated to groups?	Treatments allocated to groups?	Reference
Observational	No	No	Sect. 4.3
True experiment	Yes	Yes	Sect. 4.4.1
Quasi-experiment	No	Yes	Sect. 4.4.2

4.4.1 True experimental studies

True experiments are commonly used to answer relational RQs. An example of a true experiment is a *randomised controlled trial*, often used in drug trials.

Definition 4.4 (True experiment). In a *true experiment*, the researchers:

1. allocate treatments to groups of individuals (i.e., allocate the values of the explanatory variable to the individuals), *and*
2. determine who or what individuals are in those groups.

While the steps may not happen *explicitly*, they happen *conceptually*.

Example 4.7 (True experiment). The echinacea study (Sect. 2.11) could be designed as a *true experiment*. The researchers would allocate individuals to one of two groups, and then decide which group took echinacea and which group did not (Fig. 4.6).

These steps may happen implicitly: researchers may allocate each person at random to one of the two groups (echinacea; no echinacea). This is still a true experiment, since the researchers could decide to switch which group receives echinacea; ultimately, the decision is still made by the researchers.

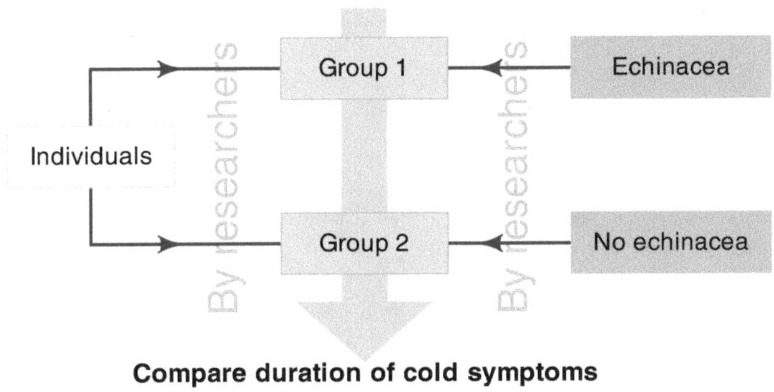

FIGURE 4.6: True experimental studies: researchers allocate individuals to groups, and treatments to groups.

4.4.2 Quasi-experimental studies

Quasi-experiments are similar to true experiments (i.e., answer relational RQs) but treatments are *allocated* to groups that *already exist* (e.g., may be naturally occurring).

Definition 4.5 (Quasi-experiment). In a *quasi-experiment*, the researchers:

1. allocate treatments to groups of individuals (i.e., allocate the values of the explanatory variable to the individuals), but
2. do *not* determine who or what individuals are in those groups.

Example 4.8 (Quasi-experiments). The echinacea study (Sect. 2.11) could be designed as a quasi-experiment. The researchers could *find* two existing groups of people (say, from Suburbs A and B), then decide to allocate people in Suburb A to take echinacea, and people in Suburb B to *not* take echinacea (Fig. 4.7).

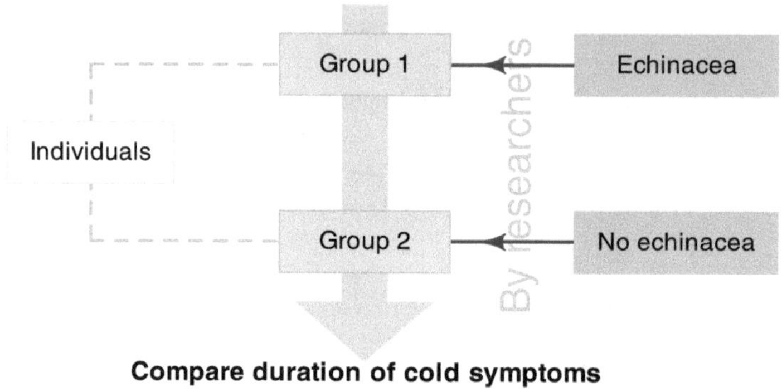

FIGURE 4.7: Quasi-experimental studies: researchers do not allocate individuals to groups, but do allocate treatments to groups. The dashed lines indicate steps not under the control of the researchers.

Example 4.9 (Quasi-experiments). A researcher wants to examine the effect of an alcohol awareness program (based on MacDonald [2008]) on the average amount of alcohol consumed per student in a university Orientation Week. She runs the program at University A only, then compares the average amount of alcohol consumed per person at two universities (A and B).

This study is a *quasi-experiment* since the researcher did not (and can not) determine the groups: the students (not the researcher) would have chosen University A or University B for many reasons. However, the researcher *did* decide whether to allocate the program to University A or University B.

4.5 Comparing study types

In *experimental* studies, researchers *create* differences in the values of the explanatory variable through allocation, and then note the effect this has on the values of the response variable. In *observational* studies, researchers *observe* differences in the values of the explanatory variable, and observe the values of the response variable.

Importantly, *only well-designed true experiments can show cause-and-effect.* Nonetheless, well-designed observational and quasi-experimental studies can provide evidence to *support* cause-and-effect conclusions, especially when supported by other evidence. Although only true experimental studies can show cause-and-effect, true experimental studies are often not possible for ethical, financial, practical and/or logistical reasons.

The advantages and disadvantages of each study type are discussed later (Sect. 8.2), after these study types are discussed in greater detail in the following chapters.

Example 4.10 (Cause and effect)**.** Many studies report that the bacteria in the gut of people on the autism spectrum is different from the bacteria in the gut of people *not* on the autism spectrum [Kang et al., 2019, Ho et al., 2020], and suggest the bacteria may contribute whether a person is autistic. These studies were observational, so the suggestion of a cause-and-effect relationship may be inaccurate.

Other studies [Yap et al., 2021] suggest that people on the autism spectrum are more likely to be 'picky eaters', which contributes to the differences in gut bacteria.

4.6 Directionality in research studies

Analytical research studies (observational; experimental) can be classified by their *directionality* (Table 4.2).

- *Forward direction* (Sect. 4.6.1): the values of the explanatory variable are obtained, and the study determines what values of the response variable occur in the future. *All experimental studies have a forward direction.*
- *Backward direction* (Sect. 4.6.2): the values of the response variable are obtained, then the study determines what values of the explanatory variable occurred in the past.
- *No direction* (Sect. 4.6.3): the values of the response and explanatory variables are obtained at the same time.

Directionality is important for understanding cause-and-effect relationships. If the explanatory variable occurs *before* the outcome is observed, a cause-and-effect relationship *may* be possible. That is, studies with a forward direction are more likely to provide evidence of causality.

TABLE 4.2: Classifying observational studies. (Experimental studies have a forward direction.)

Type	Explanatory variable	Response variable
Forward direction	When study begins	Determined in the future
Backward direction	Determined from the past	When study begins
No direction	When study begins	When study begins

Example 4.11 (Directionality)**.** In South Australia in 1988–1989, 25 cases of legionella infections (an unusually high number) were investigated. All 25 cases were gardeners.

O'Connor et al. [2007] compared 25 people with legionella infections with 75 similar people without the infection, and found that recent (past) use of potting mix was associated with an increase in the risk of contracting illness.

This study has a backward *direction*: people were identified with an infection, and then the researchers looked *back* at past activities.

Research studies are sometimes described as 'prospective' or 'retrospective', but these terms can be misleading [Ranganathan and Aggarwal, 2018] and their use not recommended [Vandenbroucke et al., 2014].

Experimental studies always have a forward direction. Observational studies may have any directionality, and may be given different names accordingly.

4.6.1 Forward-directional studies

All experimental studies have a forward direction, and include *randomised controlled trials* (RCTs) and *clinical trials.*

Observational studies with a *forward* direction are often called *cohort studies.* Both experimental studies and cohort studies can be expensive and tricky: tracking individuals (a *cohort*) into the future is not always easy, and the ability to track some individuals into the future may be lost (*drop-outs*): plants or animals may die, people may move or decide to no longer participate, etc. Forward-directional observational studies:

- may add support to cause-and-effect conclusions, since the comparison occurs *before* the outcome (only well-designed experimental studies can establish cause-and-effect).
- can examine many outcomes in one study, since the outcome(s) occur in the future.
- can be problematic for rare outcomes, as the outcome of interest may never (or rarely) appear in the future.

Example 4.12 (Forward study). Chih et al. [2018] studied dogs and cats who had been recommended to receive intermittent nasogastric tube (NGT) aspiration for up to 36 h. Some pet owners did not give permission for NGT, while some did; thus, whether the animal received NGT was *not* determined by the researchers (the study is observational). The researchers then observed whether the animals developed hypochloremic metabolic alkalosis (HCMA) in the next 36 h.

Since the explanatory variable (whether NGT was used) was recorded at the start of the study, and the response variable (whether HCMA was observed) was determined within the following 36 h, this study has a *forward direction.*

4.6.2 Backward-directional studies

Observational studies with a *backward* direction are often called *case-control* studies. The 'cases' are often individuals with a certain disease, and then the controls are those without the disease (see Example 4.11). Researchers find individuals with specific values of the response variable (cases and controls), and determine values of the explanatory variable from the past. Case-control studies:

- only allow one outcome to be studied, since individuals are chosen to be in the study based on the value of the response variable of interest.
- are useful for rare outcomes, as the researchers can purposely select large numbers with the rare outcome of interest.
- do not effectively eliminate other explanations for the relationship between the response and explanatory variables (*confounding*; Def. 3.7).
- may suffer from *selection bias* (Sect. 6.7), as researchers purposively try to locate individuals with a rare outcome.
- may suffer from *recall bias* (Sect. 9.3.2) when the individuals are people: accurately recalling the past can be unreliable.

Example 4.13 (Backwards study). Pamphlett [2012] examined patients with and without sporadic motor neurone disease (SMND), and asked about *past* exposure to metals.

The response variable (whether the respondent had SMND) is assessed when the study begins, and whether subjects had exposure to metals (explanatory variable) is determined from the *past*. This observational study has a *backward* direction.

4.6.3 Non-directional studies

Non-directional observational studies are called *cross-sectional* studies. Cross-sectional studies:

- are good for findings associations between variables (and these associations may or may not be causation).
- are generally quicker and cheaper to conduct than other types of studies.
- are not useful for studying rare outcomes.
- do not effectively eliminate other explanations for the relationship between the response and explanatory variables (*confounding*; Def. 3.7).

Example 4.14 (Non-directional study). Russell et al. [2014] asked older Australian their opinions of their own food security, and recorded their living arrangements. Individuals' responses to both the response variable and explanatory variable were gathered at the same time. This observational study is *non-directional.*

4.7 The role of research design

Choosing the *type* of study is only one part of research design; many other decisions must be made also. The purpose of these decisions is to ensure researchers can confidently study the relationship between the response and explanatory variables (*internal validity*) in the population of interest (*external validity*) from studying one the many possible samples. This is related to the idea of *bias.*

Definition 4.6 (Bias). *Bias* refers to any systematic misrepresentation of the target population or a parameter caused by the sampling or the study design.

Various types of bias are possible, some of which are studied later. Maximising internal and external validity reduces bias. Bias may occur during research design, sample selection (Sect. 6.7), data collection (selection bias; Sect. 6.8), analysis, or interpretation of results (Chap. 8). This book only discusses some possible biases.

Designing a study to maximise *internal validity* means:

- identifying *what else* might influence the values of the response variable, apart from the explanatory variable (Chap. 3).
- designing the study to be *effective* (Chap. 7).

In general, experimental studies have better internal validity than observational studies.

Designing a study to maximise *external validity* means:

- identifying who or what to study, since the whole population cannot be studied (Chap. 6).
- determining *how many* individuals to study. (We need to learn more before we can answer this critical question in Chap. 32.)

Details of the data *collection* (Chap. 9) and *ethical* issues (Chap. 5) also form part of the study design.

4.8 Chapter summary

Three types of research studies are: *descriptive studies* (for studying descriptive RQs), *observational studies* (for studying relationships *without* an intervention), and *experimental* (for studying relationships *with* an intervention).

Observational studies can be classified as having a *forward direction* (cohort studies), *backward direction* (case-control studies), or *no direction* (cross-sectional studies). Experimental studies always have a forward direction. Relational RQs with an intervention can be classified as *true experiments* or *quasi-experiments*. Cause-and-effect conclusions can only be made from well-designed *true experiments*.

Ideally studies should be designed to be *internally* and *externally* valid. In general, experimental studies have better internal validity than observational studies.

4.9 Quick review questions

Are the following statements *true* or *false*?

1. Fraboni et al. [2018] studied the 'red-light running behaviour of cyclists in Italy'. This study is most likely to be observational.
2. In a true experiment, the researchers apply treatments to groups that they have determined; in a quasi-experiment, the researchers apply treatments to groups that they have not determined.
3. In a quasi-experiment, the researchers allocate treatments to groups that they cannot manipulate.
4. True experiments generally have a higher internal validity than observational studies.
5. Observational studies generally have a higher external validity than quasi-experimental studies.

4.10 Exercises

Answers to odd-numbered exercises are given at the end of the book.

Exercise 4.1. Consider this RQ [McLinn et al., 1994]:

> In children with acute otitis media, what is the difference in the average duration of symptoms when treated with cefuroxime compared to amoxicillin?

1. Is the comparison a within- or between-individuals comparison?
2. Is this RQ descriptive, relational, repeated-measures or correlational?
3. Is there likely an intervention?
4. Is the RQ an estimation or decision-making RQ?
5. Is the study observational or experimental? If observational, what is the *direction*? If experimental, is this a quasi-experiment or true experiment?

Exercise 4.2. Khair et al. [2015] studied the time needed for organic waste to turn into compost. For some batches of compost, earthworms were added. In other batches, earthworms were *not* added to the waste. One RQ asked whether the composting times for waste with and without earthworms was the same or not.

1. Is the comparison a within- or between-individuals comparison?
2. Is this RQ descriptive, relational, repeated-measures or correlational?
3. Is there an intervention?
4. Is the RQ an estimation or decision-making RQ?
5. Is the study observational or experimental? If observational, what is the *direction*? If experimental, is this a quasi-experiment or true experiment?

Exercise 4.3. Gonzalez-Fonteboa and Martinez-Abella [2007] studied recycled concrete beams. Beams were divided into three groups, different loads were then applied to each group, then the shear strength needed to fracture the beams was measured. Is this a *quasi-experiment* or a *true experiment*? Explain.

Exercise 4.4. A research study compared the use of two different education programs to reduce the percentage of patients experiencing ventilator-associated pneumonia (VAP). Paramedics from two cities were chosen to participate. Paramedics in City A were allocated to receive Program 1, and paramedics in the other city to receive Program 2.

1. Is this RQ descriptive, relational, repeated-measures or correlational?
2. Is the comparison a within- or between-individuals comparison?
3. Is there likely an intervention?
4. Is the study observational or experimental? If observational, what is the *direction*? If experimental, is this a quasi-experiment or true experiment?

Exercise 4.5. Manzano et al. [2013] compared 'the effectiveness of alternating pressure air mattresses vs. overlays, to prevent pressure ulcers' (p. 2 099). Patients were *provided* with alternating pressure air overlays (in 2001) or alternating pressure air mattresses (in 2006). The number of pressure ulcers were recorded.

This study is experimental, because the researchers *provided* the mattresses. Is this a *true* experiment or *quasi*-experiment? Explain.

Exercise 4.6. Sacks et al. [2009] compared four weight-loss diets, using 811 overweight adults each randomly assigned to one diet. The diets used comparable foods. The authors state (p. 859):

> The primary outcome was the change in body weight after 2 years in [...] comparisons of low fat versus high fat and average protein versus high protein and in the comparison of highest and lowest carbohydrate content.

1. What is the *between*-individuals comparison?
2. What is the *within*-individuals comparison?
3. Is this study observational or experimental? Why?
4. Is this study a quasi-experiment or a true experiment? Why?
5. What are the units of analysis?
6. What are the units of observation?
7. What is the response variable?
8. What is the explanatory variable?

Exercise 4.7. Consider this initial RQ (based on Friedmann and Thomas [1985]), that clearly needs refining: 'Are people with pets healthier?'

1. Briefly describe useful and practical definitions for P, O and C.
2. Briefly describe an *experimental* study to answer the RQ.
3. Briefly describe an *observational* study to answer the RQ.

Exercise 4.8. Consider this initial RQ, that clearly needs refining: 'Are seeds more likely to sprout when a seed-raising mix is used?'

1. Briefly describe useful and practical definitions for P, O and C.
2. Briefly describe an *experimental* study to answer the RQ.
3. Briefly describe an *observational* study to answer the RQ.

Answers to *Quick review* questions: 1. True. **2.** True. **3.** True. **4.** True. **5.** False (irrelevant).

5

Ethics in research

You have learnt how to ask an RQ, and identify different types of studies to obtain data. **In this chapter**, you will learn to:

- list common ethical issues to be considered in research design.
- understand the purpose of reproducible research.

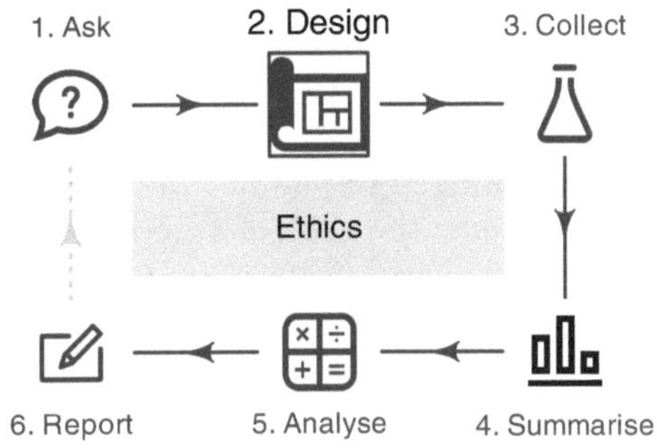

5.1 Introduction: obtaining ethical clearance

All research *must* be ethical, and *must* meet ethical guidelines, to minimise risk of harm to the environment, property and to participants, and to preserve the well-being, dignity, rights and safety of participants (including animals). Practically every university and research organisation in the world promotes and enforces ethical research practices.

Most research studies require an ethics committee to formally grant ethics approval *before* research begins. Only brief comments about research ethics are given here.

Example 5.1 (Ethics). Ethics are important for *all* studies, not just those involving people or animals. For example:

- in engineering, 238 articles published between 1945 and 2015 were retracted, mostly for unethical research practice [Rubbo et al., 2019].
- in the chemical sciences, 331 retractions were reported in 2017 and 2018 due to ethical issues, such as falsification of data and plagiarism [Coudert, 2019].

5.2 Ethical issues in research design

Ethical issues embrace many areas when designing research studies.

- *Acknowledgements*: all those who contributed to the research should be acknowledged, including those who prepare figures, take photographs, or have helped collect data.
- *Analysis*: the analysis must use appropriate methods.
- *Confidentiality*: data should be kept confidential and secure.

- *Consent*: when appropriate, people should consent to being in the study, and hence should be told what the study involves. People should also be able to withdraw from the study without penalty.
- *Economic risks*: financial loss to participants should be avoided. Reimbursements of reasonable costs for participating may be appropriate.
- *Environmental risks*: environmental impacts and damage should be avoided or minimised.
- *Funding*: sources of funding should be disclosed. Any studies funded by, or sanctioned by, companies or organisations with vested interests need to be carefully scrutinised. These may lead to, or may give the impression of, conflicts of interest.
- *Incentives to participate*: if participants are offered incentives to participate (above reimbursement of costs), these should be acknowledged as it may cause (perhaps unconsciously), or may give the impression of causing, participants to influence the results.
- *Legal risks*: participants should not be put in the position of breaking laws, and the research itself should not break any laws.
- *Plagiarism*: the work of others should be appropriately acknowledged and not claimed to be original (see Sect. 35.5.7).
- *Physical risks*: physical harm or discomfort (to researchers, participants or bystanders) should be avoided or minimised.
- *Psychological risks*: psychological harm or discomfort (to researchers, participants or bystanders) should be avoided or minimised.
- *Resourcing*: the study should not waste resources, time or money (e.g., if the answer to the RQ is already known, the study is not necessary).
- *Sample size*: the study should not use more individuals than necessary.
- *Social risks*: social harm or discomfort (to researchers, participants or bystanders) should be avoided or minimised.
- *Storage of data*: data should be stored securely, kept for the required amount of time, then (if appropriate) securely disposed.

Example 5.2 (Poor ethics). In the Tuskegee syphilis experiment, conducted between 1932 and 1972, treatments were actively withheld from men with syphilis [Corbie-Smith, 1999]. The men's wives and children were often affected, and the men were lied to about their treatments. This study was highly unethical, and could not be conducted now.

Example 5.3 (Poor ethics in analysis). In 1986, the American space shuttle *Challenger* exploded just after launch, killing all seven astronauts on board. A review [Dala et al., 1989] found the cause was partly that engineers failed to use some data that should have been used. This was unethical.

5.3 Reproducible research

One way to ensure that research results are reliable and trustworthy is through *reproducible* research: enabling someone else to repeat the study and analysis, to confirm the findings. For research to be reproducible, the methods, data, analysis methods and relevant computer code must be available [Laine et al., 2007] when possible. (Sometimes releasing data is unethical, such as when individuals may be identified, so should not be released.)

Methods for ensuring reproducible research are often discipline dependent, and beyond the scope of this book. Different journals also have different expectations regarding reproducibility. Nonetheless, the basic ideas are important.

The importance of reproducibility in the analysis phase is crucial; for example:

> There are serious medical consequences to errors attributable to the effects of spreadsheet programs and software operated through a graphical user interface [...] that could have been avoided through a reproducible research paradigm...
>
> — Simons and Holmes [2019], p. 471

Using purely point-and-click interfaces for statistical analysis (e.g., spreadsheets) is not recommended, as results are not reproducible.

Rather than using spreadsheets (see Sect. 2.12), using tools which encourage reproducible research are recommended. Statistical software packages, such as jamovi, Python, R, SAS, SPSS and Stata, are recommended as the analysis commands can be recorded (even when using the point-and-click interfaces), and hence the analysis is reproducible.

5.4 Chapter summary

Studies must be ethical, and any formal study must obtain ethical approval *before* beginning. Ethics covers issues including, but not restricted to:

- acknowledgements.
- analysis methods.
- confidentiality.
- consent.
- economic risks.
- environmental risks.
- funding.
- incentives for participants.
- legal risks.
- plagiarism.
- physical risks.
- psychological risks.
- resourcing.
- sample size.
- social risks.
- storage of data.

5.5 Quick review questions

Are the following statements *true* or *false*?

1. Ethics apply for *any* type of study.
2. Ethics only refer to the interactions of the researchers with participants in the study.
3. Ethics only apply when *people* are the individuals.
4. Ethics only apply when *people* or *animals* are the individuals.
5. Ethics can extend to storage of data.
6. Ethics only apply to the design of the study.
7. Ethics apply even to the analysis of the data.

5.6 Exercises

Answers to odd-numbered exercises are given at the end of the book.

Exercise 5.1. Consider this conundrum [Crozier and Schulte-Hostedde, 2015]:

> A research team has an extraordinarily successful long-term study of a population of bighorn sheep (*Ovis canadensis*) on Ram Mountain...
>
> The population contains marked individuals for which the research team has incredibly detailed data [...] this research has lead to numerous important publications.
>
> Recently, however, a cougar (*Puma concolor*) that has learned to specialize on these sheep is slowly but surely eating all of them. This is a study of a natural population, which includes predation, but this cougar is drastically reducing the sample size of the study.
>
> Since it is legal to hunt cougars in the region where this study is taking place, one option is to try to kill the predator; however, even if a cougar were successfully hunted, this would not ensure that it was the correct one.

What action would you recommend, from an ethical point-of-view?

Exercise 5.2. Suppose a research group is testing a new drug, with the potential to cure a debilitating illness. The researchers could (a) *use* a control group that does not receive the new drug, and so obtain stronger evidence for using the drug if it works; or (b) *not use* a control group, so that everyone in the study potentially benefits from the using the new drug.

What would you decide? Explain.

Exercise 5.3. Suppose a very deadly and highly contagious disease breaks out. Is it ethical to use a new drug to treat those affected, even though the drug is experimental and the potentially harmful side effects are unknown? Discuss your point-of-view.

Exercise 5.4. Is it ethical to lie to subjects? *Deception* is used in some disciplines, and may be approved by ethics committees under some circumstances (such as potential benefits of the study, and whether the deception may cause physical or psychological discomfort to the participants).

Is it ethical to tell participants that they are taking an active medication, when it is actually ineffective (a 'placebo')? Discuss the advantages and disadvantages.

Answers to *Quick review* questions: 1. True. **2.** False. **3.** False. **4.** False. **5.** True **6.** False. **7.** True.

6

External validity: sampling

You have learnt to ask an RQ, and identify a study design. **In this chapter**, you will learn to:

- distinguish and explain precision and accuracy.
- distinguish and explain random and non-random sampling.
- explain why random samples are preferred over non-random samples.
- identify, describe and use different sampling methods.
- identify ways to obtain samples likely to be representative.

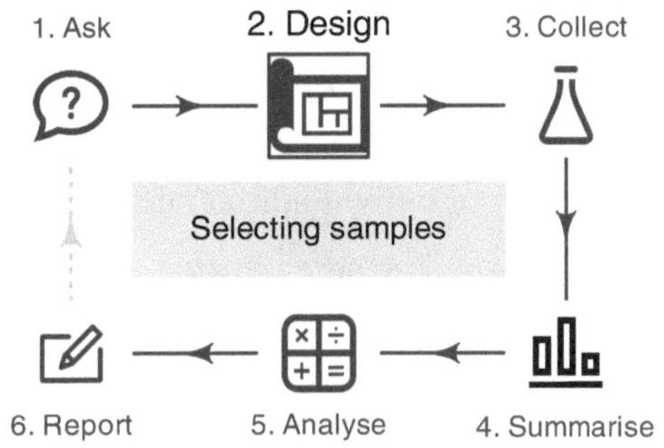

6.1 Introduction

In a research study, the researchers learn about the *population* by studying just one of the countless possible *samples*. Ideally the sample that is studied is representative of the population, so the results from the sample generalise to the population. This is called *external validity*. *External validity* does *not* mean that the results apply more widely than the intended population.

Example 6.1 (External validity). Suppose the *population* in a study is *Californian home-owners*. The sample comprises the Californian home-owners studied by the researchers. If the study is externally valid, the sample is representative of all Californian home-owners.

The results will not necessarily apply to home-owners outside of Californian, or all Californian residents. However, this *is irrelevant for external validity*. External validity concerns how the *sample* represents the intended population in the RQ, which is *Californian home-owners*.

6.2 The idea of sampling

Studying every member of a population is very rare due to cost, time, ethics, logistics and/or practicality. Instead, a subset of the population (a *sample*) is studied, and *many* different samples are possible.

⚠ The challenge of research is learning about a population from studying just one of the countless possible samples.

Example 6.2 (Samples). A study of the effectiveness of aspirin in treating headaches cannot possibly study every single human who may one day take aspirin. Not only would this be prohibitively expensive, time consuming, and impractical, but the study would not even study those not yet born who might use aspirin.

Using the whole target population is *impossible*, and a sample must be used.

Only studying one sample out of countless possible samples raises questions:

- *which* individuals should be included in the sample to be studied?
- *how many* individuals should be included in the sample to be studied?

The first issue is studied in this chapter. The second issue is studied later (Chap. 32), after learning about the implications of studying samples rather than populations.

Many samples are possible, and *every sample is likely to be different*. Hence, the results of studying a sample are likely to vary, depending on which individuals are in the studied sample. The differences between the samples, and differences in the results from each sample, are called *sampling variation*. That is, each sample has different individuals, produces different data, and may even suggest different answers to the RQ.

Example 6.3 (Number of samples). In a 'population' of just 100, the number of possible samples of size 25 is more than twice the number of people currently living on earth.

This is the challenge of research: *making decisions about populations, using just one of the many possible samples*. A lot can be learnt about the population if selecting a sample is approached correctly.

⚠ Almost always, researchers study *samples*, not *populations*. Many samples are possible, and *every sample is likely to be different*, and the *results from every sample are likely to be different*. This is called *sampling variation*.

As a result, *conclusions from a sample are never certainties*, though special techniques allow us to still learn about the *population* from a *sample*.

Example 6.4 (Sampling variation). Consider a fair pack of cards (a *population*), where 50% of cards are red. The percentage of red cards is not the same in every hand (every *sample*) of ten cards. This is a simple example of *sampling variation*.

6.3 Precision and accuracy

Two questions concerning sampling in Sect. 6.2 were: *which* individuals should be in the sample, and *how many* individuals should be in the sample. The first question addresses the *accuracy* of using a sample value to estimate a population value. The second addresses the *precision* with which a population value is estimated using a sample. An estimate that is not accurate is called *biased* (Def. 4.6).

Definition 6.1 (Accuracy). *Accuracy* refers to how close a *sample* estimate is likely to be to the *population* value, on average.

Definition 6.2 (Precision). *Precision* refers to how similar the sample estimates from different samples are likely to be to each other (that is, how much variation is likely in the sample estimates).

Using this language:

- the sampling *method* (i.e., *how* the sample is selected) impacts the *accuracy* of the sample estimate (i.e., *external validity*).
- the *size* of the sample impacts the *precision* of the sample estimate (i.e., *internal* validity).

Large samples are more likely to produce *precise* estimates, but they may or may not be accurate estimates. Similarly, random samples are likely to produce *accurate* estimates, but they may or may not be *precise*. As an analogy, consider an archer aiming at a target. The shots can be accurate, or precise, or (ideally) both (Fig. 6.1).

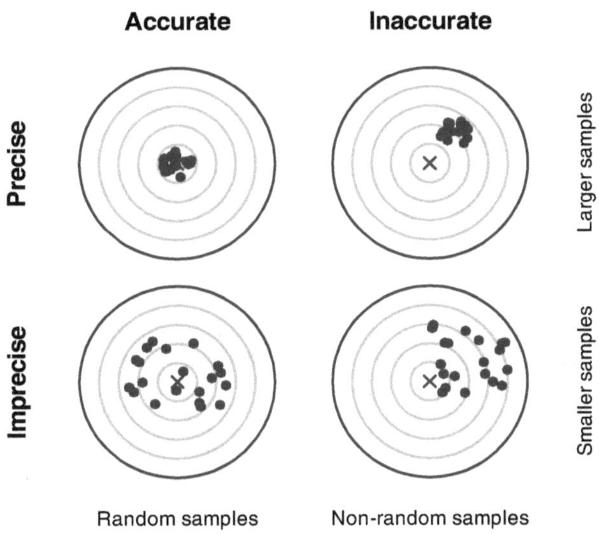

FIGURE 6.1: Precision and accuracy: each dot indicates where a shot at the target lands, and is like a sample estimate of the population value (shown by the central ×).

Example 6.5 (Precision and accuracy). To estimate the average age of *all Canadians*, 9 000 Canadian school children could be sampled.

The answer obtained from the sample will be *inaccurate* because the sample is not representative of *all* Canadians. Since the sample is large, the answer will give a *precise* answer but to a *different* question: 'What is the average age of Canadian school children?'

6.4 Types of sampling

One key to obtaining accurate estimates about the population from the sample is to ensure that the sample faithfully represents the population. So, *how* is such a sample selected from the population?

The individuals selected for the sample can be chosen using either *random sampling* or *non-random sampling.* The word *random* here has a specific meaning that is different from how it is often used in everyday use. It does *not* mean 'haphazard', 'erratically' or 'picking individuals as aimlessly as I can'.

Definition 6.3 (Random). *Random* means determined completely by impersonal chance.

6.4.1 Random sampling

In a *random sample*, both of these statements are true:

1. each individual in the population can be selected.
2. each individual is chosen on the basis of *impersonal* chance (such as using a random number generator, or a table of random numbers).

Some examples of random sampling methods appear in Table 6.1, and are explained further in Sect. 6.5.

Definition 6.4 (Random sample). In a *random* sample, each individual in the population can be selected; and each individual is chosen on the basis of *impersonal* chance.

⚠ The results obtained from a random sample are likely to generalise to the population from which the sample is drawn; that is, *random samples* are likely to produce *externally valid* and *accurate* studies.

TABLE 6.1: Comparing five types of random sampling.

Type	**Stage 1**	**Stage 2**	**Ref.**
Simple random	Individuals chosen at *random*		§6.5.1
Systematic	Start at a *random* location	Take every *n*th element thereafter	§6.5.2
Stratified	Split into a few large groups ('strata') of similar individuals	Select a *simple random sample* from *every* stratum	§6.5.3
Cluster	Split into many small groups ('clusters'); select a *simple random sample* of clusters	Select *all* individuals in the chosen clusters	§6.5.4
Multi-stage	Select a *simple random sample* from the larger collection of units	Select a *simple random sample* from those chosen in Stage 1; etc.	§6.5.5

A pot of soup can be tested randomly or non-randomly. If the soup is stirred (randomised), the small spoonful of soup can be tasted to obtain an overall impression. However, an *overall* impression is not obtained from a non-random sample (i.e., a non-stirred pot of soup).

6.4.2 Non-random sampling

A *non-random* sample is selected using personal input from the researchers. Examples include:

- *judgement samples*. Individuals are selected based on the researchers' judgement (possibly unconsciously), perhaps because the individuals are (or may appear) agreeable, supportive, easily accessible, or helpful. For example, researchers may select rats that are less aggressive, or plants that are accessible, or people that look approachable.
- *convenience samples*. Individuals are selected because they are convenient for the researcher. For example, researchers may study beaches that are nearby, or use their friends for a study.
- *voluntary response (self-selecting) samples*. Individuals participate if they wish to. For example, researchers may ask people to volunteer to take a survey.
- *cherry-picking*. Individuals are specifically chosen to reach the conclusion that the researchers want.

In non-random sampling, the individuals *in* the study are probably different from those *not in* the study. That is, *non-random samples are not likely to be externally valid.*

Researchers may use a non-random sample intentionally (e.g., to deceive) which is unethical, or unintentionally (e.g., accidentally, or due to practicality (such as meeting budgets)). Ethically, a random (or somewhat representative sample; Sect. 6.6) should be used when possible.

⚠ Using a non-random sample means that the results probably do not generalise to the intended population: they probably do not produce externally valid or accurate studies.

6.5 Methods of random sampling

6.5.1 Simple random sampling

The most straightforward idea for obtaining a random sample is a *simple random sample.*

Definition 6.5 (Simple random sample)**.** In a *simple random sample*, *every* possible sample of a given size has the *same* chance of being selected.

Selecting a simple random sample requires a list of all members of the population, called the *sampling frame*, from which to select a sample. Obtaining the sampling frame is often difficult or impossible, and so finding a simple random sample is also difficult. For example, finding a simple random sample of wombats would require having a list and location of all wombats. This is absurd; other random sampling methods, like special ecological sampling methods (e.g., Manly and Alberto [2014]), would be used instead.

Definition 6.6 (Sampling frame)**.** The *sampling frame* is a list of *all* the individuals in the population.

Selecting a simple random sample from the *sampling frame* can be performed using *random numbers* (e.g., using random number tables, or websites like https://www.random.org). Other random sampling methods avoid the need for a sampling frame, but still use randomness rather than human choice.

⚠ This book assumes simple random samples, unless otherwise noted.

Example 6.6 (Simple random sampling). Consider the letter-typing RQ again (Example 3.1, p. 34):

> For students in a large university course, is the average typing speed (in words per minute) the same for those aged under 25 ('younger') and 25 or over ('older')?

Suppose budget and time constraints mean approximately 40 students (out of 441) can be selected for the study. The *sampling frame* is the list of all students enrolled in the course. Obtaining the sampling frame is feasible here; instructors have access to this information for grading.

A simple random sample could be found using the course enrolment list, by first placing all 441 student names into rows of a spreadsheet (ordered by name, student ID, or any way). Then, using random numbers, 40 rows are selected at random (without repeating numbers) between 1 and 441 inclusive. For instance, when I used `https://random.org/integers`, the first few random numbers were: `410`, `215`, `384`, `158`, `296`.

Every student chosen using this method becomes part of the study. If a student could not be contacted or did not respond, more students could be chosen at random to ensure 40 students participated (Fig. 6.2, left panel). By chance, the sample comprises 15 younger students and 25 older students.

6.5.2 Systematic sampling

In *systematic sampling*, the first case is *randomly* selected; then, more individuals are selected at regular intervals thereafter. In general, we say that every nth individual is selected after the initial random selection.

Example 6.7 (Systematic sampling). For the study in Example 3.1, a sample of 40 students in a course of 441 is needed. To find a systematic random sample, select a random number between 1 and 441/40 (approximately 11) as a starting point; suppose the random number selected is 9 (as in Fig. 6.2, right panel).

The first student selected is the 9th person in the student list (which may be ordered alphabetically, by student ID, or other means). Thereafter, every 441/40th person, or 11th person, in the list is selected: people in rows 9, 20, 31, 42, and so on (Fig. 6.2, right panel). By chance, the sample comprises 17 younger students and 23 older students.

⚠ Care needs to be taken when using systematic samples to ensure a pattern is not hidden. Consider taking a systematic sample of every 10th residence on a long street. In many countries, odd numbers are usually on one side of the street, and even numbers usually on the other side. Selecting every 10th house (for example) would include houses all on the same side of the street, and hence with similar exposure to the sun, traffic, etc.

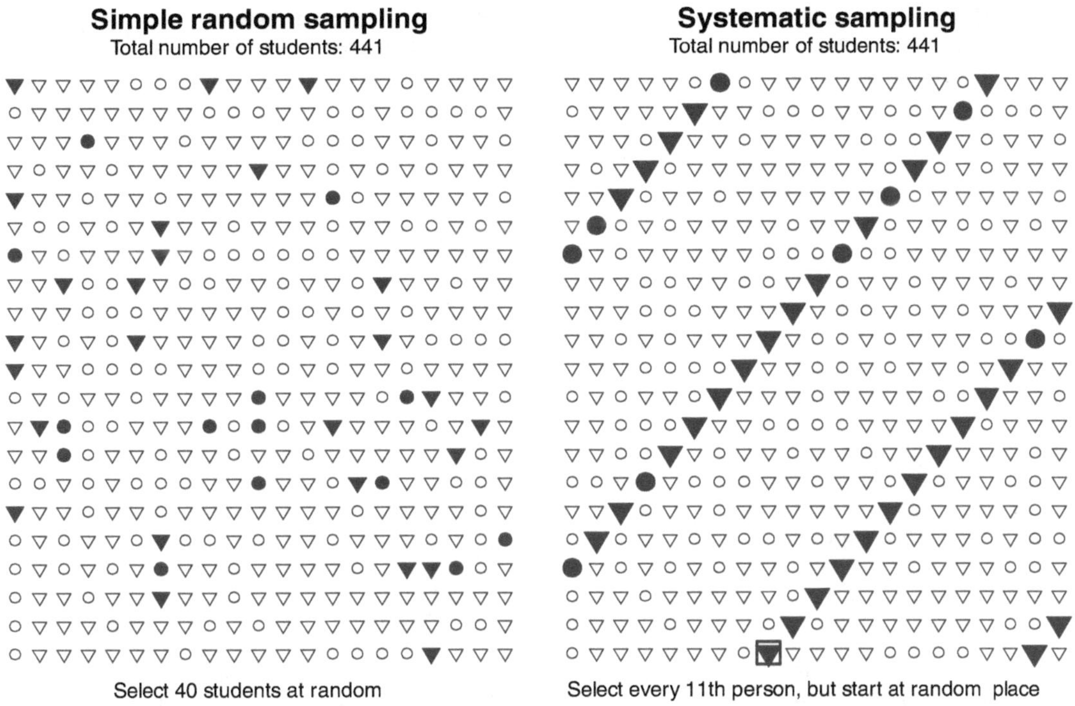

FIGURE 6.2: A simple random sample (left) and a systematic random sample (right) for obtaining a random sample of size 40 from a class of 441. Triangles ▽ represent younger students (there are 294), circles ◯ represent older students (there are 147), and filled shapes represent those individuals selected in the sample. In the right panel, the boxed individual in the bottom row is the initial, randomly-chosen person (person number nine).

Example 6.8 (Systematic sampling). Alary and Joly [1991] studied households in Quebec to determine if their hot water systems kept their water sufficiently hot to avoid Legionellae bacteria. They used a systematic random sample to select households to study (p. 2361):

> The first house was selected by using a random-number table. Thereafter, each fifth house that satisfied the [...] criteria was eligible for the study.

6.5.3 Stratified sampling

In *stratified sampling*, the population is split into a *small* number of *large* (usually similar) groups called *strata*, then cases are selected using a *simple random sample* from *each* stratum. Every individual in the population must be in one, and only one, stratum.

Example 6.9 (Stratified sampling). For the typing study in Example 3.1, 20 younger and 20 older students could be selected to obtain a sample of size 40. The sample is stratified by *age group* of the person (Fig. 6.3, left panel).

Since 66.7% of the students are younger in the population, the sample could be selected so that two-thirds of the sample of size 40 (i.e., 27 students) were younger students (Fig. 6.3, right panel). This is a *proportional* stratified sample.

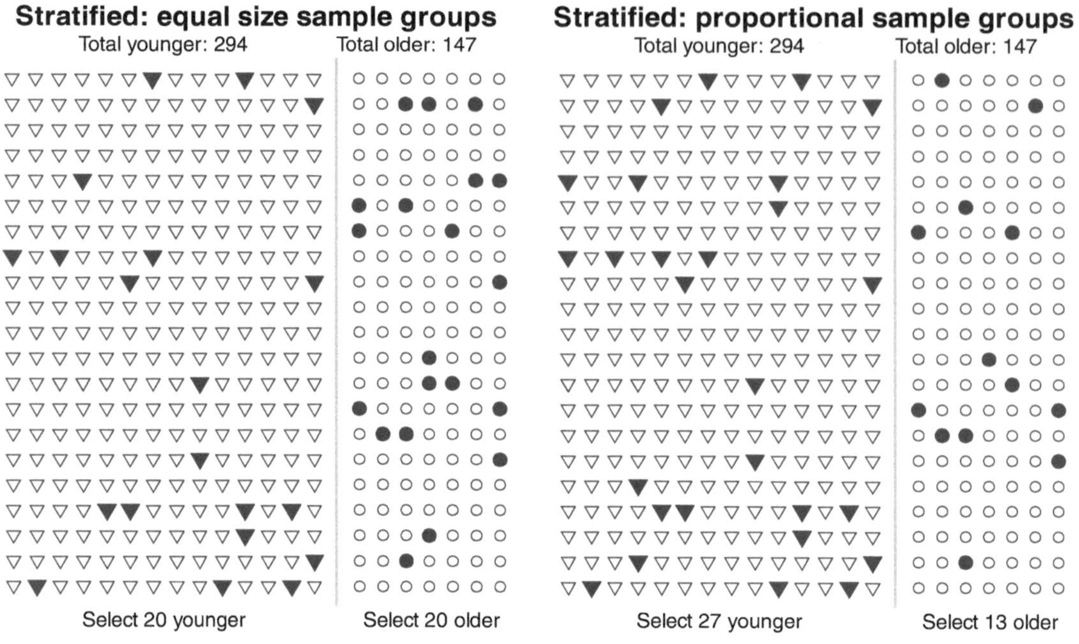

FIGURE 6.3: Two stratified sampling methods for taking a random sample of size 40 from a class of 441. Left: equal numbers of younger and older students. Right: proportional numbers of younger and older students. Triangles ▽ represent younger students, circles ○ represent older students, and filled shapes represent those individuals selected in the sample.

6.5.4 Cluster sampling

In *cluster sampling*, the population is split into a *large* number of *small* groups called *clusters*. Then, a *simple random sample* of clusters is selected, and *every* member of the chosen clusters become part of the sample. Every individual in the population must be in one, and only one, cluster.

Example 6.10 (Cluster sampling). For the study in Example 3.1, a simple random sample of (say) three of the many small-group classes for the course could be selected, and *every* student enrolled in those selected small groups constitute the sample (Fig. 6.4, left panel). By chance, the chosen classes produce a sample size of $n = 47$ (31 younger; 16 older).

6.5.5 Multi-stage sampling

In *multi-stage sampling*, larger collections of individuals are selected using a *simple random sample*, then smaller collections of individuals *within* those large collections are selected using a *simple random sample*. The simple random sampling continues for as many levels as necessary, until individuals are being selected (at random in each step).

Example 6.11 (Multi-stage sampling). For the study in Example 3.1, a *simple random sample* of ten of the many small-group classes could be selected (Stage 1), and then four students are *randomly* selected from each of these 10 selected small groups (Stage 2) (Fig. 6.4, right panel). The sample size is $10 \times 4 = 40$, comprising (by chance) 24 younger students and 16 older students.

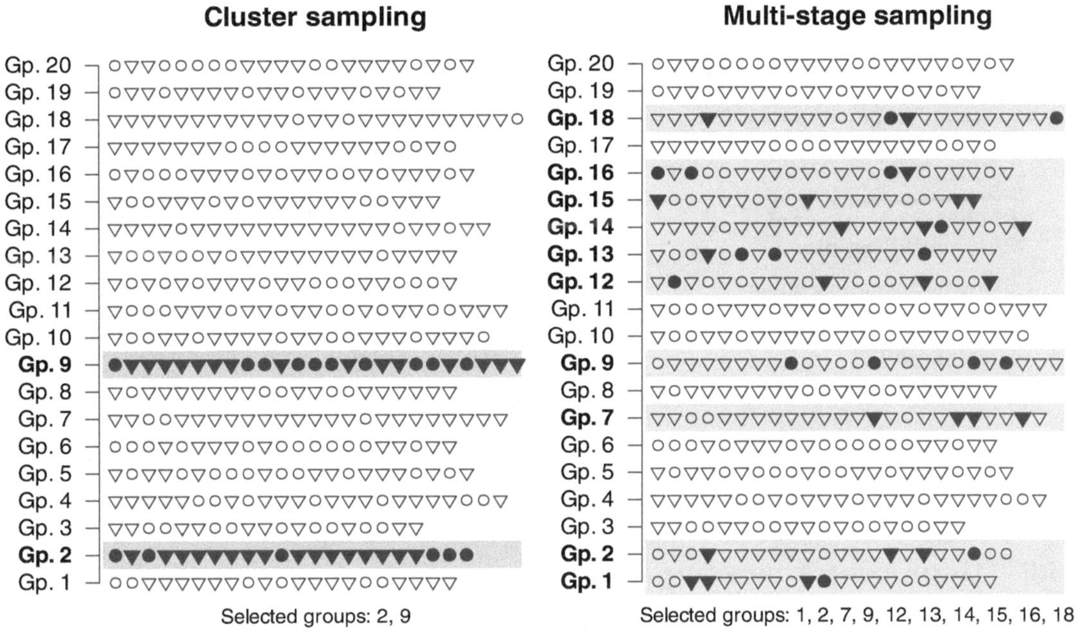

FIGURE 6.4: Cluster sampling (left) and multi-stage sampling (right) for taking a random sample of size approximately 40. Classes shown bold and shaded represent classes randomly selected to be in the sample in the first stage. Triangles ▽ represent younger students, circles ○ represent older students, and filled shapes represent those individuals selected in the sample.

Example 6.12 (Multi-stage sampling). Multi-stage sampling is often used by national statistical agencies. For example, to obtain a multi-stage random sample from a country:

- *Stage 1*: randomly select some cities in the nation.
- *Stage 2*: randomly select some suburbs in these chosen cities.
- *Stage 3*: randomly select some streets in these chosen suburbs.
- *Stage 4*: randomly select some houses in these chosen streets.

This is cheaper than simple random sampling, as data collectors can be deployed in a small number of cities (only those chosen in Stage 1).

6.5.6 Comparing the samples

The different random sampling methods produce different samples, with different proportions of younger and older students by chance (Table 6.2). Of course, repeating the random sampling processes would produce different samples each time. In all cases, only *one* of the countless possible samples is studied.

TABLE 6.2: A summary of the various random samples selected using different random sampling methods. In the population, 66.7% of students are younger students. Repeating any random sampling method is likely to produce a different sample each time.

	Number of students selected			
	Younger	Older	Total	Percentage younger
Simple random sample	26	14	40	65.0
Systematic sample	31	9	40	77.5
Stratified sample: equal	20	20	40	50.0
Stratified sample: proportional	27	13	40	67.5
Cluster sample	31	16	47	66.0
Multi-stage sample	24	16	40	60.0

6.6 Representative samples

Obtaining a truly random sample is usually hard or impossible in practice. Sometimes the best compromise is to select a sample sufficiently diverse so that it is likely to be *somewhat representative* of the diversity in the population. Specifically, those *in* the sample are not likely to be different from those *not in* the sample, at least for the variables of interest. This is often the only practical way to sample.

Definition 6.7 (Representative sample). In a *representative* sample, those *in* the sample are not likely to be different from those *not in* the sample, at least for the variables of interest. A representative sample is *not* a random sample.

As always, the results from any non-random sample *may not generalise* to the intended population (but will generalise to the population which the sample *does* represent).

Example 6.13 (Representative sample). Suppose we wish to evaluate the functionality of two types of hand prosthetics.

If a randomly-chosen group of Alaska and Texas residents is asked for their feedback, probably (but not certainly) their views would be similar to those of all Americans. No obvious reason exists for why residents of Alaska and Texas would be very different from residents in the rest of the United States, regarding their view of hand prosthetic functionality.

Even though the sample is not a random sample of all Americans, the results *may* generalise to all Americans (though we cannot be sure). This sample *may* be representative of the population.

Example 6.14 (Non-representative samples). Suppose we wish to determine the average time per day that Americans households use their air-conditioners for *cooling* in summer.

A sample of Texas residents would not be expected to represent all Americans: it would *over*-represent the average number of hours air-conditioners are used for *cooling* in summer. In this case, those *in* the sample are very different to those *not in* the sample, regarding their air-conditioners usage for cooling in summer.

In contrast, suppose a sample of Alaskans was asked the same question. This sample would not represent all Americans either (it would *under*-represent air-conditioner use). Again, those *in* the sample are likely to be very different to those *not in* the sample, regarding their air-conditioners usage for *cooling* in summer. This sample would not be representative of the population.

Sometimes, a *combination* of sampling methods is used. If the combination includes a non-random sampling method, the sampling method does *not* produce a random sample, but is probably more likely to produce an externally valid sample than a completely non-random sample.

Example 6.15 (A combination of sampling methods). In a study of pathogens present on magazines in doctors' surgeries in Dublin, some suburbs can be selected at *random*, and then (within each suburb) all surgeries are contacted, and some surgeries *volunteer* to be part of the study. This study does not use a random sample.

Sometimes, practicalities dictate how the sample is obtained, which may result in a non-random sample. Even so, the impact of using a non-random sample on the conclusions should be discussed (Chap. 8). Sometimes, simple steps can be taken to obtain a sample that is *more likely* to be representative.

⚠ Random samples are often difficult to obtain, and sometimes *representative* samples are the best that can be achieved, In a representative sample, those *in* the sample are not obviously different from those *not in* the sample. Try to ensure that a broad cross-section of the target population appears in the sample.

Even if a random or representative sample cannot be obtained, the study can still be useful. The results still apply to the population represented by the sample. If individuals in the sample are unlikely to be different from individuals *not* in the sample, for the variables important to the study, the results are likely to approximately apply to the population.

Example 6.16 (Representative sample). For the typing study in Example 3.1, only selecting students who attend the gym, or only students who are at a certain café, is unlikely to be somewhat representative of the student population. Instead, the researchers could approach students at different days, times and locations:

- at the café on Monday at 8am.
- at the gym on Tuesday at 11:30am.
- at the library on Thursdays at 2pm.

This is not a random sample, but should contain a variety of students. Ideally, *students would not be included more than once in the sample*, though this is often difficult to ensure. The students *in* the sample are probably somewhat similar to those *not* in the sample in terms of average typing speeds (there is no obvious reason why they would not be), but we cannot be sure.

To determine if the sample is somewhat representative of the population, sometimes information about the sample and population can be compared. The researchers may then be able to make some comment about whether the sample seems reasonably representative. For example, the sex and age of a sample of university students may be recorded; if the proportion of females in the sample, and the average age of students in the sample, are similar to those of the whole university population, then the sample may be considered somewhat representative of the population (though we cannot be sure).

Example 6.17 (Comparing samples and populations). Egbue et al. [2017] studied the adoption of electric vehicles (EVs) by Americans, using a sample of 121 people found through social media (such as Facebook) and professional engineering channels. This is *not* a random sample of Americans.

The authors compared some characteristics of the sample with the American population from the 2010 census. Compared to the US population, the sample contained a higher percentage of males, a higher percentage of people aged 18–44, and a higher percentage of wealthy individuals.

6.7 Sampling biases

The sample may not be representative of the population for many reasons, all of which compromise how well the sample represents the population (i.e., compromises *external* validity and accuracy). This is called *selection bias.*

Definition 6.8 (Selection, non-response and response bias). *Selection bias* is the tendency of a sample to over- or under-estimate a population quantity.

Non-response bias occurs when chosen participants do not respond: those responding may be different from those not responding.

Response bias occurs when participants provide *incorrect or false information.*

Selection bias is less common in studies with forward directionality, compared to studies that are non-directional or have backward directionality (Sect. 4.6). *Selection bias* may occur if the wrong sampling frame is used, or non-random sampling is used. The sample is biased because those *in* the sample may be different from those *not in* the sample (which may not always be obvious). Biased samples are less likely to produce externally valid studies.

Example 6.18 (Selection bias). Consider Example 6.14, about estimating the average time per day that air conditioners are used for cooling in summer. Even a *random* sample of Alaskans produces a biased sample of Americans: the sampling frame (Alaskans) does not represent the target population ('Americans'). This is *selection bias.*

Non-response bias occurs when chosen participants do not respond. Bias occurs because those who *do not* respond may be different from those who *do* respond. Non-response bias can occur because of a poorly-designed survey, using voluntary-response sampling, chosen participants refusing to participate, participants forgetting to return completed surveys, etc.

Example 6.19 (Non-response bias). Consider a study to determine the average number of hours of overtime worked by various professions. People who work a large amount of overtime may be too busy to answer the survey. Those who answer the survey may be likely to work less overtime than those who do not answer the survey. This is an (extreme) example of *non-response bias.*

Response bias occurs when participants provide *incorrect or false information.* This may be intentional (for example, respondents lie) or non-intentional (for example, the question is poorly written (see Sect. 9.3.1), personal, or misunderstood).

Example 6.20 (Poor sampling). Obtaining data using a telephone survey only includes people who own a telephone, who answer the phone, who do not hang up, who volunteer to complete the survey, and who then finish the whole survey. The people who participate in the survey must meet these criteria, and probably do not represent the population.

Obtaining data using a TV station call-in at 6:15pm only includes people watching that channel, at that time, and who are sufficiently motivated to call. These people must meet very specific criteria, and probably do not represent the population.

Randomly sampling students at your university, because it is easier than finding a random sample of all university students in your country, will only generalise to students at that university and not to students at *all* universities in your country.

6.8 Chapter summary

Almost always, the entire population of interest cannot be studied, so a *sample* (a subset of the population) must be studied. *Many* samples are possible; only one sample is studied. Samples can be obtained using *random* or *non-random* methods. Conclusions made from random samples can usually be generalised to the population (that is, they are externally valid and accurate).

Random sampling methods include *simple random samples*, *systematic samples*, *stratified samples*, *cluster samples*, and *multi-stage samples*. Random samples are likely to be *externally valid* and *accurate*.

Non-random sampling methods include *convenience samples*, *judgement samples*, *voluntary (self-selecting) samples*, and *cherry-picking*. Random samples are often very difficult to obtain, so *reasonably representative* samples are sometimes used, where those *in* the sample are unlikely to be very different from those *not in* the sample. Non-random samples *may not be externally valid* or *accurate*.

6.9 Quick review questions

Are the following statements *true* or *false*?

1. Suppose students are randomly selected and sent postal surveys from their university, but some students have moved and so never receive the survey. This is *response* bias.
2. A *large* sample is *always* better than a *random* sample.
3. *Convenience* sampling and *judgement* sampling are examples of non-random sampling.

6.10 Exercises

Answers to odd-numbered exercises are given at the end of the book.

Exercise 6.1. What is the main advantage of using a *random* sample?

a. It is easier.
b. It is more likely to produce an experimental study.
c. It is more likely to produce an externally-valid study.
d. It is more likely to produce precise estimates.

Exercise 6.2. What is the main advantage of using a *large* sample?
a. It is easier.
b. It is more likely to produce an experimental study.
c. It is more likely to produce an externally-valid study.
d. It is more likely to produce precise estimates.

Exercise 6.3. For the following scenarios, is the selected sample likely to *over*- or *under*-estimate the unknown population value, or estimate the value accurately? Explain *why* the over- or -under-estimation occurs, if relevant, and whether this is likely to be intentional or unintentional.
1. In a study by biologists to estimate biodiversity, researchers decide to focus only on easily accessible areas of a forest due to budget and time constraints.
2. A city council wishes to report the crime rate of various neighbourhoods, so employs interviewers to go door-to-door interviewing residents, between 8am and 5pm.
3. In a campaign speech, a politician reports on some large successes during her term.

Exercise 6.4. For the following scenarios, is the selected sample likely to *over*- or *under*-estimate the unknown population value, or estimate the value accurately? Explain *why* the over- or -under-estimation occurs, if relevant, and whether this is likely to be intentional or unintentional.
1. A county wants to report the number of homeless, so researchers record data from homeless shelters.
2. In a study of soil fertility, a junior researcher takes soil from the surface for testing.
3. A university has introduced a complex and time-consuming system for professors to report students suspected of cheating. When the university produces its *Annual Report*, the reported incidence of cheating is used to claim that 'reports of cheating have dropped'.

Exercise 6.5. A researcher has three months in which to collect the data for a study on car park usage at a shopping centre. Suppose the researcher wants to take a systematic sample of days, and on each of the selected days records the number of cars in the car park.

To select the days in which to collect data, she decides (by using random numbers) to start data collection on a Tuesday, and then every seventh day thereafter.
1. What problem is evident in this sampling scheme?
2. What suggestions would you give to improve the sampling?

Exercise 6.6. Suppose you need to estimate the average number of pages in physical books in a university library (with a library in each of five campuses). Describe how to select a sample of 200 books using:

1. a *simple random sample* of books.
2. a *stratified sample* of books.
3. a *cluster sample* of books.
4. a *convenience sample* of books.
5. a *multi-stage sample* of books.

Which sampling scheme would be most *practical*?

Exercise 6.7. Suppose you need a sample of residents from apartments in a large residential complex, comprising 30 floors with 15 apartments on each floor. You plan to survey the residents of these apartments. For each of the possible sampling schemes given below, first describe the sampling scheme, and then determine which methods are likely to give random (or representative) sample (explaining your answers).
1. *Randomly* select five floors, then *randomly* select four apartments from each of those five floors, and interview a randomly-chosen adult living at the apartment.
2. *Randomly* select one floor, and select the 15 apartments on that floor, then interview the oldest resident of that apartment.
3. Wait at the ground-floor elevator, and ask people who emerge to complete the survey.
4. *Randomly* select five floors, then wait by the elevator on those floors and survey residents as they arrive at the elevator.

Exercise 6.8. Suppose a researcher needs a sample of customers from a large, local shopping centre to complete a questionnaire. Four sampling schemes are listed below. For each, describe the type of sampling. Then, determine which would be the best method (explain why), and determine which (if any) produce a random sample.

1. The researcher locates themselves outside the supermarket at the shopping centre one morning, and approaches every tenth person who walks past.
2. The researcher waits at the main entrance for 30 mins at 8am every morning for a week, and approaches every fifth person.
3. The researcher leaves a pile of survey forms at an unattended booth in the shopping centre, and a locked barrel in which to place completed surveys.
4. The researcher goes to the shopping centre every day for two weeks, at a different time and location each day, and approaches someone every 15 mins.

Exercise 6.9. Ridgewell et al. [2009] investigated how children in Brisbane travel to state schools. Researchers randomly sampled four schools from a list of all Brisbane state schools, and invited every family at each of those four schools to complete a survey.

What *type* of sampling method is this? How could the researchers determine if the resulting sample was approximately representative?

Exercise 6.10. A study comparing two new malaria vaccines recruited 200 Kenyans who had contracted malaria. These recruits were obtained by approaching all patients with a confirmed malaria diagnosis who were admitted to hospitals. Patients could volunteer for the study or not. The study was then conducted to a high standard. Which of the following statements are *true*?

1. This is a voluntary response sample.
2. The study is likely to have high *external* validity.
3. The sample size is too small for the study results to provide useful information.

Exercise 6.11. Suppose a natural forest region is classified into two quite different zones. Zone A is mostly dunes and lightly vegetated, and on the coastal side of a ridge; Zone B is more densely vegetated and on the inland side of the ridge.

A sample of sugar ants (*Camponotus app*) is taken from Zone A, and another sample of sugar ants from Zone B, to study the average size of the ants. What is the best description of the *type* of sampling method being used?

Exercise 6.12. A survey in 2001 concluded (Hieger [2001], cited in Bock et al. [2010], p. 283):

> All but 2% of the home buyers have at least one computer at home, and 62% have two or more. Of those with a computer, 99% are connected to the internet.

The article later reveals the survey was conducted *online* (recall the survey was conducted in 2001). The target population is home buyers; however, home buyers *with* internet access were far more likely to complete the survey than home buyers *without* internet access.

What type of bias is this?

Exercise 6.13. Researchers are studying the percentage of farms that use a specific management technique. The researchers *randomly* select 20 regions around the country, then *randomly* select farms within each region, then ask farmers to volunteer to be in the study.

Explain why this is *not* a multi-stage sample, and what changes are necessary for the researchers to have a multi-stage sample.

Exercise 6.14. Researchers are comparing the average time that experienced school teachers and first-year school teachers spend in the sun. The researchers select schools by asking school principals to volunteer their schools, then record information for *every* teacher in those schools.

Explain why this is *not* a cluster sample, and what changes are necessary for the researchers to have a cluster sample.

Exercise 6.15. Walters et al. [2018] asked this RQ:

> What factors are preventing the adoption of household solar technologies in Santiago?

1. For this RQ, what is the *Population*?
2. The study will be *externally valid* if which of these statements is true?
   a. The sample is representative of all households in the world.
   b. The sample is representative of all solar technologies.
   c. The sample is representative of all households in Santiago.
   d. The sample is representative of all households in Chile.
3. Suppose the researchers mail surveys to all households in Santiago, and people return the survey if they wished to. What is the *best* description of this sampling method?
4. Suppose the researchers randomly select five suburbs in Santiago; then ten streets within each suburb; then ten households on each street. What is the *best* description of this sampling method?

Answers to *Quick review* questions: 1. False. **2.** False (larger: more precise; random: more accurate). **3.** True.

7

Internal validity

So far, you have learnt to ask an RQ, select a study type, and select a sample. **In this chapter**, you will learn to:

- maximise the internal validity of studies.
- manage confounding in studies.
- explain, identify and manage the Hawthorne, observer, placebo and carryover effect in studies.
- explain different types of blinding.

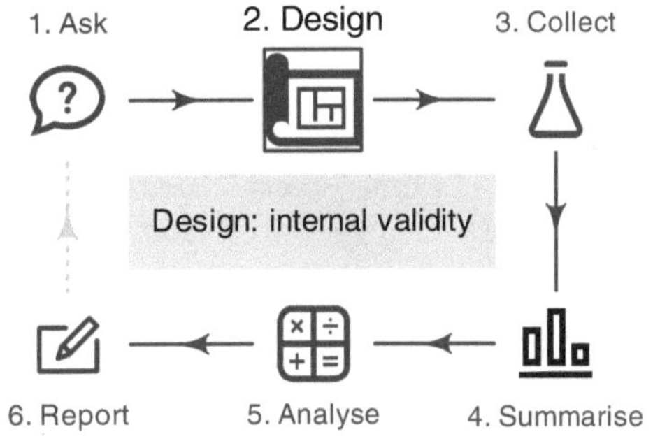

7.1 Introduction

A well-designed study is needed to draw solid conclusions: a study with high *external validity* (Sect. 3.1) and high *internal validity* (Sect. 3.2). This chapter discusses some research design decisions to maximise internal validity.

Example 7.1 (Importance of internal validity)**.** Beaman et al. [2013] describe an experiment where free fertiliser was provided to a sample of female farmers in Mali (at the recommended rate, or at half the recommended rate).

All farmers knew they were part of a study, so changed their farm management: they employed more hired labour and used more herbicide than usual. Consequently, the yields for *all* farmers improved. Knowing if changes in yield were the result of applying the fertiliser is difficult, as the study had poor *internal validity.*

Specific design strategies for maximising internal validity include:

- managing confounding (Sect. 7.2).
- managing the Hawthorne effect by blinding individuals (Sect. 7.3).
- managing the observer effect by blinding the researchers (Sect. 7.4).
- managing the placebo effect by using controls, objective measures and blinding (Sect. 7.5).
- managing the carryover effect by using washouts (Sect. 7.6).

Not all of these strategies will be relevant to every study.

7.2 Managing confounding

For this chapter, the following RQ will be used to demonstrate ideas.

Example 7.2 (Himalaya study). Consider this relational RQ (based on Bird et al. [2008]):

> Among Australians, is the average faecal weight the same for people eating provided food made from wholegrain *Himalaya 292* compared to eating provided food made from refined cereal?

Suppose that the researchers created two groups of individuals for this experimental study:

- *Group A*: women recruited from a female-only gym.
- *Group B*: men recruited from a local nursing home.

The researchers gave *Himalaya 292* to Group A, and the refined cereal to Group B. If a difference in faecal weight was detected between the two groups, many reasons may explain the difference:

- the different *diets* (the explanatory variable in the RQ) for each group.
- the different *sexes* in each group (Group A was all women; Group B was all men).
- the different *ages* in each group (Group A is likely to be younger on average than those in Group B).
- the different *overall health* in each group (Group A would generally be healthier than those in Group B).

Any difference in faecal weight detected between the two groups may not be due to the diets (Table 7.1): the study has very poor internal validity, due to poor research design.

Sex, age and overall health are *confounding variables* (Def. 3.6): they are associated with the type of diet (explanatory variable) *and* faecal weight (response variable). For example, the age of the subject may be associated with faecal weight (older people tend to eat less, and eat differently, than younger people), and the research design means older people are more likely to be consuming the refined cereal. This is an extreme case of *confounding* (Fig. 7.1); usually, confounding is more subtle (and more difficult to detect) than in this example.

TABLE 7.1: Comparing Groups A and B: extreme confounding.

Group A	Variable	Group B
Women	**Sex**	Men
Younger	**Age**	Older
Himalaya 292	**Cereal**	Refined
Fitter	**Fitness**	Less fit

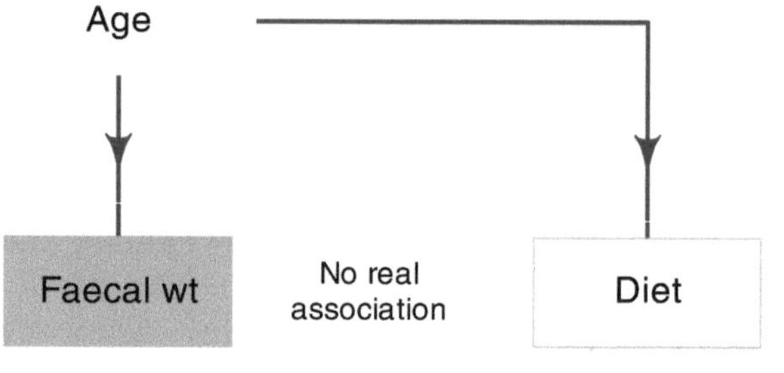

FIGURE 7.1: An extreme example of confounding.

⚠ *The groups being compared should be as similar as possible*, apart from the difference being studied.

Since the groups being compared should be as similar as possible, apart from what is being studied, researchers often compare the comparison groups on potential confounding variables (e.g., the average age of people in each comparison group).

In *experimental* studies, an excellent way to manage confounding is:

1. *Randomly allocating* individuals to the comparison groups.

Random allocation should ensure that the values of potential confounding variables are approximately evenly distributed between the comparison groups. This is true for identified potential confounders (such as age), but also for unidentified potential confounders, or variables that are hard to measure or observe (e.g., genetic conditions). One of the comparison groups is often a control group (Def. 2.17).

Example 7.3. Lian et al. [2024] studied alleviating post-operative thirst experienced by patients admitted to the intensive care unit. They compared standard procedures with the use of ice-water spray. To use random allocation of patients to the two groups, the researchers:

> ... assigned unique numbers from 1 to 56 according to [students'] admission order [...] two-digit numbers were read from the random number table's rows and columns, generating random values that were matched with the respective admission numbers [...]

Any patient assigned a number between 1 to 28 (inclusive) was allocated to the control group, while patients assigned numbers 29 to 56 were assigned to the experimental group.

Example 7.4. Witmer and Pipas [2020] studied using bear faeces to prevent bears damaging trees in an Idaho forest. The researchers painted the bear faeces on sample of trees. As a control, researchers could take observations from trees that they had not approached, and hence had no bear faeces applied. However, if a difference was found between the trees with bear faeces and trees they had not approached, the difference may have been due to the presence of humans near the trees rather than the treatment (i.e., poor internal validity).

For this reason, the control group comprised trees on which the researchers applied water. This is a better control, since trees in both groups (faeces; water) had been approached by humans. Now, if a difference was found between the faeces and water-sprayed trees, the presence-of-humans explanation has been eliminated.

Randomly allocating individuals to comparison groups is *not possible* in observational or quasi-experimental studies. For this reason, confounding is often a major threat to internal validity in these studies, as individuals who are in one comparison group may be different, in general, to those who are in another group.

Fortunately, other (though less effective) means for managing confounding also exist.

2. *Restricting* the study to a certain subgroup of the population.

Sometimes, specifically excluding or including members of the population is helpful for reducing confounding. For the *Himalaya 292* study, for example, age is a potential confounder: older people have different dietary needs, general health and gut health when compared to younger people. Hence, the researchers may decide to use an *inclusion criterion*, restricting the study to people aged from 30 to 50.

In addition, some people may have specific conditions or diseases that mean participating in the study will be problematic. For instance, coeliacs have an autoimmune disorder which results in a severe intolerance to gluten (found in wheat, barley and rye). Hence, the researchers may decide to use *exclusion criteria*, excluding coeliacs from participating in the study. Those individuals that are excluded from the population are not less important than those individuals that are included.

Inclusion and exclusion criteria may be applied for other reasons too; for example, to clarify a population of interest, to address ethical concerns (i.e., by excluding children) or to exclude rare and unusual individuals.

Definition 7.1 (Inclusion and exclusion criteria). *Inclusion criteria* are characteristics that individuals must meet explicitly to be included in the study.

Exclusion criteria are characteristics that explicitly disqualify potential individuals from being included in the study.

Exclusion and inclusion criteria clarify which individuals are explicitly included or excluded from the population for the purposes of the study, and their use should be explained when their purpose is not obvious. Exclusion and inclusion criteria are not both necessary; none, one or both may be used. These are a type of *control variable* (Def. 3.5).

Example 7.5 (Inclusion and exclusion criteria). In a strength study where the population is 'concrete test cylinders', cylinders with severe cracks may be *excluded*.

In a study of exercise regimes for people over 60, severe asthmatics may be *excluded* from the study for health reasons.

Example 7.6 (Inclusion, exclusion criteria). Mackowiak et al. [1992] studied men and women aged 18 to 40; this is the *population*. The *exclusion* criteria include people under 18 years of age and over 40 years of age; alternatively, the *inclusion* criteria are people aged between 18 and 40 years of age. Either of these can be stated; both are not needed.

Example 7.7 (Inclusion and exclusion criteria). Guirao et al. [2017] studied the walking abilities of amputees. Inclusion criteria included (p. 27):

> ... length of the femur of the amputated limb of at least 15 cm measured from the greater trochanter; use of the prosthesis for at least 12 months prior to enrollment and more than 6 h/day...

Exclusion criteria included (p. 27) people with:

> ... cognitive impairment hindering the ability to follow instructions and/or perform the tests; body weight over 100 kg...

3. *Blocking*, when units of analysis are arranged into different groups containing individuals that are similar to each another (see Sect. 29.7 for an example).

For the *Himalaya 292* study, for example, subjects may be *paired* (i.e., groups of two). That is, each person is paired with another person of the same sex and of a similar age and weight; one of each pair is given the *Himalaya 292* diet, and the other is given the refined cereal diet. Each pair is called a *block*.

Definition 7.2 (Blocking). *Blocking* occurs when units of analysis are analysed as separate groups of similar units (called *blocks*).

4. *Analysing* using special methods (beyond this book), after recording the values of potential confounding variables.

To use this approach, *recording all potential extraneous variables* is important. Most studies involving people record the participants' age and sex if possible, as these two variables are common confounders. Once a sample is obtained, recording this extra information usually requires little extra effort. Then, these extraneous variables can be included in the analysis.

Restricting and *blocking* are useful if one or two confounding variables are suspected. Multiple approaches can be used, such as randomly allocating individuals to groups, *and* recording other variables that can be managed through analysis.

Randomly allocating is superior when possible, because confounding is reduced for variables not even suspected as being confounders. Hence, *experimental* studies should use random allocation whenever possible.

For any study (but especially for observational and quasi-experimental studies), recording the values of any potential confounding variables is useful, so that special analysis methods can be used to manage confounding.

⚠ *Record* all the extraneous variables likely to be important (Sect. 7.8). This may include information about the *individuals* in the study, and the *circumstances* of the individuals in the study (that is, the circumstances the individuals find themselves in; these may not be measured on the individuals themselves).

Example 7.8 (Managing confounding: experimental study). For the *Himalaya* study, different methods can be used to manage confounding due to age.

The study could be *restricted* to people under 30. Age would be a *control variable.*

Blocking could be used by finding similar pairs of subjects (e.g., pairs of subjects of the same sex, with similar age and weight). One of each pair is given the refined cereal diet, and one given the *Himalaya 292* diet. The *differences* in faecal weight for each pair can be analysed using special methods (see Chap. 29 for example).

Information *about the individuals* could be recorded, such as age and pre-study weight. Information *about the circumstances* of the individuals could also be recorded, such as where they live. Then, special methods of *analysis* could be used to analyse the data.

Since the study is experimental, participants could be *randomly allocated* into one of two groups, so both groups would have a similar distribution of ages (and other potential confounders). Then groups could be randomly allocated to receive one of the diets (Fig. 7.2).

In the *Himalaya 292* study, individuals were randomly allocated to the diets (p. 1 033), which manages confounding due to age and other potential confounding variables also.

Example 7.9 (Managing confounding: observational study). Froud et al. [2018] studied 2 599 kiwifruit orchards using an observational study, exploring the relationship between the time since a bacterial canker was first detected (in weeks) as the explanatory variable, and the orchard productivity (in tray-equivalents per hectare) as the response variable.

The researchers also recorded potential extraneous variables such as 'whether the farm was organic', 'elevation of the orchard' and 'whether general fungicides were used'. These variables were used in their analysis to manage the potential effects of confounding.

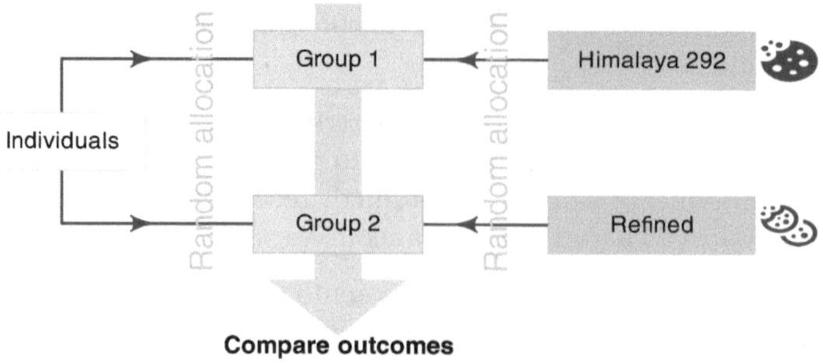

FIGURE 7.2: Random allocation can occur in two places for the *Himalaya* study.

Example 7.10 (Comparing study groups: observational study). An observational study compared the iron levels of active and sedentary women aged 18 to 35 [Woolf et al., 2009]. The active women ($n = 28$) and sedentary women ($n = 28$) were compared on a variety of characteristics (Table 7.2). The active women were similar to the sedentary women on these characteristics, but were (in general) slightly younger, slightly heavier, and slightly more likely to use hormonal contraceptives.

TABLE 7.2: The demographic information for those in the study of iron levels in women.

Characteristic	Active women	Sedentary women
Average age (in years)	20	24
Average weight (in kg)	68	62
Percentage using hormonal contraceptives	13	11

⚠ Observational studies *can* (and often do) have control groups. Indeed, one specific type of observational study is called a *case-control study* (Sect. 4.6.2). However, individuals are *not allocated to the control group* by the researchers in observational studies, so initially the control and study groups may be very different, which may explain any differences in the outcome.

Random *sampling* and random *allocation* are different concepts (Fig. 7.3) with different purposes, but are often confused.

- *Random sampling* impacts *external* validity. Its purpose is *finding individuals* to study, and is possible in both observational and experimental studies.
- *Random allocation* helps eliminate confounding, by distributing possible confounders across treatment groups, and is only possible in *experimental* studies. *Random allocation* impacts *internal* validity. Its purpose is *allocating treatments to individuals*, which does not occur in observational studies.

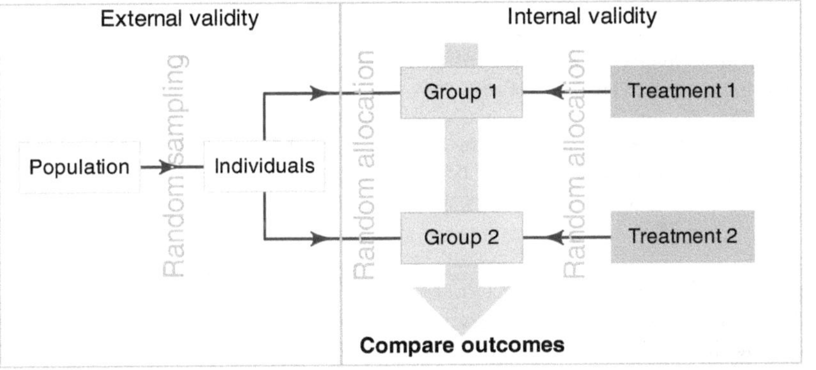

FIGURE 7.3: Comparing random allocation and random sampling.

7.3 Hawthorne effect and blinding individuals

People, and perhaps animals, may behave differently if they know (or think) they are being watched, which could compromise the internal validity of the study. This is called the *Hawthorne effect.*

Definition 7.3 (Hawthorne effect). The *Hawthorne effect* is the tendency of individuals to change their behaviour if they know (or think) they are being observed.

Example 7.11 (Hawthorne effect: observational study). Wu et al. [2018] examined hand hygiene (HH) of staff in a tertiary teaching hospital, using *covert* (secret) observers and *overt* (obvious) observers. HH compliance was higher with overt observation (78%) than with covert observation (55%).

The impact of the Hawthorne effect can be minimised by blinding the individuals, so that:

- the individuals do not know that they are *participating* in a study.
- the individuals do not know the *aims of the study.*
- the individuals do not know which *comparison group they are in.*

Any or all of these may be true, depending on the study design. Blinding individuals in all three of these ways is not always possible.

In *experimental* studies, *people* are often informed that they are in a study, due to ethics requirements (Sect. 5.2); they may not, however, know *which* treatment they have received. In *observational* studies, individuals *may* or *may not* know they are being observed. For instance, in a study where subjects' blood pressure is measured, subjects clearly know they are being observed, which has the potential to alter the subjects' behaviour (e.g., people become tense, called 'white-coat hypertension'). As far as possible, efforts should be made to ensure that individuals do not know that they are being observed (the participants are *blinded*).

Example 7.12 (Hawthorne effect: experimental study). For the *Himalaya* study (Example 7.2), the article reports that (p. 1033):

> The study was explained fully to the subjects, both verbally and in writing, and each gave their written, informed consent...

That is, the subjects knew they were in a study, and knew the aims of the study, so the Hawthorne effect may influence the results in this study. However, the subjects did not know *which* diet they were given.

Example 7.13 (Hawthorne effect: experimental study). People are more health-conscious if they know they will be examined regularly. For example, a study aiming to increase fruit and vegetable intake in young adults [Clark et al., 2019] noted that the observed increases in intake 'could be explained by the Hawthorne effect' as adults 'know they are being observed...' (p. 96).

Example 7.14 (Hawthorne effect: observational study). During the COVID-19 lockdowns in Denmark, Olesen and Feldthaus [2021] covertly observed adults entering a large mall in Copenhagen. They noticed that (p. 1)

> Almost all subjects [340/345 (99%)] wore a personal protective face mask, but only 141 (41%) made use of the hand sanitizer.

Both masks and hand sanitiser were recommended by the Danish Health Authority, but the adherence to the safety measures were very different. The authors surmised (p. 1):

> ... wearing a face mask corresponded to being observed continuously [...] hand hygiene takes moments to perform, and no one can see whether or not it has been done.

In other words, wearing a face mask is obvious (that is, others could *observe* whether the subjects was adhering to this guideline) but hand hygiene is not (so other people *could not* observe whether the subject was adhering to this guideline). The authors conclude that 'the Hawthorne effect may explain why almost all subjects wore a face mask'.

7.4 Observer effect and blinding researchers

Perhaps surprisingly, researchers' expectations or hopes may unconsciously influence how the researchers interact with the individuals and record observations. In addition, this may (unconsciously) influence the behaviour of the individuals in the study. This is called *observer effect*. (In experiments, the observer effect is sometimes called the *experimenter effect*.) This could compromise the internal validity of the study.

Definition 7.4 (Observer effect). The *observer effect* occurs when the researchers unconsciously change their behaviour to conform to expectations because they know what values of the explanatory variable apply to the individuals. This may then cause the *individuals* to change their behaviour or reporting also.

The impact of the observer effect can be minimised by blinding the *researchers*, so that they do not know which treatments the individuals are receiving. The researchers *giving* the treatment and the researchers *evaluating* the treatment can both be blinded, by using a third party. For example, the researchers may give an assistant two drugs, labelled A and B. The assistant administers the drug and evaluates the participants' response to the treatments. Later, the assistant tells the researchers whether Drug A or Drug B performed better, but only the researchers know which drugs the labels A and B refer to (Fig. 7.4).

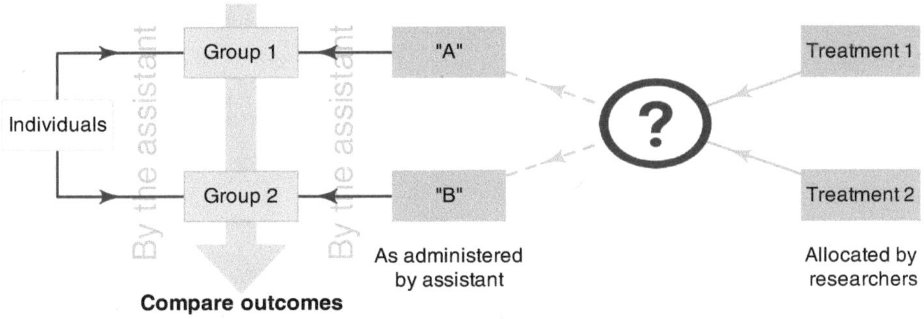

FIGURE 7.4: Using a third party to avoid the observer effect.

Example 7.15 (Observer effect: experimental study). Seo et al. [2020] examined the impact of an injection to alleviate post-operative umbilical pain, and stated (p. 392):

> ...the postoperative pain scores were gathered by a nurse practitioner who was blinded to the usage of bupivacaine to avoid observer-expectancy bias [i.e., the observer effect].

The observer effect does not just apply to situations with *people* as individuals.

Example 7.16 (Observer effect). Clever Hans was a horse that seemed to perform simple mental arithmetic. By using an experiment where the people interacting with the horse were blinded, Carl Stumpf realised that the horse was responding to involuntary (and unconscious) cues from the trainer.

The same effect has been observed in narcotic sniffer dogs [Bambauer, 2012], who may respond to their handlers' unconscious cues.

⚠ The *observer effect* is when the researcher *unconsciously* influence the individuals, and are not aware it is occurring. *Intentionally* influencing the individuals is fraud.

The observer effect can impact both observational and experimental studies. For example, consider a study measuring the blood pressure of smokers and non-smokers [Verdecchia et al., 1995]. This study is observational (individuals cannot be allocated to be a smoker or non-smoker), but if the researchers *know* an individual is a smoker when they measure blood pressure, then the observer effect could impact the results (recalling that the observer effect is an *unconscious* effect). For example, the researchers may *expect* smokers to have a high blood pressure.

The observer effect could be managed by *first* measuring the blood pressure, and *then* asking if the individual was a smoker or not. That is, the researchers may be *blinded* to whether the subject is a smoker when they measure blood pressure. This may only be partially successful; the researcher may see the subject carrying cigarettes, or can smell smoke on their breath, for example. Nonetheless, since it may prove at least partially successful and is easy to implement, this strategy should form part of the research design.

Example 7.17 (Observer effect: observational study). Zimova et al. [2020] took photos of snowshoe hares, at various stages of moulting and in various environmental conditions. Eighteen independent observers rated the moult stage from the photographs (p. 4):

> ... images were randomly named and sorted, with the dates [...] removed to minimize observer expectancy bias [i.e., the observer effect].

Blinding the observer is not always possible, but should be used when possible to improve the internal validity of the study.

7.5 Placebo effect, controls, objective data, and blinding

Perhaps surprisingly, individuals in a study may report effects of a treatment, even if they have not received an active treatment. This could compromise the internal validity of the study. This is called the *placebo effect*, which generally only impacts people as individuals.

Definition 7.5 (Placebo effect). The *placebo effect* occurs when individuals report perceived or actual effects, despite not receiving an active treatment.

For example, people who attend therapy expect a positive outcome; this expectation may result in temporary or perceived (or sometimes even real) improvements in their condition. This is the placebo effect.

To manage the placebo effect, researchers should record *objective* data (Sect. 7.9) rather than patient-reported (subjective) outcomes when possible [Enck et al., 2013]. (The operational definitions (Sect. 2.10) for the variables should make clear whether subjective or objective data are recorded.) Using a *control* group (Def. 2.17), if possible, is also useful: it acts as a benchmark for detecting changes in the outcome due to the treatment of interest. In addition, *blinding* the individuals and the researchers may help manage the placebo effect, as then the individuals cannot know which group they are in.

Example 7.18 (Placebo effect). Three active pain relievers were compared to different-coloured placebo [Huskisson, 1974] in 22 patients. The most pain relief was experienced by those taking *red* placebos (Fig. 7.5), who experienced even more pain relief than those given true pain relievers. Note that the outcome is subjective: a *patient*-reported outcome.

Since the placebo effect is concerned with individual responses to *allocated treatments*, it is not directly relevant to observational studies.

Example 7.19 (Placebo effect). In the *Himalaya* study, the individuals 'were not told the identity of the test cereal in the foods provided' (Bird et al. [2008], p. 1 033). The subjects were blinded to the diet they were exposed to. However, some may *think* they are on the refined cereal or *Himalaya* diet, and respond accordingly (perhaps unconsciously). The use of the refined cereal was acting as a control (Def. 2.17). Researchers measured faecal weight, an *objective* outcome, to minimise the placebo effect.

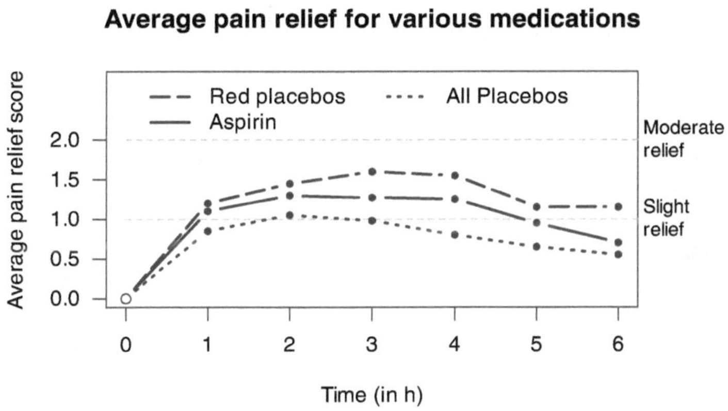

FIGURE 7.5: Pain relief, for various pain relief medicine and different-coloured placebos.

7.6 Carryover effect and washouts

In the *Himalaya* study (Example 7.2), the diet is a *between-individuals* comparison: one group of patients was given the refined cereal diet (the control), and a different group of people was given *Himalaya 292*. The study *also* used a *within-individuals* comparison: each person in the study was actually placed on both diets at different times.

Suppose all patients spent four weeks on the *Himalaya 292* diet, then the next four weeks on the refined cereal diet. Potentially, the first diet could still be impacting the subjects' faecal weight for a little while after stopping the first diet. This could compromise the internal validity of the study. This is an example of the *carryover effect*: when the influence of one treatment or condition on the response variable carries over to influence the value of the response variable for next treatment or condition. The carryover effect is only a concern for *within-individuals* comparisons.

Definition 7.6 (Carryover effect)**.** The carryover effect occurs when the influence of one treatment or condition on the response variable influences the response variable for subsequent treatments or conditions (in a repeated-measures study).

The impact of the carryover effect may be minimised by using a *washout* or similar between treatments or conditions. For example, after tasting a food sample, participants may rinse their mouth with water before tasting another food sample. For the *Himalaya* study, the participants could spend two weeks on their usual (before-study) diet, before starting each of the diets in the study. This is called a *washout period.*

Example 7.20 (Carryover effect: experimental study)**.** In the *Himalaya* study, 'there was no washout period' (Bird et al. [2008], p. 1 033) since the response variable was only recorded after individuals spent four weeks on each diet. Since faecal weight was not measured until the *end* of the four-week periods, the carryover effect is essentially irrelevant.

Sometimes, in experimental studies, researchers can *randomly allocate* the *order* in which the treatments are used (a *crossover study*). That is, some participants start by spending four weeks on the *Himalaya 292* diet, then four weeks on the refined cereal diet; meanwhile, other participants start by spending four weeks on the refined cereal diet, then four weeks on the *Himalaya 292* diet.

Example 7.21 (Carryover effect). In the *Himalaya* study (Example 7.2), subjects were allocated randomly to begin the study on the *Himalaya 292* diet *or* the refined cereal diet.

Example 7.22 (Washout periods: experimental study). MacDonald et al. [2006] required paramedics to conduct eight different tasks (such as electrical defibrillation and intravenous cannulation). Each of the paramedics began the series of tasks at a random task, to mitigate the carryover effect. A washout period between tasks (i.e., a rest time) was also used.

The carryover effect also is a potential concern to internal validity in observational studies involving a within-individuals comparison. However, since treatments are *not allocated* in observational studies, carryover effects may be difficult to prevent, as washouts cannot be imposed, and the order of the conditions cannot be imposed. However, *observing* individuals exposed to Condition A then Condition B, and other individuals exposed to Condition B then Condition A, may be possible.

Example 7.23 (Carryover effects: observational study). Norris [2005] studied the carryover effect in ecological observational studies of animals (p. 181):

> ...individuals occupying poor quality winter habitat may experience reduced reproductive success the following breeding season when compared to individuals occupying high quality winter habitat.

7.7 Describing blinding

Blinding occurs when those involved in the study do not know information about the study. The *individuals* in the study may be blinded to

- whether they are *involved in a study.*
- the *aims of the study* in which they are participants.
- *which comparison group* they are in.

Any or all of these may be true, depending on the study design.

The *researchers* and the *analysts* can be blinded to which comparison groups apply to the individuals (to help manage the observer effect).

When blinding is used in as many ways as possible, the internal validity of the study is increased and bias reduced. However, when people are the individuals, ethics requirements may mean that the individuals need to know they are in a study (especially if experimental), and the purpose of the study.

If *only* the individuals are blinded to the comparison groups, the study is called *single blind.* If *both* the researchers and participants are blinded to the comparison groups, the study is called *double-blind.* If the researchers, participants *and* the analyst are blinded to the comparison groups, the study is sometimes called *triple-blind.* Rather than using these terms, explicitly stating who or what is blinded to which parts of the study is clearer.

Blinding should be considered in all studies when possible (it is *not* always possible). *Blinding participants does not just apply to people*; it also may apply to animals (Example 7.16).

Example 7.24 (Double-blinding). Bulte et al. [2014] compared yields from modern and traditional cowpea crops in Tanzania. The two seed types ('traditional' and 'modern') were made similar in appearance, so the farmers were blinded to which group they were in (control or treatment). The seed type would eventually become obvious as the crop grew, but 'key inputs were already provided' by then (p. 817). In addition, the researchers interacting with the farmers were not informed about the type of seed distributed.

In observational studies, blinding individuals *may* be easier than in experimental studies (Sect. 7.3). Blinding the researchers may be difficult, since the researchers need to record the value of the explanatory variable.

Example 7.25 (Blinding: observational studies). Emerson et al. [2010] studied Achilles tendinopathy in gymnasts, by comparing 40 elite gymnasts with 41 similar controls who were non-gymnasts. The authors state (p. 38) that

> Although the primary investigator was blind to the clinical status of the subjects, there was no blinding to whether each subject was in the gymnast or control group during image collection [...]

When the images were reviewed, however, the article explains that the examiner was unaware of the clinical state and group of the subjects.

The paper explains who was blinded and to what parts of the study they were blinded.

7.8 Recording extraneous variables

One way to design a high-quality study is to record information about many (potential) extraneous variables. Various reasons for doing this have been given.

- To evaluate *external validity* to determine if the sample is representative of the population (Sect. 6.6), by comparing the sample and population.
- To improve *internal validity*, by helping to manage confounding:
  - by avoiding lurking variables (Sect. 3.4).
  - by determining if the comparison groups are similar (Sect. 7.2).
  - by using the information in analysis (Sect. 7.2).

⚠ *Record the values of all extraneous variables that may be important in the study!*

Example 7.26 (Poor internal validity). In the 1800s, Semmelweis recorded mortality rates of women after childbirth over many years [Dunn, 2005] at two clinics:

- in Clinic 1, with male doctors delivering babies: 9.9%.
- in Clinic 2, with female midwives delivering babies: 3.4%.

Was the difference in mortality rate (the outcome) due to the sex of the person delivering the babies (the comparison)?

One possible confounder was the clinic; however, the clinic was eliminated as an explanation. For example, Clinic 2 was actually *more* overcrowded than Clinic 1, and the climate was similar for both clinics.

However, an important *lurking variable* was present. At the time, the benefits of hand-washing were not understood, nor commonplace. Many (male) doctors performed autopsies immediately before delivering babies, without washing their hands between procedures. In contrast, autopsies were not performed by the (female) nurses.

The lurking variable was 'whether the baby was delivered by someone with clean hands', which was related to the mortality rate *and* to the sex of the person delivering the baby. The female midwives had clean hands, and hence the mortality rate was (relatively) low. The male doctors did *not* have clean hands, and hence the mortality rate was high.

After instituting hand-washing for doctors, the mortality rate in Clinic 1 reduced to a rate similar to that in Clinic 2.

7.9 Recording objective data

Recording *objective* data is often more reliable than recording *subjective* data, as subjective data can be influenced by the Hawthorne, observer or placebo effects. Perceptions are often unreliable also. However, sometimes recording objective data is not possible, and sometimes the researchers are explicitly interested in the subjective responses of people to certain treatments or conditions.

Definition 7.7 (Subjective and objective data). *Subjective* data refers to opinions, feelings, and interpretations (by the subjects or the researchers). *Objective* data refers to facts and measurable evidence.

⚠ If possible, objective data should be recorded whenever possible.

Example 7.27 (Subjective and objective data). Ueberham et al. [2019] studied cyclists using everyday routes over one week in Leipzig (Germany). Sixty-six cyclists wore sensors that *objectively* recorded particle number counts (i.e., pollution), noise, humidity and temperature. The cyclists also *subjectively* recorded similar information.

The researchers concluded that (p. 1)

> Except for heat, no significant associations between the objective and subjective data were found.

That is, the subjective and objective data generally did not agree, except for heat. The perceptions of heat may have been influenced by the Hawthorne effect (p. 7):

> ...most people are pre-informed about the daily temperature by the weather forecast and expect a certain degree of heat, which ultimately also affects their perception of it to a great extent.

Example 7.28 (Subjective and objective data). Johnson et al. [2021] asked 70 people in south-west Ireland to *subjectively* self-report their Body-Mass index (BMI) category, as 'Normal' or 'Overweight'. *Thirty-six* subjects self-reported their BMI category as 'Normal'.

The researchers also *objectively* recorded the BMI of the same subjects. *Twenty-nine* subjects were objectively categorised as 'Normal'.

7.10 Chapter summary

Designing effective studies (Fig. 7.6) requires researchers to *manage or minimise confounding* where possible, by: *restricting* the study to certain groups; *blocking* individuals into similar groups; through special *analysis* methods; and/or through *random allocation* of the units of analysis. Random allocation is only possible for experimental studies.

Well-designed studies manage the *Hawthorne effect* (e.g., by blinding *participants*); the *observer effect* (e.g., by blinding the *researchers*); the *placebo effect* (experimental studies only; e.g., by using controls, objective outcomes and blinding subjects); and the *carryover effect* (e.g., by using a washout, or randomly allocating the treatment order). Recording objective data is usually better than recording subjective data.

Often, however, not all of these strategies can be used. For instance, people usually know they are involved in an experimental study, so the Hawthorne effect may impact conclusions. In these cases, the possible impacts should be minimised as far as possible, and then the likely impact on the conclusions discussed. The impact of these issues are often reported as *limitations* (Chap. 8).

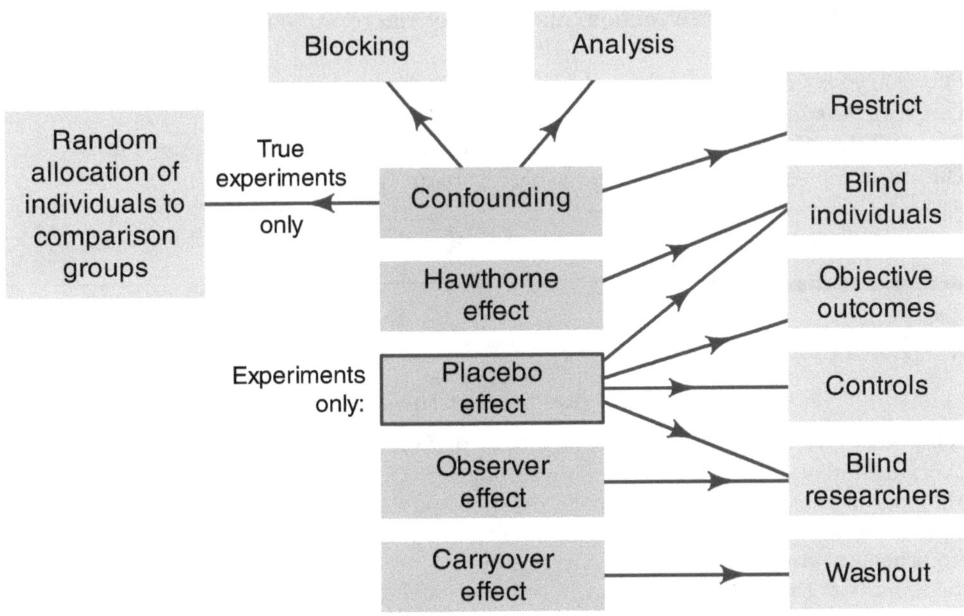

FIGURE 7.6: Design considerations for designing studies. Note: lurking variables become confounding variables when recorded in the study, and so can be managed as confounding variables. The arrows indicate the main design strategies to (perhaps partially) manage the indicated potential bias. Not all strategies are possible for every study.

Example 7.29 (Research design). Cross et al. [2019] (p. 3) compared chest compressions by student paramedics using dominant and non-dominant hands, and stated:

> ... participants were allocated randomly to one of two groups: 'dominant hand on chest' or 'non-dominant hand on chest'. Group allocation was determined by a computer-generated randomisation schedule...

The participants were blinded to the *purpose* of the study, but not to which group they were allocated. The analyst was also blinded to the group allocations. This study used many good design features.

7.11 Quick review questions

Doosti-Irani et al. [2016] wanted to determine the relationship between the depth of bruising on apples and the impact force. The researchers purposefully hit apples with three different *forces* (200, 700 and 1200 mJ) to inflict bruises on the apples. The researchers then recorded the *depth* of the bruising. The study was conducted separately for three different *regions* of the apple (lower; middle; upper), and each apple was only used once.

Are the following statements *true* or *false*?

1. The *response variable* is 'the depth of bruising'.
2. The *explanatory variable* is 'the force used on the apples'.
3. The variable 'location of the bruising' would be classified as a confounding variable'.
4. The researchers could minimise the effects of confounding by incorporating potential confounding variables in the analysis.
5. The researchers could use random allocation of the treatments to the apples to minimise confounding.
6. The *carryover* effect is likely to be a big problem in this study.
7. The *Hawthorne* effect is likely to be a big problem in this study.
8. The *placebo* effect is likely to be a big problem in this study.
9. The *observer* effect is likely to be a big problem in this study.

7.12 Exercises

Answers to odd-numbered exercises are given at the end of the book.

Exercise 7.1. Are the following statements *true* or *false*?

1. Experimental studies *must* use random samples.
2. An experimental study *must* blind the researchers.
3. Only observational studies can manage the observer effect.
4. Experimental studies *must* use a control group.
5. Using random samples is important in observational studies as a way to manage confounding.

Exercise 7.2. Which of the following statements are true?

1. Observational studies cannot have a control group.
2. Only experimental studies can use random allocation to avoid confounding.
3. An experimental study *must* blind the participants.

4. Only experimental studies can use random sampling.
5. In experimental studies, the treatments *must* be allocated by the researchers.

Exercise 7.3. Which of the following can improve internal validity in *experimental* studies?

- Blinding the individuals
- Using a control group.
- Using special methods of analysis.
- Randomly allocating treatments to groups.
- Blinding the researchers.
- Using random samples.

Exercise 7.4. Which of the following can improve internal validity in *observational* studies?

- Blinding the individuals
- Using a control group.
- Using special methods of analysis.
- Randomly allocating treatments to groups.
- Blinding the researchers.
- Using random samples.

Exercise 7.5. Is the Hawthorne effect only a (potential) issue for experiments. Explain.

Exercise 7.6. Lorenz et al. [2019] compared the efficacy of a new type of toothpaste. Participants were given either a new or an existing toothpaste, and evaluations of plaque remaining on the teeth were taken after brushing. All participants knew they were being assessed after brushing.

Would the Hawthorne effect likely impact the internal validity of this study? Explain.

Exercise 7.7. A study compared the average amount of pollen returned to the hive per bee, for two types of native Australian bees: yellow and black carpenter bees, and green carpenter bees. The researchers also recorded the size of the hive, among other variables. *Why* did they do this?

Exercise 7.8. In a study to treat septic shock, Hwang et al. [2020] used two study groups of size $n = 58$ each: one group received the treatment of interest (intravenous infusion of vitamin C and thiamine) and the other group received intravenous saline.

Explain why the researchers gave saline to 58 subjects, when it has no chance of successfully treating septic shock. Is this ethical?

Exercise 7.9. Consider a study comparing the average weight loss for patients who are *instructed* to do about 30 mins of exercise a day (Group A), to patients who are instructed to do about 60 mins of exercise a day (Group B). Which of the following statements are true?

1. This is an experimental study.
2. The extraneous variable is the amount of exercise per day (in h).
3. The response variable is the weight loss for each person.
4. The explanatory variable is whether the patient performs about 30 or 60 mins of exercise per day.
5. The response variable is the *average* weight loss.
6. The explanatory variable is the number of minutes of exercise the patient does per day.
7. Age is likely to be a lurking variable.
8. Age is likely to be an extraneous variable.
9. Age is likely to be a confounding variable.
10. Which (if any) of the following are possible *confounding* variables?
    - The sex of the patients.
    - The initial weight of the patients.
    - The names of the patients.

Exercise 7.10. Stafford et al. [2010] studied smoking in alfresco restaurants in two cities in Western Australia. The concentration of particulate matter with a diameter smaller than or equal to 2.5 (per cubic metre of air) was recorded ($PM_{2.5}$) from 12 cafés and 16 pubs. The researchers were interested in the relationship between $PM_{2.5}$ and the number of smokers. They also recorded the wind strength (calm; light breeze; windy) and the amount of cover (fully open; overhead cover only; overhead cover and enclosed sides).

1. Is this an experimental or observational study?
2. What are the response and explanatory variables?
3. What are the extraneous variables, if any?
4. Is blinding the individuals possible?
5. Is random allocation possible?

Exercise 7.11. In a study of time spent applying sunscreen [Heerfordt et al., 2018], the aim was to 'determine whether time spent on sunscreen application is related to the amount of sunscreen used' (p. 117). The study is described as follows (p. 118):

> The volunteers were asked to apply the provided sunscreen [...] the way they would normally do on a sunny day at the beach in Denmark [...] The volunteers wore swimwear during the whole session. No other information was given. Participants applied sunscreen behind a curtain and were not observed during application. Measurements of time and sunscreen weight were made without the subjects' being aware of this.

1. Is this an experimental or observational study?
2. What are the response and explanatory variables?
3. The researchers also recorded age, height, weight and body surface area of each participant. Why would they have done this?
4. The researchers also compared the average values of the response variable for males and females, and the average values of the explanatory variable for males and females. Why would they have done this?
5. What design features are evident in the quote?

Exercise 7.12. Paramedics were involved in a study to compare two treatments (A and B) for Post Traumatic Stress Disorder (PTSD), as randomly allocated to two groups of patients.

1. Is this an experimental or observational study?
2. What would be the control group?
    a. The group receiving Treatment A.
    b. A group of paramedics who do not have PTSD.
    c. The group receiving Treatment B.
    d. A group of paramedics not involved in the study.
    e. A group of people with PTSD who are not paramedics.
    f. A group receiving a pill that looks just like Treatment A and B, but has no effective ingredient.
3. The patients did not know which treatment they received. What is this called?
4. What is the purpose of blinding the participants?

Exercise 7.13. Blondal et al. [2023] studied older Icelandic adults, to (p. 632)

> ...investigate effects of six-month nutrition therapy on hospital readmissions, LOS [length of hospital stay], mortality and need for long-term care residence...

Patients in the study were all aged over 64-years-of-age, and were randomly allocated into either the intervention ($n = 53$) or control ($n = 53$) groups. The intervention group received 'nutrition therapy', which included free delivery of energy- and protein-rich foods for six months after discharge from hospital. Patients with cognitive impairment, dietary allergies and undergoing active cancer treatment were excluded from the study.

Statistical software (SPSS, version 26.0) was used to generate the random numbers for randomly allocating patients into groups, and the allocations were hidden from the researchers. Table 1 of the published article compared the two groups on (among other variables) age, percentage female, height, weight, and percentage with a higher education. Ethics approval was given by the Ethics Committee for Health Research of the National University Hospital of Iceland.

Identify good design principles used in this study, and their purpose. Can you identify any improvements that could be made to the study design?

Exercise 7.14. Braga et al. [2021] evaluated the use of body-worn cameras (BWC) by police on residents' perceptions of the police. The 40 precincts in New York City with the highest numbers complaints against New York Police Department officers were part of the study. Twenty precincts were matched with 20 other precincts (using demographics, socioeconomic characteristics, crime, and police activity). Using a computer, one of each pair of precincts was allocated to have officers wear BWC, and the other to not wear BWCs. The researchers compared the officers in each group, and found the officers in the two groups were similar 'in terms of demographics, length of service, rank, work activities, and number of citizen complaints' (p. 285).

Identify the good design principles used in this study (using the language of this chapter where possible), and their purpose. Can you identify any improvements that could be made to the study design?

Exercise 7.15. Bulte et al. [2014] compared yields from two types of cowpea in rural Tanzania. Two different studies were used for the comparison; for both studies, traditional and new cowpea seeds were randomly allocated to a sample of farmers.

In Study A, farmers knew which seeds were the newer variety, and the researchers interacting with the farmers also knew the types of seed allocated. In Study B, the farmers were blinded to which seeds were new and which were traditional (both seeds were the same size and colour), and the researchers interacting with the farmers also did *not* know the types of seed allocated.

Farmers who knew they received the modern seed (Study A) planted their seed on larger plots than farmers in Study B; that is, they behaved differently.

When the data from Study A were analysed, the new seed yield was 27% greater, compared to the traditional seed. However, in Study B, the yield of new seed was very similar to the yield for the traditional seed.

1. Discuss the blinding used in each study.
2. Explain the discrepancies from the two methods, using the language of this chapter.
3. Which method would be more likely to be recording accurate information? Why?

Exercise 7.16. Greier et al. [2021] studied 36 Austrian eighth-grade students, recording the time spent performing moderate-to-vigorous physical activity (MVPA). For each student, the time spent in MVPA was recorded in two ways: (a) using a wrist-worn accelerometer, worn for seven days; and (b) self-reported using a questionnaire at the end of the week.

MVPA reported in the questionnaire was more than double the time measured using the accelerometer; thus, 88.9% of students met PA recommendations based on self-report, but only 22.2% did so based on the accelerometer data.

1. Which method recorded *subjective* data? Which method recorded *objective* data?
2. Explain the discrepancies in the results from the methods, using the language of this chapter.
3. Which method would be more likely to be recording accurate information? Why?

Exercise 7.17. A scientist is testing whether tap water tastes the same as bottled water (based on Teillet et al. [2010]). Subjects taste either bottled or tap water from a plastic cup, and give a rating of the taste on a scale of 1 (terrible) to 5 (fantastic). You decide to address this question using an *experiment*. Describe what these might look like for this study, and which are feasible: random allocation; blinding; double-blinding; control; carryover effect; finding a random sample.

What potential problems can you identify with the research design?

Exercise 7.18. A scientist is testing whether tap water tastes the same as bottled water (based on Teillet et al. [2010]). Subjects taste either bottled or tap water from a plastic cup, and give a rating of the taste on a scale of 1 (terrible) to 5 (fantastic).

You want to answer this question using an *observational study*. Describe what these might look like for this study, and which are feasible: random allocation; blinding; double-blinding; control; carryover effect; finding a random sample.

Exercise 7.19. A scientist compares the effects of two types of fertiliser on the yield of tomatoes (based on Klanian et al. [2018]). He plants tomato seedlings, and fertilises with Fertiliser I, and later records the yield of tomatoes. He then immediately plants more tomato seedlings in the same field, fertilises with Fertiliser II, and measures the yield of tomatoes.

What potential problems can you identify with the research design?

Exercise 7.20. Skulberg et al. [2004] compared two office-cleaning methods (p. 72):

> The participants were randomly allocated to an intervention group or a control group using group level matching by sex, level of irritation symptom index, and allergy status... The participants and the field researchers were blinded to the group status of the participants...

All offices were cleaned after the employees had left the building for the evening.

The researchers compared the change in nasal congestion for the two groups (intervention: 'a comprehensive cleaning'; control: 'a superficial cleaning'), finding only small differences between the two groups. In the analysis, the researchers incorporated age and sex of the office workers.

1. How did the researchers manage confounding?
2. What other design features are evident from the quote?
3. What is the response variable?
4. What is the explanatory variable?
5. What extraneous variables are apparent?

Exercise 7.21. *Formwork* is used in construction with reinforced concrete, and can be labour intensive. Mine et al. [2015] examined the relationship between the floor area of the building (in m^2 per storey) and the number of hours of labour needed for constructing the formwork (in person-mins per storey). The researchers also recorded the average age of the workers (in years); the average years of experience of the workers (in years); and the storey height (in meters) for each of $n = 15$ multi-storey buildings in the study.

Two observers recorded the labour time by observing workers from the start to the end of the work.

1. What is the *explanatory variable*?
2. What is the *response variable*?
3. What *type* of description is appropriate for the variable 'age of the workers'?
   a. A confounding variable, since it is likely to be related to the explanatory variable only.
   b. A confounding variable, since it is likely to be related to the response variable only.
   c. An extraneous variable, because it is likely to be related to the response variable only.
   d. A lurking variable, since we don't know how it might be related to the response and explanatory variables.
4. What is the most likely way to manage confounding in this study restricting, blocking, analysis, random allocation?
5. True or false: the *carryover* effect is likely to be a big problem in this study.
6. True or false: the *Hawthorne* effect is likely to be a big problem in this study.
7. True or false: the *placebo* effect is likely to be a big problem in this study.
8. True or false: *observer* effect is likely to be a big problem in this study.

Answers to *Quick review* questions: 1. True. **2.** True. **3.** False; extraneous. **4.** True. **5.** True. **6.** False. **7.** False. **8.** False. **9.** False.

8

Research design limitations

So far, you have learnt to ask an RQ and design research studies. **In this chapter**, you will learn to identify limitations to:

- internal validity.
- external validity.
- ecological validity.

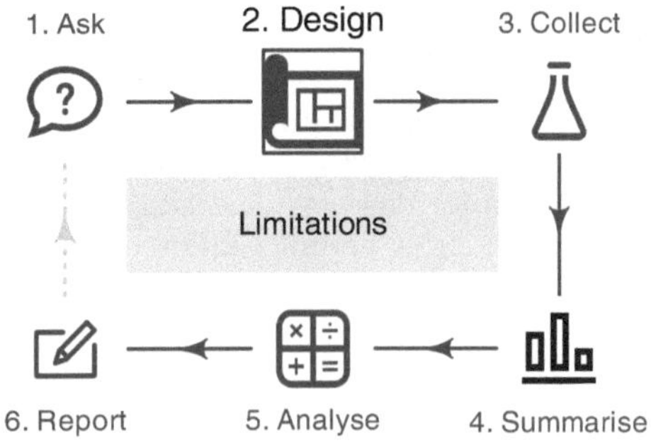

8.1 Introduction

The type of study (Chap. 4) and the research design determine how the results of the study should be interpreted. Ideally, all studies would be perfectly externally and internally valid; in practice this is very difficult to achieve. Practically *every* study has limitations. The results of a study should be interpreted in light of these limitations. Limitations are not necessarily *problems*.

Limitations generally can be discussed through three components.

- *Internal validity* (Sect. 3.1): discuss any limitations to internal validity due to the research design (such as identifying possible confounding variables). This is related to the *effectiveness* of the study within the sample (Sect. 8.2).
- *External validity* (Sect. 6.1): discuss how well the sample represents the intended population. This is related to the *generalisability* of the study to the intended population (Sect. 8.3).
- *Ecological validity*: discuss how well the study methods, materials and context approximate the real situation of interest. This is related to the *practicality* of the results to real life (Sect. 8.4).

All these issues should be addressed when considering the study limitations.

⚠ Almost every study has limitations. *Identifying* potential limitations, and *discussing the likely impact* they have on the interpretation of the study results, is important and ethical.

Different types of research studies have limitations. Experimental studies, in general, have higher internal validity than observational studies, since more of the research design in under the control of the researchers; for example, random allocation of treatments is possible to minimise confounding.

⚠ Only well-designed and well-conducted experimental studies can show cause-and-effect relationships.

However, experimental studies may suffer from poor *ecological* validity; for instance, laboratory experiments are often conducted under controlled temperature and humidity. Many experiments also require that people be told about being in a study (due to ethics), and so internal validity may be comprised (the Hawthorne effect).

Example 8.1 (Limitations due to the study type). Giandomenico et al. [2022] studied retro-fitting houses with energy-saving devices, and found large discrepancies in energy savings for observational studies (12.2%) and experimental studies (6.2%). The authors say that 'this finding reinforces the importance of using study designs with high internal validity to evaluate program savings' (p. 692).

8.2 Limitations related to internal validity

Internal validity refers to the extent to which a cause-and-effect relationship can be established in a study, eliminating other possible explanations (Sect. 3.1); that is, the *effectiveness* of the study in the sample. A discussion of the limitations of internal validity should cover, as appropriate: possible confounding and lurking variables; the impact of the Hawthorne, observer, placebo and carryover effects; the impact of any other design decisions.

If anything is likely to compromise internal validity, the implications on the interpretation of the results should be discussed. For example, if the participants were not blinded, this should be clearly stated, and the conclusion should indicate that the individuals in the study may have behaved differently than usual.

Example 8.2 (Study limitations). Axmann et al. [2020] randomly allocated Ugandan farmers to receive, or not receive, hybrid maize seeds. One potential threat to internal validity was that farmers receiving the hybrid seeds could share seeds with their neighbours. Hence, the researchers contacted the 75 farmers allocated to receive the hybrid seeds; none of the contacted farmers reported selling or giving seeds to other farmers. This extra step increased the internal validity of the study.

Maximising internal validity in *observational studies* is more difficult than in experimental studies (since random allocation is not possible). However, the internal validity of *experimental studies* involving people is often compromised because people must be informed that they are participating in a study.

Example 8.3 (Internal validity). In a study of the hand-hygiene practices of paramedics [Barr et al., 2017], *self*-reported hand-hygiene practices were very different from what was reported by *peers*. That is, how people self-report their behaviours may not align with how they actually behave, which influenced the internal validity of the study.

8.3 Limitations related to external validity

External validity refers to the ability to *generalise* the findings made from the sample to the entire *intended* population (Sect. 6.1). For a study to be externally valid, it must first be internally valid: that is, if the study is not effective in the sample studied (i.e., internally valid), the results may not apply to the intended population either.

⚠ External validity refers to how well the sample is likely to represent the *intended population* in the RQ.

If the population is Californians, then the study is externally valid if the sample is representative of Californians. The results *do not* have to apply to people in the rest of the United States to be externally valid (though this can be commented on too). The intended population is *Californians.*

External validity depends on *how* the sample was obtained. Results from random samples (Sect. 6.5) are likely to generalise to the population and be externally valid. (The analyses in this book assume all samples are *simple random samples.*) Furthermore, results from approximately representative samples (Sect. 6.6) *may* generalise to the population and be externally valid if those *in* the study are not obviously different from those *not in* the study.

Any inclusion criteria, exclusion criteria or control variables may also limit the external validity of the study.

Example 8.4 (External validity). Gammon et al. [2012] identified (for well-documented reasons) a *population* of interest: 'women of South Asian origin living in New Zealand' (p. 21). The women in the *sample* were 'women of South Asian origin [...] recruited using a convenience sample method throughout Auckland' (p. 21).

The results may not generalise to the *intended* population (*all* women of South Asian origin living in New Zealand) because all the women in the sample came from Auckland, and the sample was not a *random* sample from this population anyway. The study was still useful however, since we have still learnt information about the population that is represented by the sample, which may be *similar* to the intended population.

Example 8.5 (Using biochar). Farrar et al. [2018] studied growing ginger using biochar on one farm at Mt Mellum, Australia. The results may only generalise to growing ginger at Mt Mellum, but since ginger is usually grown in similar types of climates and soils, the results *may* apply to other ginger farms also.

8.4 Limitations related to ecological validity

The likely *practicality* of the study results in the real world should also be discussed. This is called *ecological validity.*

Definition 8.1 (Ecological validity). A study is *ecologically valid* if the study methods, materials and context closely approximate the real situation of interest.

Studies don't *need* to be ecologically valid to be useful; much can be learnt under special conditions, as long as the potential limitations are understood when applying the results to the real world. The ecological validity of experimental studies may be compromised because the experimental conditions are sometimes highly controlled (for good reason).

Example 8.6 (Ecological validity). Consider a study to determine the proportion of people that buy coffee in a reusable cup. People could be *asked* about their behaviour. This study may not be ecologically valid, as what people *do* may not align with what they *say* they will do (i.e., subjective data).

An alternative study could *watch* people buy coffee at various coffee shops, and record what people *do* in practice. (i.e., objective data). This second study is more likely to be *ecologically valid*, as real-world behaviour is observed.

8.5 Chapter summary

The limitations in a study need to be identified, and may be related to:

- *internal validity* (effectiveness); how well the study is conducted within the sample, isolating the relationship of interest.
- *external validity* (generalisability); how well the sample results are likely to apply to the intended population.
- *ecological validity* (practicality); how well the results may apply to the real-world situation of interest.

8.6 Quick review questions

Are the following statements *true* or *false*?

1. When interpreting the results of a study, the steps taken to maximise internal validity should be evaluated.
2. If studies are not externally valid, then they are not useful.
3. When interpreting the results of a study, the steps taken to maximise external validity do not need to be evaluated.
4. When interpreting the results of a study, ecological validity is about the impact of the study on the environment.

8.7 Exercises

Answers to odd-numbered exercises are given at the end of the book.

Exercise 8.1. Gentile [2022] examined how people can save energy through lighting choices. The study states (p. 9) that the results 'are limited to the [sample] and cannot be easily projected to other similar settings'.

What type of validity is being discussed here?

Exercise 8.2. Fill the blanks with the correct word: *internal*, *external* or *ecological*.

When interpreting the results of studies, we consider the practicality (__________ validity), the generalisability (__________ validity) and the effectiveness (__________ validity).

Exercise 8.3. A student project asked if 'the average word retention is higher in male students than female students at UniX'. When discussing *external validity*, the students stated:

> We cannot say whether or not the general public have better or worse word retention compared to the students that we will be studying.

Why is the statement *not relevant* in a discussion of external validity?

Exercise 8.4. Yeh et al. [2018] conducted an experimental study to 'determine if using a parachute prevents death or major traumatic injury when jumping from an aircraft'.

The researchers randomised 23 volunteers into one of two groups: wearing a parachute, or wearing an empty backpack. The response variable was a measurement of death or major traumatic injury upon landing. From the study, death or major injury was the same in both groups (0% for each group). However, the study used 'small stationary aircraft on the ground, suggesting cautious extrapolation to high altitude jumps' (p. 1).

Discuss the internal, external and ecological validity based on this information.

Exercise 8.5. Delaney et al. [2018] examined how well hospital patients sleep at night. The researchers state that 'convenience sampling was used to recruit patients' (p. 2). Later, the researchers state that (p. 7):

> ... patients requiring hospitalization will likely require some daytime nap periods. This study looks at sleep only in the night-time period 22:00–07:00h, without the context of daytime sleep considered.

Discuss the internal, external and ecological validity based on this information.

Exercise 8.6. Botelho et al. [2019] examined the food choices made when subjects were asked to shop for ingredients to make a last-minute meal. Half were told to prepare a 'healthy meal', and the other half told just to prepare a 'meal'. The authors emphasise that the purchases were 'simulated' (p. 436):

> As participants did not have to pay for their selection, actual choices could be different. Participants may also have not behaved in their usual manner since they were taking part in a research study...

Discuss the internal, external and ecological validity based on this information.

Exercise 8.7. Johnson et al. [2018] studied the use of over-the-counter menthol cough-drops in people with a cough. One conclusion from the *observational* study of 548 people was that, taking 'too many cough drops [...] may actually make coughs more severe', as one author explained (University of Wisconsin, March 2018). Critique this statement.

Exercise 8.8. Suppose a group of students was studying this RQ:

> Among Australians, is the average serum cholesterol concentration different for smokers and non-smokers?

The students gave the following information about their study. Explain *why* each of these statements is incorrect.

1. The design is observational, as we cannot manipulate each person's serum cholesterol.
2. The outcome is 'the average serum cholesterol concentration for smokers and non-smokers'.
3. The study is not externally valid, as the results may not apply to all people in the world.
4. The response variable is serum cholesterol.
5. In this experiment, the population is 'Australians'.
6. The data file will have two columns: one for smokers, and one for non-smokers.
7. 'Whether the person owns a cat' is likely to be a confounding variable.
8. The observer effect is not relevant, as the participants will know they are involved in a study.

Exercise 8.9. Delarue et al. [2019] discuss studies where subjects rate the taste of new food products. They note that taste-testing studies should be externally and internally valid (p. 78): However, even with good internal and external validity, these studies often result in a 'high rate of failures of new launched products'.

Discuss the internal, external and ecological validity based on this information.

Answers to *Quick review* questions: 1. True. **2.** False. **3.** False. **4.** False.

Part III

Collecting data

9

Collecting data

So far, you have learnt to ask an RQ and design the study. **In this chapter**, you will learn to:

- record the important steps in data collection.
- describe study protocols.
- ask questionnaire questions.

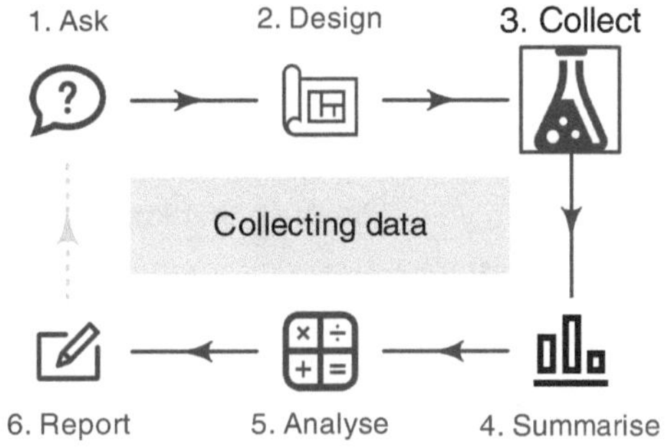

9.1 Introduction

If the RQ is well-constructed, terms are clearly defined, and the study is well-designed and explained, then the process for collecting the data should be easy to describe. Data collection is often time-consuming, tedious and expensive, so collecting the data correctly first time is important, hence an accurate description of the data collection process is essential.

ⓘ Data collection is often tedious, time-consuming and expensive: you usually get one chance to collect data. In contrast, data (once collected) can be analysed as many times as necessary. Design the study properly the first time!

9.2 Protocols

Before collecting the data, a plan should be established and documented that explains exactly *how* the data will be obtained, which will include *operational definitions* (Sect. 2.10). This plan is called a *protocol*, and allows results to be confirmed and compared.

Definition 9.1 (Protocol). A *protocol* is a procedure documenting the details of the design and implementation of studies, and for data collection.

A protocol usually has at least three components that describe:

1. how individuals are chosen from the population (i.e., external validity).
2. how data are collected from the individuals (i.e., internal validity).
3. the types of analyses and software (including version) used.

Data collection often encounters problems or challenges, which should be documented also.

Example 9.1 (Protocol). Romanchik-Cerpovicz et al. [2018] made cookies using puréed green peas in place of margarine (to increase the nutritional value of cookies). They assessed the acceptance of these cookies to college students.

The protocol discussed *how the individuals were chosen* (p. 4):

> ...through advertisement across campus from students attending a university in the southeastern United States.

This voluntary sample comprised 80.6% women, a higher percentage than in the general population, and in the college population. (Other extraneous variables were also recorded.)

Exclusion criteria were also applied, excluding people 'with an allergy or sensitivity to an ingredient used in the preparation of the cookies' (p. 5). The researchers also described *how the data were obtained*, including these steps (p. 5):

- tasters sat at individual tables, so they could not be influenced by other tasters.
- cookies were presented individually, on individual plates.
- the cookies were presented in a randomised order (fat substituted with green pea puree at 25%, 0%, 50%, 100% and 75%).
- between tastes, tasters 'cleansed their palates' by drinking distilled water at 25°C.
- the tasters recorded (p. 5):

> characteristics of color, smell, moistness, flavor, aftertaste, and overall acceptability, for each sample of cookies...

Thus, internal validity was addressed using random allocation (to manage confounding), blinding individuals (to partially manage the Hawthorne effect), and washouts (to manage the carryover effect). Details are also given of how the cookies were made, and how objective measurements (such as moisture content) were determined. Subjects were not blinded to being in a study, but were blinded to which substitution percentage was in each cookie.

The *type of analyses and software used* were also given.

Example 9.2 (Operational definitions). In a study where the sex of a person is the explanatory variable, an *operational definition* for the sex of the person is needed. That is, a description is needed for *how* the researchers determine the sex of the person.

One option is to *ask* each individual their sex. Another option is to make an *informed assessment*, based on clothing and body shape, for instance. A third option is to have subjects indicate their sex in a *multiple-choice question*.

Different types of studies may require different means for determining the sex of a person. The study protocol should explain *how* the sex of the person will be established.

Unforeseen complications are not unusual, so often a *pilot study* (or a *practice run*) is conducted before the actual data collection, to:

- determine the feasibility of the data collection protocol.
- identify unforeseen challenges.
- obtain data to determine appropriate sample sizes (Sect. 32).
- identify ways to potentially save time and money.

The pilot study may suggest changes to the protocol.

Definition 9.2 (Pilot study). A *pilot study* is a small test run of the study used to check that the protocol is appropriate and practical, and to identify (and hence fix) possible problems with the research design or protocol.

The data can be collected once the protocol has been finalised. Protocols ensure studies are reproducible (Sect. 5.3), so others can confirm or compare results, and others can understand exactly what was done, and how. Protocols should indicate how design aspects (such as blinding the individuals, random allocation of treatments, etc.) will happen. The final *protocol*, without pedantic detail, should be reported. Diagrams can be useful to support explanations. All studies should have a well-established protocol for describing how the study was done.

9.3 Collecting data using questionnaires

9.3.1 Writing questions

Collecting data using *questionnaires* is common for both observational and experimental studies. Questionnaires are very difficult to do well: question wording is crucial, and surprisingly difficult [Fink, 1995]. Pilot testing questionnaires is crucial.

Definition 9.3 (Questionnaire). A questionnaire is a set of questions for respondents to answer.

ⓘ A *questionnaire* is a set of questions to obtain information from individuals. A *survey* is an entire methodology, that includes gathering data using a questionnaire, finding a sample, and other components.

Questions in a questionnaire may be *open-ended* (respondents can write their own answers) or *closed* (respondents select from a small number of possible answers, as in multiple-choice questions). Open-ended and closed questions both have advantages and disadvantages. Answers to open-ended questions more easily lend themselves to qualitative analysis and closed questions more to quantitative research. This section briefly discusses writing questions.

Example 9.3 (Open and closed questions). German students were asked a series of questions about microplastics [Raab and Bogner, 2021], including:

1. Name sources of microplastics in the household.
2. In which ecosystems are microplastics in Germany? Tick the answer (multiple ticks are possible). *Options*: (a) sea; (b) rivers; (c) lakes; (d) groundwater.
3. Assess the potential danger posed by microplastics. *Options*: (a) very dangerous; (b) dangerous; (c) hardly dangerous; (d) not dangerous.

The first question is *open-ended*: respondents provide their own answers. The second question is *closed*, where *multiple* options can be selected. The third question is *closed*, where only *one* option can be selected.

Writing good questionnaire questions is difficult. Good practice requires:

- *avoiding leading questions* that may lead respondents to answer a certain way.
- *avoiding ambiguity* by avoiding unfamiliar or technical terms.
- *avoiding asking the uninformed*; avoid asking respondents about issues they don't know about. Many people will give a response even if they do not understand (and such responses are worthless). For example, people may give directions to places that do not even exist [Collett and O'Shea, 1976].
- *avoiding complex and double-barrelled questions*, which are hard to understand and the answers hard to interpret.
- *avoiding problems with ethics*, such as questions about people breaking laws, or revealing confidential or private information. In special cases and with justification, ethics committees may allow such questions.
- *ensuring clarity* in question wording.
- (for closed questions) ensuring options are *mutually exclusive* (responses fit into only one category) and *exhaustive* (categories cover *all* options).

Example 9.4 (Poor question wording). Consider a questionnaire asking these questions:

1. Because bottles from bottled water create enormous amounts of non-biodegradable landfill and hence threaten native wildlife, do you support banning bottled water?
2. Do you drink more water now?
3. Are you more concerned about Coagulase-negative *Staphylococcus* or *Neisseria pharyngis* in bottled water?
4. Do you drink water in plastic and/or glass bottles?
5. Do you have a water tank installed illegally, without permission?
6. Do you avoid purchasing water in plastic bottles unless it is carbonated, unless the bottles are plastic but not necessarily if the lid is recyclable?

Question 1 is *leading* because the expected response is obvious. Better would be: 'Do you support or not support banning bottled water?'

Question 2 is *ambiguous*: it is unclear what 'more water now' is being compared to.

Question 3 is unlikely to give sensible answers, as most people will be *uninformed*. Many people will still give an opinion, but the data will be effectively useless (though the researcher may not realise).

Question 4 is *double-barrelled*, and would be better asked as two separate questions (one asking about plastic bottles, and one about glass bottles).

Question 5 is unlikely to be given *ethical approval* or to obtain truthful answers, as respondents are unlikely to admit to breaking rules.

Question 6 is *unclear*, since knowing what a *yes* or *no* answer means is confusing.

Example 9.5 (Question wording). Question *wording* can be important. In the 2014 *General Social Survey* (https://gss.norc.org), when white Americans were asked for their opinion of the amount America spends on *welfare*, 58% of respondents answered 'Too much'. However, when white Americans were asked for their opinion of the amount America spends on *assistance to the poor*, only 16% of respondents answered 'Too much' [Jardina, 2018].

Example 9.6 (Mutually exclusive options). In a study to determine the time doctors spent with patients (from Chan et al. [2008]), doctors were given the options:

- 0–5 mins.
- 5–10 mins.
- more than 10 mins.

This is a poor question, because a respondent does not know which option to select for an answer of '5 mins'. The options are not *mutually exclusive.*

9.3.2 Challenges using questionnaires

Using questionnaires presents myriad challenges.

- *Non-response bias* (Sect. 6.7): non-response bias is common with questionnaires, as they are often used with voluntary-response samples (Sect. 6.4.2). The people who *do not* respond to the survey may be different from those who *do* respond.
- *Response bias* (Sect. 6.7): people do not always answer truthfully; for example, what people *say* may not correspond with what people *do* (Example 8.6). Sometimes this is unintentional (e.g., poor questions wording), due to embarrassment, or because questions are controversial. Sometimes, respondents repeatedly provide the same answer (without reading the question) to a series of multiple-choice questions (i.e., always select 'Always').
- *Recall bias*: people may not be able to accurately recall past events clearly, or recall when they happened.
- *Question order*: the order of the questions can influence the responses.
- *Interpretation*: phrases and words such as 'Sometimes' and 'Somewhat disagree' may have different meanings to different people.

Many of these can be managed with careful questionnaire design, but discussing the methods are beyond the scope of this book.

Example 9.7. Alharthy et al. [2023] studied the knowledge, attitude, and level of confidence of paramedics when managing patients with visual or hearing problems. They used a questionnaire, which was sent to (p. 5)

> ... 372 potential participants with the expectation that at least 310 questionnaires [would] be returned (83% assumed return rate). However, only 97 out of 372 participants completed the questionnaire, resulting in actual return rate of 26%.

Response rates from questionnaires are often very low (and unrepresentative).

9.3.3 Preparing software for questionnaire data

Care is needed when preparing software for data collected using a questionnaire. Sometimes, of course, the data are collected by computer (e.g., online) and are supplied to the researchers already formatted and in electronic format.

Data from open questions are usually text-based (such as words, sentences or paragraphs of text). These can generally be included in the data worksheet (though there may be a limit to the length of such data), but cannot be analysed using the quantitative methods described in this book.

Closed questions are easily included in a data worksheet. In closed questions where respondents can select *one* option only, one column is needed for the question that records which option was selected. In closed questions where respondents can select *all* options that apply, each option requires its own column that records each respondents' answer for that option.

Example 9.8 (Open and closed questions: software). In Example 9.3, three questionnaire questions are given that were asked of German students about microplastics [Raab and Bogner, 2021]. Some (artificial) data are shown entered in Fig. 9.1.

The first question requires open-ended, text-based answers (`Sources`). For the second (closed) question, students could select *multiple* options, so each option needs one column in the data worksheet (`WhereSeas` to `WhereGroundwater`). The third (closed) question required students to select *one* option from a given list, so one column (`Danger`) is needed to record responses. As usual (Sect. 2.12), each row represents one unit of analysis (student).

Sources	WhereSeas	WhereRivers	WhereLakes	WhereGroundwater	Danger
dont know	No	Yes	Yes	No	Dangerous
water	No	No	Yes	No	Not dangerous
Food chain.	Yes	Yes	Yes	Yes	Very dangerous

FIGURE 9.1: The data worksheet for some example data, for the microplastics study.

9.4 Chapter summary

Having a detailed procedure for collecting the data (the *protocol*) is important. Using a *pilot study* to trial the protocol can reveal unexpected changes necessary for a good protocol. Creating good questionnaires questions is difficult, but important.

9.5 Quick review questions

1. What is the biggest problem with this question: 'Do you have bromodosis?'
2. What is the biggest problem with this question: 'Do you spend too much time connected to the internet?'
3. What is the biggest problem with this question: 'Do you eat fruits and vegetables?'
4. *True* or *false*: A well-defined protocol allows the researchers to make the study externally valid.
5. *True* or *false*: This question is likely to be a *leading* question. 'Do you support a ban on drinks sold in unrecyclable plastic bottles?'

9.6 Exercises

Answers to odd-numbered exercises are given at the end of the book.

Exercise 9.1. What is the problem with this question?

What is your age? (Select one option)

- Under 18.
- Over 18.

Exercise 9.2. What is the problem with this question?

> How many children do you have? (Select one option)
> - None.
> - 1 or 2.
> - 2 or 3.
> - More than 4.

Exercise 9.3. Which of these questionnaire questions is better? Why?

1. Should concerned cat owners vaccinate their pets?
2. Should domestic cats be required to be vaccinated or not?
3. Do you agree that pet-owners should have their cats vaccinated?

Exercise 9.4. Which of these questionnaire questions is better? Why?

1. Do you own an environmentally-friendly electric vehicle?
2. Do you own an electric vehicle?
3. Do you own or do you not own an electric vehicle?

Exercise 9.5. Falk and Anderson [2013] studied sunscreen use, and asked participants questions, including these:

- how often do you sun bathe with the intention to tan during the summer in Sweden? (Possible answers: never, seldom, sometimes, often, always).
- how long do you usually stay in the sun between 11am and 3pm, during a typical day-off in the summer (June–August)? (Possible answers: < 30 mins, 30 mins–1 h, 1–2 h, 2–3 h, > 3 h).

Critique these questions. What biases may be present?

Exercise 9.6. Morón-Monge et al. [2021] studied primary-school children's knowledge of their natural environment. They were asked three questions:

1. Do you usually visit Guadaira Park?
    - No, I don't like parks.
    - No, I don't usually visit it.
    - Yes, once per week.
    - Yes, more than once a week
2. How many times have you visited nature (the beach, countryside, mountains, etc.) in the last month?
    - Never.
    - Once.
    - Two to three times.
    - More than three times.
3. Which is your favourite natural place?
    - Write a story.
    - Draw a picture.

Which questions are *open* and which are *closed*? Which questions will produce *qualitative* data? Critique the questions.

> **Answers to *Quick review* questions: 1.** *Language*: most people do not know what 'bromodosis' is. **2.** *Ambiguous*: 'too much', compared to what? **3.** *Double-barrelled*: some people may eat fruits but not vegetables, for example.
> Over what time frame? **4.** False. **5.** True.

Part IV

Classifying and summarising data

10

Classifying data and variables

So far, you have learnt to ask an RQ, design a study, and collect the data. **In this chapter**, you will learn to:

- identify and distinguish qualitative and quantitative variables.
- identify and distinguish nominal and ordinal qualitative variables.
- identify and distinguish continuous and discrete quantitative variables.

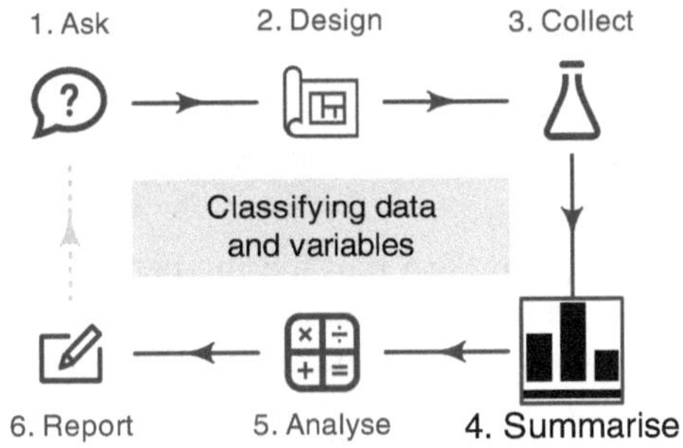

10.1 Introduction

Understanding the type of data collected is essential before summarising or analysing, because the *type* of data determines how to proceed. Broadly, data may be classified as either *quantitative* data (Sect. 10.2) or *qualitative* data (Sect. 10.3). The *data* are the recorded *values* of the variables, so we also talk about quantitative and qualitative *variables*. Quantitative variables record quantitative data; qualitative variables record qualitative data.

Example 10.1 (Variables and data). 'Age' is a *variable* because age varies from individual to individual (Def. 2.9). The *data* may include values like 13 months, 21 years and 76 years.

⚠ *Quantitative research* summarises and analyses data using numerical methods (Sect. 1.3). *Quantitative research* can involve both *quantitative* and *qualitative* data, because both can be summarised numerically (Chaps. 11 and 12 respectively) and analysed numerically.

10.2 Quantitative data: discrete and continuous data

Quantitative data are mathematically numerical. Most data arising from counting or measuring are quantitative. Quantitative data often (but not always) have measurement units (such as *kg* or *cm*). Be careful: numerical data are not necessarily quantitative; only *mathematically* numerical data are quantitative (numbers with numerical *meanings*).

Definition 10.1 (Quantitative data). *Quantitative data* are *mathematically* numerical: the numbers have numerical meaning, and represent quantities or amounts. Quantitative data generally arise from counting or measuring.

Example 10.2 (Quantitative data). The weight of numbats, the thickness of sheet metal, and blood pressure are all *measured*, and are quantitative variables.

The number of power failures per year, the number of solar panels per home, and the number of tangelos per tree are all *counts*, and are quantitative variables.

Australian postcodes are four-digit numbers, but are *not* quantitative; the numbers are labels. A postcode of 4556 isn't one 'better' or 'more' than a postcode of 4555. The values do not have numerical *meanings*. Indeed, alphabetic postcodes could have been chosen. For example, the postcode of Caboolture (Queensland) is 4510, but could have been QCAB.

Quantitative data may be further classified as *discrete* or *continuous*. *Discrete* quantitative data have possible values that can be *counted*, at least in theory. Sometimes, the possible values may have no theoretical upper limit, yet are still considered 'countable'. *Continuous* quantitative data have values that cannot, at least in theory, be recorded exactly: another value can always be found between any two given values of the variable, if we *measure* to a greater number of decimal places. In practice, though, values must be rounded to a reasonable number of decimal places.

Definition 10.2 (Discrete data). *Discrete* quantitative data has a countable number of possible values between any two given values of the variable.

Example 10.3 (Discrete quantitative data). These quantitative variables are *discrete*:

- the *number* of people in passenger vehicles being driven on a certain road. Possible values: 1, 2, ..., with an upper limit of perhaps 8.
- the *number* of cracked eggs in a carton of 12. Possible values: 0, 1, 2, ... 12.
- the *number* of orthotic devices a person has used. Possible values: 0, 1, 2, ...
- the *number* of turbine cracks after 750 h use. Possible values: 0, 1, 2, ...

Definition 10.3 (Continuous data). *Continuous* quantitative data have (at least in theory) an infinite number of possible values between any two given values.

Height is continuous: between the heights of 179 cm and 180 cm, many heights exist, depending on how many decimal places are used to record height. In practice, however, heights are usually rounded to the nearest centimetre for convenience. All continuous data are rounded.

Example 10.4 (Continuous quantitative data). These quantitative variables are *continuous*:

- the *weight* of 6-year-old Fijian children. Values exist between any two given values of weight, by measuring to more decimal places of a kilogram. However, weights are usually reported to the nearest kilogram.
- the *energy consumption* of houses in London. Values exist between any two given values of energy consumption, by measuring to more and more decimal places of a kiloWatt-hour (kWh). Consumption would usually be given to the nearest kWh.

- the *time* spent in front of a computer each day for employees in a given industry. Values exist between any two given times, by measuring to more decimal places of a second. The values may be reported to the nearest minute, or the nearest 15 mins.

Sometimes, discrete quantitative data with a very large number of possible values may be treated as continuous.

Example 10.5 (Treating discrete data as continuous). Annual income is discrete, since no income is between \$80 000.00 and \$80 000.01. However, annual incomes are much larger than cents, and vary at scales much greater than cents, and so are often treated as continuous.

10.3 Qualitative data: nominal and ordinal data

Qualitative data has distinct labels or categories, and are not mathematically numerical. Be careful: *numerical* data may be qualitative if those numbers don't have numerical *meanings*. The categories of a qualitative variable are called the *levels* or the *values* of the variable.

Definition 10.4 (Qualitative data). *Qualitative data* are not *mathematically* numerical data: they comprise mutually exclusive (and usually exhaustive) categories or labels.

Definition 10.5 (Levels). The *levels* (or the *values*) of a qualitative variable refer to the names of the distinct categories.

Example 10.6 (Qualitative data). 'Brand of mobile phone' is a variable (as 'brand' varies from phone to phone) that is qualitative. Many levels (i.e., brands) are possible, but could be simplified by using the levels as 'Apple', 'Samsung', 'Google' and 'Other'.

Example 10.7 (Qualitative data). Social Security Numbers (SSN) in the US are nine-digit numbers unique to each individual. The first three digits represent geographic regions; the next two digits are assigned to groups in that region. The last four digits are unique to individuals.

Although the SSN is a nine-digit number, SSN is a qualitative variable.

Example 10.8 (Clarity in variables). 'Age' is a *continuous quantitative* variable, since age could be measured to many decimal places of a second. Age is usually rounded down to the number of completed years, for convenience. However, the age of young children may be given as '3 days' or '10 months'.

Sometimes *Age group* is used (such as Under 20; 20 to under 50; 50 or over) instead of Age. 'Age group' is *qualitative*. Ensure you are clear about which is used.

Example 10.9 (Levels). The levels of a variable depend on how the variable is defined. For example, the variable 'How does the person commute to work' may have two levels: 'Using public transport' and 'Not using public transport'.

Alternatively, the variable could be written as 'Does the person use public transport to commute to work?' For this variable, the levels are 'Yes' and 'No'.

Qualitative data can be further classified as *nominal* or *ordinal*.

Definition 10.6 (Nominal qualitative variables). A *nominal* qualitative variable is a qualitative variable where the levels *do not* have a natural order.

Definition 10.7 (Ordinal qualitative variables). An *ordinal* qualitative variable is a qualitative variable where the levels *do* have a natural order.

Example 10.10 (Nominal and ordinal data). *Blood type* is qualitative with four levels: Type A; Type B; Type AB; Type O. These levels have no natural order; they can be ordered alphabetically, or by prevalence. *Blood type* is nominal.

Age group could be listed with levels Under 20; 20 to under 50; 50 or over. These levels have a natural order: youngest to oldest. *Age group* is ordinal.

Example 10.11 (Ordinal data). Consider this questionnaire question:

> Please indicate the extent to which you agree or disagree with this statement: 'Vaping should be banned'.
>
> Strongly disagree; Disagree; Neither agree nor disagree; Agree; Strongly agree.

The responses will be *ordinal* with five levels. Giving the levels in the given order (or the reverse order) makes sense; giving the levels in alphabetical order, for example, would be very confusing. The levels have a natural order.

Example 10.12 (Types of variables). Consider a study to determine if the weight of 500 g bags of pasta actually weigh 500 g (or more) on average. One approach is to record the weight of pasta in each bag (a *quantitative* variable), and compare the *average* weight to the target weight of 500 g.

Another approach is to record whether each bag of pasta was underweight using a balance scale. This variable would be *qualitative*, with two *levels* (underweight; not underweight). The *percentage* of underweight bags could be reported.

Most *statistical* software requires variables to be classified as quantitative or qualitative (and perhaps discrete or continuous; ordinal or nominal). This enables the software to produce appropriate output and suggest appropriate analyses.

10.4 Example: water access

López-Serrano et al. [2022] studied three rural communities in Cameroon, and recorded information about their access to water. The study could be used to determine contributors to the incidence of diarrhoea in young children (85 households had children under 5 years of age). The variables in the `WaterAccess` dataset are classified in Tables 10.1 and 10.2.

TABLE 10.1: The qualitative variables in the water-access dataset.

Qualitative variable	Type	Levels
Region	Nominal	Mbeng; Mbih; Ntsingbeu
Education	Ordinal	Primary or less; Secondary or higher
Distance to water source	Ordinal	Under 100 m; 100 m to 1000 m; over 1000 m
Queuing time at water source	Ordinal	Under 5 mins; 5 to 15 mins; Over 15 mins
Household has a garden	Nominal	Yes; No
Household keeps livestock	Nominal	Yes; No
Water source	Nominal	Well; Bore; Tap; River
How often water container washed	Ordinal	Before each fill; Once per week; Once per month
Diarrhoea in children under 5	Nominal	Yes; No

TABLE 10.2: The quantitative variables in the water-access dataset.

Quantitative variable	Type	Extra information
Household coordinator's (woman's) age	Continuous	Rounded to nearest year
Number of people in household	Discrete	
Number of children under 5 in household	Discrete	

10.5 Chapter summary

The *type* of data collected determines the types of summaries and analyses that are needed. Data and variables can be classified as either:

- *quantitative* (*discrete* or *continuous*) if mathematically numerical.
- *qualitative* (*nominal* or *ordinal*) if not mathematically numerical.

10.6 Quick review questions

Benetou et al. [2020] studied school-aged adolescents in Greece. Among other variables, for each child they recorded the body-mass index (weight, divided by height-squared), diet quality (poor; moderate; good), the region where they lived (Attica; Thessaloniki; Other), the number of days they performed physical exercise in the last week, and school grade.

Are the following statements *true* or *false*?

1. 'Body-mass index' is a quantitative discrete variable.
2. 'Diet quality' is a qualitative ordinal variable.
3. 'Region of residence' is a qualitative nominal variable.
4. 'Number of days the child performed physical exercise in the last week' is a quantitative discrete variable.
5. 'School grade' is a quantitative continuous variable.

10.7 Exercises

Answers to odd-numbered exercises are given at the end of the book.

Exercise 10.1. Classify these variables as quantitative (discrete or continuous) or qualitative (nominal or ordinal).

1. The knee-flex angle after treatment.
2. Whether laser drilling of small holes in concrete is successful.
3. Length of time between arrival at an emergency department, and admission.
4. Telephone numbers.

Exercise 10.2. Classify these variables as quantitative (discrete or continuous) or qualitative (nominal or ordinal).

1. Number of eggs laid by female brush turkeys.
2. Whether a child eats the recommended serving of fruit each day.
3. Bar code numbers on supermarket products.
4. The breed of dog used for koala detection.

Exercise 10.3. True or false: these variables are *qualitative* and *nominal*.

1. The age group of respondents to a survey.
2. Whether a cyclist is wearing a helmet or not.
3. The dosage of a medication applied: 40, 60 or 80 mg per day.

Exercise 10.4. True or false: these variables are *qualitative* and *ordinal*.

1. The brand of fertiliser being applied.
2. The age of trees.
3. Highest level of education (never finished school; primary school; secondary school; beyond secondary school).

Exercise 10.5. A study recorded whether people (who were not swimming) were wearing head-protection at the beach. The results were recorded as None; Cap; or Hat. Which of the following words could be used to classify this variable:

Nominal; qualitative; continuous; quantitative; ordinal.

Exercise 10.6. Schepaschenko et al. [2017] studied lime trees (*Tilia cordata*), and recorded these variables for 385 trees in Russia: the foliage biomass (in kg); the tree diameter (in cm); the age of the tree (in years); and the origin of the tree (one of Coppice, Natural, or Planted).

Classify the variables in the study using the language of this chapter.

Exercise 10.7. A study is comparing the proportion of females and males who wear hats between 10am and 2pm. Which one of these could be the *explanatory* variable?

- The sex of the person.
- 'Female' and 'male'.
- The percentage of people who are female.

Exercise 10.8. A study is comparing the proportion of older women (aged 40+) and younger women (under 40) who work full-time. Which one of these could be the explanatory variable?

- 'Full-time' and 'part-time'.
- The percentage of women who are aged under 40.
- Whether a woman is aged under 40.
- 'Yes' and 'No'.

Exercise 10.9. Are these variables quantitative (discrete or continuous; what units of measurement), or qualitative (nominal or ordinal, and with what levels?)?

1. Systolic blood pressure.
2. Diet (vegan; vegetarian; neither vegan nor vegetarian).
3. Socioeconomic status (low income; middle income; high income).
4. Number of times a person visited the doctor last year.

Exercise 10.10. Alley et al. [2017] studied body-mass index and its relationship with use of social media, and recorded these variables (among others) from a group of 1 140 participants:

1. age (under 45; 45 to 64; 65 or over).
2. gender (male; female).
3. location (urban; rural).
4. social media use (none; low; high).
5. total sitting time, in minutes per day.

For each variable, classify the *type* of variable: quantitative (discrete or continuous; what units of measurement?), or qualitative (nominal or ordinal; what levels?).

Exercise 10.11. The *Behavioral Risk Factor Surveillance System* (BRFSS; Centers for Disease Control and Prevention (CDC) [2021–2023]) survey collects data annually in all 50 US states, the District of Columbia and three US territories, from more than 400 000 adults each year. The following questions, among many others, appear in the 2023 BRFSS survey.

1. Do you own or rent your home? (Options: Own, Rent; Other.)
2. How many children less than 18 years of age live in your household?
3. How many cell (mobile) phones do you have for personal use? (Options: 1; 2; 3; 4; 5; 6 or more.)
4. Have you ever served on active duty in the United States Armed Forces? (Options: Yes; No.)
5. About how much do you weigh without shoes?

Classify the type of data collected from each question.

Exercise 10.12. The *National Health and Nutrition Examination Survey* (NHANES; Centers for Disease Control and Prevention (CDC) [2024]):

> ... examines a nationally representative sample of about 5 000 persons each year...

The following questions, among many others, appear in the 2021–2023 NHANES survey, and are asked about the person selected in the household (SP) to complete the questionnaire.

1. Do you consider SP now to be overweight, underweight, or about the right weight?
2. How many rooms are in SP's home? (Count the kitchen and do not count any bathrooms, or an unfinished basement, or a laundry room.)
3. How many people who live in SP's home smoke cigarettes, cigars, little cigars, pipes, water pipes, hookah, or any other tobacco product?
4. Has SP ever been told by a doctor or other health professional that SP had asthma? (Options: Yes; No; Don't know.)
5. Overall, how would SP rate the health of SP's teeth and gums? (Options: Excellent; Very good; Good; Fair; Poor.)

Classify the type of data collected from each question.

Exercise 10.13. Swinnen et al. [2018] studied the use of ankle-foot orthoses in children with cerebral palsy. Table 10.3 give the data for the 15 subjects. (GMFCS is the Gross Motor Function Classification System describing the impact of cerebral palsy on motor function; *lower* levels mean *better* functionality.) Classify the variables in the study using the language of this chapter.

Exercise 10.14. Lane [2002] studied fertiliser use, and recorded the soil nitrogen after applying different fertiliser doses. These variables were recorded for each field:

1. the *fertiliser dose*, in kilograms of nitrogen per hectare;
2. the *soil nitrogen*, in kilograms of nitrogen per hectare; and
3. the *fertiliser source*; one of 'inorganic' or 'organic'.

Classify the variables in the study.

Exercise 10.15. Brunton et al. [2019] recorded the response of kangaroos to overhead drones (one of 'No vigilance', 'Vigilance', 'Flee < 10 m', or 'Flee > 10 m') and the altitude of the drone (30 m, 60 m, 100 m or 120 m). The mob size and sex of the kangaroo was also recorded. Classify the variables in the study.

Exercise 10.16. Dokur et al. [2018] studied people who died while taking selfies, and recorded the data in Table 10.4. Which of the following are the *variables* in the table? For each that is a variable, classify the variable.

TABLE 10.3: The orthoses dataset.

Gender	Age (years)	Height (cm)	Weight (kg)	GMFCS
M	9	136	34.5	1
M	7	106	16.2	2
M	7	129	21.1	1
M	12	152	40.4	1
M	11	146	39.3	2
M	5	113	18.1	1
M	6	112	16.7	2
M	8	112	19.1	1
M	8	138	28.6	1
M	6	116	19.3	1
F	7	113	17.6	1
M	11	141	34.9	1
M	7	136	34.5	1
F	9	128	21.9	1
F	8	133	23.0	1

1. The location.
2. The number of people who died at each location.
3. The percentage of people who died at each location.

TABLE 10.4: Locations of people dying while taking selfies.

	Number	Percentage
Nature, associated environments	48	43.2
Train, railway, associated structures	22	19.9
Buildings, associated structures	17	15.3
Road, bridge, associated structures	12	10.8
Dams, associated structures	7	6.3
Fields, farms, associated structures	4	3.6
Others	1	0.9

Answers to *Quick review* questions: 1. False (quantitative continuous). **2.** True. **3.** True. **4.** True **5.** False (qualitative ordinal).

11

Summarising quantitative data

So far, you have learnt to ask an RQ, design a study, collect the data, and classify the data. **In this chapter**, you will learn to:

- summarise quantitative data using the appropriate graphs.
- summarise quantitative data using shape, average, variation and unusual features.

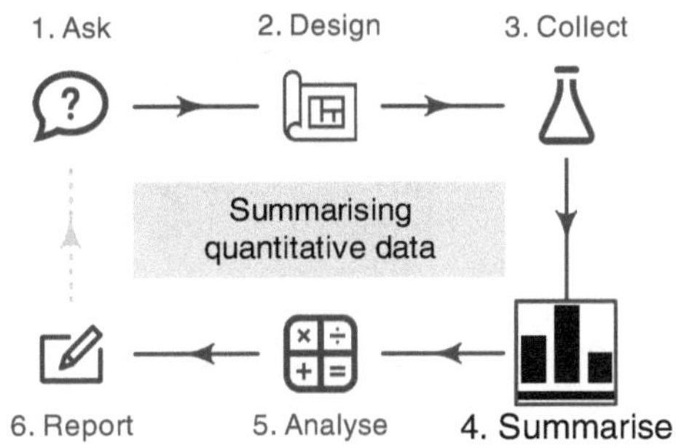

11.1 Introduction

Many quantitative research studies involve quantitative variables. Except for very small amounts of data, understanding the data is difficult without a summary. Quantitative data can be summarised by knowing how often various values of the variable appear. This is called the *distribution* of the data.

Definition 11.1 (Distribution). The *distribution* of a variable describes what values are present in the data, and how often those values appear.

The distribution can be displayed using a frequency table (Sect. 11.2) or a graph (Sect. 11.3). The distribution of quantitative data can be described by the shape (Sect. 11.5), and summarised numerically by computing the average value (Sect. 11.6), computing the amount of variation (Sect. 11.7), and identifying outliers (Sect. 11.8).

11.2 Frequency tables for quantitative data

Quantitative data can be collated in a *frequency table* by grouping the variables into appropriate intervals ('bins'). The intervals should be *exhaustive* (cover all values) and *mutually exclusive* (observations belong to one and only one category). While not essential, usually the categories have equal width.

A frequency table for *discrete* quantitative data uses bins defined to contain single discrete values, or a small number of discrete values.

Example 11.1 (Frequency table: discrete data). The data in Table 11.1 show the number of severe cyclones in the Australian region, for each year from 1969 to 2005.

A frequency table can be constructed by binning each discrete value individually (Table 11.2, left table) or grouped in pairs (Table 11.2, right table). The table gives the number of years, and the corresponding percentages, that recorded the given number of cyclones.

TABLE 11.1: The first five and last five observations (of 37) of the number of severe cyclones recorded in the Australian region for each year.

Year	Cyclones recorded
1969	3
1970	3
1971	9
1972	6
1973	4
⋮	⋮

Year	Cyclones recorded
⋮	⋮
2001	3
2002	3
2003	5
2004	5
2005	8

TABLE 11.2: Two frequency tables for the severe cyclone data.

Cyclones recorded	Num. of years	Percentage of years
3 cyclones	8	22
4 cyclones	10	27
5 cyclones	3	8
6 cyclones	5	14
7 cyclones	2	5
8 cyclones	4	11
9 cyclones	4	11
10 cyclones	0	0
11 cyclones	1	3

Cyclones recorded	Number of years	Percentage of years
3 or 4 cyclones	18	49
5 or 6 cyclones	8	22
7 or 8 cyclones	6	16
9 or 10 cyclones	4	11
11 cyclones	1	3

For *continuous* data, care is needed when creating frequency tables. Bins must be carefully constructed, since all continuous data are rounded. The bins should be defined to ensure no values lie on the border between bins, and hence creating ambiguity.

Example 11.2 (Frequency table: continuous data). Table 11.3 gives the weights of babies born in a hospital on one day [Dunn, 1999, Steele, 1997], plus the gender of each baby, and the number of minutes after midnight of the birth (shown in birth order).

To display the distribution of birth weights, the weights can be grouped into clearly-defined weight intervals (Table 11.4, left column). Alternatively, the breaks between the bins can be given to one more decimal place than the data to avoid observations landing exactly on the bin divisions (final column).

The table also gives percentage of births in each bin; for example, the percentage of babies over 4.0 kg is $1/44 \times 100 = 2.27\%$, or about 2%. Most babies in the sample are between 3 and 4 kg at birth.

Sometimes trial and error is needed to find useful intervals for continuous data. Usually, but not universally, the intervals *include* values at the lower end of the interval, but *exclude* values at the upper end (as in Table 11.4).

TABLE 11.3: The first nine observations (of $n = 44$) of the baby-births data: the number of babies born in a Brisbane (Australia) hospital on one specific day. The 'birth time' is the number of minutes after midnight.

Gender	Weight (kg)	Birth time
Female	3.8	5
Female	3.3	64
Male	3.6	78
Male	3.8	115
Male	3.6	177

Gender	Weight (kg)	Birth time
Female	2.2	245
Female	1.7	247
Male	2.8	262
Male	3.2	271
⋮	⋮	⋮

TABLE 11.4: The baby-weights data, displayed in a frequency table. The first and last columns show two different (but equivalent) ways to group the data.

Weight group	Number of babies	Percentage of babies	Alterative weight group
1.5 kg to under 2.0 kg	1	2	1.45 kg to 1.95 kg
2.0 kg to under 2.5 kg	4	9	1.95 kg to 2.45 kg
2.5 kg to under 3.0 kg	4	9	2.45 kg to 2.95 kg
3.0 kg to under 3.5 kg	17	39	2.95 kg to 3.45 kg
3.5 kg to under 4.0 kg	17	39	3.45 kg to 3.95 kg
4.0 kg to under 4.5 kg	1	2	3.95 kg to 4.45 kg

11.3 Graphs for quantitative data

The graphs in this section are appropriate for *continuous* quantitative data, and sometimes for *discrete* quantitative data if many values are possible. Sometimes, *discrete* data with very few recorded values are better displayed using graphs designed for qualitative data (Sect. 12.3).

Graphs used to display the distribution of one quantitative variable include:

- *histograms* (Sect. 11.3.1), which are best for moderate to large amounts of data.
- *stemplots* (Sect. 11.3.2), which are best for small amounts of data, and are only sometimes useful.
- *dot charts* (Sect. 11.3.3), which are used for small to moderate amounts of data.

⚠ The purpose of a graph is to display the information in the clearest, simplest possible way, to facilitate understanding the message(s) in the data.

11.3.1 Histograms

Histograms are a series of boxes, where the width of the box represents an interval of *values* of the variable being graphed, and the height of the box represents the *number* (or *percentage*) of observations within that range of values.[1] The height of the histogram bars indicate the number (or percentage) in each category (often called 'bins'). A histogram is essentially a

[1]Actually, the *area* of the box is proportional to the number of observations. We only consider histograms where the boxes have the same width, so the statements are equivalent.

picture of a frequency table. The vertical axis can be counts (labelled as 'Counts of trees', 'Number of frogs', 'Frequency of people', or similar) or percentages.

When the quantitative variable is *discrete*, the labels usually are placed on the axis aligned with the centre of the bar (see Example 11.3).

Example 11.3 (Histograms: discrete data)**.** Consider again the number of severe cyclones in the Australian region (Table 11.1). A histogram can be constructed from either frequency table in Table 11.2; see Fig. 11.1. For example, the left histogram shows there were eight years in which three severe cyclones were recorded.

Notice that different bin locations and widths change the appearance of the distribution.

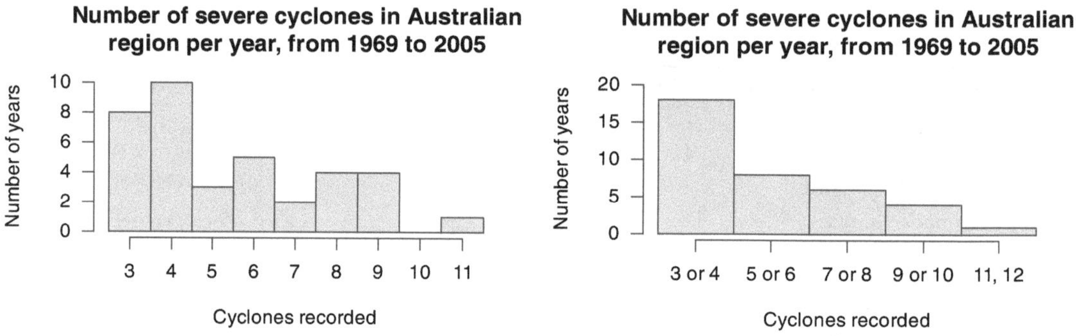

FIGURE 11.1: Two histograms of the severe-cyclone data.

⚠ The axis displaying the counts (or percentages) should *start from zero*, since the height of the bars visually implies the frequency of those observations (see Example 17.3).

When the quantitative variable is *continuous*, care is needed when constructing the histogram. Since the data are continuous, the data must be rounded. (For instance, the birthweights in Table 11.3 are rounded to one decimal place of a kilogram.) This means care is needed when defining boundaries between bins, and ensuring clarity about which bin contains observations when they lie on (or near) a boundary. One way to do this is to define the boundaries between bins to one more decimal place than the given data (as in the final column of Table 11.4).

The choice of bin size and bin boundaries can substantially change how a histogram displays the data (Examples 11.4 and 11.6). For large datasets, these choices tend to matter less.

When observations lie on the boundary of the boxes, some software includes these observations in the lower box (which is common) and some in the higher box.

Example 11.4 (Histograms: continuous data)**.** Consider again the weights (in kg) of babies born in a Brisbane hospital in one day (Table 11.3). A histogram can be constructed for these data; Fig. 11.2 shows the histogram in the process of being constructed.

An observation on a boundary between the bins may be placed in the *higher* box (i.e., 2.5 kg is in the '2.5 to 3.0 kg' box, not the '2.0 to 2.5 kg' box): see the left panel. Alternatively, a boundary observation may be placed in the *lower* box; see the centre panel. This histogram is a picture of the frequency table in Table 11.4.

To avoid confusion, the boundaries can be defined to one more decimal place than the data (right panel), which is equivalent to counting the observations in the lower box (as in the left panel). Notice that the choice impacts the appearance of the histogram.

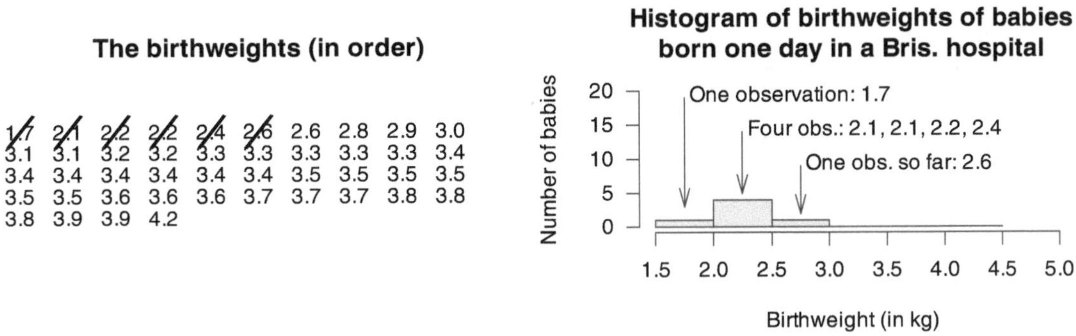

FIGURE 11.2: Making the histogram for the baby-birth data: the first six observations added.

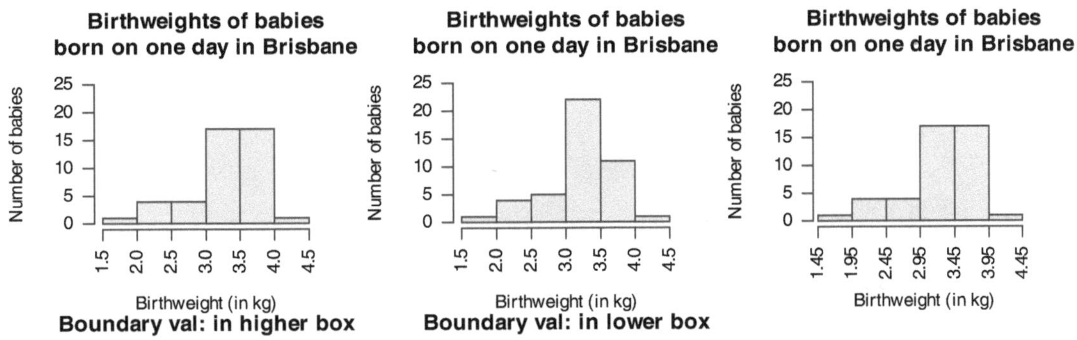

FIGURE 11.3: Histograms can be constructed in different ways to manage observations on the boundary of bins. Left: boundary values counted in the higher box. Centre: boundary values counted in the lower box. Right: defining boundaries with one more decimal place than the data may be clearer.

Example 11.5 (Histograms). Mages et al. [2017] recorded the length of 'brain freezes' after consuming cold food or drink. A histogram of the data (Fig. 11.4), shows 11 people experience symptoms less than 5 s in length; nine people experienced symptoms for at least 5 but less than 10 s; and 1 person experienced symptoms for at least 35 s but under 40 s.

Software tries to use sensible default choices for the number of bins, and width of the bins. However, the bin size can substantially change the appearance of the histogram. Software makes it easy to try different bin sizes to find one that displays the overall distribution well.

Example 11.6 (Histograms: bin width). A histogram for the time between eruptions [Härdle et al., 1991] of the *Old Faithful* geyser in Yellowstone National Park (USA) (Fig. 11.5) shows the shape of the distribution depends on the bin width.

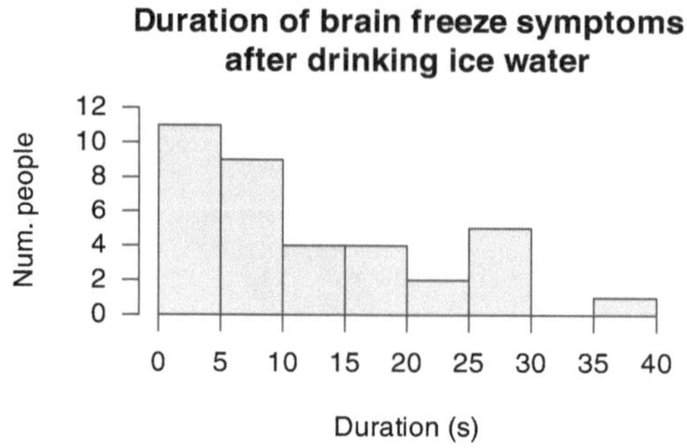

FIGURE 11.4: Histogram of the duration of brain-freeze symptoms after drinking ice water. Boundary observations are counted in the lower box.

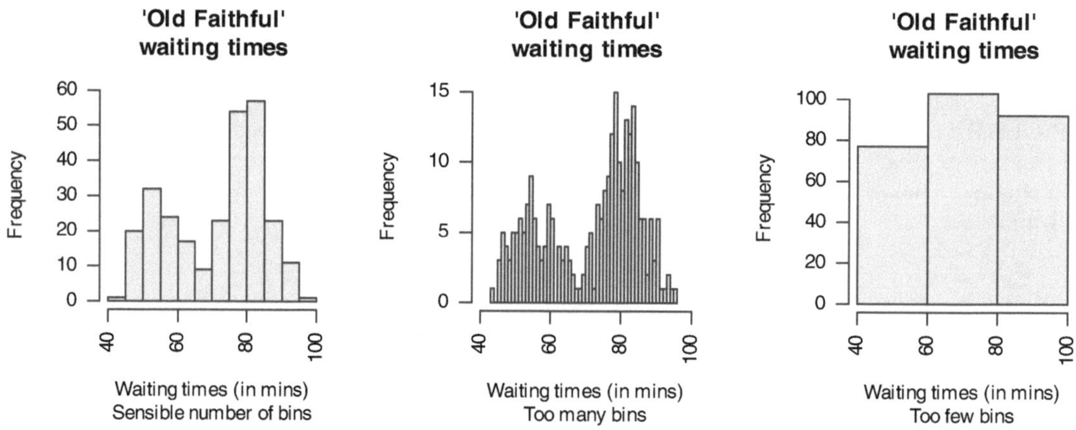

FIGURE 11.5: Histograms of the waiting times between eruptions for the *Old Faithful* geyser; changing the bins can change the impression of the distribution. Left: sensible number of bins. Centre: too many bins. Right: too few bins. For all histograms, boundary values are counted in the lower box.

11.3.2 Stemplots

Stemplots (or *stem-and-leaf plots*) are best described and explained using an example. Consider again the data in Table 11.3 and Fig. 11.2: the weights of babies born in a Brisbane hospital on one day.

In a stemplot, part of each number is placed to the left of a vertical line (the *stems*), and the rest of each number to the right of the line (the *leaves*). The weights in Table 11.3 are given to one decimal place of a kilogram, so the whole number of kilograms is placed to the left of the line (as the *stem*), and the first decimal place is placed on the right of the line (as a *leaf*). Figure 11.6 shows the stemplot in the process of being built, and Fig. 11.7 shows the final stemplot. The first weight, of 1.7 kg, is entered with the 1 to the left of the line, and the 7 to the right: 1 | 7. Similarly, 2.1 kg is entered as 2 | 1 and 2.2 kg is entered as 2 | 2, sharing the same stem as for 2.1 kg. The plot shows that most birthweights are 3-point-something kilograms.

The birthweights (in order)

~~1.7~~	~~2.1~~	~~2.2~~	2.2	2.4	2.6	2.6	2.8	2.9	3.0
3.1	3.1	3.2	3.2	3.3	3.3	3.3	3.3	3.3	3.4
3.4	3.4	3.4	3.4	3.4	3.4	3.5	3.5	3.5	3.5
3.5	3.5	3.6	3.6	3.6	3.7	3.7	3.7	3.8	3.8
3.8	3.9	3.9	4.2						

Stemplot of birthweights

1	7
2	12**2**
3	
4	

2 | 6 means 2.6 kg

FIGURE 11.6: Starting to make the stemplot for the baby-weight data: the first four observations added. The data are on the left; the stemplot during construction on the right.

Stemplot of birthweights

1	7
2	12246689
3	0112233333444444555555666777888899
4	2

2 | 6 means 2.6 kg

FIGURE 11.7: The completed stemplot for the baby-weight data.

For stemplots:

- the original data remain visible.
- place the left-most digit(s) (e.g., kilograms) on the left (stems).
- place the right-most digit (e.g., first decimal of a kilogram) on the right (leaves).
- some data do not work well for a stemplot.
- data may sometimes need suitable rounding before creating the stemplot (the baby weights were originally given to three decimal places).
- the numbers in each row should be evenly spaced, with the numbers in the columns under each other, so the length of each stem is proportional to the number of observations.

- the observations are *ordered* within each stem, so patterns in the data can be seen.
- add an explanation for reading the stemplot. For example, the stemplot for the baby-birth data says '2 | 6 means 2.6 kg' (rather than, say, 0.26 kg, or 2 lb 6 oz).

Example 11.7 (Stemplots). Wright et al. [2021] recorded the chest-beating rate of gorillas. The stemplot (Fig. 11.8) for gorillas aged under 20 years of age shows a lot of variation in the chest-beating rate.

The chest-beating data (in increasing order)

0.7	0.9	1.3	1.5	1.5
1.5	1.7	1.7	1.8	2.6
3.0	4.1	4.4	4.4	

Stemplot of chest-beating rates in young gorillas

0	7 9
1	3 5 5 5 7 7 8
2	6
3	0
4	1 4 4

1|7 means 1.7 beats per 10 hs

FIGURE 11.8: The stemplot (right) for the gorilla chest-beating data (shown on the left).

11.3.3 Dot charts (quantitative data)

Dot charts show the data on a single (usually horizontal) axis, with each observation represented by a dot (or other symbol). Sometimes, observations are identical, or nearly so; to avoid points being plotted on top of other points (called *overplotting*), the points are *jittered* (placed with some added randomness in the vertical direction) or *stacked* (placed above each other).

Example 11.8 (Dot charts). Consider the weights (in kg) of babies born in a Brisbane hospital (Table 11.3). A dot chart (Fig 11.9, left panel) shows that most babies were born between 3 and 4 kg. The points have been *jittered*.

Example 11.9 (Dot charts). The chest-beating rate of young gorillas (Example 11.7) can be displayed using a dot chart (Fig. 11.9, right panel). The points have been *stacked*.

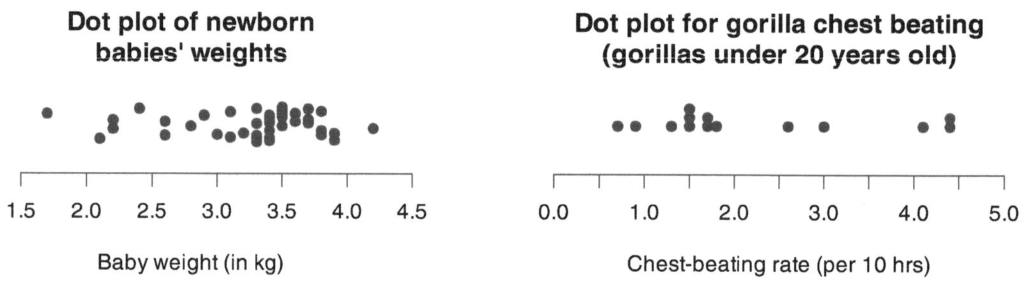

FIGURE 11.9: Left: a dot chart of the baby-weight data (with similar observations jittered). Right: a dot chart of the gorilla chest-beating rates (with similar observations stacked).

11.3.4 Describing the distribution

Graphs are constructed to help readers understand the data. Hence, after producing a graph, the *distribution* of the data should be described, focusing on four features:

1. The *shape* of the distribution. That is, are most of the values smaller or larger, or about evenly distributed between smaller and larger values?
2. The *average* of the data. What is an average, central or typical value?
3. The *variation* in the bulk of the data.
4. Any *outliers* (unusually large or small observations) or unusual features.

These can be described in rough terms. The average, variation and outliers are usually described numerically, too (Sect. 11.6 to Sect. 11.8).

Example 11.10 (Describing quantitative data). The weights of babies (Example 11.4) are typically between about 2.5 kg and 3 kg (the *average*), with most between 1.5 kg and 4.5 kg (*variation*). A few babies have very low weights (*shape*), probably premature births. No unusual values are present.

11.4 Parameters and statistics

The purpose of describing *sample* data is to understand the *population* that the sample comes from, and which the RQ asks about. Any computed numerical quantities (such as averages) are computed from the *sample*, even though the *population* is of interest. As a result, distinguishing *parameters* and *statistics* is important.

Definition 11.2 (Parameter). A *parameter* is a number, usually unknown, describing some feature of a population.

Definition 11.3 (Statistic). A *statistic* is a number describing some feature of a sample (to estimate the unknown value of the population *parameter*).

A statistic is a numerical value estimating an unknown population value. However, countless samples are possible (Sect. 6.2), and so countless possible values for the statistic—all of which are estimates of the value of the parameter—are possible. The observed value of the statistic depends on which one of the countless possible samples is selected.

⚠ The RQ identifies the population, but in practice only one of the many possible samples is studied. *Statistics* are estimates of *parameters*, and the value of the *statistic* is not the same for every possible *sample*. We only observe one value of the statistic from our single observed sample.

11.5 Describing shape

The *shape* of a distribution may be able to be described using some common terminology.

- In *right* (or *positively*) skewed distributions, most data are smaller, with some larger values.
- In *left* (or *negatively*) skewed distributions, most data are larger, with some smaller values.
- In symmetric distributions, the left and right sides of the graph are roughly similar.
- In bimodal distributions, the distribution has two peaks.

Figure 11.10 shows typical shapes. Sometimes, no short descriptions (as above) are suitable.

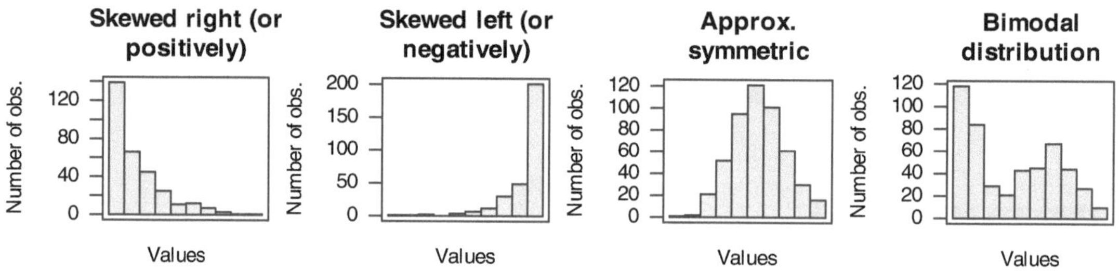

FIGURE 11.10: Some common shapes of the distribution of qualitative data.

Example 11.11 (Bimodal data)**.** The *Old Faithful* geyser in Yellowstone National Park (USA) erupts regularly [Härdle et al., 1991]. The histogram for the time between eruptions (Fig. 11.5, left panel) is bimodal, with peaks near 55 mins and 80 mins.

The weight of babies born in Brisbane (Fig. 11.3) are slightly skewed left.

11.6 Numerical summary: averages

The average (or *location*, or *central value*) for *quantitative sample data* can be described numerically in many ways. The two most common are:

- the *sample mean* (or *sample arithmetic mean*), which estimates the unknown population mean (Sect. 11.6.1).
- the *sample median*, which estimates the unknown population median (Sect. 11.6.2).

In both cases, the parameter is *estimated* by a statistic. Understanding whether to use the mean or median is important.

ⓘ 'Average' can refer to means, medians or other measures of centre. Use the precise term 'mean' or 'median', rather than 'average', when possible.

Example 11.12 (Averages). Consider the *daily* river flow volume ('streamflow') at the Mary River (Queensland, Australia) from 01 October 1959 to 17 January 2019 [Marshman and Dunn, 2025]. The 'average' daily streamflow in February could be described using either:

- the sample *mean* daily streamflow, which is 1 123.2 ML.
- the sample *median* daily streamflow, which is 146.1 ML.

Both are an 'average value', and of the same data, yet give *very* different values. This implies the mean and median measure the 'average' differently, and have different meanings. Which is the best 'average' to use? To decide, both averages need to be studied.

11.6.1 Average: the mean

The mean of the population is denoted by μ, and its value is almost always unknown. The mean of the population is *estimated* by the mean of the sample, denoted $\bar{x}$. In this context, the value of the unknown *parameter* is μ, and the value of the *statistic* is $\bar{x}$.

The sample mean *estimates* the population mean, and every sample is likely to give a different value for the sample mean. We usually only have one sample.

The Greek letter μ is pronounced 'mew' (rhymes with 'chew'). $\bar{x}$ is pronounced 'ex-bar'.

Example 11.13 (Estimating a population mean). Consider a small dataset for answering this descriptive RQ: 'For gorillas aged under 20, what is the average chest-beating rate?' The population mean rate (denoted μ) is to be estimated.

Every gorilla cannot be studied, so a *sample* is studied. The unknown population mean μ is estimated using the sample mean ($\bar{x}$) of $n = 14$ young gorillas (Fig 11.8, left panel). Of course, a different sample would likely give a different value for $\bar{x}$.

The sample mean is the 'balance point' of the observations, as shown in Fig. 11.11 (left panel) for the gorilla data. Also, the positive and negative distances (the 'deviations') of the observations from the mean add to zero (Fig. 11.11, right panel). Both of these explanations seem reasonable for identifying an 'average' value for the data.

Definition 11.4 (Mean). The *mean* is one way to measure the 'average' value of quantitative data. The *arithmetic mean* is the 'balance point' of the data. The positive and negative distances from the mean add to zero.

To find the *value* of the sample mean, *add* (denoted by $\sum$) all the observations (denoted by x) then *divide* by the number of observations (denoted by n). In symbols:

$$\bar{x} = \frac{\sum x}{n}.$$

Example 11.14 (Computing a sample mean). For the chest-beating data (Fig 11.8, left panel), an *estimate* of the population mean (i.e., the sample mean) chest-beating rate is

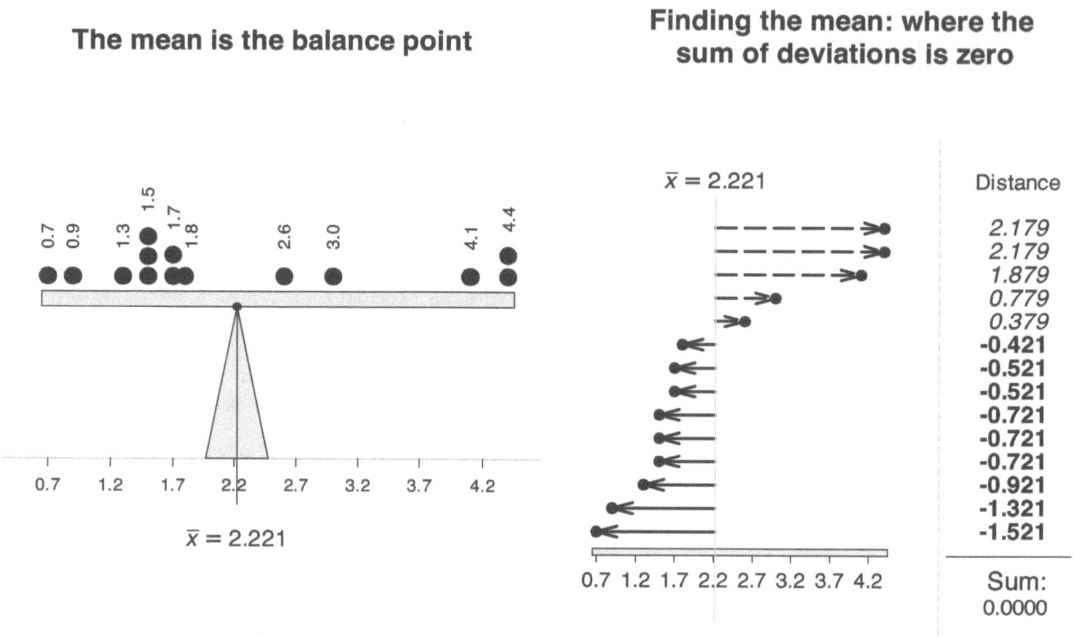

FIGURE 11.11: Two ways to understand the (arithmetic) mean. Left: the mean is the balance point of the data. Right: the mean is the value such that the positive and negative distances sum to zero.

found by summing all $n = 14$ observations then dividing by $n = 14$:

$$\bar{x} = \frac{\sum x}{n} = \frac{0.7 + 0.9 + \cdots + 4.4}{14} = \frac{31.1}{14} = 2.221429.$$

The sample mean, the best estimate of the population mean, is 2.22 beats per 10 h.

ⓘ The sample mean is usually calculated using statistical software for large amounts of data, or a calculator for small amounts of data. However, knowing *how* the mean is computed is helpful.

Software and calculators often produce numerical answers to many decimal places, not all of which may be meaningful or useful. A simple, but often useful, rule-of-thumb is to round to one or two more significant figures than the original data. Software usually does not add measurement units to the answer either.

The chest-beating data are given to one decimal place, so the *sample mean* rate is given as $\bar{x} = 2.22$ beats per 10 h.

Example 11.15 (Computing a sample mean). Griffin et al. [1960] recorded the distance at which flies (*Drosophila*) were detected by bats for $n = 11$ detections (Table 11.5). The population mean distance is estimated by the sample mean as $\bar{x} = 532/11 = 48.4$ cm.

TABLE 11.5: The distance at which small fruit flies were detected by bats.

Detection distance (in cm)										
62	52	68	23	34	45	27	42	83	56	40

11.6.2 Average: the median

A median is a value separating the largest 50% of the data from the smallest 50% of the data. In a dataset with n values, the median is *ordered observation number* $(n+1) \div 2$. (The value of the median is *not* $(n+1) \div 2$, and the median *not* necessarily halfway between the minimum and maximum values in the data.)

ⓘ Many calculators cannot find the median. The median has no commonly-used symbol, though $\tilde{\mu}$ and $\tilde{x}$ are sometimes used for the population and sample medians respectively.

Definition 11.5 (Median). The *median* is one way to measure the 'average' value of data. A *median* is a value such that half the values are larger than the median, and half the values are smaller than the median.

Example 11.16 (Computing a sample median). To find a sample median for the chest-beating data (Fig 11.8, left panel), first arrange the data *in numerical order* (Table 11.6). The median separates the larger seven numbers from the smaller seven numbers. With $n = 14$ ordered observations, the median is at position $(14+1)/2 = 7.5$ (the *median itself is not* 7.5). This means that the median is located between the seventh and eighth ordered observations.

Thus, the sample median, an estimate of the *population* median, is between 1.7 (ordered observation 7) and 1.7 (ordered observation 8). Since these values are the same, the sample median is 1.7 beats per 10 h.

TABLE 11.6: The chest-beating rate of young gorillas, in increasing order.

Chest-beating rate, per 10 h													
0.7	0.9	1.3	1.5	1.5	1.5	1.7	1.7	1.8	2.6	3.0	4.1	4.4	4.4

To clarify:

- if the sample size n is *odd* (see Example 11.17), the median is the middle number when the observations are ordered.
- if the sample size n is *even* (such as the chest-beating data; Example 11.16), the median is halfway between the two middle numbers, when the observations are ordered.

Software may use slightly different rules when n is even, producing slightly different values for the median.

⚠ The sample median *estimates* the population median, and every sample is likely to have a different value for the sample median. We usually only have one sample.

Example 11.17 (Computing a sample median)**.** For the bat data (Table 11.5), the estimate of the population *median* distance at which bats detect the flies is the *sample* median. With $n = 11$, the median is the $(11 + 1)/2 = 6$th ordered value, which is 45 cm.

11.6.3 Which average to use?

Consider the daily streamflow at the Mary River again (Example 11.12): the sample *mean* daily streamflow is 1 123 ML, and the sample *median* daily streamflow is 146.1 ML. Which is 'best' for measuring the average streamflow?

For these data, 86% of observations are *smaller* than the mean, but 50% of the observations are smaller than the median (by definition). The mean is hardly a *central* value.

A dot chart of the daily streamflow (Fig. 11.12; jittered) shows that the data are *very* highly right-skewed, with many *very* large outliers (presumably flood events).

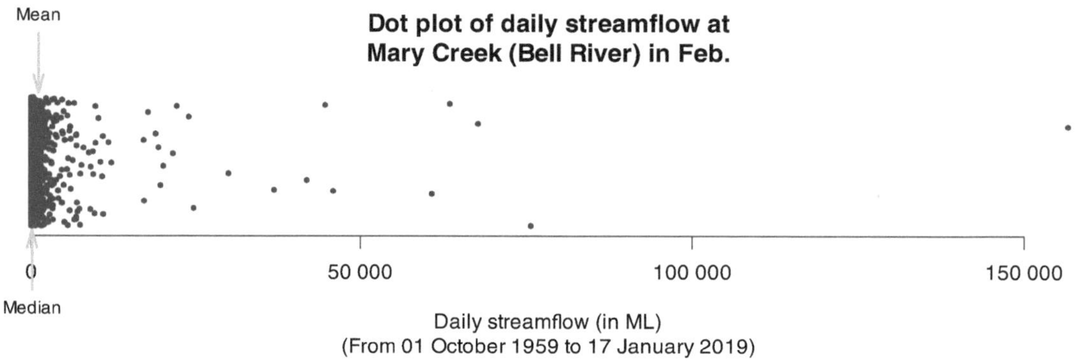

FIGURE 11.12: A dot plot of the daily streamflow at Mary River from 1960 to 2017, for February ($n = 1\,650$). Many very large outliers exist. Note: values have been jittered in the vertical direction, but overplotting is still present near 0.

The streamflow data are *very* right skewed, which is important for explaining the difference between the values of the sample mean and the sample median:

- *means* are best used for approximately symmetric data, because the mean is influenced by outliers and skewness.
- *medians* are best used for data that are highly skewed or contain outliers, because the median is *not* influenced by outliers or skewness.

Means tend to be too large if the data contain large outliers or severe right skewness, and too small if the data contain small outliers or severe left skewness. The Mary River data contains extremely large outliers—and many of them—so the mean is much larger than the median. *The median is the better measure of average for these data.* However, understanding the variation is probably more important than understanding the average value (Sect. 11.7), and the data may even be better described using percentiles (Sect. 11.7.4).

The mean is generally used rather than the median if possible (for practical and mathematical reasons), and is the most commonly-used measure of location. However, the mean is not always appropriate (as the mean is influenced by outliers and skewness). The mean and median are similar in approximately symmetric distributions. Sometimes, quoting *both* the mean and the median may be appropriate.

11.7 Numerical summary: variation

For quantitative data, the amount of *variation* in the bulk of the data should be described. Many ways exist to measure the variation in a dataset, including:

- the *range*, which is very simple and simplistic so is not often used (Sect. 11.7.1).
- the *standard deviation*, which is commonly used (Sect. 11.7.2).
- the *interquartile range (or IQR)*, which is commonly used (Sect. 11.7.3).
- *percentiles*, which are useful in specific situations (Sect. 11.7.4), especially for very skewed data.

As always, a value computed from a *sample* (the statistic) estimates the unknown value in the *population* (the parameter), and every sample can produce a different estimate.

11.7.1 Variation: the range

The range is the simplest and easiest-to-compute measure of variation.

Definition 11.6 (Range)**.** The range is the maximum value *minus* the minimum value.

The range is not often used as it only uses two values in a data set, both of which are extreme observations. As a result, the range is highly influenced by outliers. Sometimes, the *range* is given by stating both the maximum and the minimum value in the data instead of the *difference* between these values. The range is measured in the same measurement units as the data, and is usually quoted with the median.

Example 11.18 (The range)**.** For the chest-beating data (Table 11.6), the largest value is 4.4, and the smallest value is 0.7; hence

$$\text{Range} = 4.4 - 0.7 = 3.7.$$

The range of the chest-beating rate is 3.7 beats per 10 h.

11.7.2 Variation: the standard deviation

The population standard deviation (a parameter) is denoted by σ and is estimated by the sample standard deviation s (a statistic). The standard deviation is the most commonly-used measure of variation. It is tedious to compute manually, so is usually calculated using statistical software for large amounts of data, or a calculator for small amounts of data.

The *standard deviation* is (approximately) the mean distance that observations are from the mean. This seems like a reasonable way to measure the amount of variation in data.

The Greek letter σ is pronounced 'sigma'.

⚠ The sample standard deviation *estimates* the population standard deviation, and every sample is likely to have a different value for the sample standard deviation. We usually only have one sample.

Definition 11.7 (Standard deviation). The *standard deviation* is, approximately, the mean distance of the observations from the mean.

Even though *you do not have to use the formula* to calculate s (use software), we will demonstrate to show exactly what s calculates. The formula for computing the value of s is

$$s = \sqrt{\frac{\sum(x - \bar{x})^2}{n-1}},$$

where $\bar{x}$ is the sample mean, x represents the individual data values, n is the sample size, and the symbol '$\sum$' means to *add* (Sect. 11.6.1). Using the formula requires these steps.

1. Calculate the sample mean: $\bar{x}$.
2. Calculate the *deviation* of each observation x from the sample mean: $x - \bar{x}$.
3. Square these deviations (to make them all *positive* values): $(x - \bar{x})^2$.
4. Add these squared deviations: $\sum(x - \bar{x})^2$.
5. Divide the answer by $n - 1$.
6. Take the (positive) square root of the answer (to 'undo' the squaring of the data).

⚠ *You do not need to use the formula!* You should know how to use software or a calculator to find the value of the standard deviation, what the standard deviation measures, and how and when to use it.

Example 11.19 (Computing a sample standard deviation). For the chest-beating data (Table 11.6), the squared *deviations* of each observation from the mean of 2.2214 (using four decimal places in calculations) are shown in Fig. 11.13. The sum of the squared distances is 20.9636. Then, the sample standard deviation is:

$$s = \sqrt{\frac{20.9636}{14-1}} = \sqrt{1.612585} = 1.269876.$$

The sample standard deviation of the chest-beating rate is 1.27 per 10 h.

The sample standard deviation s is:

- positive (unless all observations are the same, when $s = 0$; that is, *zero* variation).
- best used for (approximately) symmetric data.
- usually quoted with the mean.
- the most commonly-used measure of variation.
- measured in the same units as the data.
- influenced by skewness and outliers, like the mean.

11.7.3 Variation: the interquartile range (IQR)

The standard deviation uses the value of $\bar{x}$, so is impacted by skewness and outliers just like the sample mean. A measure of variation *not* affected by skewness and outliers is the interquartile range, or IQR. To understand the IQR, understanding *quartiles* is necessary first.

Definition 11.8 (Quartiles). *Quartiles* describe the shape of the data.

The sum of the squared distances of each observation from the mean

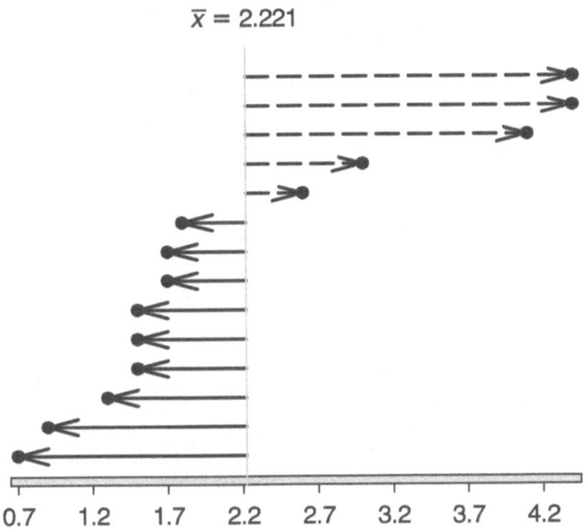

Deviation	$(\text{Dev.})^2$
2.1786	4.7462
2.1786	4.7462
1.8786	3.5290
0.7786	0.6062
0.3786	0.1433
-0.4214	0.1776
-0.5214	0.2719
-0.5214	0.2719
-0.7214	0.5205
-0.7214	0.5205
-0.7214	0.5205
-0.9214	0.8490
-1.3214	1.7462
-1.5214	2.3147
0.0000	20.9636

FIGURE 11.13: The standard deviation is related to the sum of the squared-distances from the mean. The chest-beating data are used. The sum of the deviations is *always* zero.

- The first quartile Q_1 is a value separating the smallest 25% of observations from the largest 75%. The Q_1 is like the median of the *smaller* half of the data, halfway between the minimum value and the median.
- The second quartile Q_2 is a value separating the smallest 50% of observations from the largest 50%. (This is also the *median.*)
- The third quartile Q_3 is a value separating the smallest 75% of observations from the largest 25%. The Q_3 is like the median of the *larger* half of the data, halfway between the median and the maximum value.

Quartiles divide the data into four parts of approximately equal numbers of observations. The *interquartile range* (or *IQR*) is the difference between Q_3 and Q_1.

Definition 11.9 (IQR)**.** The *IQR* is the range in which the middle 50% of the data lie: the difference between the third and the first quartiles.

⚠ The sample IQR *estimates* the population IQR, and every sample is likely to have a different value for the sample IQR. We usually only have one sample.

For the chest-beating data (Table 11.6), where $n = 14$ (an *even* number of observations), the median is 1.7 (Example 11.16). The data then can be split into *smaller* and *larger* halves, each with seven values:

- Smaller half: 0.7 0.9 1.3 1.5 1.5 1.5 1.7
- Larger half: 1.7 1.8 2.6 3.0 4.1 4.4 4.4

Since each half has seven observations (an *odd* number), the median of each half is the $(7+1)/2 = 4$th ordered value. Hence:

- Q_1, the *first quartile*, is the median of the smaller half; $Q_1 = 1.5$. About 25% of observations are smaller than 1.5.
- Q_2, the *second quartile* or *median*, is 1.7, so 50% of observations are smaller than 1.7.
- Q_3, the *third quartile*, is the median of the larger half; $Q_3 = 3.0$. About 75% of observations are smaller than 3.0.

To divide the data into four parts of equal numbers of observations, each part would need $14/4 = 3.5$ observations, which is not possible. Hence, we say the values of Q_1 and Q_3 are 'about' the values given. (Software sometimes uses a different method for computing the quartiles.) Using these values, the IQR is $Q_3 - Q_1 = 3.0 - 1.5 = 1.5$, as shown in Fig. 11.14.

Since the IQR measures the range of the central 50% of the data, the IQR is not influenced by outliers. The IQR is measured in the same measurements units as the data.

Software often uses different rules to compute quartiles (and medians) that may produce slightly different answers. In large datasets, the differences are usually minimal.

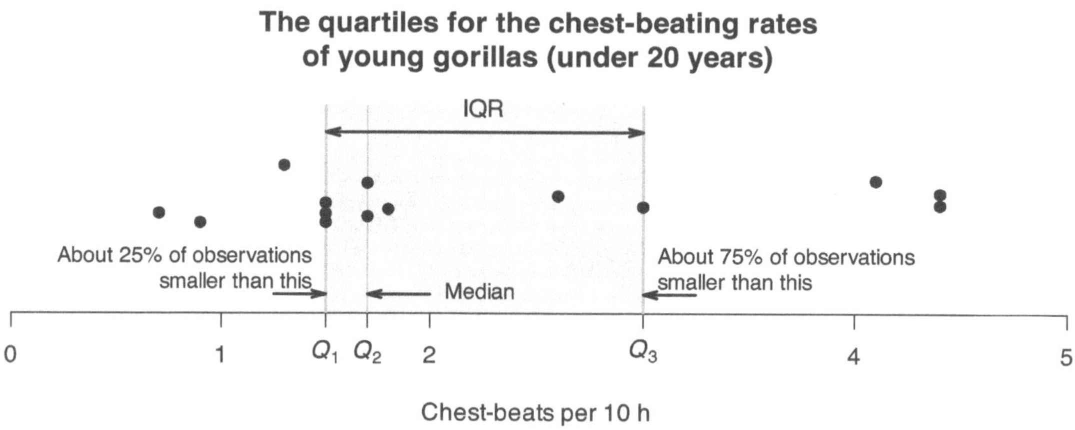

FIGURE 11.14: A dot chart (with jittering) for the chest-beating data for young gorillas, showing the IQR.

When n is odd, the median *may* or *may not* be included in each of these halves when computing Q_1 and Q_3; we decide *not* to include the median in each half.

Example 11.20 (Computing the IQR for n odd). The bat data (Table 11.5) has $n = 11$ observations. The smaller and larger halves of the data, *without* the median of 45, are:

- Smaller half: 23 27 34 40 42: the median is $Q_1 = 34$.
- Larger half: 52 56 62 68 83: the median is $Q_3 = 62$.

Hence, the IQR is $62 - 34 = 28$ cm. (If the median *is* included in each half, the IQR is $59 - 37 = 22$ cm.)

11.7.4 Variation: percentiles

Percentiles are like quartiles; in fact, quartiles are a special case of percentiles.

Definition 11.10 (Percentiles)**.** The pth percentile of the data is a value separating the smallest $p\%$ of the data from the rest.

For example:

- the 12th percentile separates the smallest 12% of the data from the rest.
- the 67th percentile separates the smallest 67% of the data from the rest.
- the 94th percentile separates the smallest 94% of the data from the rest.

This means that the first quartile Q_1 is the 25th percentile, the second quartile Q_2 is the 50th percentile (and median), and the third quartile Q_3 is the 75th percentile.

Software uses various rules to compute percentiles. In large datasets, the differences are usually minimal.

Percentiles are especially useful for very skewed data in certain applications. For instance, scientists who monitor rainfall and stream heights, and engineers who use this information, are more interested in extreme weather events rather than the 'average' event. Structures need to be designed to withstand 1-in-100 year events (the 99th percentile) or similar, rather than 'average' events. Percentiles are measured in the same measurements units as the data.

Example 11.21 (Percentiles)**.** For the streamflow data at the Mary River (Example 11.12), the February data are highly right-skewed (Fig. 11.12). The median (50th percentile) is 146.1 ML. However, the 95th percentile is 3 480 ML and the 99th percentile is 19 043 ML. Constructing infrastructure for the *median* streamflow would be highly inadequate.

11.7.5 Which measure of variation to use?

Which is the 'best' measure of variation for quantitative data? As with measures of location, the answer depends on the data.

Since the standard deviation formula uses the mean, it is impacted in the same way as the mean by outliers and skewness. Hence, the standard deviation is best used with approximately symmetric data. The IQR is best used when data are skewed or outlier are present. Sometimes, both the standard deviation and the IQR can be quoted.

11.8 Numerical summary: identifying outliers

Outliers are 'unusual' observations: those quite different from the bulk of the data (larger or smaller). Outliers are 'unusual', but not necessarily 'incorrect' or 'bad' observations. Rules for deciding if an observation is an outlier are always arbitrary.

Definition 11.11 (Outliers)**.** An *outlier* is an observation that is 'unusual' (either larger or smaller) compared to the bulk of the data. Rules for identifying outliers are arbitrary.

Two rules for identifying outliers are:

- the *standard deviation rule*, which is only useful when the data have an approximately symmetric distribution (Sect. 11.8.1).
- the *IQR rule*, which is useful in most situations (Sect. 11.8.2).

11.8.1 The standard deviation rule

The standard deviation rule uses the mean and the standard deviation, so is suitable for approximately symmetric distributions (when means and standard deviations are sensible numerical summaries). The rationale behind this rule is explained in Sect. 20.3.

Definition 11.12 (Standard deviation rule for identifying outliers). For approximately symmetric distributions, an observation more than three standard deviations from the mean may be considered an outlier.

All rules for identifying outliers are arbitrary, and sometimes the standard deviation rule is given slightly differently. For example, outliers may be identified as observations more than 2.5 standard deviations from the mean. Both rules are acceptable, since the definition is arbitrary.

Example 11.22 (Standard deviation rule for identifying outliers). An engineering project [Hald, 1952] studied a new building material, to estimate the average permeability. Permeability time (the time for water to permeate the sheets) was measured from 81 pieces of material (in seconds).

For these data, the mean is $\bar{x} = 43.162$ and the standard deviation is $s = 27.358$. Using the standard deviation rule, outliers are observations *smaller* than $43.162 - (3 \times 27.358)$ or *larger* than $43.162 + (3 \times 27.358)$; that is, *smaller* than -38.9 s (which is clearly not appropriate here, as the data must be positive values), or *larger* than 125.2 s. This rule is shown in Fig. 11.15; two observations are identified as outliers using the standard deviation rule.

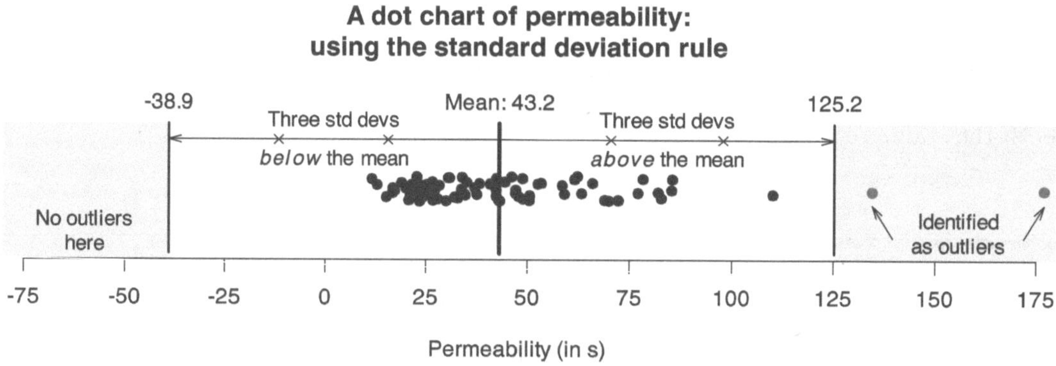

FIGURE 11.15: Outliers identified using the standard deviation rule for the permeability data.

11.8.2 The IQR rule

Since the standard deviation rule for identifying outliers relies on the mean and standard deviation, it is not appropriate for non-symmetric distributions. Another rule is needed for identifying outliers in these situations: the IQR rule.

Definition 11.13 (IQR rule for identifying outliers). The IQR rule identifies mild and extreme outliers.

- *Extreme outliers*: observations $3 \times$ IQR more unusual than Q_1 or Q_3.
- *Mild outliers*: observations $1.5 \times$ IQR more unusual than Q_1 or Q_3 (that are not extreme outliers).

This definition is easier to understand using an example.

Example 11.23 (IQR rule for identifying outliers). Using the permeability data seen in Example 11.22, a computer shows that $Q_1 = 24.7$ and $Q_3 = 50.6$, so IQR $= 50.6 - 24.7 = 25.9$. Then, *extreme* outliers are observations $3 \times 25.9 = 77.7$ more unusual than Q_1 or Q_3. That is, *extreme* outliers are observations are:

- more unusual than $24.7 - 77.7 = -53.0$ (that is, *less* than -53.0); or
- more unusual than $50.6 + 77.7 = 128.3$ (that is, *greater* than 128.3).

Mild outliers are observations $1.5 \times 25.9 = 38.9$ more unusual than Q_1 or Q_3 (that are not extreme outliers). That is, *mild* outliers are

- more unusual than $24.7 - 38.9 = -14.2$ (that is, *less* than -14.2); or
- more unusual than $50.6 + 38.9 = 89.5$ (that is, *greater* than 89.5).

Three observations are identified as outliers using the IQR rule (Fig. 11.16): two extreme outliers, and one mild outlier.

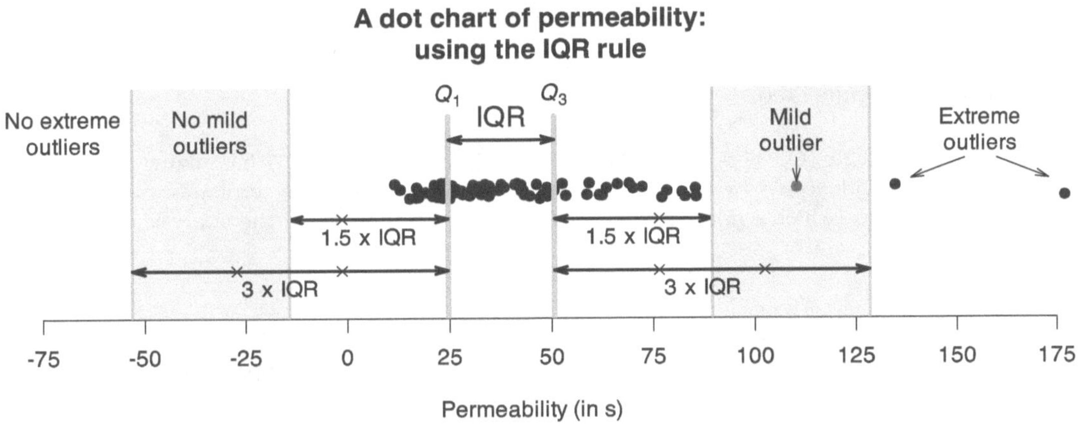

FIGURE 11.16: Mild and extreme outliers, using the IQR rule, for the permeability data.

11.8.3 Which outlier rule to use?

The standard deviation rule is most appropriate for *approximately symmetric distributions*; the IQR rule can be used for *any distribution*, but primarily for those skewed or with outliers.

⚠ Remember: all rules for identifying outliers are arbitrary.

11.8.4 What to do with outliers?

What should be done if outliers are identified in data? Deleting or removing outliers simply because they are identified as outliers is *very poor practice*. After all, the outliers were obtained from the study like all other observations; they belong in the data as much as any other observation. In addition, the rules for identifying outliers are *arbitrary*: some observations may be identified as outliers using one rule, but not by another.

⚠ Outliers are *unusual* observations; they are not necessarily *mistakes*.

Managing outliers depends on *why* they occur (Dunn and Smyth [2018], p. 138):

- *the outlier is clearly a mistake* (e.g., an age of 222). If the mistake cannot be fixed (e.g., the completed questionnaire form is lost), the observation can be *deleted*. Similarly, if the outlier comes from an error or mistake in the data collection (e.g., too much fertiliser was accidentally applied), the observation can be deleted.
- *the outlier represents a different population*. Suppose an outlier is identified in a study of students, corresponding to a student aged 65. If the next oldest student in the data is aged 39, the outlier can be removed, since it belongs to a different population ('students aged over 40') than the other observations ('students aged 40 and under'). The remaining observations can be analysed, but the results only apply to students aged under 40 (which should be clearly communicated).
- *the reason for the outlier is unknown*. In these cases, *discarding outliers routinely is not recommended*; the outliers are probably real observations that are just as valid as the others. Perhaps a different analysis is necessary (e.g., using medians rather than means). Furthermore, very large datasets are expected to have a small number of observations identified as outliers using the above arbitrary rules.

In all cases, whenever observations are removed from a dataset, this should be clearly explained and documented.

Example 11.24 (Outliers). The Mary River dataset (Sect. 11.6) has many *extremely* large outliers identified by software, but each is reasonable. They probably correspond to flood events (which could be confirmed). Removing these from the analysis would be inappropriate.

Example 11.25 (Outliers). The permeability data (Example 11.22) has large outliers, but all seem reasonable. Removing these from the analysis would be inappropriate.

11.9 Numerical summary tables

In studies with quantitative variables, the quantitative variables should be summarised in a table. The table should include, as a minimum, measures of average, variation and the sample sizes. An example is given in the next section (Table 11.7).

11.10 Example: water access

López-Serrano et al. [2022] recorded data about access to water for three rural communities in Cameroon. Three quantitative variables are recorded. Part of understanding the data requires summarising the quantitative variables; histograms are shown in Fig. 11.17, and a summary table in Table 11.7.

Many households are coordinated by women in their late 50s. The number of people and number of children under 5 years of age are both right-skewed. One household has over 30 people, and has 10 children in that household. (These are identified as outliers, but are unlikely to be mistakes.) Some observations are missing for some variables, explaining the differences in sample sizes in Table 11.7.

TABLE 11.7: Summarising the quantitative data in the water-access study.

	n	Mean	Median	Std dev.	IQR	Min.	Max.
Woman's age (years)	120	41.6	40.5	14.56	30.25	19	61
Household size	121	7.0	6.0	4.80	4.00	0	32
Children aged under 5	120	1.6	1.0	1.65	2.00	0	10

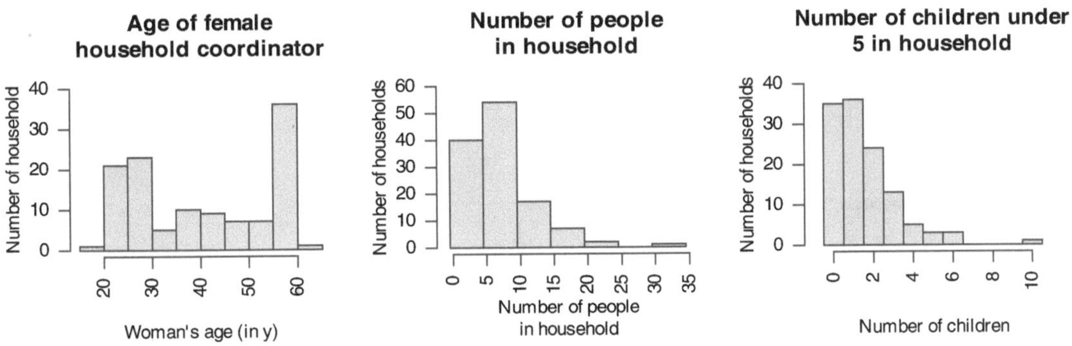

FIGURE 11.17: The age of the female household coordinator, the number of people in the household, and the number of children in the household aged under 5 years, for the water-access study.

11.11 Chapter summary

Quantitative data can be graphed using a histogram, stemplot (in special circumstances), or dot charts. Quantitative data can be summarised numerically; the most common techniques are indicated in Table 11.8. The *mean* and *standard deviation* are usually used whenever possible, for practical and mathematical reasons. Sometimes quoting both the mean and median (and the standard deviation and IQR) may be appropriate.

TABLE 11.8: Summarising quantitative data.

	For distributions with a shape that is:	
Feature	**Approximately symmetric**	**Not symmetric, or has outliers**
Average	Mean	Median
Variation	Standard deviation	IQR
Outliers	Standard deviation rule	IQR rule

11.12 Quick review questions

Are the following statements *true* or *false*?

1. The IQR measures the amount of variability in data.
2. The mean and the median can both be called an 'average'.
3. The mean and the median are not always the same value.
4. The range is a simple measure of variation in a set of data.
5. The standard deviation measures the amount of variability in data.
6. Another name for the median is Q_2.
7. Q_3 is the median of the largest half of the data.
8. The IQR is a useful measure of variation in data that are skewed.
9. The IQR is the difference between the first and second quartiles.
10. Another name for the 75th percentile is Q_3.
11. The units of the standard deviation and the IQR are the same as for the original data.

11.13 Exercises

Answers to odd-numbered exercises are given at the end of the book.

Exercise 11.1. The *Australian Bureau of Statistics* (ABS) records the age at death of Australians. The histogram of the age of death for females in 2012 is shown in Fig. 11.18 (left panel). Describe the distribution.

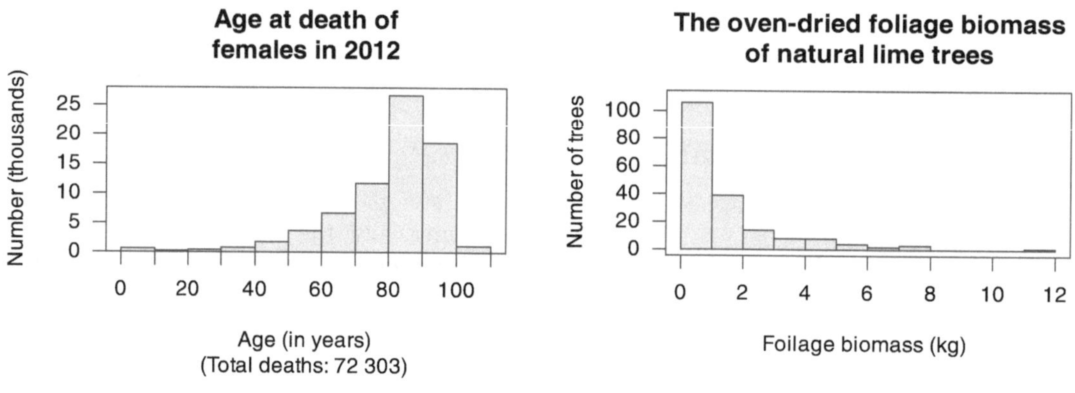

FIGURE 11.18: Left: histograms of age at death for female Australians in 2012. Right: the oven-dried foliage biomass for naturally-grown lime trees.

Exercise 11.2. [*Dataset*: `Lime`] Schepaschenko et al. [2017] measured the oven-dried foliage biomass of lime trees grown in natural environments. The histogram of the foliage biomass is shown in Fig. 11.18 (right panel). Describe the distribution.

Exercise 11.3. [*Dataset*: `NHANES`] The histogram of the direct HDL cholesterol concentration from the American National Health and Nutrition Examination Survey (NHANES) [Pruim, 2015] from 1999–2004 is shown in Fig. 11.19 (left panel).

1. Should the mean or median be used to measure the 'average' HDL cholesterol concentration? Explain.
2. Describe the distribution.

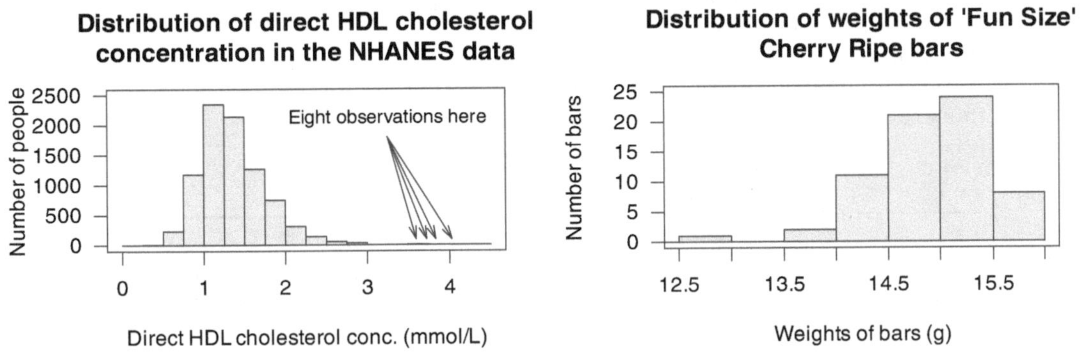

FIGURE 11.19: Left: the histogram of direct HDL cholesterol concentration from the NHANES study (large outliers exist but are hard to see, as the sample size is very large). Right: the weights of 'Fun Size' *Cherry Ripe* chocolate bars.

Exercise 11.4. [*Dataset*: `CherryRipe`] The histogram of the weights of 'Fun Size' *Cherry Ripe* chocolate bars between 2016 and 2019 is shown in Fig. 11.19 (right panel).

1. Should the mean or median be used to measure the 'average' weight of a 'Fun Size' *Cherry Ripe* bar? Explain.
2. Describe the distribution.

Exercise 11.5. Levenson [2005] recorded the number of fatalities at amusement rides in the US from 1994 to 2003 (Table 11.9). Using software or a calculator, compute:

1. the sample mean number of fatalities per year over this period.
2. the sample median number of fatalities per year over this period.
3. the sample standard deviation of the number of fatalities per year over this period.
4. the sample IQR of the number of fatalities per year over this period.

TABLE 11.9: Fatalities at amusement park rides in the US.

	1994	**1995**	**1996**	**1997**	**1998**	**1999**	**2000**	**2001**	**2002**	**2003**
Fatalities:	2	4	3	4	7	6	1	3	2	5

Exercise 11.6. Furness and Bryant [1996] studied fulmars (a seabird). The mass of the female birds were (in grams): 635; 635; 668; 640; 645; 635.

1. Construct a stemplot (using the first two digits as the stems).
2. Using your calculator, find the value of the *sample* mean.
3. Using your calculator, find the value of the *sample* standard deviation.
4. Find the value of the *sample* median.
5. Find the value of the *population* standard deviation.

Exercise 11.7. Draw a stemplot of the average monthly SOI (from the Australian *Bureau of Meteorology*) in August from 1995 to 2003 (Table 11.10). Then, use your calculator (where possible) to calculate the:

1. sample mean
2. sample median.
3. range.
4. sample standard deviation.
5. sample IQR.

TABLE 11.10: The average monthly SOI values in August from 1995 to 2003.

	1995	**1996**	**1997**	**1998**	**1999**	**2000**	**2001**	**2002**	**2003**
Monthly average SOI:	0.8	4.6	−19.8	9.8	2.1	5.3	−8.2	−14.6	−1.8

Exercise 11.8. [*Dataset*: **FriesWt**] Wetzel [2005] weighed orders of french fries to determine how they matched the target weight of 171 g (Table 11.11).

1. Produce graphs to summarise the data.
2. Use software to produce numerical summary information.

Do you think the weights meet the target weight, on average?

TABLE 11.11: The weight of servings of french fries.

Weight of large orders of fries (in g)										
117.0	132.0	134.0	139.0	141.0	143.0	146.0	152.0	154.0	155.0	157.0
126.0	133.0	137.0	139.0	142.0	143.5	146.0	152.0	154.5	156.0	176.0
128.0	133.0	138.0	140.0	142.5	145.0	151.0	154.0	154.5	156.5	117.0

Exercise 11.9. [*Dataset*: **Orthoses**] Swinnen et al. [2018] studied the influence of using ankle-foot orthoses in children with cerebral palsy. The data for the 15 subjects is shown in Table 10.3.

1. Compute the values of the sample mean, sample median, sample standard deviation and sample IQR for the heights.
2. What are the values of the population mean, population median, population standard deviation and population IQR for the heights?
3. Produce a stemplot of the children's heights.
4. Produce a dot chart of the children's heights.
5. Produce a histogram of the children's heights.
6. Describe the distribution of the children's heights.

Exercise 11.10. An article studied patients who had been admitted to Castle Hill Hospital [Jenner et al., 2022]. The total number of microplastics found in the lungs of each patient are shown in Table 11.12. For these patients:

1. Draw a stemplot, using the numbers as (say) 8.0, with the decimals as the leaves.
2. What are the values of the sample mean and sample median number of microplastics?
3. What are the values of the population mean and population median number of microplastics?
4. What is the value of the sample standard deviation of the number of microplastics?
5. What is the value of the sample IQR of the number of microplastics?

TABLE 11.12: The number of microplastics found in 11 patients.

Number of microplastics										
8	3	5	2	0	2	1	7	5	1	0

Exercise 11.11. Describe the histogram in Fig. 11.4 for the brain-freeze data.

Exercise 11.12. The standard deviation for Dataset A in Fig. 11.20 is $s = 2$. Will the standard deviation of Dataset B be *smaller* than or *greater* than 2? Why?

Exercise 11.13. [*Dataset*: **Jeans**] Diehm and Thomas [2018] recorded the size of pockets in men's and women's jeans, including the minimum heights of the front pockets (Fig. 11.21, left panel).

1. What proportion of jeans in the sample have a minimum height less than 17 cm, for men's and women's jeans?

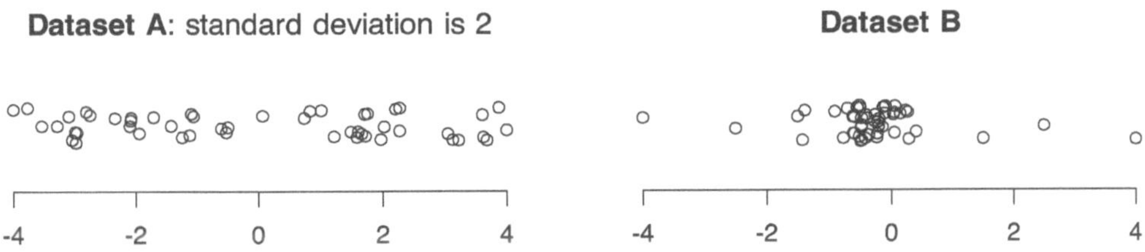

FIGURE 11.20: Dotplots of two datasets (with jittering).

2. What proportion of jeans in the sample have a minimum height less than 13.25 cm, for men's and women's jeans?

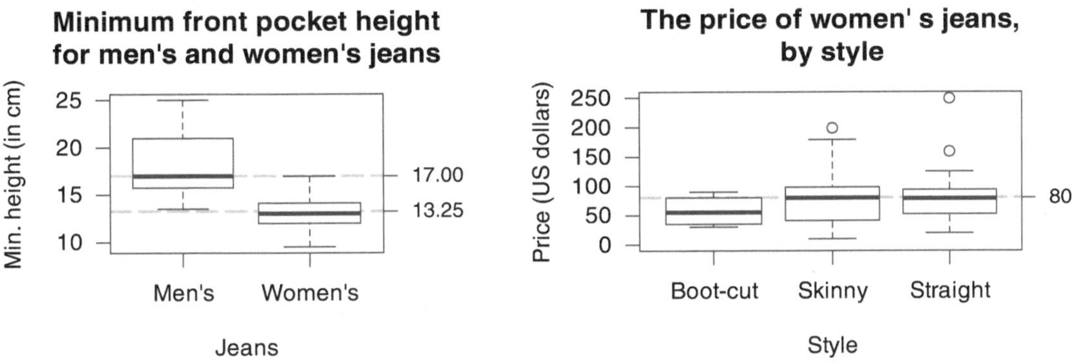

FIGURE 11.21: Left: the minimum height of the height of front pockets in jeans. Right: the price of different styles of women's jeans.

Exercise 11.14. Diehm and Thomas [2018] recorded data on the price of different styles of women's jeans (Fig. 11.21, right panel).

1. What proportion of boot-cut jeans in the sample cost less than $80?
2. What proportion of skinny jeans in the sample cost less than $80?
3. What proportion of straight jeans in the sample cost less than $80?
4. In general, which type of jeans are the cheapest?

Exercise 11.15. A professor has recorded the marks (as a percentage) for all students in her four classes for an assignment. All classes have the same number of students. The corresponding histograms are shown in Fig. 11.22.

1. In which class would the median be the largest?
2. In which class would the median be the smallest?
3. In which class would the standard deviation be the largest?
4. In which class would the standard deviation be the smallest?

Exercise 11.16. Consider the four histograms in Fig. 11.23. Which histogram is most likely to describe the *shape* of the following data? Why?

1. The time that students remain in an examination room for a *short*, *easy* two-hour examination.
2. The heights of females at a local adults' dance club.
3. The *starting* salaries of new science graduates employed full-time.
4. The volume of drink in 375 mL cans of soft drink.
5. The time that students remain in the examination room for a *hard*, *long* two-hour examination.

(The first bar of the histogram is not necessarily at zero; it is the *shape* of the histogram that is of interest here: right skewed, left skewed, symmetric, etc.)

Answers to *Quick review* questions: Only Statement 9 is false.

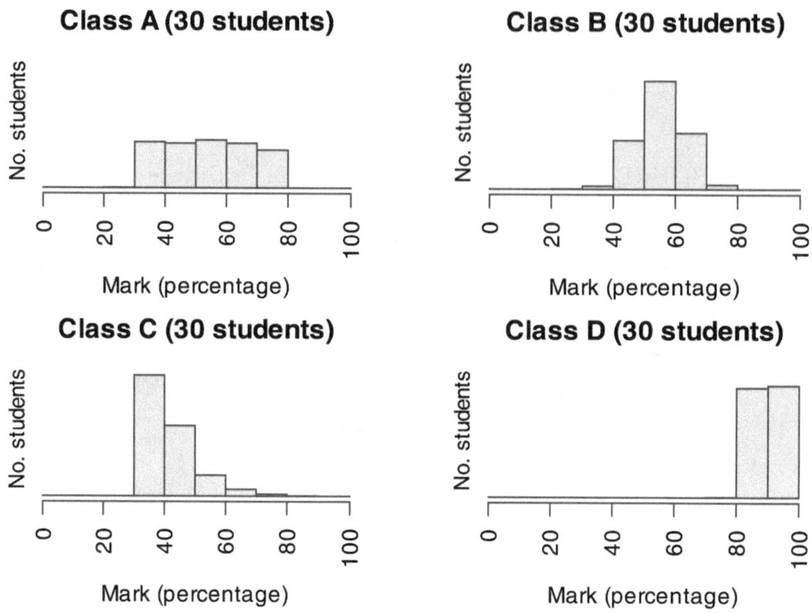

FIGURE 11.22: Histogram of marks for four classes.

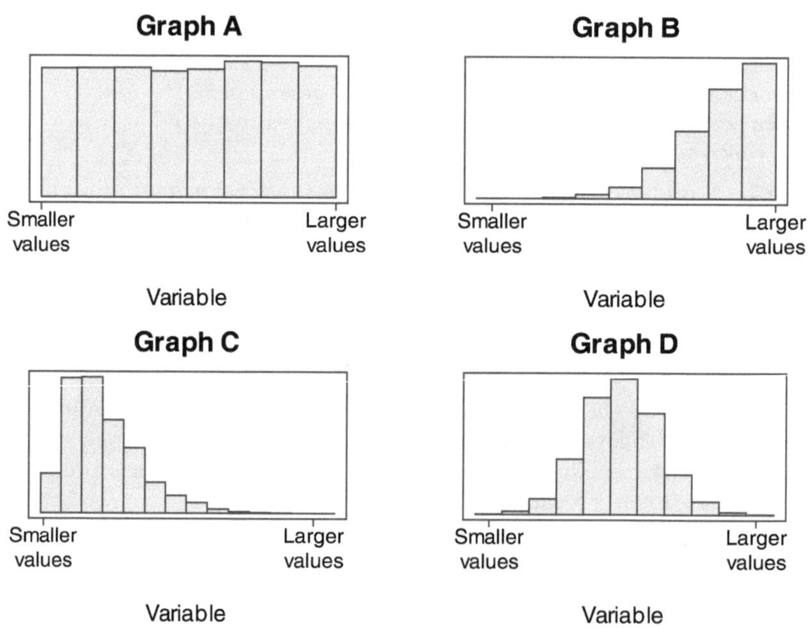

FIGURE 11.23: Four histograms: where would they be useful?

12

Summarising qualitative data

So far, you have learnt to ask an RQ, design a study, collect the data, classify the data, and summarise quantitative data. **In this chapter**, you will learn to:

- summarise qualitative data using the appropriate graphs.
- summarise qualitative data using, for example, medians, proportions and odds.

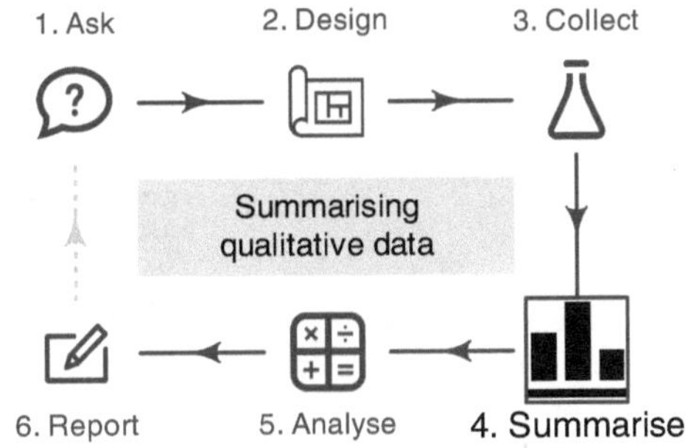

12.1 Introduction

Many quantitative research studies involve qualitative variables. Except for very small amounts of data, understanding the data is difficult without a summary. As with quantitative data, qualitative data can be understood by knowing how often values of the variables appear. This is called the *distribution* of the data (Def. 11.1).

The distribution can be displayed using a frequency table (Sect. 12.2) or a graph (Sect. 12.3). Qualitative data can be summarised by finding modes or, for ordinal qualitative data, using medians (Sect. 12.6). The distribution of qualitative data can be summarised numerically by computing proportions, percentages (Sect. 12.4) or odds (Sect. 12.5).

12.2 Frequency tables for qualitative data

Qualitative data are typically collated in a *frequency table*. The rows (or the columns) should list the *levels* of the variable, and these should be *exhaustive* (cover all levels) and *mutually exclusive* (observations belong to only one level). The number of observations or the percentage of observations (or both) are then given for each level.

For *nominal* data, the levels of the variables can be displayed in alphabetical order, in order of size, in order of personal preference, or in any other order: use the order most likely to be useful to readers. For *ordinal* data, the natural order of the levels should almost always be used.

Example 12.1 (Opinions of AV vehicles). Pyrialakou et al. [2020] surveyed 400 residents of Phoenix (Arizona) about their opinions of autonomous vehicles (AVs). Demographic information (Table 12.1) and respondents' opinions of sharing roads with AVs (Table 12.2) were recorded.

The gender of the respondent is *nominal* (two levels), while the age group is *ordinal* (six levels). The levels are shown in the rows. The three questions about safety (Table 12.2) all yield *ordinal* responses (five levels, in columns).

TABLE 12.1: Demographic information for the AV data for 400 respondents.

	Number	Percentage
Gender ($n = 400$)		
Female	204	51
Male	196	49
Age group ($n = 400$)		
18 to 24	52	13
25 to 34	76	19
35 to 44	76	19
45 to 54	72	18
55 to 64	56	14
65+	68	17

TABLE 12.2: Responses to three scenarios for the AV data for 400 respondents (rows sum to $n = 400$).

	Unsafe		Somewhat unsafe		Neutral		Somewhat safe		Safe	
	n	%	n	%	n	%	n	%	n	%
Driving near an AV	58	14	79	20	96	24	97	24	70	18
Cycling near an AV	77	19	104	26	87	22	76	19	56	14
Walking near an AV	63	16	86	22	103	26	82	20	66	16

12.3 Graphs for qualitative data

Three options for graphing qualitative data include:

- *dot charts* (Sect. 12.3.1), which are usually a good choice.
- *bar charts* (Sect, 12.3.2), which are usually a good choice.
- *pie charts* (Sect. 12.3.3), which are only useful in special circumstances, and can be hard to interpret.

Sometimes these graphs are used for *discrete* quantitative data with a small number of possible options.

⚠ The purpose of a graph is to display the information in the clearest, simplest possible way, to facilitate understanding the message(s) in the data.

12.3.1 Dot charts (qualitative data)

Dot charts indicate the counts (or corresponding percentages) in each level using dots (or some other symbol). The levels can be on the horizontal or vertical axis, and the counts or percentages on the other. Placing the levels on the vertical axis often makes for easier reading, and space for long labels.

Example 12.2 (Dot plots)**.** For the AV study in Example 12.1, a dot chart of the age group of respondents is shown in Fig. 12.1 (top left panel).

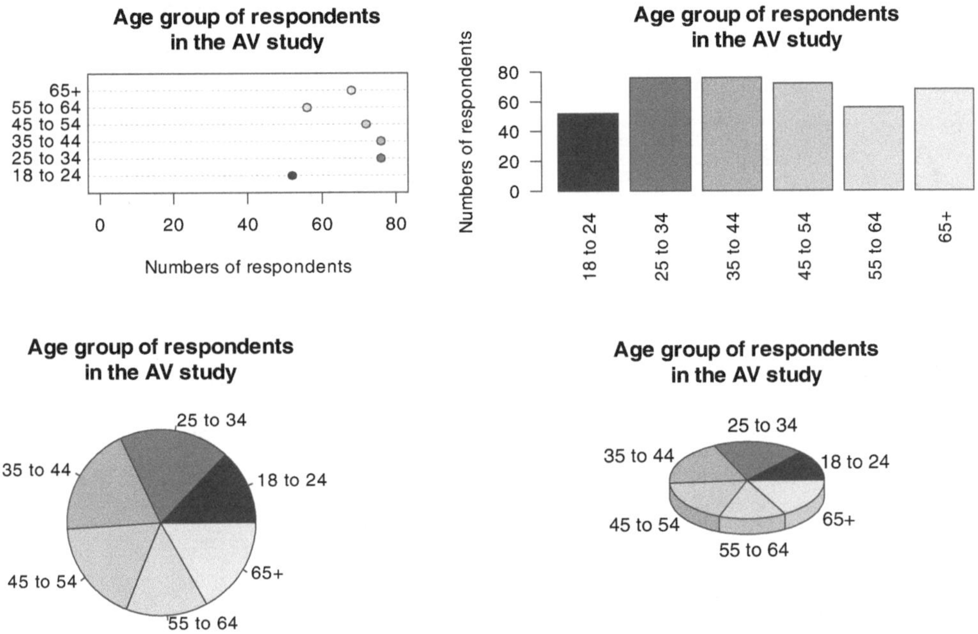

FIGURE 12.1: The age group of respondents in the AV study. All graphs present the same data.

For dot charts:

- place the qualitative variable on the horizontal or vertical axis (and label with the levels of the variable).
- use counts or percentages on the other axis.
- for nominal data, *think about the most helpful order* for the levels.

⚠ The axis displaying the counts (or percentages) should *start from zero*, since the distance of the dots from the axis visually implies the frequency of those observations (see Example 17.3).

12.3.2 Bar charts

Bar charts use bars to represent the number (or percentage) of observations in each level. As with dot charts, the levels can be on the horizontal or vertical axis, but placing the level names on the vertical axis often makes for easier reading, and room for long labels.

Example 12.3 (Bar plots)**.** For the AV study in Example 12.1, a bar chart of the age group of respondents is shown in Fig. 12.1 (top right panel).

For bar charts:

- place the qualitative variable on the horizontal or vertical axis (and label with the levels of the variable).
- use counts or percentages on the other axis.
- for nominal data, levels can be ordered any way: *think about the most helpful order.*
- bars have gaps between bars, as the bars represent distinct categories.

In contrast to bar charts, bars in histograms are butted together (except when an interval has a zero count), as the variable-axis usually represents a continuous numerical scale.

⚠ The axis displaying the counts (or percentages) should *start from zero*, since the height of the bars visually implies the frequency of those observations (see Example 17.3).

12.3.3 Pie charts

In pie charts, a circle is divided into segments proportional to the number in each level of the qualitative variable.

Example 12.4 (Pie charts)**.** For the AV study in Example 12.1, a pie chart of the age group of respondents is shown in Fig. 12.1 (bottom left panel).

Using pie charts may present challenges (see Sect. 17.2.4):

- pie charts only work when graphing parts of a whole.
- pie charts only work when *all* options are present ('exhaustive').
- pie charts are difficult to use with levels having zero or small counts (see Example 17.4).
- pie charts are difficult to interpret when many categories are present.
- pie charts are hard to read, as humans compare *lengths* (bar and dot charts) better than *angles* (pie charts) [Friel et al., 2001].

Example 12.5 (Pie chart unsuitable)**.** Consider studying the percentage of people who use Firefox, Chrome, and Safari as web browsers. A pie chart is *not suitable* for displaying the data, as people can use more than one of these browsers (i.e., the options are not *mutually exclusive*) nor *exhaustive* (i.e., other options exist).

12.3.4 Comparing dot, bar and pie charts

In the pie chart (Fig. 12.1, bottom left panel), determining *which* age groups have the fewest and most respondents is hard. The equivalent bar chart or dot chart makes the comparison easy. The *tilted* pie chart makes this comparison even harder (Fig. 12.1, bottom right panel).

Recall that the *purpose of a graph is to display the information in the clearest, simplest possible way, to facilitate understanding the message(s) in the data.* A pie chart often makes the message hard to see [Siegrist, 1996].

12.4 Numerical summary: proportions and percentages

Qualitative data can be summarised numerically by using the *proportion* or *percentage* of individuals in each level. These can be given instead of, or with, the counts (Tables 12.1 and 12.2).

Definition 12.1 (Proportion). A *proportion* is a fraction out of a total, and is a number between 0 and 1.

Definition 12.2 (Percentages). A *percentage* is a proportion, multiplied by 100. In this context, percentages are numbers between 0% and 100%.

Population proportions are almost always unknown. Instead, the *population* proportion (the parameter), denoted p, is estimated by a *sample* proportion (a statistic), denoted by $\hat{p}$.

The symbol $\hat{p}$ is pronounced 'pee-hat', and refers to a *sample* proportion. The caret above the p is called a 'hat'.

As always, only one possible sample is studied. *Statistics* are estimates of *parameters*, and the value of the *statistic* is not the same for every possible *sample*.

Example 12.6 (Proportions and percentages). Consider the AV data in Table 12.1, summarising results from a sample of $n = 400$ respondents. The *sample proportion* of respondents aged 25 to 34 is $76 \div 400$, or 0.19. The *sample percentage* of respondents aged 25 to 34 is 0.19×100, or 19%, as in the table.

12.5 Numerical summary: odds

For the AV data in Table 12.1, the number of females is slightly larger than the number of males. Specifically, the *ratio* of females to males is $204 \div 196 = 1.04$; that is, there are 1.04 *times* as many females as males. This value of 1.04 is the *odds* that a respondent is female in the sample. An alternative interpretation is that there are $1.04 \times 100 = 104$ females for every 100 males in the sample.

While proportions and percentages are computed as the number of results of interest divided by the *total number*, the *odds* are computed as the number of results of interest divided by *the remaining number* (Fig. 12.2).

Definition 12.3 (Odds). The *odds* are the number (or proportion, or percentage) of results of interest, divided by the remaining number (or proportion, or percentage) of results:

$$\text{Odds} = \frac{\text{Number of results of interest}}{\text{Remaining number of results}}$$

or (equivalently)

$$\text{Odds} = \frac{\text{Proportion of results of interest}}{\text{Remaining proportion of results}} = \frac{\text{Percentage of results of interest}}{\text{Remaining percentage of results}}.$$

The *odds* are how many *times* the result of interest *occurs* compared to the number of times the results of interest does *not occur*.

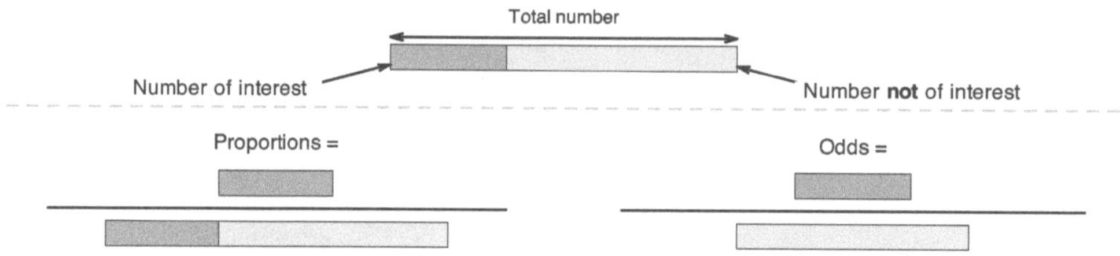

FIGURE 12.2: Proportions (left) are the number of interest divided by the total number. Odds (right) are the number of interest divided by the rest.

Example 12.7 (Interpreting odds)**.** The AV data (Table 12.1) includes 204 females and 196 males. The *odds* that a respondent is female is 1.04. The odds are greater than one, as there are more females than males. Alternatively, there are 104 females for every 100 males.

The *odds* that a respondent is male is $196/204 = 0.96$; there are 0.96 *times* the number of males as females. The odds are less than one, as there are fewer males than females. Alternatively, there are 96 males for every 100 females.

When interpreting odds:

- odds *greater* than 1 mean the result of interest is *more* likely to happen than not.
- odds *equal to* 1 mean the result of interest is *equally likely* to happen as not.
- odds *less* than 1 mean the result of interest is *less* likely to happen than not.

Example 12.8 (Odds and percentages)**.** Consider the AV data in Table 12.1, summarising results from a sample of $n = 400$ respondents.

The percentage of respondents aged 18 to 24 is $52/400 \times 400 = 13\%$. The *odds* that a respondent is aged 18 to 24 is $52/(400 - 52) = 0.15$. This means the number of respondents aged 18 to 24 is 0.15 times (i.e., less than) the number of respondents aged over 24.

The *odds* that a respondent is aged 18 to 54 is $(52 + 76 + 76 + 72)/(56 + 68) = 2.23$. This means the number of respondents aged 18 to 54 is 2.23 times (i.e., greater than) the number of respondents aged 55 or over.

The *population* odds (the parameter) are almost always unknown, and are estimated by the *sample* odds (the statistic). No symbol is commonly used to denote odds.

Take care: proportions and odds are similar, but are different ways of numerically summarising quantitative data (Fig. 12.2).

12.6 Describing the distribution: modes and medians

Graphs are constructed to help readers understand the data, so any important features in the graph should be described. One simple way is to identify the level (or levels) with the *most* observations. This is called the *mode*.

Definition 12.4 (Mode). A *mode* is the level (or levels) of a qualitative variable with the most observations.

Example 12.9 (Modes). Consider the data in Tables 12.1 and 12.2:

- the *mode* for gender is 'Female' (with 204 respondents, or 51%).
- the *mode* age groups are 25 to 34 and 35 to 44 (each with 19 respondents, or 4.8%).
- the *modal* response to the question about *driving* near AVs is 'Somewhat safe'.
- the *modal* response to the question about *cycling* near AVs is 'Somewhat unsafe'.
- the *modal* response to the question about *walking* near AVs is 'Neutral'.

Medians can be found for *ordinal* data (but *not* nominal data), since ordinal data have levels with a natural order. The *median* is the level in which the middle response is located, when the levels from all individuals are placed in order. The sample median estimates the unknown *population* median.

⚠ Medians can be used to summarise *quantitative data* and *ordinal* data, but *never* nominal data.

Example 12.10 (Medians). Consider the data in Tables 12.1 and 12.2. 'Gender' is *nominal* qualitative, so medians are not appropriate. However, the other variables are *ordinal*, so medians could be used to describe each variable. Since $n = 400$, the median response will be halfway between the location of the 200th and 201st response when ordered:

- the *median* age group is 35 to 44.
- the *median* response to the driving-near-AVs question is 'Neutral'.
- the *median* response to the cycling-near-AVs question is 'Neutral'.
- the *median* response to the walking-near-AVs question is 'Neutral'.

For each variable, ordered observations 200 and 201 both fall into the indicated level.

Importantly, all these numerical quantities are computed from a sample (i.e., are statistics; Def. 11.3), even though the whole population is of interest (i.e., the parameter; Def. 11.2).

Means (Sect. 11.6.1) are generally not suitable for numerically summarising qualitative data. However, *ordinal* data *may be* numerically summarised like quantitative data in *rare and very special circumstances.* Means may be appropriate if both of these are true:

- the levels are considered equally spaced.
- assigning a number to each level is appropriate (for example, using a mid-point for numerical age groups).

We will not consider means for ordinal data further.

12.7 Numerical summary tables

Qualitative variables should be summarised in a table. The table should include, as a minimum, numbers and/or percentages for each level. While useful in other contexts (see Chap. 15), odds are usually not given in the summary table. Examples are shown in Tables 12.1 and 12.2, and in the next section.

12.8 Example: water access

López-Serrano et al. [2022] recorded data about access to water for three rural communities in Cameroon (see Sect. 11.10). Numerous qualitative variables are recorded; some are displayed in Fig. 12.3, and summarised in Table 12.3. Notice that the levels of the two ordinal variables are displayed in their natural order.

The distance to the nearest water source is usually less than 1 km, and the wait is often over 15 mins. The most common water source (i.e., the mode) is a bore (68.6%). The median (and mode) distance to the water source was 100 m to 1000 m; the median wait time was 5 to 15 mins (the mode wait time was under 5 mins).

TABLE 12.3: Summarising some qualitative data in the water-access study. Left: the ordinal variables. Right: the nominal variable.

	Number	**%**	**Odds**
Distance to water source ($n = 121$)			
Under 100 m	55	45.5	0.83
100 m to 1000 m	57	47.1	0.89
Over 1000 m	9	7.4	0.08
Wait time at water source ($n = 120$)			
Under 5 mins	50	41.7	0.71
5 to 15 mins	28	23.3	0.30
Over 15 mins	42	35.0	0.54

	Number	**%**	**Odds**
Water source ($n = 121$)			
Tap	7	5.8	0.06
Bore	83	68.6	2.18
Well	16	13.2	0.15
River	15	12.4	0.14

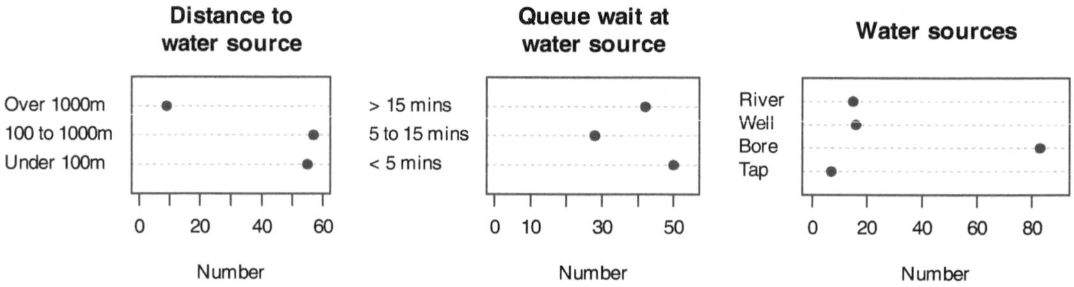

FIGURE 12.3: The distance to the water source (left), the wait time at the water source (centre), and the water sources (right) for the water-access study.

12.9 Chapter summary

Qualitative data can be graphed with a dot chart, bar chart or (in special circumstances) pie chart. Qualitative data can be described using the mode or (for *ordinal* data only) a median. Qualitative data can be numerically summarised using *proportions*, *percentages* or *odds*.

12.10 Quick review questions

Are the following statements *true* or *false*?

1. Nominal data can be summarised using a median.
2. Ordinal data can be summarised using a mode.
3. Odds are the ratio of how often a result of interest occurs, to how often it does *not* occur.
4. Proportions and percentages are the same.

12.11 Exercises

Answers to odd-numbered exercises are given at the end of the book.

Exercise 12.1. A study of spider monkeys [Chapman, 1990] examined the types of social groups present (Table 12.4).

1. Construct a suitable plot, and explain what the data reveal.
2. Determine, if appropriate, the median and mode social group.

TABLE 12.4: Social groups for spider monkeys.

Social group	Percentage	Social group	Percentage
Solitary	8	Mixed + no young	1
All males	3	One female + offspring	23
Female + no young	2	Many females + offspring	48
Mixed young	15		

Exercise 12.2. Czarniecka-Skubina et al. [2021] studied how Poles prepared and consumed coffee using a sample of 1 500 Poles. Some data are shown in Table 12.5.

1. Classify the variables as quantitative, nominal or ordinal.
2. Sketch appropriate graphs for the three variables.
3. Summarise the three variables.
4. Where appropriate, compute the median and mode for each variable.

Exercise 12.3. Henderson and Velleman [1981] recorded the number of cylinders in many models of cars: eleven cars had four cylinders, seven cars had six cylinders, and fourteen cars had eight cylinders. The *number* of cylinders is quantitative discrete, but with so few different values, the data could be plotted with a graph used for qualitative data. For these data:

1. Produce a dot chart.
2. Produce a histogram.
3. Produce a bar chart.
4. Produce a pie chart.

TABLE 12.5: Location of coffee consumption, brewing temperature and brewing time, from 1 500 Poles.

Where consumed	n
Home	1432
Canteen	687
Cafe	922
Others' homes	994
Work	1196

Brew Temp.	n
100°C	748
98°C	269
93°C	453
Unknown	30

Brew time	n
Under 3 mins	226
About 3 mins	267
About 4 mins	114
About 5 mins	82
About 6 mins	30
Unknown	781

What graph do you think is best? Why?

Exercise 12.4. A survey of voice assistants (such as Amazon Echo; Google Home; etc.) conducted by Nielsen asked respondents to indicate how they used their voice assistant. The options were:

- listening to music;
- listen to news;
- use alarms, timer;
- search for real-time information (e.g., traffic; weather);
- search for factual information (e.g., trivia; history);
- chat with voice assistant for fun.

Respondents could select all options that applied. What would be the best graph for displaying respondents answers? Would a pie chart be suitable? Explain your answer.

Exercise 12.5. Gębski et al. [2019] studied the taste of bread with varying salt and fibre content. Information was recorded from 300 subjects, including the subjects' responses to the statement 'Rolls with lower salt content taste worse than regular ones', on a five-point ordinal scale from 'Strongly Agree' to 'Strongly Disagree'; see Table 12.6.

1. Identify the variables, then classify them as nominal or ordinal.
2. For which variables is a mode an appropriate summary (if any)?
3. For which variables is a median an appropriate summary (if any)?
4. Compute the above statistics where appropriate.
5. Compute and interpret the odds of a respondent coming from a city background.
6. Compute and interpret the odds of a respondent agreeing *or* strongly agreeing with the statement.
7. Compute and interpret the odds of a respondent being male.

TABLE 12.6: The bread-tasting data ($n = 300$).

	Number	Percentage
Gender		
Female	150	50
Male	150	50
Place of residence		
Rural	49	16
City up to 20 000 residents	38	13
City 20 000 to 100 000 residents	83	28
City > 100 000 residents	130	43
Response to statement		
Strongly agree	30	10
Agree	84	28
Neutral	78	26
Disagree	66	22
Strongly disagree	42	14

Exercise 12.6. López-Serrano et al. [2022] asked 231 farmers what they considered to be the advantages and disadvantages of using reclaimed water on the farm. The responses are shown in Table 12.7 (not all farmers responded).

1. Produce two bar charts to display the data.
2. Produce two dot charts to display the data.
3. Produce two pie charts to display the data.
4. Determine the mode for both the advantages and disadvantages.
5. Compute the percentages for both the advantages and disadvantages.
6. Compute the odds of a farmer stating 'high price' as a disadvantage, among *all* farmers.
7. Compute the odds of a farmer stating 'high price' as a disadvantage, among farmers who listed a disadvantage.
8. What is the difference in the meaning of the last two statements?

TABLE 12.7: The advantages and disadvantages of using reclaimed water, reported by 231 farmers. (Not all farmers responded.)

Advantage	No. farmers
Water reutilization	15
Availability	27
Sustainability	16

Disadvantage	No. farmers
High price	40
Growing conductivity	12
Lack of proper filtering	21

Exercise 12.7. Henning et al. [2020] studied 284 university students in Joinville, Brazil, tabulating how students got to campus (Table 12.8; each student could select one option only).

1. What is the mode type of active transport? What about motorised transport?
2. What is the mode type of transport overall?
3. Are medians appropriate? If so, compute the median for active transport types, and motorised transport types.
4. Compute the proportions for each option, out of the total sample.
5. Compute the odds that a randomly-chosen student uses motorised transport to get to campus. Explain what this means.
6. Compute the odds that a student walks to campus. Explain what this means.
7. Construct appropriate plots to display the data.

TABLE 12.8: Modes of transport for students getting to campus.

	Number: active methods
Bicycle	29
Walking	35

	Number: motorised methods
Car	70
Bus	117
Other	33

Exercise 12.8. [*Dataset*: `BabyBoom`] Table 11.3 shows the gender of 44 babies born in a hospital on one day [Dunn, 1999, Steele, 1997]. The data are given in the order in which the births occurred.

1. What is the mode sex?
2. If appropriate, compute the median sex.
3. Compute the percentages for each sex.
4. Compute the odds that a randomly-chosen baby from the sample is female. Explain what this means.
5. Construct appropriate plots to display sex of the baby.

Exercise 12.9. [*Dataset*: `LungCap`] Tager et al. [1979] studied the lung volume of 654 children in East Boston in the 1970s (Table 12.9).

1. Construct suitable plots for all variables.
2. For each qualitative variable, determine the mode.
3. For each qualitative variable, compute the percentage and odds of one of the levels occurring in the data.
4. Compute appropriate statistics for each quantitative variable.

TABLE 12.9: The lung volume (FEV) for youth in East Boston in the 1970s; the first six observations in the dataset ($n = 654$).

Age	FEV	Height	Gender	Smoking
3	1.072	46	F	No
4	0.839	48	F	No
4	1.102	48	F	No
4	1.389	48	F	No
4	1.577	49	F	No
4	1.418	49	F	No
⋮	⋮	⋮	⋮	⋮

Exercise 12.10. Swinnen et al. [2018] studied the influence of using ankle-foot orthoses in children with cerebral palsy. The data in Table 10.3 give the data for the 15 subjects. (GMFCS is the Gross Motor Function Classification System) used to describe the impact of cerebral palsy on their motor function; where *lower* levels mean *better* functionality.)

1. Construct suitable plots for all variables.
2. For each qualitative variable, determine the mode.
3. For each qualitative variable, compute the percentage and odds of one of the levels occurring in the data.
4. Compute appropriate statistics for each quantitative variable.

Answers to *Quick review* questions: 1. False. **2.** True. **3.** True. **4.** Percentages are proportions multiplied by 100, so similar (but different).

13

Comparing quantitative data within individuals

So far, you have learnt to ask an RQ, design a study, collect the data, describe the data and summarise the data. **In this chapter**, you will learn to:

- summarise within-individual changes in quantitative data using appropriate graphs.
- summarise within-individual changes using summary tables.

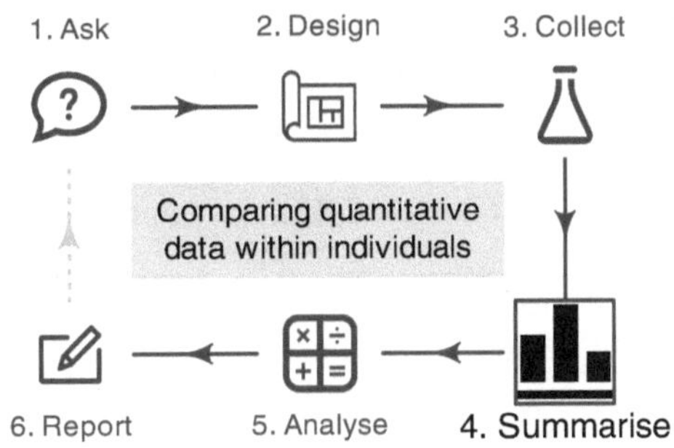

13.1 Introduction

Sometimes the same quantitative variable is measured on each individual more than once (i.e., *within*-individual changes for each unit of analysis) but only a small number of times. Examples of this type of data include:

- measurements of weekly household water consumption for many households, *before* and *after* installing water-saving devices.
- blood pressure recorded for people at 8am, 1pm and 8pm each day.

In both cases, the same variable is measured multiple times for each individual. This chapter studies summarising within-individuals changes in *quantitative* variables.

13.2 Numerical summary: mean differences

When each individual has *two* observations, the difference between the observations can be computed for each individual. Then, the appropriate numerical summary is to summarise these *differences* over all individuals; for example, using the *mean* of these *differences*.

When *more than two* observations are made for each individual, the changes from the first observation (the *reference level*) may be computed (e.g., if the first observation is pre-intervention, or a benchmark, observation); for example, using the *mean change* (see Sect. 13.5).

Example 13.1 (Within-individual comparisons). Lothian et al. [2006] studied children with atopic asthma, and measured the immunogoblin E concentrations (IgE) before and after an intervention for each child (Table 13.1), plus the *reduction* in IgE for each child. The child is the *individual.*

For the IgE data, the numerical summary table is shown in Table 13.2. The direction of the difference is implied by the word '*reduction*'.

⚠ In the numerical summary table, the information for the differences is *not* found by subtracting the information in one row from the other. In Table 13.2, for example, the number of differences is not $11 - 11 = 0$; the standard deviation of the differences is *not* $1615.53 - 1354.4 = 261.13\,\mu\text{g.L}^{-1}$. These statistics are computed from the differences (i.e., the **Reductions** in Table 13.1).

TABLE 13.1: The IgE before and after an intervention, and the change in IgE (in μg. L^{-1}).

Before (in μg. L^{-1})	**After (in μg. L^{-1})**	**Reduction (in μg. L^{-1})**	**Before (in μg. L^{-1})**	**After (in μg. L^{-1})**	**Reduction (in μg. L^{-1})**
83	83	0	1668	1000	668
292	292	0	1960	1626	334
293	292	1	2877	2502	375
623	542	81	2961	2711	250
792	709	83	5504	4504	1000
1543	1000	543			

TABLE 13.2: A numerical summary of the IgE data in μg. L^{-1}).

	Mean	**Std dev.**	**Sample size**
Before	1690.5	1615.53	11
After	1387.4	1354.28	11
Reduction	303.2	325.28	11

13.3 Graphs for the differences

For within-individual changes for a *quantitative* variable, options for plotting include:

- *histograms of differences* (Sect. 13.3.1), which are useful for changes in *pairs* of measurements or observations for each individual.
- *case-profile plots* (Sect. 13.3.2), which are useful when the same individuals are measured or observed a small number of times.

13.3.1 Histogram of differences

Sometimes the same variable is measured on each unit of analysis twice, when the *changes* (or *differences*) for each individual can be produced, and a histogram of the changes or differences can be constructed. The direction of the differences should be clear (e.g., first measurement minus second, or second measurement minus first).

Issues relevant for constructing histograms (Sect. 11.3.1), such as bin widths and boundary values, also apply here.

⚠ The axis displaying the counts (or percentages) should *start from zero*, since the height of the bars visually implies the frequency of those differences (see Example 17.3).

Example 13.2 (Within-individual comparisons). For the IgE data (Table 13.1), the *reduction* in IgE for each child can be shown using a histogram (Fig. 13.1, left panel).

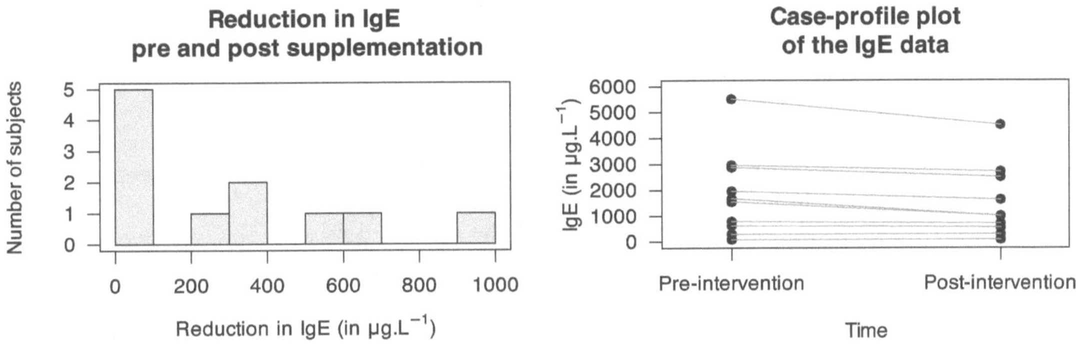

FIGURE 13.1: The IgE data. Left: a histogram of the differences (reductions); boundary observations are counted in the lower box. Right: a case-profile plot, where each line joins each person's pre-intervention to their post-intervention measurement.

13.3.2 Case-profile plots

Sometimes the variable is measured or recorded more than twice, and so a single set of differences cannot be produced. In these cases, the values for each individual can be plotted using a case-profile plot: the measurements are shown on one axis (usually the vertical), and the various points at which measurements are taken are shown on the other axis. A case-profile plot is still useful for paired data, of course.

⚠ The axis displaying the counts (or percentages) *need not start from zero*, since the distance from the axis to the lines *do not* visually imply any quantity of interest. Rather, the *relative changes* represented by the lines display the quantity of interest.

Example 13.3 (Case-profile plot). For the IgE data (Table 13.1), the measurements of IgE for each child at both times can be shown in a case-profile plot (Fig. 13.1, right panel). Each line corresponds to a unit of analysis (i.e., a child).

Example 13.4 (Case-profile plot). Runners use wearable devices to measure many performance indicators, including vertical oscillation (VO). VO contributes to running economy and injury risk, so reliable VO measurements are crucial. Smith et al. [2022] compared four devices, and obtained data from video analysis for $n = 150$ athletes; that is, each participant had the same runs measured using five methods. The case-profile plot (Fig. 13.2) shows the means for each method using a solid point; each line represents one runner. NOVA and Footpod give smaller VO measurements in general.

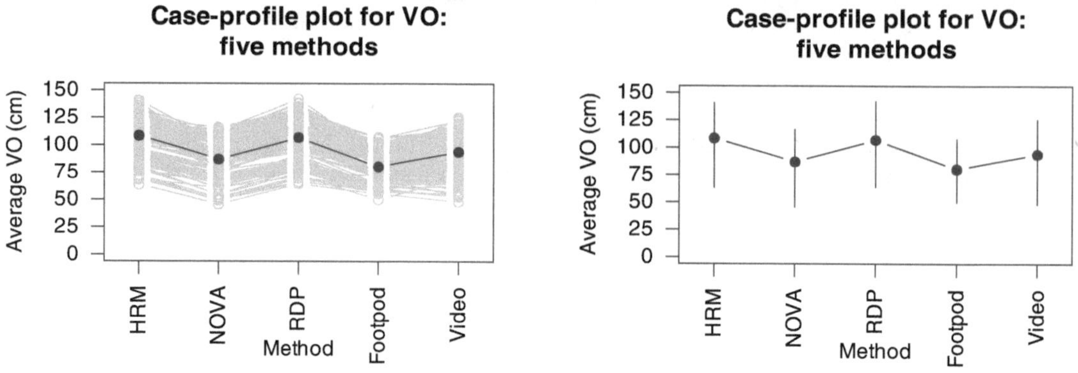

FIGURE 13.2: Vertical oscillation (VO) measured using five methods for 15 runners. The solid black points represent the means for each method. Left: a line is plotted for each individual. Right: only the means are shown, with vertical lines from the minimum value to the maximum value for each method.

As in Example 13.4, the case-profile plot is hard to read with large numbers of individuals, and so sometimes the mean (or median, as appropriate) is shown, with some measure of the variation of the observations (Fig. 13.2 shows the minimum and maximum values for each method, for instance).

13.4 Example: invasive plants

Skypilot (*Polemonium viscosum*) is a native alpine wildflower growing in the Colorado Rocky Mountains (USA). In recent years, a willow shrub (*Salix*) has been encroaching on skypilot territory and, because willow often flowers early, researchers [Kettenbach et al., 2017] are concerned that the willow may 'negatively affect pollination regimes of resident alpine wildflower species' (p. 6 965). One RQ was:

> In the Colorado Rocky Mountains, what is the mean difference between first-flowering day for the native skypilot and the encroaching willow?

Data for both species was collected at 25 different sites (Table 13.3). The site is the *individual*; the data are *paired* (Sect. 29.1), a form of blocking (Sect. 7.2). The 'first-flowering day' is the number of days since the start of the year (e.g., January 12 is 'day 12') when flowers were first observed.

Since the data are available, the data should be summarised graphically (Fig. 13.4) and numerically (Table 13.4), using software output (Fig. 13.3).

TABLE 13.3: The day of the year of first flowering by encroaching willow and native skypilot.

	First-flowering day			First-flowering day	
Site	**Willow**	**Skypilot**	**Site**	**Willow**	**Skypilot**
1	201	201	14	209	209
2	178	179	15	221	221
3	189	189	16	179	188
4	189	189	17	174	179
5	196	203	18	172	166
6	207	203	19	196	196
7	199	199	20	173	173
8	178	182	21	180	173
9	178	178	22	181	179
10	191	191	23	186	186
11	187	192	24	194	209
12	190	197	25	197	197
13	190	190			

Descriptives

	N	Missing	Mean	Median	SD	Minimum	Maximum
Willow	25	0	189.4000	189	12.1997	172	221
Skypilot	25	0	190.7600	190	13.0616	166	221
Difference	25	0	1.3600	0	4.6982	-7	15

FIGURE 13.3: Software output for the flowering-day data.

13.5 Example: pain-relieving tape

Naugle et al. [2021] studied the effect of using Kinesio Tape to alleviate pain in athletes. Pain was measured by applying a slow constant rate of pressure on the left arm, and subjects pressed a button when the sensation moved from pressure to pain. The pressure at which this occurred was recorded. This was repeated 5 mins before applying the tape, 5 mins after applying the tape, and again 15–20 mins after applying the tape.

Figure 13.5 shows the reported pain for 16 subjects. A summary table is shown in Table 13.5. The pain thresholds are increasing slightly as time progresses.

TABLE 13.4: The day of first flowering for encroaching willow and native skypilot.

	Mean	Std dev.	Sample size
Willow (encroaching)	189.4	12.20	25
Skypilot (native)	190.8	13.06	25
Differences	1.4	4.70	25

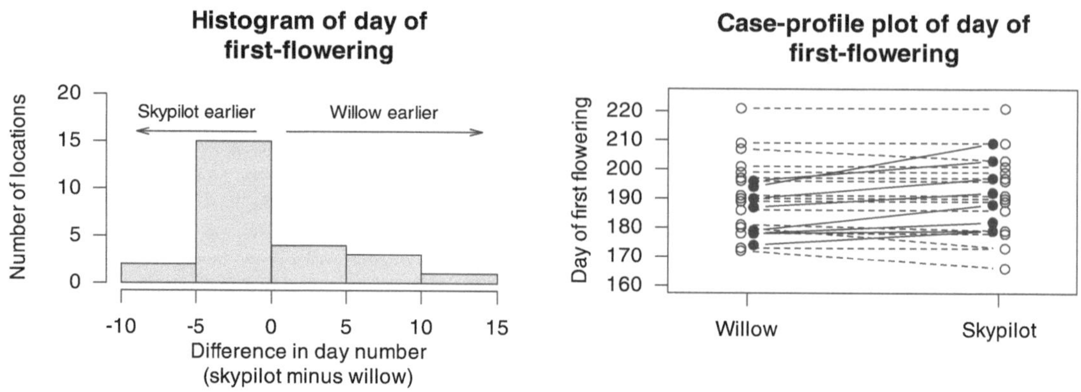

FIGURE 13.4: The flowering-day data. Left: a histogram of the difference between the first-flowering days (skypilot minus willow). Right: a case-profile plot of days of first flowering (unfilled points and dashed lines indicate earlier or same dates (smaller or equal values) for willow).

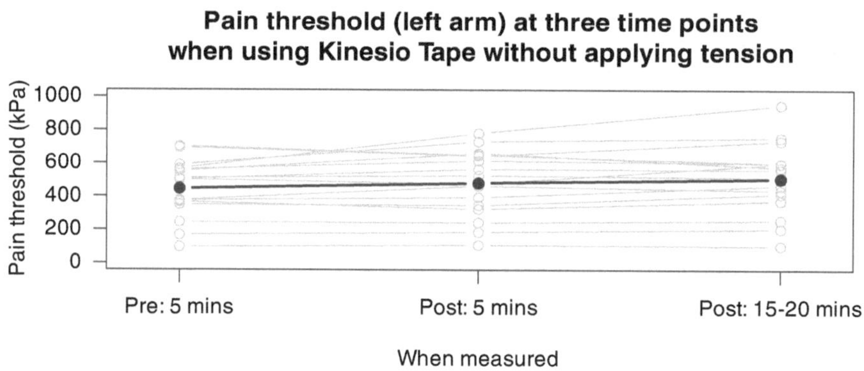

FIGURE 13.5: Pain threshold (left arm) at three time points when using Kinesio Tape, without applying tension, for $n = 16$ subjects. The filled, black points represent the means for each time point.

TABLE 13.5: A numerical summary of the tape data: pain thresholds in kPa.

	Mean	Std dev.	Sample size	Mean of change from Pre	Std dev. of change from Pre
Pre: 5 mins	446.5	175.18	16		
Post: 5 mins	479.6	199.61	16	33.1	73.93
Post: 15 − 20 mins	506.9	214.36	16	60.4	102.72

13.6 Chapter summary

Quantitative data measured within individuals can be summarised using a histogram of differences when the variable is measured or observed twice, or a case-profile plot (with two or more measurements or observations per individual). A summary table should show the numerical summaries for the quantitative variable at each measurement or observation and for appropriate changes.

13.7 Quick review questions

Are the following statements *true* or *false*?

1. A histogram of the differences is only appropriate with two within-individuals measurements or observations.
2. A case-profile plot is only appropriate for showing changes for two within-individuals measurements or observations.
3. The median and IQR are *not* appropriate for summarising differences.
4. Explaining *how* the differences are computed is important.

13.8 Exercises

Answers to odd-numbered exercises are given at the end of the book.

Exercise 13.1. [*Dataset*: `Insulation`] The *Electricity Council* in Bristol wanted to determine if a certain type of wall-cavity insulation reduced average energy consumption in winter [The Open University, 1983, Hand et al., 1996]:

> In Bristol homes, what is the *mean reduction* in energy consumption after adding home insulation?

1. What are the individuals (units of analysis)?
2. Explain why this study uses a within-individuals comparison.
3. Use the collected data (Table 13.6) to sketch a case-profile plot.
4. Use the data to sketch a histogram of the differences.
5. Use software or a calculator to prepare a summary table.

TABLE 13.6: The house insulation data: energy consumption before and after adding insulation, and the energy saving (all in MWh).

Home	Before	After	Saving
A	12.1	12.0	0.1
B	11.0	10.6	0.4
C	14.1	13.4	0.7
D	13.8	11.2	2.6
E	15.5	15.3	0.2

Home	Before	After	Saving
F	12.2	13.6	−1.4
G	12.8	12.6	0.2
H	9.9	8.8	1.1
I	10.8	9.6	1.2
J	12.7	12.4	0.3

Exercise 13.2. [*Dataset*: **Captopril**] In a study of hypertension [Hand et al., 1996, MacGregor et al., 1979], 15 patients were given a drug (Captopril) and their systolic blood pressure measured (in mm Hg) immediately before and two hours after being given the drug (Table 13.7).

1. Explain why this study uses a within-individuals comparison.
2. Construct a histogram of the differences.
3. Construct a case-profile plot for the data.

TABLE 13.7: The Captopril data: before after systolic blood pressures (in mm Hg).

Before	After	Differences	Before	After	Differences
210	201	9	173	147	26
169	165	4	146	136	10
187	166	21	174	151	23
160	157	3	201	168	33
167	147	20	198	179	19
176	145	31	148	129	19
185	168	17	154	131	23
206	180	26			

Exercise 13.3. [*Dataset*: **PainRelief**] Augustino et al. [2023] measured the reported pain of new mothers in Dodoma (Tanzania) at four times: near giving birth, then 20, 40 and 60 mins after giving birth. Mothers were administered either paracetamol or a cold pack as pain relief. Pain was recorded using a 'numeric rating scale represented by the horizontal line marked from zero to ten', where higher scores mean greater pain.

Since the number of individuals is large ($n = 912$), use the summary data in Table 13.8 to sketch a plot of the means and the range, like that in Figure 13.5.

TABLE 13.8: A summary table of reported pain for mothers after giving birth.

		At birth	After 20 mins	After 40 mins	After 60 mins
Paracetamol	Mean	7.44	6.89	4.69	2.84
($n = 456$)	Standard deviation	2.01	1.83	1.49	1.19
	Minimum	2.00	2.00	2.00	0.00
	Maximum	10.00	10.00	9.00	7.00
Cold pack	Mean	8.63	5.67	3.19	0.99
($n = 455$)	Standard deviation	1.40	2.03	1.63	0.99
	Minimum	4.00	0.00	0.00	0.00
	Maximum	10.00	9.00	6.00	4.00

Exercise 13.4. [*Dataset*: **Stress**] The concentration of beta-endorphins in the blood is a sign of stress. One study (Hand et al. [1996], Dataset 232; Hoaglin et al. [2011]) measured the beta-endorphin concentration for 19 patients about to undergo surgery.

Each patient had their beta-endorphin concentrations measured 12–14 h before surgery, and also 10 mins before surgery. A numerical summary (from software output) is in Table 13.9.

TABLE 13.9: The numerical summary for the presurgical stress data.

	Mean	Std deviation	Sample size
12–14 h before surgery	8.35	4.397	19
10 min before surgery	16.05	12.509	19
Increase	7.70	13.519	19

1. Explain why this study uses a within-individuals comparison.
2. Explain why the standard deviation for the *increase* is not the difference between the two individuals time-point standard deviations.
3. Using the data file and software, construct a histogram of the differences.
4. Using the data file and software, construct a case-profile plot for the data.

Exercise 13.5. [*Dataset*: `Running`] Create a summary table for the data in Example 13.4.

Exercise 13.6. [*Dataset*: `WCTennis`] Alberca et al. [2022] recorded the push-time for French wheelchair tennis players, while holding and not holding a racquet (Table 13.10; Alberca [2022]).

1. What do the differences mean (as given in the table)?
2. Create a plot of the data.
3. Create a numerical summary table for the data.

TABLE 13.10: The wheelchair-tennis data. One observation is missing.

	Push-time (in s)		
Person	**With racquet**	**Without racquet**	**Difference (in s)**
1	0.2625	0.1833	0.0792
2	0.2375	0.2250	0.0125
3	0.2583	0.2042	0.0542
4	0.1917	0.1875	0.0042
5	0.1875	0.1708	0.0167
6	0.2542	0.1750	0.0792
7	0.2333	0.1917	0.0417
8	0.1917	0.1708	0.0208
9	0.2208	0.2208	0.0000
10	0.2583	0.2750	−0.0167
11	0.2083	0.1750	0.0333
12	—	0.2042	—
13	0.2208	0.2292	−0.0083

Exercise 13.7. [*Dataset*: `Jumping`] Hébert-Losier et al. [2023] recorded double-legged jumping distance for 80 healthy people, when they wore shoes and were barefoot (Table 13.11).

1. What do the differences mean (as given in the table)?
2. Create a plot of the data.
3. Create a numerical summary table for the data.

TABLE 13.11: Jumping distances for athletes, with and without shoes (the first four and the last four observations).

Distance (in cm)		
With shoes	**Barefoot**	**Difference**
42.73	42.23	0.50
41.00	39.47	1.53
27.37	30.40	−3.03
46.80	36.60	10.20
⋮	⋮	⋮

Distance (in cm)		
With shoes	**Barefoot**	**Difference**
⋮	⋮	⋮
32.73	33.90	−1.17
56.50	55.10	1.40
33.57	32.07	1.50
27.77	33.57	−5.80

Answers to *Quick review* questions: 1. True. **2.** False; a case-profile plot can be used for *two or more* within-individual comparisons. **3.** False; use whatever numerical summaries are appropriate. **4.** True.

14

Comparing quantitative data between individuals

So far, you have learnt to ask an RQ, design a study, collect the data, describe the data, and summarise data. **In this chapter**, you will learn to:

- compare quantitative data between individuals using the appropriate graphs.
- compare quantitative data between individuals in summary tables.

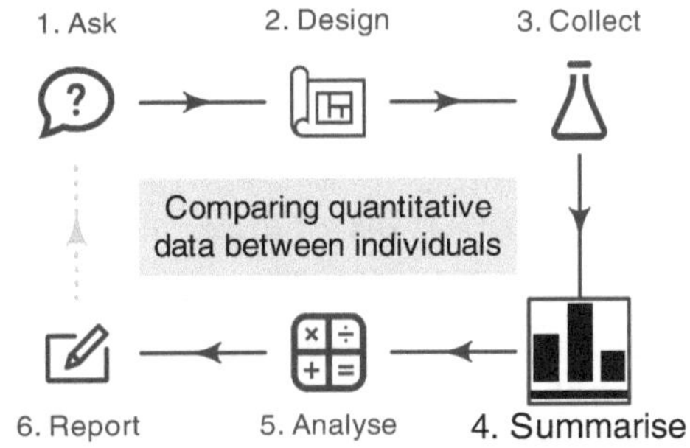

14.1 Introduction

Relational RQs compare groups. This chapter considers how to compare *quantitative* variables in different groups. Graphs are useful this purpose, and a table of the numerical summaries usually is produced also.

14.2 Numerical summary: difference between means

When comparing quantitative variables in different groups, the data should be summarised for each group. If two groups are being compared, the *difference* between the means and/or medians of the two groups must also be computed. If more than two groups are being compared, the *differences* between one of the group means/medians (the first, or the benchmark, or the initial situation as the reference level) and the other group means/medians are also usually computed.

Example 14.1 (Numerical summary table)**.** Wright et al. [2021] recorded the number of chest-beats by gorillas (Table 14.1), for gorillas under 20 years old ('younger') and 20 years and over ('older'). A summary of the data can be tabulated as in Table 14.2. Notice that no standard deviation or sample size is provided for the *difference*; these make no sense.

TABLE 14.1: The chest-beating rate of gorillas (in beats per 10 h).

Younger							**Older**					
0.7	1.3	1.5	1.7	1.8	3.0	4.4	0.0	0.2	0.4	0.8	1.1	4.0
0.9	1.5	1.5	1.7	2.6	4.1	4.4	0.1	0.3	0.6	0.9	1.6	

TABLE 14.2: A numerical summary of the gorillas data.

	Mean (in beats per 10 h)	Standard deviation (in beats per 10 h)	Sample size
Younger	2.22	1.270	14
Older	0.91	1.131	11
Difference	1.31		

14.3 Graphs for the comparison

When a *quantitative* variable is measured or observed in different groups (i.e., between individuals), the distribution of each variable can be graphed separately. However, to *compare* the quantitative variable in the groups, appropriate graphs include:

- *back-to-back stemplots* (Sect. 14.3.1), which are best for small amounts of data (and only possible for comparing *two groups*).
- *2-D dot charts* (Sect. 14.3.2), which are the best choice for small to moderate amounts of data.
- *boxplots* (Sect. 14.3.3), which are the best choice except for small amounts of data.

These situations have one quantitative variable being compared in different groups (defined by *one qualitative variable*).

14.3.1 Back-to-back stemplot

Back-to-back stemplots are two stemplots (Sect. 11.3.2) sharing the same stems; one group has the leaves emerging left-to-right from the stem, and the second group has the leaves emerging right-to-left from the stem. Back-to-back stemplots can only be used when *two* groups are being compared. Again, one advantage of using stemplots over other plots is that the original data are retained. Disadvantages are that only two groups can be compared, and not all data work well with stemplots.

Example 14.2 (Back-to-back stemplots)**.** A back-to-back stemplot for comparing the chest-beating rate of gorillas (Fig. 14.1) has the leaves for younger gorillas right-to-left, and the leaves for older gorillas left-to-right, sharing the same stems. The younger gorillas have a faster chest-beating rate in general. One older gorilla has a much faster rate than the other older gorillas (a potential outlier).

14.3.2 2-D dot charts

A two-dimensional (2-D) dot chart places a dot for each observation, separated for each level of the qualitative variable (also see Sect. 12.3.1). Any number of groups can be compared.

⚠ The axis displaying the counts (or percentages) *need not start from zero*, since the distance from the axis to the these numbers *do not* visually imply any quantity of interest. Rather, how the dots *compare* in the groups is the main feature of interest.

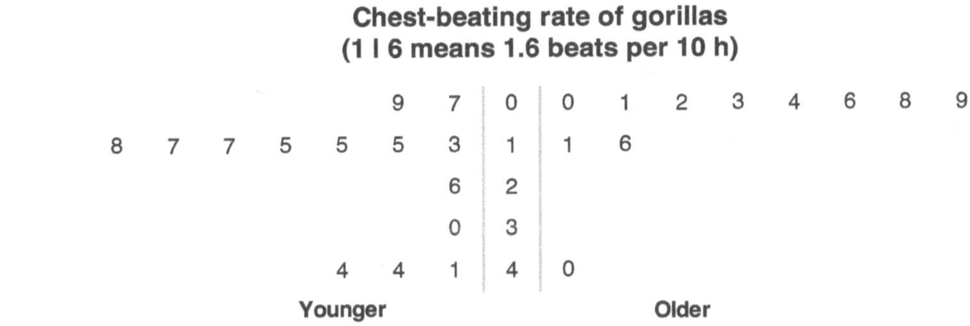

FIGURE 14.1: Stemplot for the chest-beating rate for gorillas.

Example 14.3 (Boxplots). For the chest-beating data seen in Example 14.2, a dot chart is shown in Fig. 14.2. Many observations are the same, so some points would be *overplotted* if points were not *stacked* (left panel), or *jittered* (right panel).

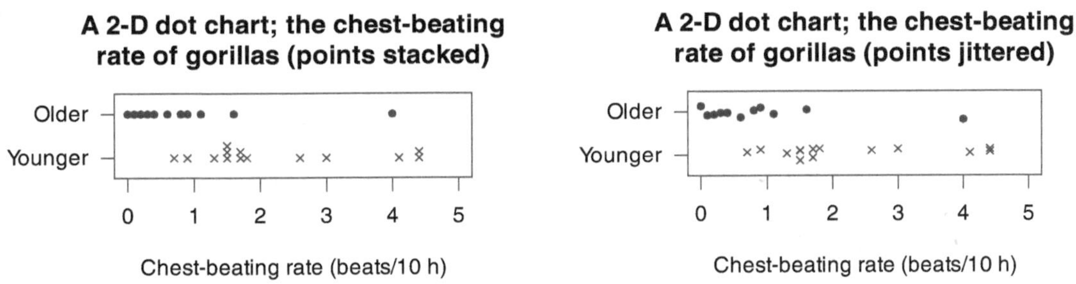

FIGURE 14.2: Two variations of a 2-D dot chart for the chest-beating data to avoid overplotting: stacking (left) and jittering (right).

14.3.3 Boxplots

A boxplot is a picture of the quantiles (Sect. 11.7.3) for each group, drawn side-by-side on the same plot (and so are sometimes called *parallel* boxplots or *side-by-side* boxplots). Any number of groups can be compared using a boxplot. Sometimes, the mean of each groups is added to the boxplot using, for example, a solid dot.

The distribution for each group is summarised by five numbers: the minimum value; the first quartile (Q_1); the median (Q_2); the third quartile (Q_3); and the maximum value. Outliers, identified using the IQR rule (Sect. 11.8.2), are usually shown too. The values of Q_1, the median, and Q_3 for each group can be used to compare the distributions. Different software may use different rules for computing quartiles, and hence may produce slightly different boxplots.

⚠ The axis displaying these five numbers *need not start from zero*, since the distance from the axis to the these numbers *do not* visually imply any quantity of interest. Rather, the boxes display the values of these five numbers for each group *relative* to each other, which is of interest.

⚠ Boxplots summarise data with only five numbers (sometimes called the five-number summary), so details of the distributions are lost. For this reason, boxplots are excellent for *comparing* distributions, but histograms are better for displaying the distribution of a single quantitative variable.

Example 14.4 (Boxplots)**.** The boxplot for the chest-beating data (Example 14.2) is shown in Fig. 14.3. No outliers are identified for younger gorillas; one large outlier is identified for the older gorillas. The boxplot shows a distinct difference between the chest-beating rates of older and younger gorillas.

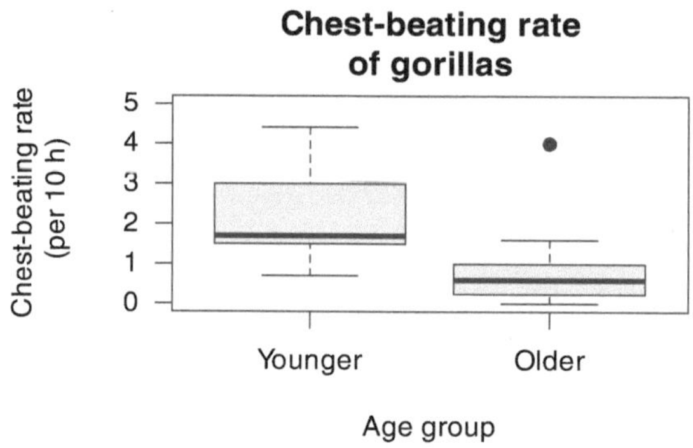

FIGURE 14.3: The boxplot for the chest-beating data.

The detail of the boxplots are explained in Fig. 14.4. Firstly, focus on just the boxplot for the *younger* gorillas (i.e., the left box). Boxplots have five horizontal lines.

1. *Top line*: the *fastest* chest-beating rate (largest value) is 4.4 per 10 h.
2. *Second line from top*: 75% of observations are smaller than about 3 per 10 h, represented by the line at the top of the central box. This is the *third quartile* (Q_3).
3. *Middle line*: 50% of observations are smaller than about 1.7 per 10 h, represented by the line inside the central box. This is the median value, the *second quartile* (Q_2).
4. *Second line from bottom*: 25% of observations are smaller than about 1.5 per 10 h, represented by the line at the bottom of the central box. This is the *first quartile* (Q_1).
5. *Bottom line*: the *slowest* chest-beating rate (smallest value) is 0.7 per 10 h.

The box for the *older* gorillas (Fig. 14.3, right box) is slightly different: one observation is identified with a point, *above* the top line. Computer software identifies this observation as an *extreme outlier* using the IQR rule (Sect. 11.8.2), and has plotted this point separately.

The values of Q_1, the median and Q_3 are all substantially larger for the younger gorillas, suggesting that younger gorillas have, in general, faster chest-beating rates.

Example 14.5 (Boxplots)**.** Boxplots can be plotted horizontally too, which leaves space for long labels of the qualitative variable. In Fig. 14.5 (based on Silva et al. [2016]), the three dental cements are very different regarding their push-out forces.

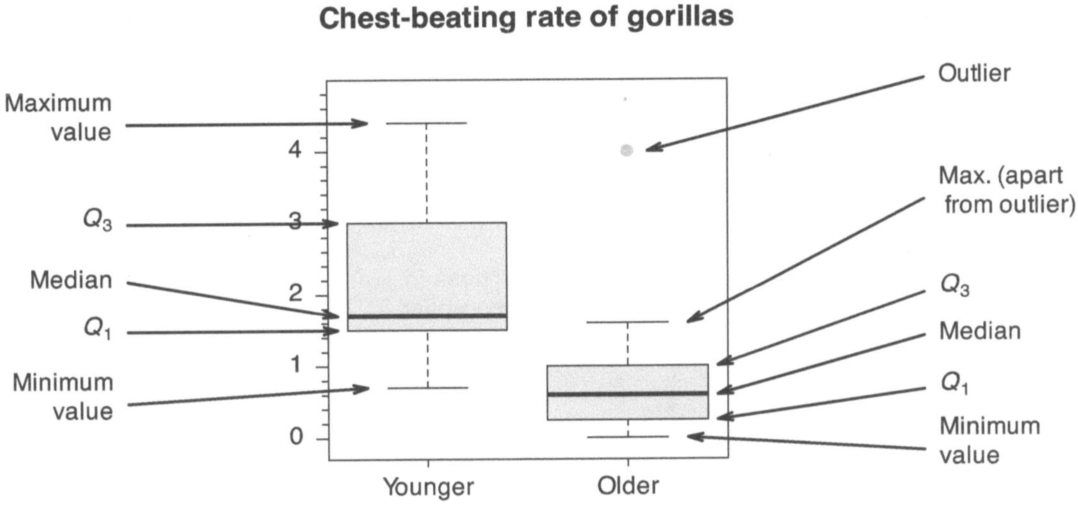

FIGURE 14.4: Explaining the boxplots for the chest-beating data.

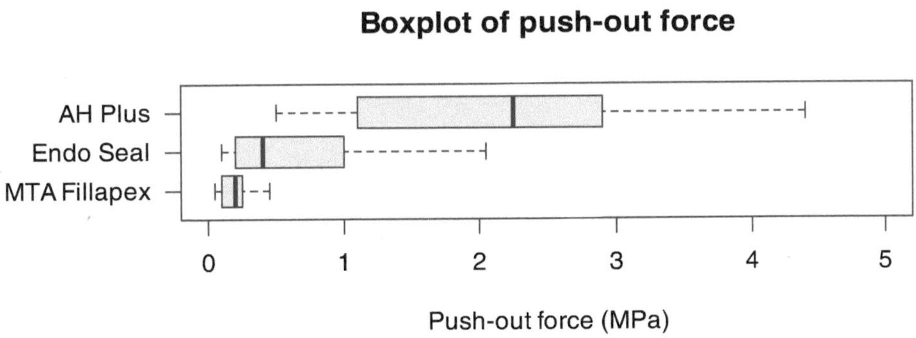

FIGURE 14.5: Comparing three push-out values for three dental cements.

14.4 Example: water access

López-Serrano et al. [2022] recorded data about access to water in three rural communities in Cameroon (Sect. 11.10). The study could be used to determine associations to the incidence of diarrhoea in young children (85 households had children under 5 years of age).

The graphs (Fig. 14.6) and summary (Table 14.3) show that households in which diarrhoea was found in the last two weeks in children had older household coordinators, more people in the household, and more children under 5 years of age in the household. These may be expected: older female coordinators probably have more children, hence more children in the household under 5, and so more children (and so people) are in the household in general.

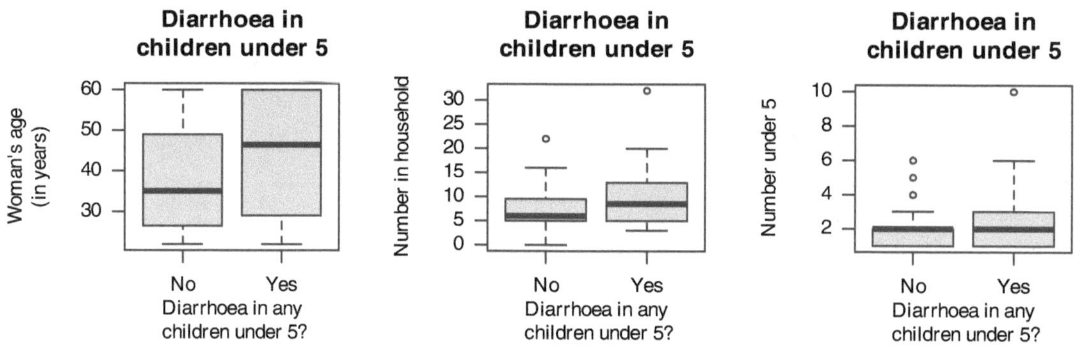

FIGURE 14.6: Three plots for the water access data in 85 households (59 household reported no diarrhoea in children under 5 years of age; 26 reported diarrhoea in children under 5 years of age).

TABLE 14.3: A summary of the quantitative variables in the water-access study, according to whether diarrhoea had been observed in the last two weeks in children under 5 years of age, for those household with children under 5 years of age.

	n	Mean	Median	Std dev.	IQR
Woman coordinator's age (in years)					
All households with children	85	40.2	37.0	13.90	28.00
Incidents of diarrhoea	26	45.0	46.5	14.04	28.50
No incidents of diarrhoea	59	38.1	35.0	13.44	22.50
Difference		6.8			
Household size					
All households with children	85	8.4	7.0	4.93	6.00
Incidents of diarrhoea	26	10.5	8.5	6.51	7.75
No incidents of diarrhoea	59	7.5	6.0	3.78	4.50
Difference		2.9			
Children under 5 in household					
All households with children	85	2.2	2.0	1.56	2.00
Incidents of diarrhoea	26	2.8	2.0	2.01	1.75
No incidents of diarrhoea	59	1.9	2.0	1.26	1.00
Difference		0.8			

14.5 Chapter summary

Quantitative data can be compared between different groups (between individuals comparisons) using a back-to-back stemplot, boxplot or 2-D dot chart. A summary table should show the numerical summaries for the levels of the quantitative variable, and the between-group differences.

14.6 Quick review questions

Are the following statements *true* or *false*?

1. A boxplot is an appropriate graph for comparing a quantitative variable in two *or more* groups.
2. A back-to-back stemplot is an appropriate graph for comparing a quantitative variable in two *or more* groups.
3. A case-profile plot is an appropriate graph for comparing a quantitative variable in two *or more* groups.
4. When comparing a quantitative variable in two groups, the difference between the two sample sizes should be included.

14.7 Exercises

Answers to odd-numbered exercises are given at the end of the book.

Exercise 14.1. Hale et al. [2009] studied two different engineering project delivery methods (Fig. 14.7, left panel): Design/Build and Design/Bid/Build. The grey, horizontal line is where the projected costs are the same as the actual cost.

1. What does the plot reveal about the two methods?
2. What is the median for each method (approximately)?
3. What is the IQR for each method (approximately)?

Exercise 14.2. [*Dataset*: `AISsub`] Telford and Cunningham [1991] studied athletes at the *Australian Institute of Sport* (AIS). Numerous physical and blood measurements were taken from high performance athletes. Figure 14.7 (right panel) compares the heights of females in two similar sports: basketball and netball. (Netball was derived from basketball.)

1. What does the plot reveal about the heights of the females in each sport?
2. What is the median for each sport (approximately)?
3. What is the IQR for each sport (approximately)?

Exercise 14.3. Consider the histograms and boxplots in Fig. 14.8.

1. Match the histogram with the corresponding boxplot.
2. For which datasets would the mean and standard deviation be the appropriate numerical summary? For which datasets would the median and IQR be the appropriate numerical summary?

Exercise 14.4. Lunn and McNeil [1991; Hand et al., 1996] compared the dimensions of jellyfish at two sites at Hawkesbury River, NSW (Dangar Island; Salamander Bay) to determine the difference

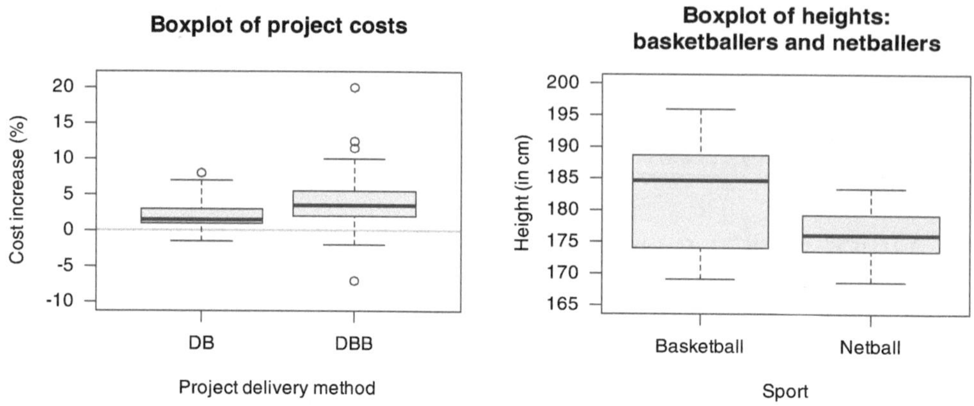

FIGURE 14.7: Left: cost increases for two different building project delivery methods: Design/Build and Design/Bid/Build (the grey, horizontal line is where the projected costs are the same as the actual cost). Right: the heights of female basketball and netball players attending the AIS.

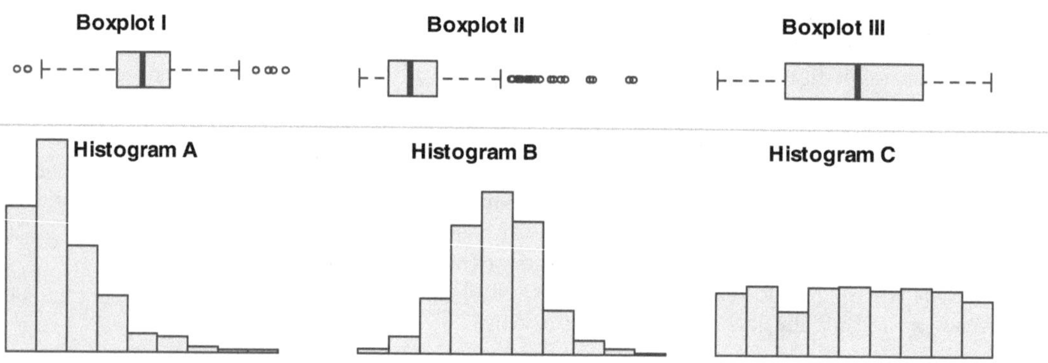

FIGURE 14.8: Match the histogram with the boxplot.

between the jellyfish at each site. A histogram of the breadth of jellyfish at Dangar Island Bay is shown in Fig. 14.9 (left panel).

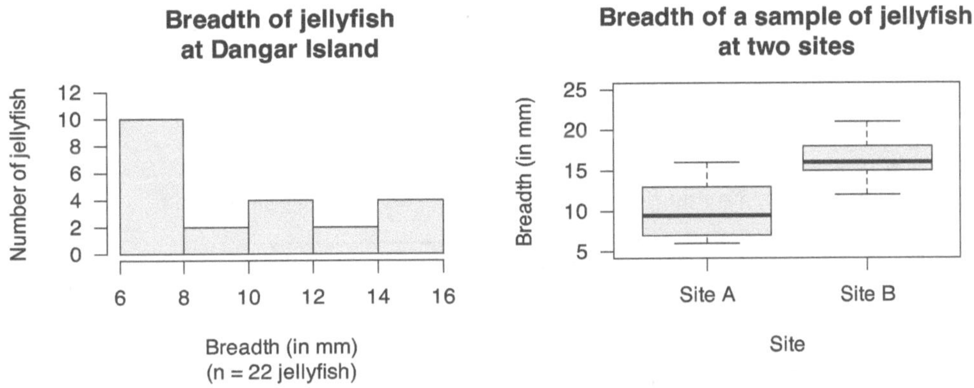

FIGURE 14.9: Left: a histogram of the breadth of jellyfish at Dangar Island. Right: a boxplot of the breadth of jellyfish at two sites.

1. Two students are arguing about the median breadth.

   Student 1 says: *The bars in the histogram have heights of* 10, 2, 4, 2 *and* 4. *When these numbers are put in order, they are:* 2, 2, 4, 4, *and* 10. *The median breadth is the median of these numbers, so the median breadth is the middle one:* 4 *mm is the median.*

   Student 2 responds: *You have the correct answer, but for the wrong reason! There are five bars, and the middle bar is the third bar. Since the third bar has a height of* 4, *the median breadth is* 4 *mm.*

   Which student, if either, is correct?
2. Describe the histogram.
3. A boxplot comparing the breadths of jellyfish at Dangar Island and Salamander Bay is shown in Fig. 14.9 (right panel). Describe and compare the breadths of the jellyfish.
4. What is the median breadth for the jellyfish at each location?
5. Which box in the boxplot represents the Dangar Island jellyfish (in Fig. 14.9, left panel)?

Exercise 14.5. Gatti et al. [2013] studied the productivity of construction workers, recording (among other variables) the rate at which concrete panels could be installed by workers. Data for three different female workers in the study are shown in Table 14.4, gathered over four installation periods of 50 mins each.

1. Compute the IQR for each worker.
2. Construct the boxplot for comparing the three workers.
3. Draw the approximate histograms for each worker.
4. What do you learn about the workers?

TABLE 14.4: The productivity of three workers installing concrete panels (in panels per minute).

	Worker 1	Worker 2	Worker 3
Mean	1.24	1.73	1.36
Minimum	0.59	1.13	0.86
1st quartile	0.88	1.51	1.16
Median	1.35	1.70	1.38
3rd quartile	1.49	1.91	1.58
Maximum	1.88	3.00	2.17

Exercise 14.6. In a study of the temperature in offices, Paul and Taylor [2008] compared the temperature in three offices (during working hours) at Charles Sturt University (Australia; CSU); the data are summarised in Table 14.5.

1. Compute the IQR for the temperatures in each office.
2. Construct the boxplot for comparing the temperatures in the three offices.
3. Draw the approximate histograms for each office.
4. What do you learn about the offices?

TABLE 14.5: A summary of the temperature (in degrees C) in three offices at CSU during working hours. Office A is on the ground floor; Offices B and C are on the top floor.

	Office A	**Office B**	**Office C**
Mean	24.1	25.3	25.7
Minimum	16.4	15.9	20.1
Q_1	22.8	23.8	24.6
Median	24.4	25.5	26.1
Q_3	25.5	26.9	27.2
Maximum	27.4	31.0	30.3

Exercise 14.7. [*Dataset*: `NHANES`] Consider this RQ:

> Among Americans, is the mean direct HDL cholesterol different for current smokers and non-smokers?

Data to answer this RQ are available from the American *National Health and Nutrition Examination Survey* (NHANES) [Pruim, 2015].

1. What would be an appropriate graph to display the comparison?
2. Use the software output (Fig. 14.10) to construct an appropriate table showing the numerical summary relevant to the RQ.

Descriptives

	SmokeNow	N	Missing	Mean	Median	SD	Minimum	Maximum
DirectChol	No	1668	77	1.3924	1.3200	0.4279	0.3900	3.8300
	Yes	1388	78	1.3077	1.2400	0.4235	0.5400	3.7200

FIGURE 14.10: Software output for the NHANES data.

Exercise 14.8. [*Dataset*: `ForwardFall`] Wojcik et al. [1999] compared the lean-forward angle in younger and older women. An elaborate set-up was constructed to measure this angle, using a harness. Consider the RQ:

> Among healthy women, what is difference between the mean lean-forward angle for younger women compared to older women?

The data are shown in Table 14.6.

1. What is an appropriate graph to display the comparison?
2. Construct an appropriate numerical summary from the software output (Fig. 14.11).

Exercise 14.9. [*Dataset*: `Speed`] Ma et al. [2019] studied adding additional signage to reduce vehicle speeds on freeway exit ramps. At one site (Ningxuan Freeway), speeds were recorded for 38 vehicles before the extra signage was added, and then for 41 different vehicles after the extra signage was added (Table 14.7).

The researchers are hoping that the addition of extra signage will *reduce* the mean speed of the vehicles. The RQ is:

TABLE 14.6: Lean-forward angles (in degrees) for older women ($n = 10$) and younger women ($n = 5$).

Younger women					Older women				
29	34	33	27	28	18	15	23	13	12
32	31	34	32	27					

Descriptives

	Group	N	Missing	Mean	Median	SD	Minimum	Maximum
LeanAngle	Younger	10	0	30.7000	31.5000	2.7508	27.0000	34.0000
	Older	5	0	16.2000	15.0000	4.4385	12.0000	23.0000

FIGURE 14.11: Software output for the lean-forward angles data.

> At this freeway exit, how much is the mean vehicle speed *reduced* after extra signage is added?

1. Using the software output in Fig. 14.12, summarise the data numerically, then construct a suitable summary table.
2. Produce a boxplot of the data (use a computer if necessary).

Exercise 14.10. [*Dataset*: `Deceleration`] Ma et al. [2019] studied adding additional signage to reduce vehicle speeds on freeway exit ramps. At one site (Ningxuan Freeway), speeds were recorded at various points on the freeway exit for 38 vehicles before the extra signage was added, and then for 41 vehicles after the extra signage was added.

From this data, the *deceleration* of each vehicle was determined (Table 14.8) as the vehicle left the 120 km.h^{-1} speed zone and approached the 80 km.h^{-1} speed zone. The RQ is:

> At this freeway exit, what is the difference between the mean vehicle deceleration, comparing the times before the extra signage is added and after extra signage is added?

In this context, the researchers are hoping that the extra signage might cause cars to slow down *faster* (i.e., they will decelerate more, on average, after adding the extra signage).

1. Using the software output in Fig. 14.13, summarise the data numerically, then construct a suitable summary table.
2. Produce a boxplot of the data (use a computer if necessary).
3. What does a *negative* deceleration value represent?

TABLE 14.7: Vehicle speeds (in km.h^{-1}) before and after adding extra signage.

Speeds before signage added					Speeds after signage added				
90.0	108.0	127.1	102.9	86.4	98.2	98.2	93.9	102.9	69.7
83.1	102.9	72.0	113.7	83.1	102.9	93.9	93.9	90.0	113.7
93.9	108.0	80.0	108.0		93.9	98.2	120.0	98.2	102.9
113.7	83.1	86.4	93.9		98.2	67.5	93.9	108.0	72.0
120.0	72.0	80.0	98.2		80.0	98.2	77.1	86.4	80.0
108.0	98.2	90.0	108.0		86.4	93.9	86.4	113.7	
98.2	102.9	98.2	108.0		102.9	98.2	98.2	72.0	
90.0	108.0	90.0	102.9		83.1	63.5	108.0	98.2	
90.0	120.0	102.9	102.9		98.2	83.1	74.5	93.9	

Descriptives

	When	N	Missing	Mean	Median	SD	Minimum	Maximum
Speed	Before	38	0	98.016	98.200	13.194	72.000	127.100
	After	41	0	92.341	93.900	13.134	63.500	120.000

FIGURE 14.12: Software output for the speed data.

TABLE 14.8: Vehicle deceleration (in m.s^{-2}) before and after adding extra signage.

Deceleration before signage added					**Deceleration after signage added**				
0.108	−0.062	0.023	0.043	0.044	0.134	0.093	0.057	0.147	0.046
0.064	0.063	0.028	0.096	0.048	−0.113	0.095	0.076	0.089	0.007
0.002	0.107	0.080	0.151		−0.052	0.054	0.167	0.074	0.063
0.029	0.081	0.061	0.114		0.093	0.118	0.076	0.085	0.087
0.167	0.102	0.031	0.154		0.080	0.154	0.107	0.079	0.096
0.042	0.054	0.071	0.107		0.044	0.076	0.044	0.119	
0.113	0.003	0.113	0.173		0.043	0.113	0.175	0.014	
0.053	0.042	0.126	0.084		0.064	0.083	0.085	0.093	
0.035	0.070	0.084	0.126		0.074	0.064	0.037	0.095	

Descriptives

	When	N	Missing	Mean	Median	SD	Minimum	Maximum
Deceleration	After	41	0	0.0765	0.0800	0.0521	-0.1130	0.1750
	Before	38	0	0.0745	0.0705	0.0494	-0.0620	0.1730

FIGURE 14.13: Software output for the deceleration data.

Exercise 14.11. [*Dataset*: **Typing**] The **Typing** dataset contains information about the typing speed and accuracy for students, from an online typing test [Pinet et al., 2022]. The four variables are: typing speed (**mTS**), typing accuracy (**mAcc**), age (**Age**), and sex (**Sex**) for 1 301 students.

1. Produce appropriate numerical summaries for the quantitative variables.
2. Produce appropriate numerical summaries for *comparing* the quantitative variables for different values of the qualitative variable.
3. What do you learn from these numerical summaries?

Exercise 14.12. [*Dataset*: **Dental**] Woodward and Walker [1994] recorded the sugar consumption and the average number of decayed, missing or filled teeth (DMFT) in 29 industrialised countries and 61 non-industrialised countries.

1. Produce appropriate numerical summaries for the two quantitative variables.
2. Produce appropriate numerical summaries for *comparing* the two quantitative variables for industrialised countries and non-industrialised countries.
3. What do you learn from these numerical summaries?

Exercise 14.13. [*Dataset*: **Snakes**] Some Mexican garter snakes (*Thamnophis melanogaster*) live in habitats with no crayfish, while some live in habitats with crayfish and use crayfish as a food source. Manjarrez et al. [2017] were interested in whether the snakes in these regions were different:

> For female Mexican garter snakes, is the mean snout–vent length (SVL) different for those in regions with crayfish and without crayfish?

Two different groups of snakes are studied (so the study uses a between-individuals comparison). (The data are shown in Table 30.1.) Boxplots of the data are shown in Fig. 14.14.

1. Describe the boxplot displaying the SVL for the two regions, for *all* crayfish (left panel). Compare the means for the two regions.
2. Describe the boxplot displaying the SVL for the two regions, for *female* crayfish (centre panel). Compare the means for the two regions.
3. Describe the boxplot displaying the SVL for the two regions, for *male* crayfish (right panel). Compare the means for the two regions.
4. How would you describe the variable 'Sex of the snake': extraneous, confounding, lurking, response or explanatory?

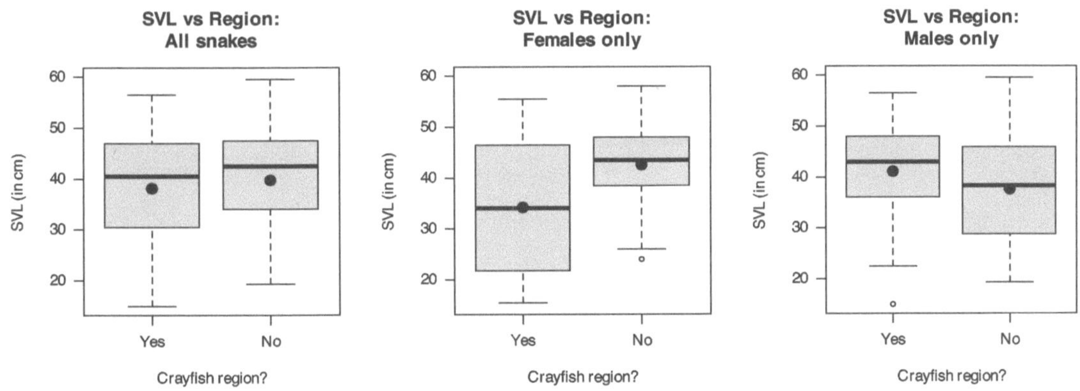

FIGURE 14.14: The snount–vent length (SVL) for Mexican garter snakes, living in crayfish or non-crayfish regions. The solid dots represent the means.

Answers to *Quick review* questions: 1. True. **2.** False; only compares two groups. **3.** False; a case-profile plot is appropriate for *within*-individual changes. **4.** False; difference between sample sizes is meaningless.

15

Comparing qualitative data between individuals

So far, you have learnt to ask an RQ, design a study, collect the data, describe the data and summarise the data. **In this chapter**, you will learn to:

- compare qualitative data between groups of individuals using the appropriate graphs.
- compare qualitative data between groups of individuals using the difference in proportions, odds ratios and summary tables.

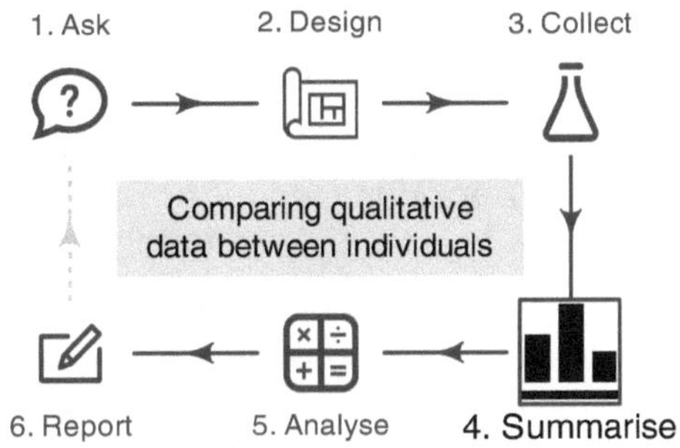

15.1 Introduction

Relational RQs compare groups. This chapter considers how to compare *qualitative* variables in different groups. Graphs are useful for this purpose, and a table including odds, odds ratios and proportions is usually produced also.

15.2 Two-way tables

When more than one qualitative variable is recorded for each individual, the data can be collated into a table. When *two* qualitative variables are cross-tabulated, the resulting table is called a *two-way table*. The categories for each variable should be *exhaustive* (cover all levels) and *mutually exclusive* (observations belong to one and only one level). Usually, the levels of the explanatory variable are in the rows of the table.

Example 15.1 (Two-way tables). To compare two treatments for kidney stones, Charig et al. [1986] collected data from 700 UK patients on two qualitative variables:

- the treatment method ('A' or 'B'), the explanatory variable.
- the result of the procedure ('success' or 'failure'), the response variable.

Both variables are *qualitative* with two *levels*, and each treatment was used on 350 patients. Treatment A was used from 1972–1980, and Treatment B from 1980–1985; that is, treatments were *not randomly allocated*, and so *confounding* may be present. For this reason, the researchers also recorded the *size* of the kidney stone ('small' or 'large') as one possible confounding variable. Firstly, consider just the *small stones* [Julious and Mullee, 1994], displayed in the two-way table in Table 15.1.

TABLE 15.1: *Counts* for two procedures with *small* kidney stones.

	Success	Failure	Total
Method A	81	6	87
Method B	234	36	270
Total	315	42	357

15.3 Summary tables by rows and columns

Each variable in a two-way table can be analysed separately, using percentages or proportions (Sect. 12.4) or odds (Sect. 12.5). For example, the two variables in Table 15.1 (Method; Result) can be analysed separately. For overall results:

- the proportion of procedures that were successful is $315/357 = 0.882$ (or 88.2%).
- the odds that a procedure was successful is $315/42 = 7.5$; that is, there were 7.5 times as many successful procedures as unsuccessful procedures.

However, to *compare* Methods A and B, the proportions (or percentages) and odds of successful results need to be computed for each row separately.

Example 15.2 (Small kidney stones)**.** The data in Table 15.1 can be summarised by computing proportions or percentages by *row*. Each row refers to a different method, so row percentages will compute success percentages for the two methods.

For the small kidney stones (Table 15.1), the *row percentages* (Table 15.2, left table) give the percentage of successes for each *Method*, since the rows represent the counts for Methods A and B. *Row* proportions (or percentages) allow the proportions (or percentages) *within the rows* (i.e., for each Method) to be compared:

- with Method A, $81 \div 87 = 0.931$ (or 93.1%) of operations in the sample were successful.
- with Method B, $234 \div 270 = 0.867$ (or 86.7%) of operations in the sample were successful.

For small kidney stones, Method A is slightly more successful (93.1%) than Method B (86.7%) in the *sample*. These percentages are collated in Table 15.2 (left table).

Odds can also be computed:

- with Method A, the odds of success is $81 \div 6 = 13.5$; there are 13.5 times as many successful procedures than failures for Method A.
- with Method B, the odds of success is $234 \div 36 = 6.5$; there are 6.5 times as many successful procedures than failures for Method B.

The odds of a success is far greater for Method A than Method B in the sample.

TABLE 15.2: Two procedures with *small* kidney stones. Left: *row* percentages. Right: *column* percentages (from Table 15.1). Proportions could be used rather than percentages.

	Success	Failure	Total
Method A	93.1	6.9	100.0
Method B	86.7	13.3	100.0

	Success	Failure
Method A	25.7	14.3
Method B	74.3	85.7
Total	100.0	100.0

Rather than comparing *methods* (in the rows), the procedure *results* can be compared (i.e., the columns).

Example 15.3 (Comparing by column). For the small kidney stones (Table 15.1), the *column percentages* (Table 15.2, right table) give the percentage of successes within each column (i.e., for successes and for failures), since the columns contain the procedure results. *Column* percentages (or proportions) allow the percentages (or proportions) within *columns* to be compared:

- the proportion of the *successful* procedures from Method A is $81 \div 315 = 0.257$ (or 25.7%).
- the proportion of the *failed* procedures from Method A is $234 \div 315 = 0.143$ (or 14.3%).

Odds can also be computed:

- the odds of a *success* coming from Method A is $81/234 = 0.346$; there are 0.346 times as many Method A procedures than Method B procedures among the successes.
- the odds of *failure* coming from Method A is $6/36 = 0.167$; there are 0.167 times as many Method A procedures than Method B procedures among the failures.

The odds of a success being a Method A procedure is quite different from the odds of a success being a Method B procedure.

Comparing rows (i.e., using row percentages and row odds) seems more intuitive than column proportions here: they compare the success percentages and odds for each method.

15.4 Graphs for the comparison

When a *qualitative* variable is compared across different groups (i.e., comparing between individuals), options for plotting include:

- *stacked bar charts* (Sect. 15.4.1).
- *side-by-side bar charts* (Sect. 15.4.2).
- *dot charts* (Sect. 15.4.3).

15.4.1 Stacked bar charts

The data can be graphed by using a bar for each level of one variable, and *stacking* the bars for the levels of the second variable. Bars indicate the counts (or percentages) in each category. The levels can be on the horizontal or vertical axis, but placing the level names on the vertical axis often makes for easier reading, and room for long labels.

⚠ The axis displaying the counts (or percentages) should *start from zero*, since the height of the bars visually implies the frequency of those observations (see Example 17.3).

Example 15.4 (Stacked bar charts). For the small kidney-stone data in Example 15.1, a stacked bar chart can be created by producing a bar for each method, and *stacking* the successes and failures for each method (Fig. 15.1, top left panel).

Rather than using *numbers*, the *percentages* separately within each group can be used too (Fig. 15.1, bottom left panel). This makes comparing the *relative* proportions easier.

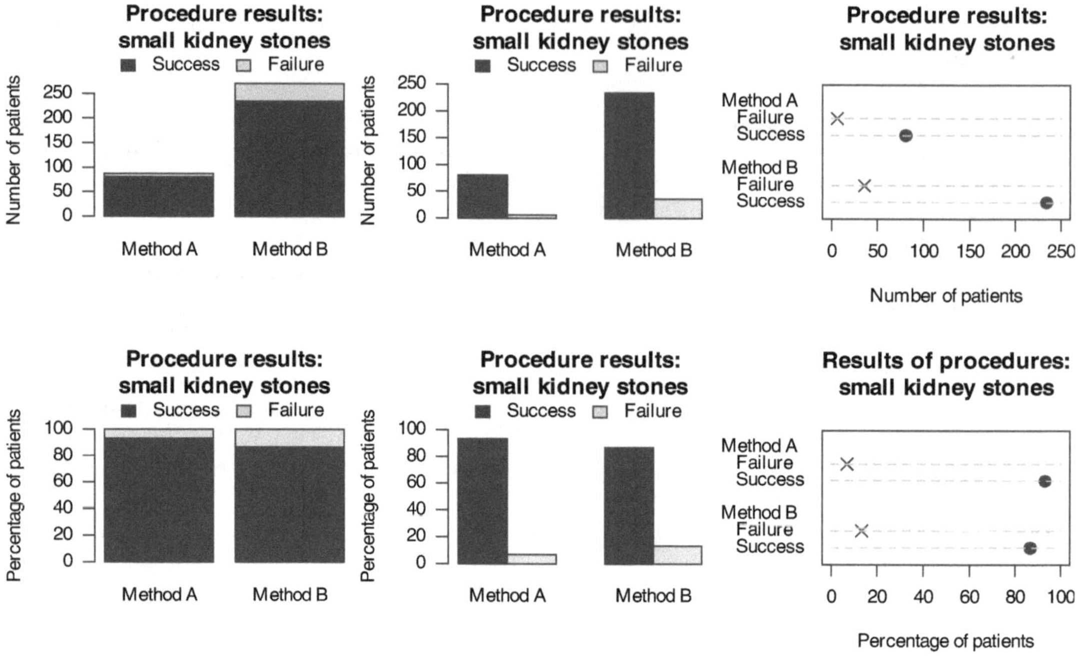

FIGURE 15.1: Six plots for the small kidney-stone data. Top plots: displaying the numbers for each method. Bottom plots: displaying the percentages for each method. Left: stacked bar chart. Centre: side-by-side bar charts. Right: dot charts.

15.4.2 Side-by-side bar charts

Instead of stacking the success and failures bars on top of each other, these bars can be placed *side-by-side* for each method. Bars indicate the counts (or percentages) in each category. The levels can be on the horizontal or vertical axis, but placing the level names on the vertical axis often makes for easier reading, and room for long labels.

⚠ The axis displaying the counts (or percentages) should *start from zero*, since the height of the bars visually implies the frequency of those observations (see Example 17.3).

Example 15.5 (Side-by-side bar charts). For the small kidney-stone data in Example 15.1, a side-by-side bar chart can be created by producing two bars for each method (one for failures; one for successes), and placing these side-by-side (Fig. 15.1, centre panels). Again, numbers or percentages within each method can be graphed.

15.4.3 Dot charts

Instead of bars, dots (or other symbols) can be used in place of the bars in a side-by-side bar chart to create a dot chart.

⚠ The axis displaying the counts (or percentages) should *start from zero*, since the distance of the dots from the axis visually implies the frequency of those observations (see Example 17.3).

Example 15.6 (Dot charts). For the data in Example 15.1, a dot chart can be created by placing plotting symbols for each result (one for failures; one for successes) side-by-side for each method (Fig. 15.1, right panels). Again, numbers or percentages can be used.

15.4.4 Other variations

Many variations of these charts are possible, by making different choices:

- using a stacked bar chart, side-by-side bar chart, or dot chart.
- using percentages or counts on one axis. (The percentages can be percentages of the total, or within the total for each level of the variable, as in the bottom plots in Fig. 15.1.)
- using the counts (or percentage) on either the horizontal or vertical axis.
- deciding which variable can be used as the first division of the data.

The guiding principle remains: *the purpose of a graph is to display the information in the clearest, simplest possible way, to facilitate understanding the message(s) in the data.*

Using a computer to create graphs is recommended, and using a computer makes it easy to try different variations to find the graph that best displays the message in the data.

15.5 Numerical summary: difference between proportions

The difference between the success-rates of the two methods for the small kidney-stone data (Table 15.1) can be summarised using the difference between the respective proportions:

- for *Method A*, the *sample* proportion of successful procedures is $\hat{p}_A = 0.931$.
- for *Method B*, the *sample* proportion of successful procedures is $\hat{p}_B = 0.867$.

The *difference* between these proportions is $\hat{p}_A - \hat{p}_B = 0.064$ (i.e., the success rate is higher for Method A). The difference between the proportions is a *statistic*, and the (unknown) difference between the population proportions (i.e., $p_A - p_B$) is a *parameter*.

15.6 Numerical summary: odds ratios

The small kidney-stone data (Table 15.1) also can be summarised using the odds of success for each method:

- for *Method A*, the odds of success are 13.5 (13.5 *times* as many successes as failures).
- for *Method B*, the odds of success are 6.5 (6.5 *times* as many successes as failures).

The odds of success for Method A and Method B are very different. In the sample, the odds of success for Method A is many *times* greater than for Method B. In fact, in the sample, the odds of success for Method A is $13.5 \div 6.5 = 2.08$ *times* the odds of a success for Method B. This value is the *odds ratio* (OR). The sample OR is a *statistic*, and the (unknown) population OR is a *parameter*. There is no commonly-used symbol for odds ratios.

Definition 15.1 (Odds Ratio (OR)). The *odds ratio* (often written OR) is the ratio of the odds of a result of interest in one group, compared to the odds of the *same* result in a *different* group:

$$\text{Odds ratio (OR)} = \frac{\text{Odds of a result in Group A}}{\text{Odds of the same result in Group B}}.$$

Example 15.7 (Odds ratios). For the small kidney-stone data, the odds of a success for Method A is $81 \div 6 = 13.5$. The odds of a success for Method B is $234 \div 36 = 6.5$. The OR is then computed as $13.5 \div 6.5 = 2.08$. The odds have been computed *with the rows.*

This means that the odds of a success for Method A is about 2.08 times the odds of a success for Method B.

Most software computes the OR from a two-way table by using the values in the *first* row and *first* column on the *top* of the fractions when computing the odds and the odds ratio. In Example 15.7, for instance, the odds for both methods were computed with the Column 1 values on the top of the fraction (81 and 234), and the OR comparing the *rows* was computed with the Row 1 odds (13.5) on top of the fraction.

However, the OR could also be computed using the odds within the columns (i.e., comparing the *columns*), rather than within the rows.

The OR can be interpreted in *either* of these ways (i.e., both are correct):

- the *odds* in each column compares Row 1 counts (top) to Row 2 counts (bottom). The *OR* then compares the Column 1 odds (top) to the Column 2 odds (bottom).
- the *odds* in each row compares Column 1 counts to Column 2 counts. The *OR* then compares the Row 1 odds to the Row 2 odds.

Odds and ORs are computed with the *first row* and *first column* values on the *top* of the fraction. While both are correct, the levels of the explanatory variable are usually the rows of the table (as in Table 15.1), so usually the *second* interpretation makes more sense (as in Example 15.7).

The OR compares the odds of the same result (e.g., success) in two groups (e.g., Method A and Method B). This means a 2×2 table can be summarised with one number: the OR.

When interpreting ORs:

- ORs *greater than* 1 mean the odds of the result is *larger* for the group on top of the fraction compared to the group on the bottom.
- ORs *equal to* 1 mean the odds of the result is the *same* for both groups (on the top and the bottom of the fraction).
- ORs *less than* 1 mean the odds of the result is *smaller* for the group on the top of the fraction compared to the group on the bottom.

The numerical summary information for comparing qualitative variables can be collated in a table. The data should be summarised by one of the qualitative variables, producing proportions (or percentages) and odds for the other. The summary table also requires the differences between the proportions (or percentages) and the odds ratio.

Example 15.8 (Numerical summary table). For the small kidney-stone data, the summary of the data can be tabulated as in Table 15.3, using percentages and odds.

TABLE 15.3: Numerical summary of the small kidney-stone data: odds and percentage of a successful procedure.

	Percentage success	Odds of success	Sample size
Method A	93.1	13.50	87
Method B	86.7	6.50	270
	Difference: 6.4	*OR:* 2.08	

15.7 Example: large kidney stones

The data in Table 15.1 are for procedures on *small* kidney stones. Data were also recorded for the *large* kidney stones (Table 15.4, left table). As for small kidney stones, the *success proportions* can be computed for both methods:

- for *Method A*, the success proportion for *large* kidney stones: $192/263 = 0.730$.
- for *Method B*, the success proportion for *large* kidney stones: $55/80 = 0.688$.

For large kidney stones, then, *Method A* has a higher success proportion than Method B, just as with the small kidney stones.

So, could the data for small (Table 15.1) and large kidney stones (Table 15.4, left table) be combined, to produce a single two-way table of just Method and Result (Table 15.4, right table)? From this table of small and large stones combined:

- for *Method A*, the success proportion for *all* kidney stones: $273/350 = 0.780$.
- for *Method B*, the success proportion for *all* kidney stones: $289/350 = 0.826$.

TABLE 15.4: The kidney stones data. Left: numbers for *large* stones only. Right: numbers for *all* kidney stones combined, without separating by the size of the kidney stone.

Large stones only			
	Success	**Failure**	**Total**
Method A	192	71	263
Method B	55	25	80

Large and small stones combined			
	Success	**Failure**	**Total**
Method A	273	77	350
Method B	289	61	350

When all kidney stones are combined, *Method A* has a *lower* success proportion than Method B. To summarise:

- *Method A* is more successful for *small* stones (0.931 vs 0.867).
- *Method A* is more successful for *large* stones (0.730 vs 0.688).
- *Method B* is more successful for *all* stones combined (0.780 vs 0.826).

That seems strange: Method A performs better for small *and* for large kidney stones, but Method B performs better when combining all kidney stones. The explanation is that the *size of the stone* is a *confounding variable* (Fig. 15.2). Size is associated with both the method (small stones are treated more often with Method B) *and* with the result (small stones have a higher success proportion for *both* methods). Method B was used more often on smaller kidney stones, for which a success is more likely (due to their smaller size).

This confounding could have been avoided by randomly allocating a treatment method to patients. However, random allocation was not possible in this observational study, so the researchers used a different method to manage confounding: *recording* the size of the kidney stones to use in the analysis (see Sect. 7.2).

In this example, incorporating information about a potential confounder (the size of the kidney stone) is important, otherwise the wrong (opposite) conclusion is reached: Method B would be incorrectly considered better if the size of the stones was ignored, when the better method really is Method A.

This is called *Simpson's paradox*. If the size of the kidney stone had not been recorded, size would be a *lurking variable*, and the incorrect conclusion would have been reached.

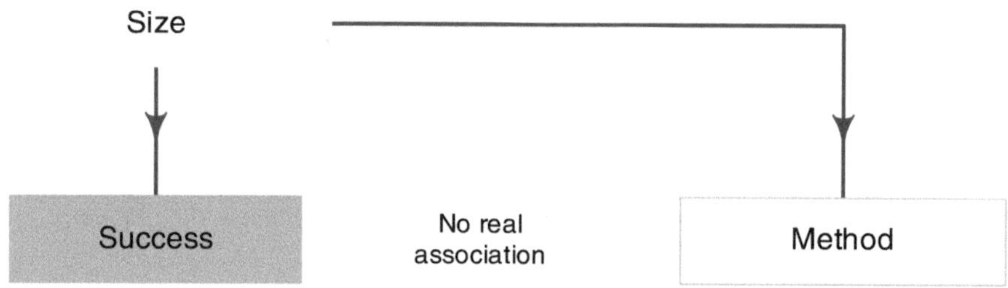

FIGURE 15.2: The size of the stones is associated with the success percentage and method.

15.8 Example: water access

López-Serrano et al. [2022] recorded data about access to water for three rural communities in Cameroon (see Sects. 11.10 and 12.8). The study could be used to determine associations to the incidence of diarrhoea in young children (85 households had children under 5). A cross-tabulation (Table 15.5) shows the relationship with keeping livestock; the numerical summary table (Table 15.6) may suggest a difference in the percentage of children with diarrhoea in households that do and do not keep livestock. The comparison in Fig. 15.3 includes some categories with small sample sizes, so the percentages shown may not be precise estimates of the population values.

As usual, the data come from one of countless possible samples, but the RQ is about the population, so making a definitive decision about the population is difficult.

TABLE 15.5: Cross-tabulation of having livestock in the household, and children under 5 years of age having diarrhoea in the household in the last two weeks.

	No diarrhoea reported in children	Diarrhoea reported in children
Household does not have livestock	17	3
Household has livestock	42	23

TABLE 15.6: Numerical summary of the water-access data: odds and percentage of children with diarrhoea in the last two weeks (comparing those without livestock to those with).

	Percentage children having diarrhoea	Odds children having diarrhoea	Sample size
Household does not have livestock	15.0	0.176	20
Household has livestock	35.4	0.548	65
	Difference: −20.4	*OR:* 0.322	

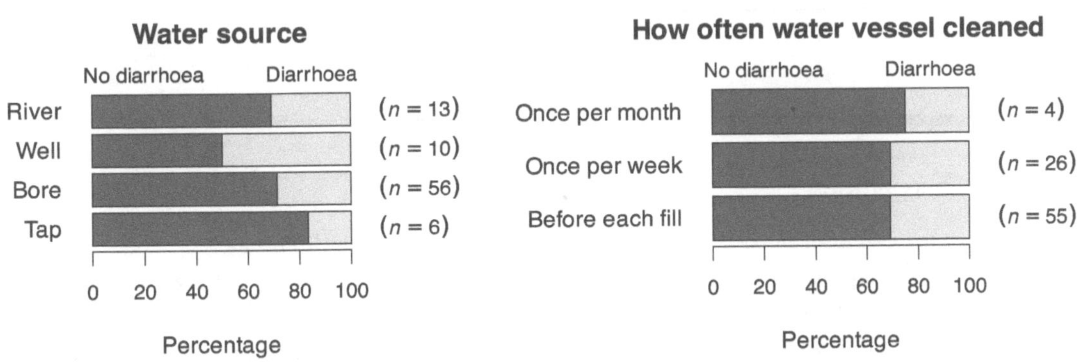

FIGURE 15.3: Percentage of children with and without diarrhoea in the last two weeks, by water source (left) and how often the water vessel was cleaned (right).

15.9 Chapter summary

Qualitative data can be compared between different groups (between-individuals comparisons) using a stacked bar chart, side-by-side bar chart or a dot chart. The data can be displayed in a two-way table, then summarised numerically by comparing proportions (or percentages) and odds. The odds ratio (OR) and the difference between the proportions (or percentages) can be used to compare the two different groups.

15.10 Quick review questions

Alley et al. [2017] examined social media use (Table 15.7), using a representative sample of Queenslanders at least 18 years of age (from the 2013 *Queensland Social Survey*).

Are the following statements *true* or *false*?

1. The *sample proportion* of *urban* residents who use social media is $416/984 = 0.423$.
2. The *sample proportion* of *rural* residents who use social media is $89/167 = 0.533$.
3. The *sample odds* of *urban* residents who use social media is $416/568 = 0.732$.
4. The *sample odds* of *rural* residents who use social media is $78/89 = 0.876$.
5. The *sample OR* of using social media, comparing *urban* to *rural* residents is $1.365/1.141 = 1.196$.
6. The *sample difference between the proportions* using social media, comparing *urban* to *rural* residents, is $0.577 - 0.533 = 0.044$.

TABLE 15.7: The number of Queenslanders using and not using social media (SM) in rural and urban locations in 2013 in a sample.

	Doesn't use SM	**Does use SM**	**Total**
Urban residents	568	416	984
Rural	89	78	167

15.11 Exercises

Answers to odd-numbered exercises are given at the end of the book.

Exercise 15.1. Suppose the sample OR has a value of one. What will be value of the difference between the sample proportions? Explain.

Exercise 15.2. Suppose the sample OR (Row 1 divided by Row 2) has a value *smaller* than one. Will the difference between the sample proportions (Row 1 minus Row 2) be a positive or a negative value? Explain carefully.

Exercise 15.3. Köchling et al. [2019] studied hangovers and recorded, among other information, when people vomited after consuming alcohol. Table 15.8 shows how many people vomited after consuming beer followed by wine, and how many people vomited after consuming only wine.

1. Compute the *row proportions*. What do these mean?
2. Compute the *column percentages*. What do these mean?
3. Compute the *overall percentage* of drinkers who vomited.
4. Compute the *sample odds* that a wine-only drinker vomited.
5. Compute the *sample odds* that a beer-then-wine drinker vomited.
6. Compute the *sample OR*, comparing the odds of vomiting for wine-only drinkers to beer-then-wine drinkers.
7. Compute the *sample OR*, comparing the odds of vomiting for beer-then-wine drinkers to wine-only drinkers.
8. Compute the difference between the *sample proportions* of people vomiting, comparing beer-then-wine drinkers to wine-only drinkers.
9. What do the data suggest about the relationship?

TABLE 15.8: How many people vomited and did not vomit, by type of alcohol consumed.

	Beer then wine	Wine only
Vomited	6	6
Didn't vomit	62	22

Exercise 15.4. Stirrat [2008] recorded the sex of adult and young wallabies at the East Point Reserve, Darwin. In December 1993, 91 males and 188 female *adult* wallabies were recorded, and 13 male and 22 female *young* wallabies were recorded.

1. Create the two-way table of counts.
2. For *adult* wallabies, what *proportion* of adult wallabies were males?
3. For *adult* wallabies, what are the *odds* that a female was observed?
4. For *young* wallabies, what *percentage* of wallabies were males?
5. For *young* wallabies, what are the *odds* that a female was observed?
6. What is the OR of observing an adult wallaby, comparing females to males?
7. What is the difference between the sample proportions of females wallabies, comparing adults to young?
8. Create a summary table.
9. Sketch a graph to display the data.
10. What do the data suggest about the relationship?

Exercise 15.5. [*Dataset*: `EmeraldAug`] The *Southern Oscillation Index* (SOI) is a standardised measure of the air pressure difference between Tahiti and Darwin, shown to be related to rainfall in some parts of the world [Stone et al., 1996], and especially Queensland, Australia [Stone and Auliciems, 1992, Dunn, 2001].

The rainfall at Emerald (Queensland) was recorded for Augusts between 1889 and 2002 inclusive [Dunn and Smyth, 2018], for months when the monthly average SOI was positive and non-positive (zero or negative); see Table 15.9.

1. Compute the *percentage* of Augusts with no rainfall.
2. Compute the *percentage* of Augusts with no rainfall, in Augusts with a *non-positive SOI*.
3. Compute the *percentage* of Augusts with no rainfall, in Augusts with a *positive SOI*.
4. Compute the *odds* of no August rainfall.
5. Compute the *odds* of no August rainfall, in Augusts with a *non-positive SOI*.
6. Compute the *odds* of no August rainfall, in Augusts with a *positive SOI*.
7. Compute the *OR* of no August rainfall, comparing Augusts with *non-positive SOI* to Augusts with a *positive SOI*.
8. Interpret this OR.
9. Create a summary table.
10. Sketch a graph to display the data.

Exercise 15.6. Haselgrove et al. [2008] asked boys and girls in Western Australia about back pain from carrying school bags (Table 15.10).

TABLE 15.9: The SOI, and whether rainfall was recorded in Augusts between 1889 and 2002.

	Non-positive SOI	Positive SOI
No rainfall recorded	14	7
Rainfall recorded	40	53

1. Compute the *percentage* of boys reporting back pain from carrying school bags.
2. Compute the *percentage* of girls reporting back pain from carrying school bags.
3. Among the boys, compute the *odds* of reporting back pain from carrying school bags.
4. Among the girls, compute the *odds* of reporting back pain from carrying school bags.
5. Compute the *odds* of a child reporting back pain.
6. Compute the *OR* of reporting back pain, comparing boys to girls.
7. Interpret this OR.
8. Create a summary table.
9. Sketch a graph to display the data.

TABLE 15.10: The number of boys and girls reporting back pain from carrying school bags.

	Males	Females
No back pain	330	226
Back pain	280	359

Exercise 15.7. Using the information in Table 12.2, create a stacked bar chart to *compare* the responses to the three questions.

Exercise 15.8. Russell et al. [2009] studied road-kill possums in the northern suburbs of Sydney (Table 15.11).

1. Identify the two variables, and classify them as nominal or ordinal.
2. Sketch some graphs to display the data.
3. What is the main message in the data? What graph shows this best?

TABLE 15.11: The number of possums found as road kill, by sex and season.

	Unknown sex	Male	Female
Autumn	75	25	21
Winter	74	27	22
Spring	71	10	18
Summer	58	10	12

Exercise 15.9. The data in Table 15.12 come from a study of Iranian children aged 6–18 years old [Kelishadi et al., 2017].

1. Compute the *proportion* of females who skipped breakfast.
2. Compute the *proportion* of males who skipped breakfast.
3. Compute the *odds* of a female skipping breakfast.
4. Compute the *odds* of a male skipping breakfast.
5. Compute the *OR* comparing the odds of skipping breakfast, comparing females to males.
6. Interpret this OR.
7. Construct a summary table.

Exercise 15.10. Yonekura et al. [2020] studied Japanese women and their coffee drinking habits (Table 15.13).

1. Compute the *proportion* of coffee drinkers who are smokers.
2. Compute the *proportion* of non-coffee drinkers who are smokers.
3. Compute the *odds* of a coffee drinker being a smoker.

TABLE 15.12: The number of Iranian children aged 6 to 18 who skip and do not skip breakfast.

	Skips breakfast	Doesn't skip breakfast	Total
Females	2 383	4 257	6 640
Males	1 944	4 902	6 846

4. Compute the *odds* of a non-coffee drinker being a smoker.
5. Compute the *OR* comparing the odds of being a smoker, comparing coffee drinkers to non-coffee drinkers.
6. Interpret this OR.
7. Construct a summary table.

TABLE 15.13: The number of Japanese women who smoked, and drank at least one cup of coffee per day.

	Smokers	Non-smokers
Coffee drinkers	10	66
Non-coffee drinkers	2	84

Exercise 15.11. Oostema et al. [2018] studied how well emergency dispatchers recognised signs of stroke (Table 15.14).

1. Sketch a side-by-side or stacked bar chart to display the data.
2. Of the *female* patients, what *percentage* had stroke symptoms suspected by the dispatcher?
3. Of the *male* patients, what *percentage* had stroke symptoms suspected by the dispatcher?
4. For *female* patients, what are the *odds* they had stroke symptoms suspected by the dispatcher?
5. For *male* patients, what are the *odds* they had stroke symptoms suspected by the dispatcher?
6. What is the *OR* that a patient had stroke symptoms suspected by the dispatcher, comparing *females* to *males*?
7. What is the *OR* that a patient had stroke symptoms suspected by the dispatcher, comparing *males* to *females*?
8. Construct a numerical summary table.

TABLE 15.14: The number of strokes suspected and missed by dispatchers.

	Suspected stroke	Missed stroke
Female patient	97	67
Male patient	39	43

Exercise 15.12. Soccer is a unique in that one aspect is 'the purposeful use of the unprotected head for controlling and advancing the ball' [Kirkendall et al., 2001]. Some researchers suspect that repeatedly 'heading' the ball may impair brain function. Kirkendall et al. [2001] studied (p. 157)

> ...whether long-term or chronic neuropsychological dysfunction (i.e., concussion) was present in collegiate soccer players

Data were collected from 240 college students for two variables:

- the student type, where each student was classified as a 'soccer player' (63 students), a 'non-soccer athlete' (96 students), or a 'non-athlete' (81 students).
- the number of head concussions, where each student was asked about the number of head concussions they had experienced; 'zero' (158 students), 'one' (45 students), or 'two or more' (37 students) concussions.

Use the study data (Table 15.15) to answer the following questions.

TABLE 15.15: The number of concussions experienced by college students.

	Number of concussions			
	0	1	2 or more	Total
Soccer players	45	5	13	63
Non-soccer athletes	68	25	3	96
Non-athletes	45	15	21	81
Total	158	45	37	240

1. Classify the two variables as nominal or ordinal.
2. Compute the percentage of college students in the *sample* who have received exactly one concussion.
3. Among the *non-athletes*, compute the odds of receiving two or more concussions. Interpret what this means.
4. Among the *soccer players*, compute the odds of receiving two or more concussions. Interpret what this means.
5. Compute the OR comparing the odds of a non-athlete player receiving two or more concussions to the odds of a soccer player receiving two or more concussions.
6. Create a table of *column* percentages. What do these tell you?
7. Create a table of *row* percentages. What do these tell you?
8. Which one of these tables is probably more sensible, and why?

Exercise 15.13. [*Dataset*: `PremierL`] In the 2019/2020 Premier League season, Chelsea had 4 wins from 10 games at home, and 7 wins from 11 wins away from home. What is the OR of a win (comparing home games and away games)?

Answers to *Quick review* questions: 1. True. **2.** False. **3.** True. **4.** True. **5.** False. **6.** False.

16

Correlations between quantitative variables

So far, you have learnt to ask an RQ, design a study, collect the data, describe the data, and summarise data. **In this chapter**, you will learn to:

- describe the relationships between two quantitative variables.
- compute and interpret correlation coefficients and R^2.

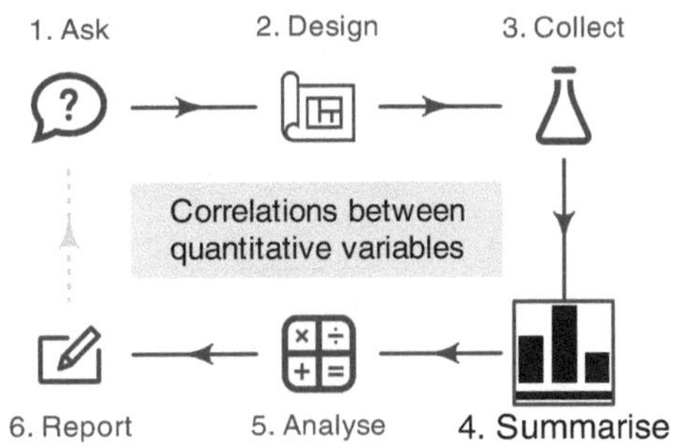

16.1 Introduction

Correlational RQs ask about the relationship between two quantitative variables. Scatterplots are useful for this purpose, and the relationship is usually described numerically using a correlation coefficient or R^2.

16.2 Graphs for the relationship

Scatterplots display the relationship between *two quantitative variables*. Conventionally, and when appropriate, the response variable (denoted y) is shown on the vertical axis, and the explanatory variable (denoted x) is shown on the horizontal axis. Two quantitative variables are measured on each individual, and a point is placed on the scatterplot for each individual (unit of analysis) to indicate the values of the two variables. In some cases, which variable is denoted x and which is y is not important (e.g., see Exercise 33.15.)

As with any graph, describing the message in the graph is important, because the purpose of a graph is to display the information in the clearest, simplest possible way.

Example 16.1 (Red-deer data). Holgate [1965] examined the relationship between the age of $n = 78$ male red deer and the weight of their molars. The data (Table 16.1) comprises two *quantitative* variables, and both measurements are made on the same individuals (i.e., male red deer).

The scatterplot (Fig. 16.1) shows one dot for each deer (individual). The response variable is the molar weight, which is on the vertical axis and denoted y. The explanatory variable is the deer age, which is on the horizontal axis and denoted x.

For instance, one deer is just over 4 years of age (so x has a value a bit larger than 4), and has a molar weight of 2.42 g (so that $y = 2.42$). This is the first deer listed in Table 16.1.

TABLE 16.1: Molar weight and age of male red deer: the first five and the last five observations are shown.

Age (in years)	**Molar weight (in g)**	**Age (in years)**	**Molar weight (in g)**
4.4	2.42	⋮	⋮
4.4	4.45	12.4	2.72
4.4	5.24	12.8	1.71
4.4	3.19	13.4	2.14
4.4	3.90	13.4	2.76
⋮	⋮	14.4	1.57

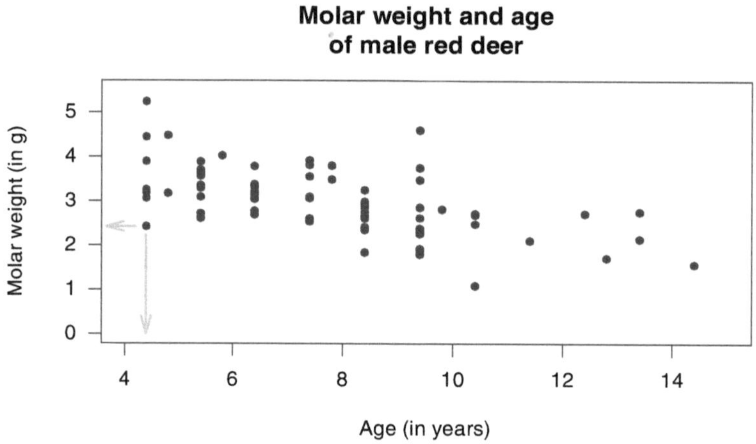

FIGURE 16.1: A plot of the red-deer data. The indicated point is the first observation in Table 16.1, where $x = 4.4$ and $y = 2.42$.

16.3 Describing scatterplots

The purpose of a graph is to facilitate understanding of the data. For a scatterplot, the *form*, *direction*, and *variation in the relationship* (or the *strength of the relationship*) are described.

1. *Form*: the overall *form* or structure of the relationship (e.g., linear; curved upwards; etc.).
2. *Direction*: the *direction* of the relationship (sometimes not relevant if the relationship is non-linear):
   - a *positive* association exists if *high* values of one variable accompany *high* values of the other variable, in general.
   - a *negative* association exists if *high* values of one variable accompany *low* values of the other variable, in general.

3. *Variation*: the amount of *variation* in the relationship. A small amount of variation in the response variable for given values of the explanatory variable means the relationship is strong; a lot of variation in the response variable for given values of the explanatory variable means the relationship is weak. Describing the variation can be difficult; an objective, numerical way to do so is explained in Sect. 16.4.

Anything unusual or noteworthy should also be discussed. These features explain the *type* of relationship (*form*; *direction*), and the *strength* of that relationship (*variation*). Examples are shown in Fig. 16.2.

⚠ The axes do not need to *start from zero*, since the distance of the dots from the axes visually do not imply any quantity of interest.

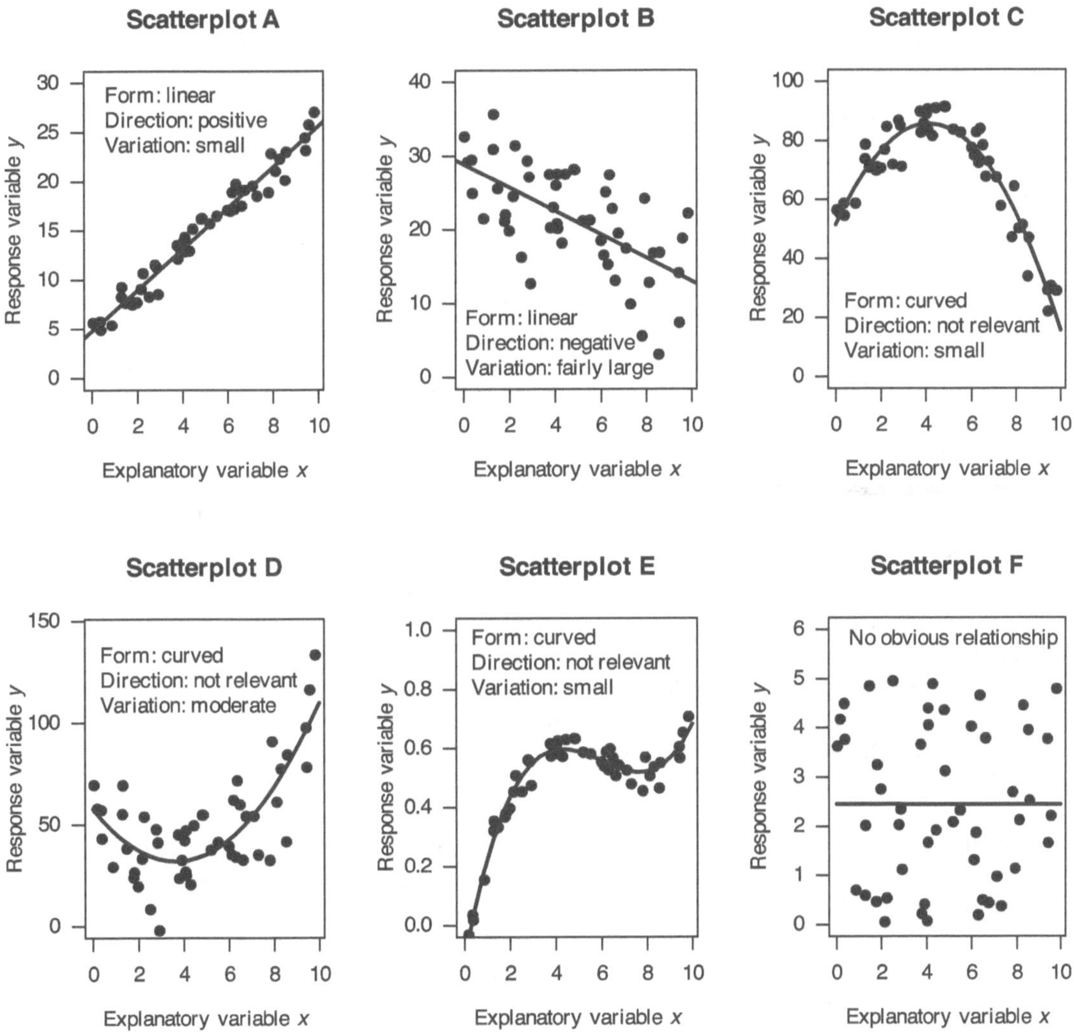

FIGURE 16.2: Some example scatterplots. The dark lines show the overall relationship between the variables.

Example 16.2 (Scatterplots). For the red-deer data (Fig. 16.1), the relationship is approximately linear (form) with a negative direction (*older* deer generally have *lighter* teeth); the *variation* is, perhaps, moderate.

Example 16.3 (Describing scatterplots). Tager et al. [1979] (cited by Kahn [2005]) measured the lung capacity of children in Boston (using forced expiratory volume, FEV, in litres). The scatterplot (Fig. 16.3) is curved (*form*), where older children have larger FEVs, in general (*direction*). The *variation* in FEV gets larger for taller youth.

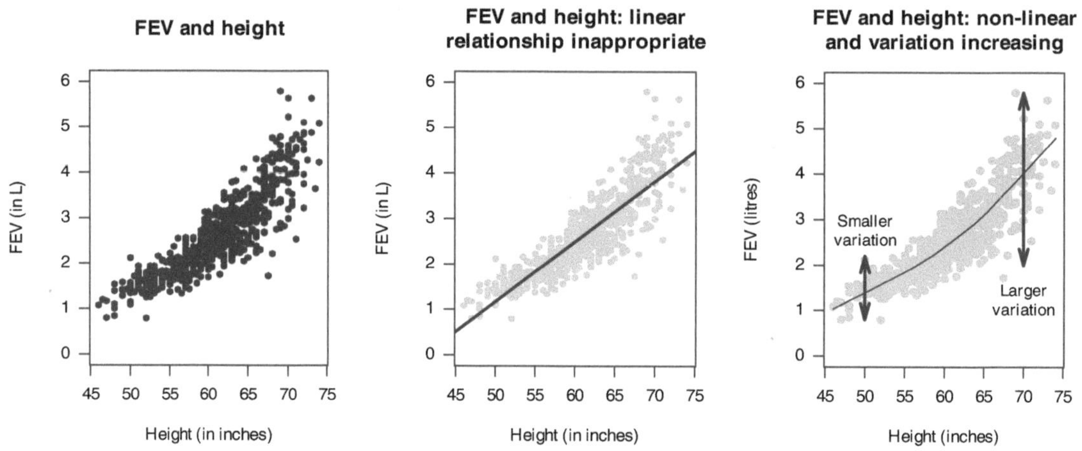

FIGURE 16.3: FEV plotted against height for children in Boston.

16.4 Numerical summary: correlation coefficient and R^2

16.4.1 Correlation coefficients

In general, summarising the relationship between two quantitative variables is difficult, because the possible relationships vary greatly (consider the variety in the scatterplots shown in Fig. 16.2). However, if we focus only on approximately *linear* relationships, the best way to numerically summarise the relationship between the variables is to use a *correlation coefficient*. Both quantitative variables can also be numerically summarised individually.

Definition 16.1 (Correlation coefficient). The Pearson correlation coefficient measures the *strength* and *direction* of the *linear* relationship between two quantitative variables. Its value is always between -1 and $+1$.

Pearson correlation coefficients only apply if both of these are true:

- the form is approximately *linear*.
- the variation in the values of y is reasonably constant for all values of x.

Hence, checking the scatterplot first is important.

Only the *Pearson* correlation coefficient is discussed in this book (and usually referred to as the 'correlation coefficient'), but other correlation coefficients also exist (such as the

Spearman or *Kendall* correlation coefficients), which may be used for increasing-only or decreasing-only *non-linear* relationships).

> ⚠ The Pearson correlation coefficient only makes sense if the relationship is approximately linear.

In the *population*, the unknown value of the correlation coefficient is denoted ρ ('rho'); in the *sample*, the value of the correlation coefficient is denoted r. As usual, r (the *statistic*) is an estimate of ρ (the *parameter*), and the value of r is likely to be different in every sample (that is, *sampling variation* exists).

> The symbol ρ is the Greek letter 'rho', pronounced 'row', as in 'row your boat'.

The values of ρ and r are *always* between -1 and $+1$. The *sign* indicates whether the relationship has a positive or negative linear association, and the *value* of the correlation coefficient describes the *strength* of the relationship, as follows.

- $r = -1$ indicates a *perfect, negative* relationship. By 'perfect', we mean that each value of x always produces the same value of y; the negative value means *larger* values of y are associated with *smaller* values of x.
- values of r between -1 and 0 indicates a *negative* relationship. Each value of x produces a range of values of y, and *larger* values of y are associated with *smaller* values of x (in general).
- $r = 0$ indicates *no linear relationship* between the variables: knowing how the value of x changes tells us nothing about how the corresponding value of y changes. The best prediction of y for *any* value of x would be the mean of y; i.e., the value of $\bar{y}$.
- values of r between 0 and $+1$ indicates a *positive* relationship. Each value of x produces a range of values of y, and *larger* values of y are associated with *larger* values of x (in general).
- $r = +1$ indicates a *perfect, positive* relationship. By 'perfect', we mean that each value of x always produces the same value of y; the positive value means *larger* values of y are associated with *larger* values of x.

Almost all values of r seen in practice are between the extremes of $r = -1$ and $r = +1$. Guessing the values of the correlation coefficient from a scatterplot is very difficult.

Example 16.4 (Correlation coefficients). Numerous example scatterplots were shown in Sect. 16.3. A correlation coefficient is not relevant for Plots C, D or E, as those relationships are not linear. For the others:

- *Plot A*: the correlation coefficient is *positive*, and reasonably close to one.
- *Plot B*: the correlation coefficient is *negative*, but not near -1.
- *Plot F*: the correlation coefficient is close to zero.

Example 16.5 (Correlation coefficients). Leuchtenberger et al. [2022] and Nishizaki et al. [2022] explored the relationship between water temperature and fertilisation rates for sand dollars (Fig. 16.4). The correlation coefficient is $r = -0.71$ (left panel), which might suggest that *higher* temperatures result in *lower* fertilisation rates. However, a *curved* relationship is apparent (right panel), and so the relationship is more complex: the fertilisation rate increases up to about 18°C, and then starts falling again.

A Pearson correlation coefficient is not suitable for describing the relationship.

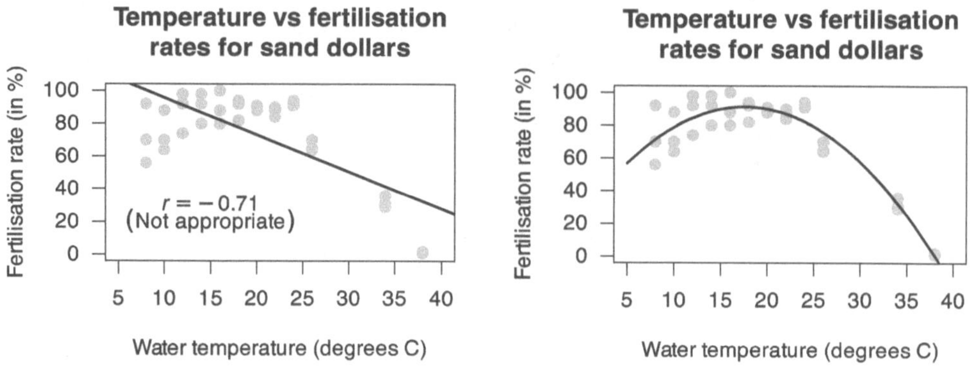

FIGURE 16.4: Water temperature vs fertilisation rates for sand dollars. Left: an inappropriate linear relationship. Right: the appropriate curved relationship.

Formulas exist to compute the value of r, but are tedious to use manually. We will use software output to obtain values of r.

Example 16.6 (Correlation coefficients)**.** For the red-deer data (Fig. 16.1), the relationship is approximately linear, and the software output (Fig. 16.5) shows that $r = -0.584$. The value of r is *negative* because, in general, *older* deer (x) are associated with *smaller* weight molars (y). The relationship may be described as 'moderately strong' perhaps.

Example 16.7 (Correlation coefficients)**.** Tager et al. [1979] studied the lung capacity (forced expiratory volume; FEV) of children in Boston [Kahn, 2005]. The scatterplot in Fig. 16.3 is not linear, so a correlation coefficient is inappropriate.

Correlation Matrix

		Age	Weight
Age	Pearson's r	—	
	p-value	—	
Weight	Pearson's r	-0.584	—
	p-value	< .001	—

FIGURE 16.5: Software output for correlation for the red-deer data.

16.4.2 R-squared (R^2)

While r describes the strength and direction of the linear relationship, knowing exactly what the value *means* is tricky. Interpretation is easier using R^2: the square of the value of r.

Definition 16.2 (R^2)**.** The value of R^2 is how much the unexplained variation in the values of y is reduced (usually expressed as a percentage) due to using the extra information in the values of x.

R^2 is pronounced 'r-squared'.

The value of R^2 is *never* negative, and is usually multiplied by 100 and expressed as a percentage.

The value of R^2 is never negative! However, you need to be careful using your calculator. On most calculators, entering `-0.5^2` returns an answer of `-0.25`. The calculator interprets your input as meaning `-(0.5^2)`.

Use brackets; `(-0.5)^2` gives the correct answer of `0.25` (or 25%).

Example 16.8 (Values of R-squared). For the red-deer data (Fig. 16.1), the value of R^2 is $R^2 = (-0.584)^2 = 0.341$, usually written as a percentage: 34.1%. The value of R^2 is positive, even though the value of r is negative.

This means a reduction of about 34.1% in the unexplained variation of the molar weights, due to using the information in the age of the deer (see Example 16.9). The rest of the variation in molar weights is due to chance, and to extraneous variables such as weight, diet, amount of exercise, genetics, etc.

R^2 measures the reduction in the unexplained variation in values of y because the value of x is known. If the values of x were unknown, the best summary of the y-values is the mean of the y-values (i.e., $\bar{y}$). However, if a relationship exists between the values of x and y then better estimates of the value of y could be made by knowing the value of x. That means that less variation should be left unexplained.

When expressed as a percentage, R^2 measures how much the unexplained variation reduces due to our knowledge of the linear relationship. If R-squared is zero, then the amount of unexplained variation has not reduced at all, and exploring the relationship between x and y has no value.

Example 16.9 (Unknown variation in y). For the red-deer data, the unexplained variation in the values of y (molar weight), without knowing anything about the age of the deer, is the variation in the *distances from the mean* to each observation (Fig. 16.6, left panels). Effectively, the unexplained variation is the standard deviation of the molar weights ($s = 0.7263$).

If the age of the deer (x) is used, the unexplained variation in the values of y is now the variation in the *distances from the line explaining the relationship* to each observation (Fig. 16.6, right panels). The distances are shorter, in general, showing a decrease in the *unexplained* variation. Effectively, the unexplained variation is the standard deviation of the distances from the line to the observations ($s = 0.5895$).

Hence, the reduction in the *square* of the standard deviations is $(0.7263^2 - 0.5895^2)/0.7263^2 = 0.341$, or 34.1%. This is the value of R^2.

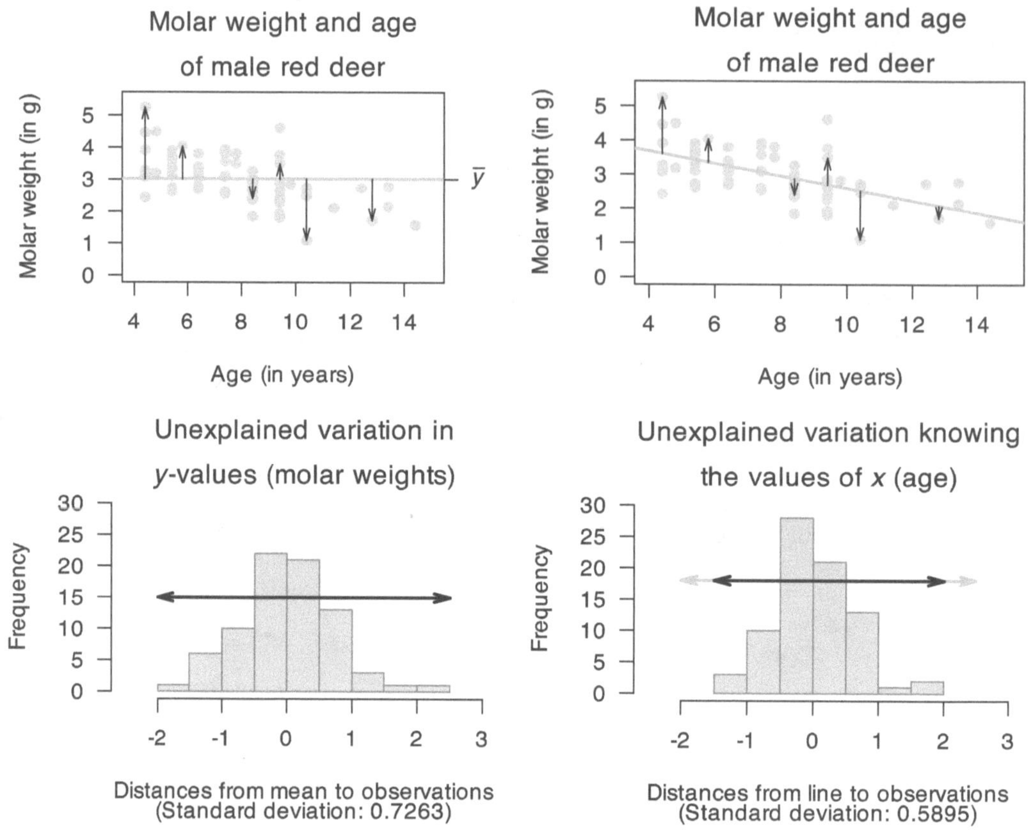

FIGURE 16.6: The unexplained variation for the red-deer data. Left panels: when no information about the age of the deer is used, the mean (the horizontal grey regression line in the top panel) is the best summary of the molar weight. Right panels: when information about the age of the deer is used (as shown by the grey line in the top panel), the distances are shorter in general. R^2 is a measure of how much smaller.

16.5 Numerical summary tables

In general, numerically summarising the relationship between two quantitative variables is difficult because of the many types of possible relationships (Sect. 16.3). However, for *linear* relationships, both quantitative variables can be summarised, and the correlation coefficient can be given (Table 16.2).

TABLE 16.2: A numerical summary of the red-deer data.

	Mean	Standard deviation	Sample size	Correlation
Age (in years)	7.7	2.34	78	−0.584
Molar weight (in g)	3.0	0.73	78	

16.6 Example: removal efficiency

In wastewater treatment facilities, air from biofiltration is passed through a membrane and dissolved in water, and is transformed into harmless by-products. The removal efficiency y (in %) may depend on the inlet temperature (in °C; x). Chitwood and Devinny [2001] asked:

> In treating biofiltation wastewater, is the removal efficiency linearly associated with the inlet temperature?

A scatterplot of $n = 32$ observations [Devore and Berk, 2007] suggests an approximately linear relationship (Fig. 16.7, left panel). The direction is positive: larger inlet temperatures are associated with a larger removal efficiency, in general. The variation is always hard to describe in words, but is perhaps 'reasonably small'.

A more precise way to measure the strength of the linear association is to use the correlation coefficient. Using software output (Fig. 16.7, right panel), $r = 0.891$, and so $R^2 = (0.891)^2 = 79.4\%$. This means that about 79.4% of the variation in removal efficiency can be explained by knowing the inlet temperature.

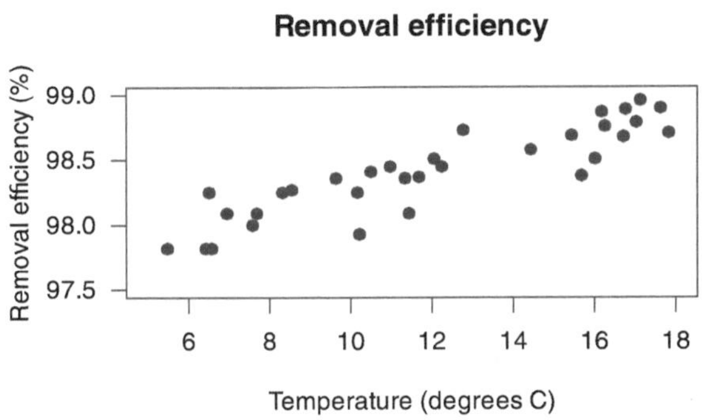

Correlation Matrix

		Removal	Temp
Removal	Pearson's r	—	
	df	—	
	p-value	—	
	95% CI Upper	—	
	95% CI Lower	—	
Temp	Pearson's r	0.8909	—
	df	30	—
	p-value	< .0001	—
	95% CI Upper	0.9458	—
	95% CI Lower	0.7865	—

FIGURE 16.7: The relationship between removal efficiency and inlet temperature. Left: scatterplot. Right: software output.

16.7 Chapter summary

A *scatterplot* displays the relationship between two quantitative variables (the response denoted y; the explanatory denoted x). The relationship is described by the *form* (linear, or otherwise), the *direction* of the relationship (sometimes not relevant if the graph is not linear), and the *variation* in the relationship (or the *strength* of the relationship).

Linear relationships are measured numerically using the correlation coefficient and R^2. *Correlation coefficients* (denoted r in the sample; ρ in the population) are always between -1 and $+1$. *Positive* values denote *positive* relationships between the two variables: as the values of one variable get larger, the values of the other tend to get larger too. *Negative* values denote *negative* relationships between the two variables: as the values of one variable get larger, the values of the other tend to get *smaller*. Values close to -1 or $+1$ are very strong relationships; values near zero shows very little linear relationship between the values of the two variables.

Sometimes, R^2 is used to describe the relationship: it measures how much the unexplained variation in the values of y is reduced due to using the extra information in the values of x.

16.8 Quick review questions

A study of onion growth [Mead, 1970] produced the scatterplot shown in Fig. 16.8.

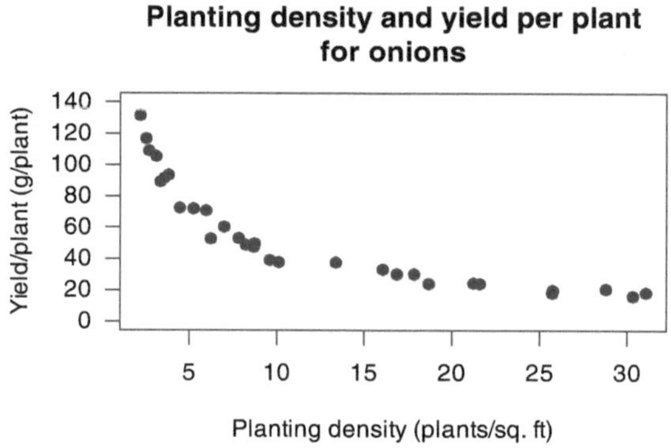

FIGURE 16.8: Onion yield plotted against planting density.

Are the following statements *true* or *false*?

1. The x-variable is 'planting density'.
2. The best description for the *form* of the relationship is 'curved'.
3. The best description for the *direction* of the relationship is 'negative'.
4. The best description for the *variation* in the relationship is 'small'.

16.9 Exercises

Answers to odd-numbered exercises are given at the end of the book.

Exercise 16.1. Draw a scatterplot with:

1. a negative correlation coefficient, with r very close to (but not equal to) -1.
2. a positive correlation coefficient, with r very close to (but not equal to) $+1$.
3. a correlation coefficient very close to 0.

Exercise 16.2. Estimate the correlation coefficients from scatterplots in Fig. 16.9, when appropriate. (You can only give very rough estimates!)

Exercise 16.3. [*Dataset*: **Peas**] Hacisalihoglu et al. [2021] studied the nutritional content of peas (*Pisum sativum*), and measured the quantities of various minerals. In these plots, it does not matter which of the pair of variables is used on the horizontal axis and which is used on the vertical axis. From Fig. 16.10 (*left* panel), estimate the value of r.

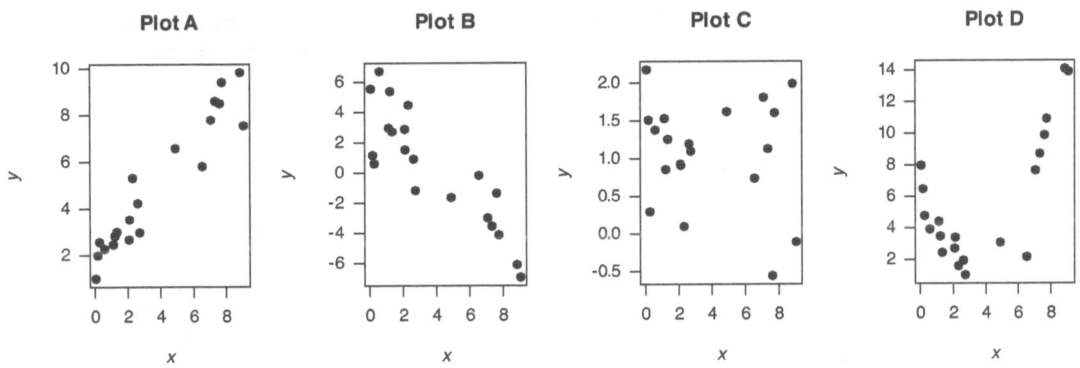

FIGURE 16.9: Four plots: estimate the correlation coefficients.

Exercise 16.4. [*Dataset*: `Peas`] Hacisalihoglu et al. [2021] studied of the nutritional content of peas (*Pisum sativum*), and measured the quantities of various minerals. In these plots, it does not matter which of the pair of variables is used on the horizontal axis and which is used on the vertical axis. From Fig. 16.10 (*right* panel), estimate the value of r.

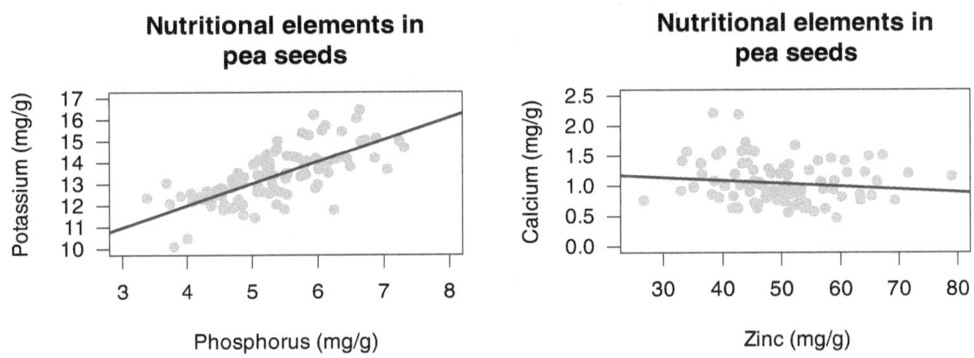

FIGURE 16.10: The relationship between some minerals in pea seeds.

Exercise 16.5. [*Dataset*: `Lime`] Schepaschenko et al. [2017] measured the diameter and the age of 385 small-leaved lime trees (Fig. 16.11).

1. What does each point on the scatterplot represent?
2. Describe the scatterplot.
3. Would a correlation coefficient be appropriate? Explain.

Exercise 16.6. [*Dataset*: `BoneQuality`] Kim and Kim [2022] measured numerous data from 969 South Korean subjects, including 517 females. A scatterplot of the height and age for females is shown in Fig. 16.12.

1. What does each point on the scatterplot represent?
2. Describe the relationship.
3. Would a correlation coefficient be appropriate? Explain.
4. Does the scatterplot suggest that people become less tall, on average, as they age? Explain.

Exercise 16.7. [*Dataset*: `SDrink`] Montgomery and Peck [1992] examined the time taken to deliver soft drinks to vending machines.

1. Describe the relationship (Fig. 16.13, left panel).
2. What does each point represent?
3. Would a correlation coefficient be appropriate? Explain.

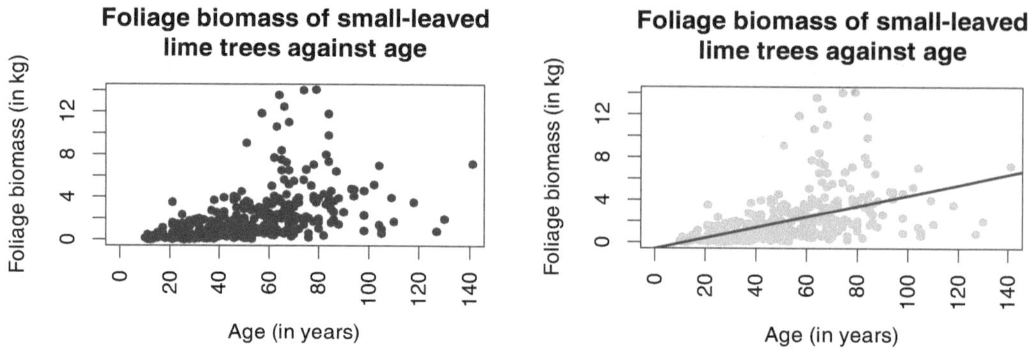

FIGURE 16.11: The age and foliage biomass of small-leaved lime trees grown in Russia ($n = 385$). The solid line on the right panel displays the linear relationship.

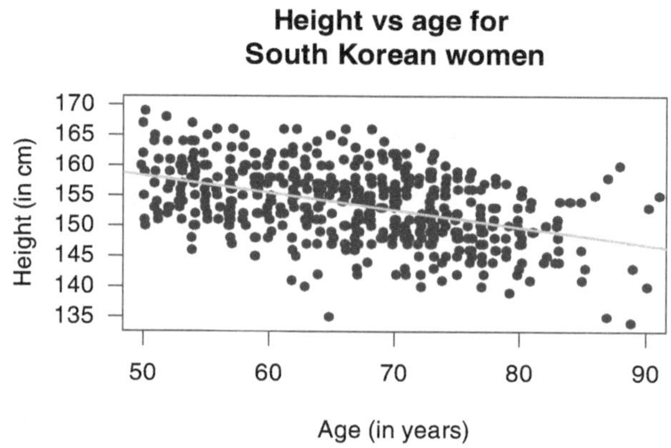

FIGURE 16.12: The age and height of South Korean females ($n = 517$). The solid line shows the linear relationship.

Exercise 16.8. [*Dataset*: `Mandible`] Royston and Altman [1994] examined the mandible length and gestational age for 167 foetuses from the 12th week of gestation onward. Describe the relationship (Fig. 16.13, right panel).

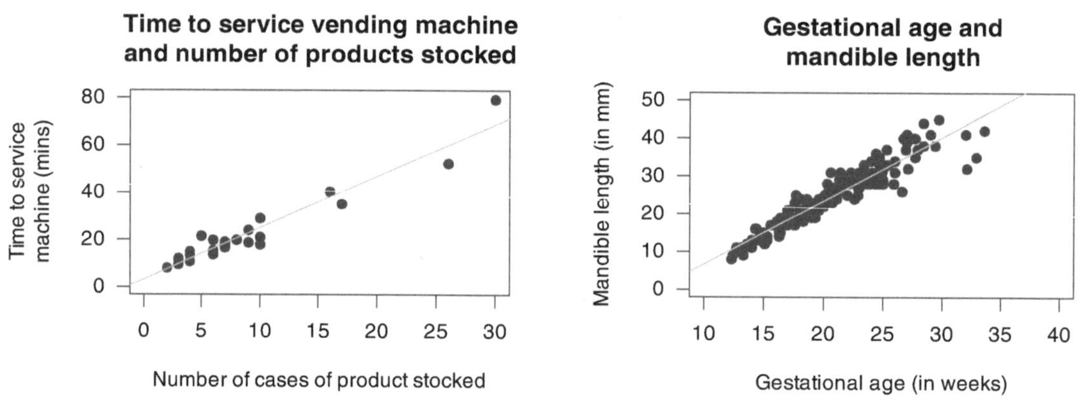

FIGURE 16.13: Two scatterplots. Left: the time taken to deliver soft drinks to vending machines. Right: the relationship between gestational age and mandible length. In both plots, the solid grey line displays the linear relationship.

Exercise 16.9. [*Dataset*: `Gorillas`] Wright et al. [2021] recorded the chest-beating rate of 25 gorillas, and the gorillas' size (measured by the breadth of the gorillas' backs). Describe the relationship (Fig. 16.14, left panel).

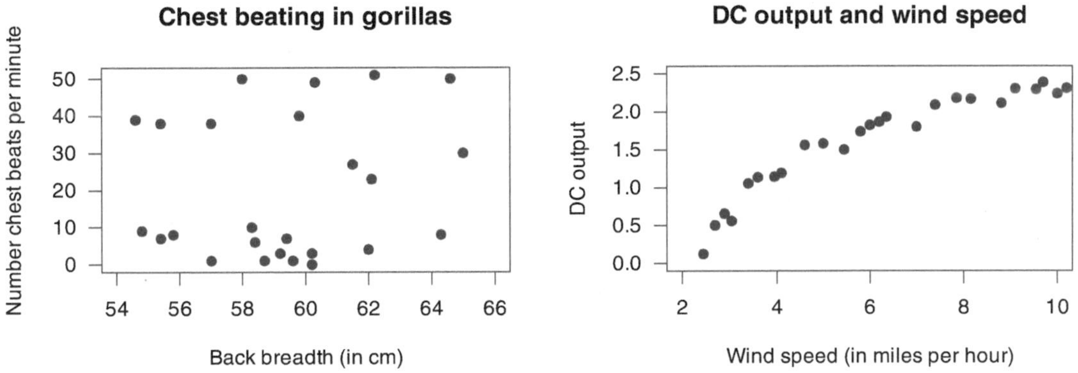

FIGURE 16.14: Two scatterplots. Left: chest beating in gorillas. Right: the relationship between DC output and wind speed.

Exercise 16.10. [*Dataset*: `Windmill`] Joglekar et al. [1989] examined the relationship between direct current (DC) generated by a windmill and wind speed [Hand et al., 1996]. Describe the relationship (Fig. 16.14, right panel).

Exercise 16.11. [*Dataset*: `SoilCN`] Lambie et al. [2021] recorded the percentage carbon (C) and the percentage nitrogen (N) in 28 irrigated farming plots (Fig. 16.15, left panel).

1. Describe the relationship.
2. Does it matter which variable is x and which is y? Explain.
3. What does each point represent?

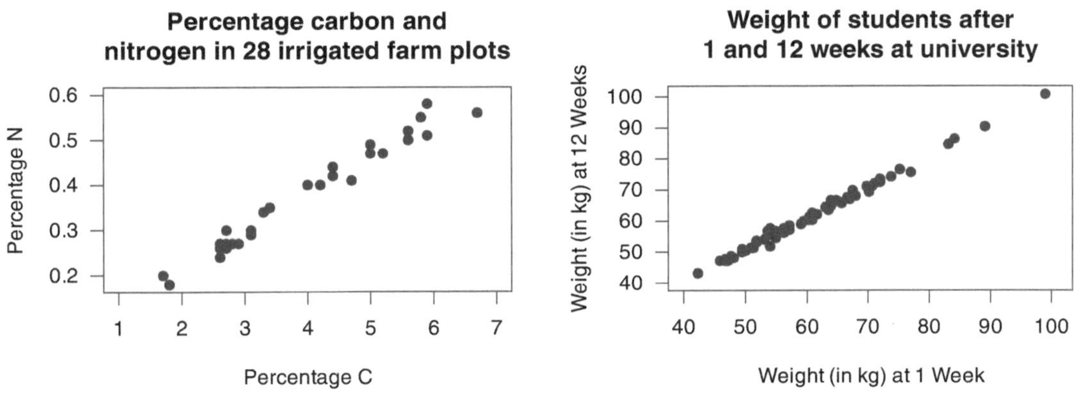

FIGURE 16.15: Left: the percentage N and percentage C in irrigated plots. Right: the weight of students in Week 1 and Week 12 of the university semester.

Exercise 16.12. [*Dataset*: `StudentWt`] The weights of students starting at university (Week 1) and in Week 12 are shown in Fig 16.15 (right panel).

1. Describe the relationship.
2. What does each point represent?

Exercise 16.13. [*Dataset*: **`Cyclones`**] The relationship [Dunn and Smyth, 2018] between the number of cyclones y in the Australian region each year from 1969 to 2005, and a climatological index called the *Ocean Niño Index* (ONI, x), is shown in Fig. 16.16.

From software, $r = -0.682$. What is the value of R^2? What does it mean?

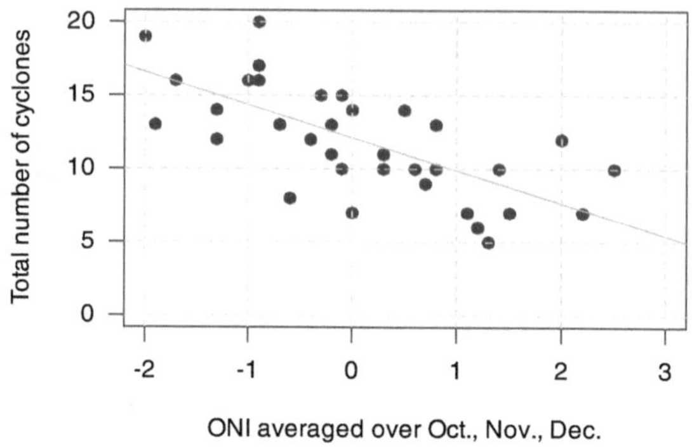

FIGURE 16.16: The number of cyclones in the Australian region each year from 1969 to 2005, and the ONI for October, November, December.

Answers to *Quick review* questions: 1. True. **2.** True. **3.** True. **4.** True.

17

More details about tables and graphs

So far, you have learnt to ask an RQ, design a study, collect the data, describe the data, and summarise data. **In this chapter**, you will learn to:

- construct clear and informative graphs.
- construct clear and informative tables.

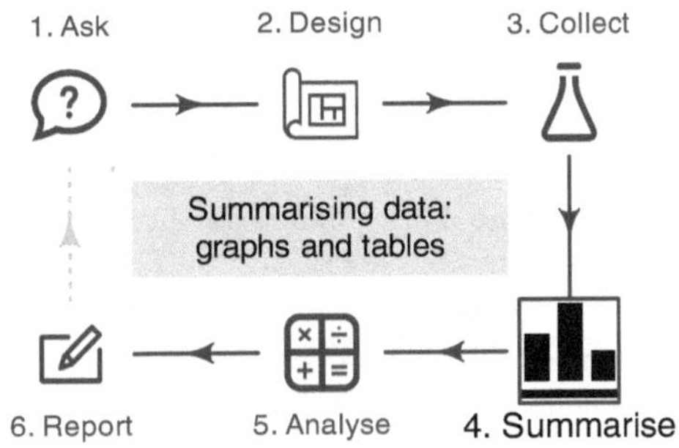

17.1 Introduction

A summary of the data is important for understanding the data, and for planning the direction of the analysis. In this chapter, we make some general comments for constructing graphs and tables. Always remember:

⚠ The purpose of a graph and a table is to display the information in the clearest, simplest possible way, to facilitate understanding the message(s) in the data.

17.2 More details about preparing graphs

Helping readers to understand the data is the goal of producing a graph. You should be able to sketch graphs by hand, but *usually software is used to produce graphs*. Using a computer makes it easy to try different graphs, to change features of graphs, and to produce the best graph possible. When creating graphs, ensure you:

- *do* make graphs clear and well-labelled.
- *do* add informative titles and axis labels.
- *do* add units of measurement where necessary.
- *do* add informative captions *below* the figure.
- *do* add units of measurement and axis labels where appropriate.
- *do* make sure text and details are easy to read.
- *do* ensure the axis scales are appropriate.
- *do* add any necessary explanations.
- *do* make it easy for readers to easily make the important comparisons, as far as possible.

- *do not* add artificial third dimensions, or other 'chart junk' [Su, 2008]; see Sect. 17.2.1.
- *do not* add optical illusions, such as an artificial third dimension.
- *do not* use distracting colours and fonts; only use different colours and fonts if necessary, and explain that purpose if it is not clear.
- *do not* make errors.

Some specific problems to be aware of are discussed in the subsections that follows.

17.2.1 Avoid unnecessary third dimensions

Graphs should focus on clear communication. One barrier to clear communication is using an unnecessary third dimension. This is poor: such graphs can be misleading and hard to read [Siegrist, 1996].

Example 17.1 (Two- and three-dimensional plots). In the NHANES study [Centers for Disease Control and Prevention (CDC), 2024], the age and sex of each participant were recorded. Using Fig. 17.1 (left panel), can you easily determine if more females or more males are present in each age group?

The artificial third dimension makes determining the heights of the bars hard. In contrast, a side-by-side bar graph (Fig. 17.1, right panel) makes it clear whether each age group has more females or more males.

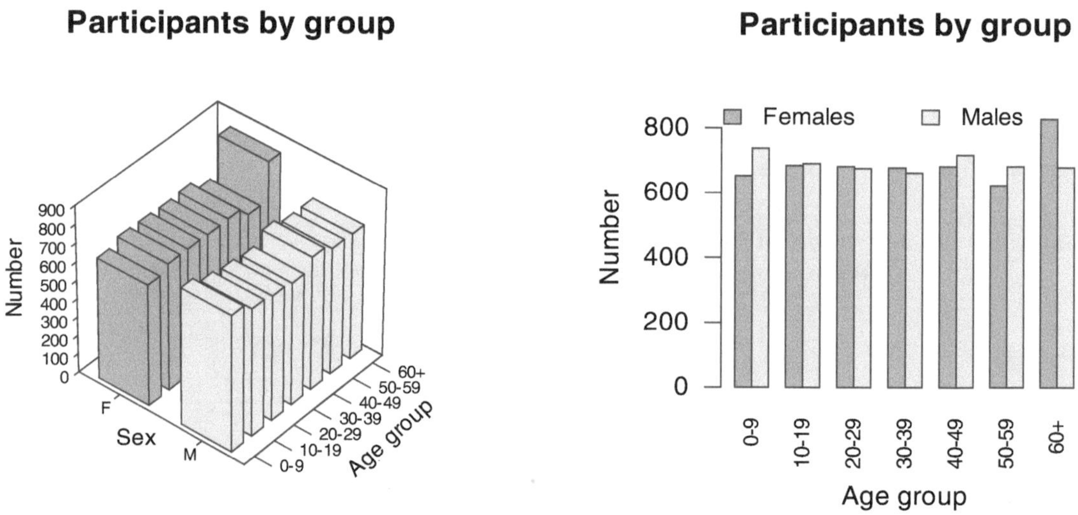

FIGURE 17.1: Two plots of the NHANES participants, divided by age group and sex. Left: a three-dimensional bar chart. Right: a side-by-side bar chart.

17.2.2 Avoid overplotting

Some plots, such as dot charts and scatterplots, may suffer from *overplotting*: when multiple observations have the same (or nearly the same) values, and these cannot be distinguished on the graph. Overplotting can especially be a problem when plotting *discrete* quantitative data. In many cases (such as dot charts), points can be *jittered* by adding a small amount of randomness to the observations, or *stacked*; see Example 14.3. Jittering is the best option for scatterplots. Overplotted points can change readers' impression of the data, since some observations are obscured and are effectively 'lost' to the reader.

17.2.3 Take care when truncating axes

One common optical illusion occurs when the frequency (or percentage) axis does not start at zero. This is a problem in graphs where the distance represented visually implies the frequency of those observations, as with the count (or percentage) axis in bar charts, dot charts, or histograms. This is *not* a problem in, for example, boxplots and scatterplots, where the distance of points from zero do not visually imply any quantity of interest.

Sometimes, the axes may be truncated intentionally so the differences are easier to see. In these cases, the reader should be alerted that the axes have been truncated for this reason.

Example 17.2 (Truncating is not appropriate)**.** Consider data recording the number of lung cancer cases in Fredericia in various age groups [Andersen, 1977].

Figure 17.2 (left panel) shows a good bar chart with the count (vertical) axis starting at zero; the counts in each age group look similar. In contrast, if the vertical axis starts at 9, the counts look very different (Fig. 17.2, right panel) for two age categories, suggesting large difference between the number of lung cancer cases. The graph is visually misleading when the graph does not start at a count of zero, since the height of the bars from the axis visually implies the frequency of those observations.

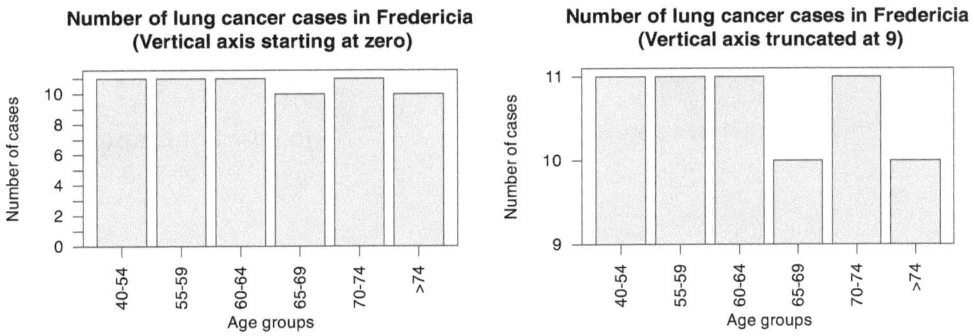

FIGURE 17.2: The same data presented in two bar charts, without truncating the vertical axis (left) and truncating the vertical axis (right).

Example 17.3 (Truncating is appropriate)**.** Consider data recording the body temperature of $n = 130$ people (Mackowiak et al. [1992], Shoemaker [1996]). A histogram of the data (Fig. 17.3, left panel) clearly shows the distribution of body temperatures.

The vertical axis, displaying the counts, must start at zero since the bar heights visually imply a quantity of interest. However, the horizontal axis starts at 35.5°C, which does not create any problems as the *distances* from a temperature of 0°C do not visually imply any quantity of interest.

In contrast, starting the horizontal axis at a temperature of 0°C (Fig. 17.3, right panel) makes any details in the histogram difficult to see; the histogram is pointless.

17.2.4 Take care when using pie charts

As noted in Sect. 12.3.3, pie charts may be hard to read: humans compare *lengths* (bar and dot charts) better than *angles* (pie charts) [Friel et al., 2001]. Pie charts are also difficult to use with levels having zero or small counts.

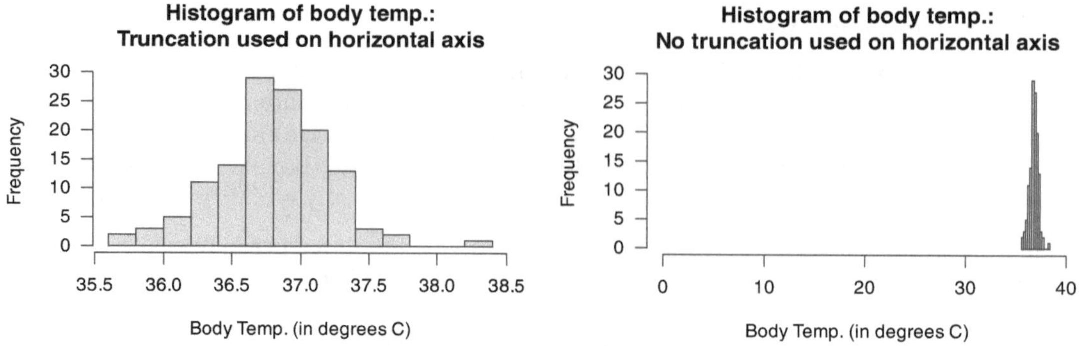

FIGURE 17.3: Two histograms displaying the body temperature of 130 people. Left: a well-constructed histogram. Right: a poorly-constructed histogram.

Example 17.4 (Pie charts with small counts). Solomon et al. [2002] studied the use of ginkgo for memory enhancement. Caregivers blinded to the treatment (ginkgo or placebo) reported the impact on subjects' memory. The bar chart (Fig. 17.4, left panel), for subjects on the placebo, shows that four of the available categories had zero responses, and one had a very small number of responses (two). The pie chart (right) make the small category difficult to see, and the categories with zero counts impossible to see.

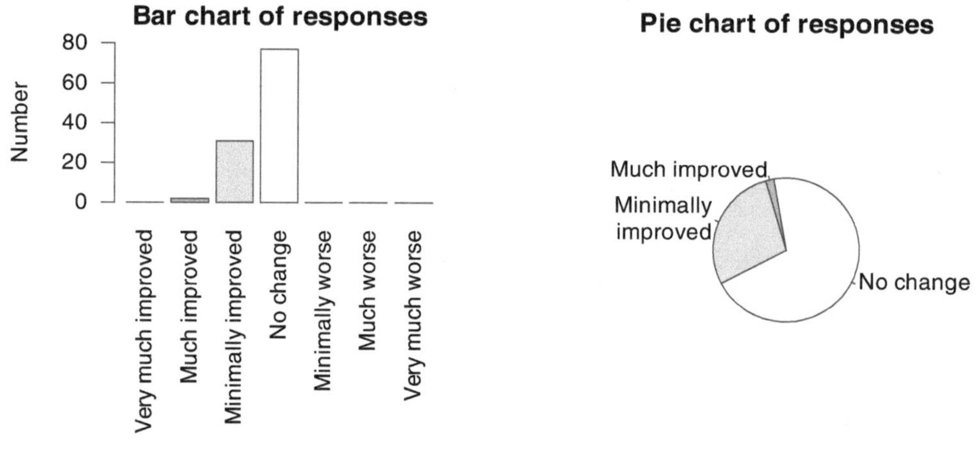

FIGURE 17.4: Data with zeros and small counts are easy to see in a bar chart (left panel) and dot chart, but difficult to see in a pie chart (right panel).

17.3 More details about preparing tables

A computer is helpful for constructing tables. Using a computer also makes it easy to try different orientations or layouts. As with graphs, the purpose of tables is to help readers understand the data. When creating numerical summary tables, ensure you:

- *do* make tables clear and well-labelled.
- *do* use clear and informative row and column labels (as necessary).

- *do* add units of measurement where necessary.
- *do* add informative captions *above* the table.
- *do* add units of measurement and value labels where appropriate.
- *do* make sure text and details are easy to read.
- *do* round numbers appropriately (don't necessarily use all figures provided by software).
- *do* align numbers in the table by decimal point if possible, for easier reading and comparing.
- *do* construct the table to allow readers to easily make the important comparisons, as far as possible (space restriction may take precedence, for example).
- *do not* use distracting colours and fonts; only use different colours and fonts if necessary, and explain that purpose if it is not clear.
- *do not* use vertical lines (in general), and use *very few* horizontal lines. Horizontal lines can be used to group columns (for example, see Table 17.1).

17.4 Example: water access

López-Serrano et al. [2022] recorded data about access to water in three rural communities in Cameroon (see Sects. 12.8 and 11.10). The study could be used to determine associations to the incidence of diarrhoea in young children (85 households had children under 5 years of age). Relationships between the incidence of diarrhoea and some other variables appear in Figs. 15.3 and 14.6. A summary table of information can also be constructed (Table 17.1).

In this table, note that:

- quantitative and qualitative variables are summarised differently, but appropriately.
- units of measurements are given where appropriate (i.e., only for age).
- numbers in columns are aligned for easier reading and comparing.

The table summarises the *sample*, but RQs are about the *population*. For example, one relational RQ could be:

> Is the percentage of households with children under 5 years of age having diarrhoea the same for households that do and do not keep livestock?

Since the observed sample is one of countless possible samples that may have been selected, answering RQs about the population is not straightforward. In the observed sample, 85.0% of households that *did not* keep livestock reported diarrhoea in children under 5, while 64.6% of households that *did* keep livestock reported diarrhoea in children under 5. That is, a difference is seen *in the sample*; but RQs ask about the *population*.

Broadly, two possible reasons could explain why the *sample* percentages of households reporting diarrhoea in children are different:

1. *The **population** percentages are the same.* The *sample* percentages are *different* simply because of the households selected in this particular sample. Another sample, with different households, might produce different sample percentages. *Sampling variation explains the difference in the sample percentages.*
2. *The **population** percentages are different.* The difference between the *sample* percentages reflects this difference between the *population* percentages.

TABLE 17.1: Numerical summary of the water-access data in 85 households with children. 'All households' are broken into those that reported, and did not report, diarrhoea in children under 5 years of age in the last two weeks.

	All households		Diarrhoea		No diarrhoea	
	n	Summary	n	Summary	n	Summary
Age (in years)[a]	85	37.0 (28.0)	59	35.0 (22.5)	26	46.5 (28.5)
Household size[a]	85	7.0 (6.0)	59	6.0 (4.5)	26	8.5 (7.8)
Under 5s in household[a]	85	2.0 (2.0)	59	2.0 (1.0)	26	2.0 (1.8)
Region[b]						
Mbeng	26	30.6%	14	53.8%	12	46.2%
Mbih	28	32.9%	19	67.9%	9	32.1%
Ntsingbeu	31	36.5%	26	83.9%	5	16.1%
Water source[b]						
Tap	6	7.1%	5	83.3%	1	16.7%
Bore	56	65.9%	40	71.4%	16	28.6%
Well	10	11.8%	5	50.0%	5	50.0%
River	13	15.3%	9	69.2%	4	30.8%
Education[b]						
Primary or less	38	44.7%	27	71.1%	11	28.9%
Secondary or higher	47	55.3%	32	68.1%	15	31.9%
Has livestock[b]						
No	20	23.5%	17	85.0%	3	15.0%
Yes	65	76.5%	42	64.6%	23	35.4%

[a] Quantitative variables are summarised using medians and IQR.
[b] Qualitative variables are summarised using counts and percentages.

The difficulty is knowing which of these reasons ('hypotheses') is the most likely explanation for the difference between the sample percentages. This question is of prime importance as it answers the RQ. Tools for answering these questions are considered later in this book.

17.5 Quick review questions

Are the following statements *true* or *false*?

1. Graphs usually have their captions *under* the figure.
2. Graphs should use as many colours as possible.
3. Graphs should usually be carefully created using computer software.
4. Tables should have plenty of horizontal and vertical lines.
5. Tables usually have their captions *under* the table.

17.6 Exercises

Answers to odd-numbered exercises are given at the end of the book.

Exercise 17.1. What would be the best graph for displaying the data for these situations?

1. Researchers record the pH of water and the temperature of the water, in various creeks in the north island of New Zealand, to explore the relationship between pH and temperature.
2. Researchers measure the difference between each swimmers' fastest 100 m time and their fastest 200 m time. The researchers were interested in the average time *difference*.
3. A research study examined the way in which students usually came to university (bus; private car; carpooling; etc.) and their program of study.

Exercise 17.2. What would be the best graph for displaying the data for these situations?

1. Researchers record the number of times a specific recycling bin is used each day at a shopping centre, over many days.
2. Researchers measure the difference between heart rate before and two hours after drinking a cup of coffee. The researchers were interested in the average increase in heart rate.
3. A research study recorded the diet of students (vegan; vegetarian; other) and the cost of groceries in the previous week, for many students. The researchers were exploring if there was any relationship between diet and cost of groceries.

Exercise 17.3. Schepaschenko et al. [2017] recorded these variables for 385 lime trees in Russia: the foliage biomass (in kg; the response variable); the tree diameter (in cm; the explanatory variable); the age of the tree (in years); and the origin of the tree (one of Coppice, Natural, or Planted).

The purpose of the study is to estimate the foliage biomass from the tree diameter, in the presence of some extraneous variables. What graphs would be useful?

Exercise 17.4. Lane [2002] recorded the soil nitrogen after applying different fertiliser doses. The researchers recorded:

- the fertiliser dose, in kilograms of nitrogen per hectare;
- the soil nitrogen, in kilograms of nitrogen per hectare; and
- the fertiliser source; one of 'inorganic' or 'organic'.

What graphs would be useful for understanding the data?

Exercise 17.5. Maron [2007] counted the number of noisy miners (an Australian bird) and eucalyptus trees in random quadrats. Critique the graph (not given by Maron [2007]!) of the data (Fig. 17.5, left panel).

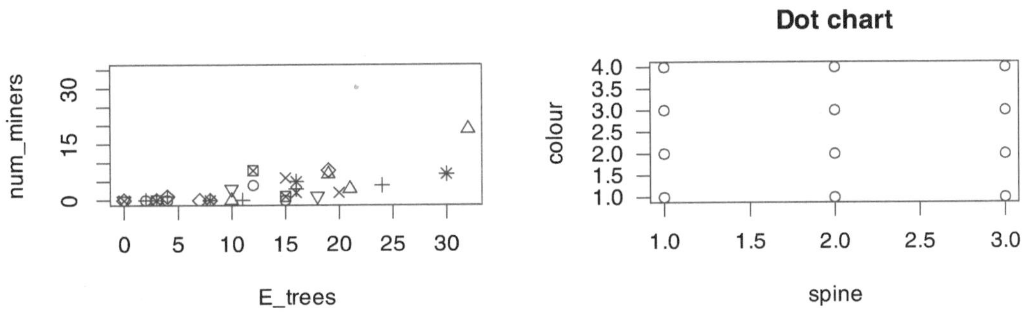

FIGURE 17.5: Left: the number of noisy miners and the number of eucalyptus trees. Right: a scatterplot of the colour of female horseshoe crabs and the condition of their spines.

Exercise 17.6. Brockmann [1996] recorded, among other variables, the colour of the carapace ('Light medium', 'Medium', 'Dark medium' or 'Dark') and the condition of the carapace ('Both OK', 'One OK', 'None OK') of $n = 173$ female horseshoe crabs. Critique the scatterplot (Fig. 17.5, right panel) used to explore the data.

Exercise 17.7. Danielsson et al. [2014] examined the change in MADRS (a *quantitative* scale measuring level of depression) and treatment group (whether each person was treated using: exercise; body awareness; or advice).

1. What is the response variable?
2. What is the explanatory variable?
3. What graphs would be useful for exploring the data and the relationships of interest?

Exercise 17.8. A study of high-performance athletes at the *Australian Institute of Sport* (AIS) [Telford and Cunningham, 1991] recorded numerous variables about athletes. A plot for the sports played by the athletes is shown in Fig. 17.6. How would you describe the data: left skewed, right skewed, approximately symmetrical? Or something else?

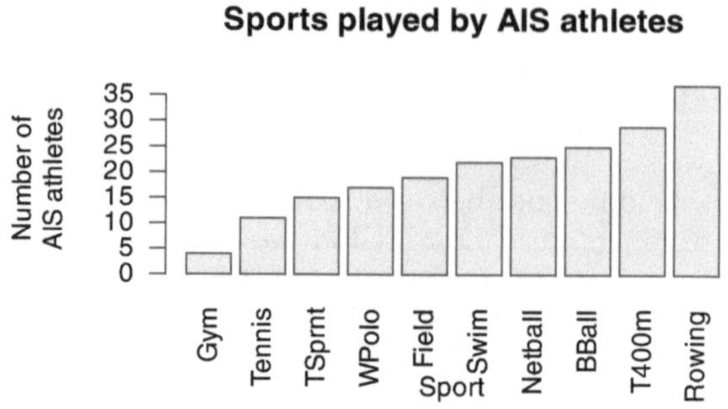

FIGURE 17.6: Sports played by athletes in the AIS study.

Exercise 17.9. [*Dataset*: **`Typing`**] The **`Typing`** dataset [Pinet et al., 2022] contains four variables: typing speed (**`mTS`**), typing accuracy (**`mAcc`**), age (**`Age`**), and sex (**`Sex`**) for 1 301 students. Produce graphs necessary for understanding the data, making sure to explain what they reveal.

Does the mean typing speed or mean accuracy appear to differ by the age or sex of the student? What other questions could be useful to ask about the data?

Exercise 17.10. [*Dataset*: **`NHANES`**] Consider the NHANES data. In preparing a paper about this study, suppose Fig. 17.7 and Tables 17.2 were produced. Critique these.

TABLE 17.2: A table of results.

	Mean	**Std dev.**
Current smoker	206.6	46
Current non-smoker	214.64	48.7945
Difference	8.03	

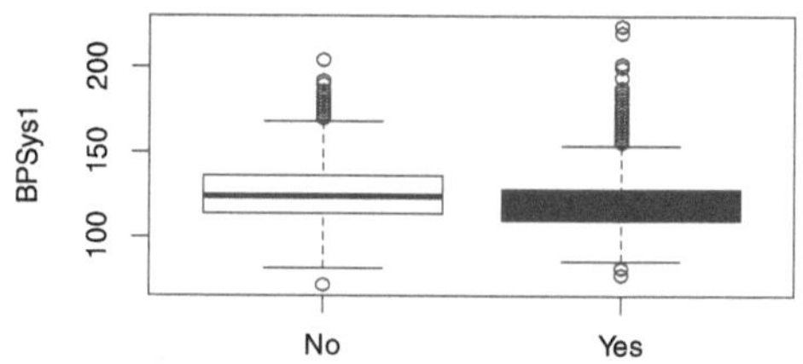

FIGURE 17.7: A boxplot.

Answers to *Quick review* questions: 1. True. **2.** False. Use different colours only if they have a purpose (and explain that purpose if it is not clear). **3.** True. **4.** False: very few vertical lines (if any); minimum of horizontal lines. **5.** False.

Part V

Tools for answering RQs

18

Probability

So far, you have learnt to ask an RQ, design a study, and describe and summarise the data. **In this chapter**, you will learn to:

- explain probability.
- identify and apply various approaches to computing probability.
- determine the probability of events described using **and**, **or** and **not** in simple situations.
- identify events that are independent.
- compute simple conditional probabilities.

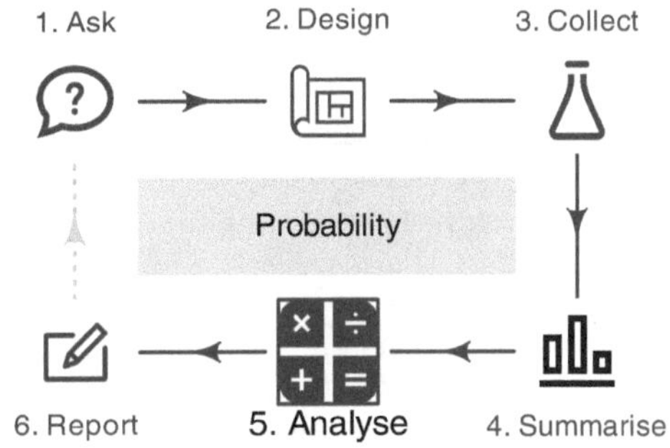

18.1 Introduction

This chapter briefly discusses *probability*. *Probability* quantifies the chance of observing a specific, unknown result (an 'event'). Before discussing probability, some associated terms need defining.

Definition 18.1 (Random procedure). A *random procedure* is a sequence of well-defined steps that (a) can be repeated, in theory, indefinitely under essentially identical conditions; (b) has well-defined results; and (c) has results that are unpredictable for any individual repetition.

Using this definition, the result of rolling a die is a 'random procedure', with possible results ⚀, ⚁, ⚂, ⚃, ⚄ and ⚅. Similarly, tossing a coin is a random procedure with two possible results: **Heads** Ⓗ or **Tails** Ⓣ.

18.2 Sample spaces, events and probability

A list of all mutually exclusive possible results from one instance of a random procedure is the *sample space*. A *simple event* is any element of the sample space.

Definition 18.2 (Sample space). The *sample space* is a list of all possible and mutually exclusive (distinct) results after administering a random procedure once.

Definition 18.3 (Simple event). A *simple event* is a single element of the sample space.

Example 18.1 (Sample spaces). Consider rolling a fair, six-sided die (the random procedure). We do not know what face will be uppermost until we roll the die.

However, the *sample space* for this procedure can be listed: ⚀, ⚁, ⚂, ⚃, ⚄ and ⚅. These are all mutually exclusive (or distinct) results and cover all possible results (exhaustive) from a single roll. The sample space is *discrete* (see Sect. 10.2).

The event 'rolling a ⚀' is a simple event.

Combinations of the elements in the sample space are usually of more interest than simple events. These are called *compound events.*

Definition 18.4 (Compound event). A *compound event* is any combination of simple events.

Example 18.2 (Events). Some *events* that can be defined using the sample space in Example 18.1 include:

- rolling a ⚃. This *simple event* includes one element of the sample space: ⚃.
- rolling an odd number. This *compound event* includes three elements of the sample space: ⚀, ⚂ and ⚄.
- rolling a number larger than ⚁. This *compound event* includes four elements of the sample space: ⚂, ⚃, ⚄ and ⚅.

The sample space is *discrete* (see Sect. 10.2).

Example 18.3 (Sample spaces and events). Consider the distance you can throw a baseball (the random procedure). We do not know beforehand what distance your next throw will be, but the *sample space* (i.e., the throwing distance) is a number greater than 0 m. This sample space is *continuous.*

Many *compound events* can be defined using this sample space; for example:

- throwing more than 50 m.
- throwing between 10 and 40 m.

Because the sample space is continuous, throwing an *exact* distance (such as *exactly* 10 m) is technically not possible (see Sect. 10.2).

Events are often defined using **and, or, not**. Consider two events called A and B. Then, 'A **and** B' is the event comprising events only occurring in *both* events A and B. 'A **or** B' is the event comprising all events in A, all events in B, and events in both. The event '**not** A' comprises all the events in the sample space that are *not* in Event A.

Example 18.4 (Defining events). Consider rolling a fair, six-sided die again (Example 18.1). Suppose these two (compound) events are defined:

- Event A is 'roll a number divisible by 2'.
- Event B is 'roll a number divisible by 3'.

Event A comprises the simple events 'roll a ⚁', 'roll a ⚃' and 'roll a ⚅'. Event B comprises the simple events 'roll a ⚂' and 'roll a ⚅'.

Then, the Event 'A **and** B' includes all events only occurring in both A and in B; that is, 'A **and** B' comprises the single simple event 'roll a ⚅'.

Event '*A* **or** *B*' include the events in *A*, the events in *B*, and those in both; that is, '*A* **or** *B*' comprises the four simple events 'roll a ⚁', 'roll a ⚂', 'roll a ⚃' and 'roll a ⚅'.

The event '**not** *A*' comprises the three simple events 'roll a ⚀', 'roll a ⚂' and 'roll a ⚄'.

Using these definitions, a *probability* can be defined.

Definition 18.5 (Probability). A *probability* is a number between 0 and 1 inclusive (or between 0% and 100% inclusive) that quantifies the likelihood that a certain event will occur.

A probability of 0 (or 0%) means the event is 'impossible' (will *never* occur), and a probability of 1 (or 100%) means that the event is *certain* to happen (will *always* occur). Most events have a probability between the extremes of 0% and 100%.

Example 18.5 (Probabilities). Consider these examples:

- the probability of receiving negative rainfall in London next year is 0; it is impossible.
- the probability of receiving some rain in London next year is 1; it is certain.
- the probability of receiving rain on 01 January next year in London is between 0 and 1 inclusive.

18.3 Determining probabilities

Three different ways to think about probability are:

- the *classical approach* (Sect. 18.3.1).
- the *relative frequency approach* (Sect. 18.3.2).
- the *subjective approach* (Sect. 18.3.3).

These approaches help determine, or approximate, values for probabilities.

18.3.1 Classical approach

What is the probability of rolling a ⚃ on a die? The sample space has six possible outcomes (see Example 18.1) that are *equally likely* to occur (i.e., no reason exists to expect one event to occur more often than the others), and the event 'rolling a ⚃' comprises just *one* of those events. Thus,

$$\text{Probability of rolling a ⚃} = \frac{\text{The number of rolls that are a ⚃}}{\text{The number of possible events in the sample space}} = \frac{1}{6}.$$

This approach to computing probabilities is called the *classical* approach to probability, and is only appropriate when all events in the sample space are *equally likely*.

Definition 18.6 (Classical approach to probability). In the *classical approach to probability*, the probability of an event occurring is the number of elements of the sample space included in the event, divided by the total number of elements in the sample space, *when all outcomes are equally likely*.

By this definition:

$$\text{Prob. of an event} = \frac{\text{Number of simple events in the event of interest}}{\text{Total number of possible equally-likely events}}.$$

We can say that 'the probability of rolling a ⚃ is 1/6', or 'the probability of rolling a ⚃ is 0.1667'. The answer can also be expressed as a *percentage*: 'the probability of rolling a ⚃ is 16.67%'. The answer could also be interpreted as 'the *expected* proportion of rolls that are a ⚃ is 0.1667'. That is, about 16.67% of a very large number of future rolls are likely to be a ⚃.

The probability of rolling a ⚃ is 0.1667, but any single roll of the die either *will* or *will not* produce a ⚃, and we don't know which will occur.

Example 18.6 (Probabilities for compound events). Consider rolling a standard six-sided die. With six equally-likely results (Example 18.1), the probability of rolling an even number is 3/6, since there are three even numbers in the sample space.

Example 18.7 (Describing probability). Consider rolling a standard six-sided die.

- The *probability* of rolling an even number is $3 \div 6 = 0.5$.
- The *percentage* of rolls expected to be even is $3 \div 6 \times 100 = 50\%$.
- The *odds* of rolling an even number is $3 \div 3 = 1$.

Example 18.8 (Probabilities). Consider the probabilities of the events in Example 18.2.

- The probability of rolling a ⚃ is 1/6 (or about 0.1667).
- The probability of rolling an odd number is 3/6, or 1/2 (or 0.5).
- The probability of rolling a number larger than ⚁ is 4/6, or 2/3 (or about 0.6667).

18.3.2 Relative frequency approach

What is the probability that a newborn baby will be male? The sample space could be listed as: *male* and *non-male*. Since the sample space has two elements, the classical approach suggests the probability is $1 \div 2 = 0.5$. However, this approach is appropriate *only if* males and non-males are *equally likely* to be born. But are they?

In 2021 in Australia, 289 603 live births occurred, with 148 636 male births, 140 944 female births, and 23 others (or 'not stated'). The *proportion* of males born in the 2021 sample is $148\,636 \div 289\,603 = 0.513$, or about 51.3%. An *estimate* of the probability that the next birth will be male is about 0.513 (or 51.3%), based on using past data.

This is the *relative frequency* approach to calculating probabilities: based on past data. The relative frequency method can only ever produce an *approximate* probability, as it is based on a limited number of past observations. An actual probability would require an infinite number of observations.

Definition 18.7 (Relative frequency approach to probability). In the *relative frequency approach to probability*, the probability of an event is *approximately* the number of times the outcomes of interest has appeared in the past, divided by the number of 'attempts' in the past. This produces an *approximate* probability.

Example 18.9 (Relative frequency probability). Based on the earlier information, the *odds* that a new baby will be a boy is *approximately* $0.513 \div (1 - 0.513) = 1.053$ (i.e., 105.3 boys per 100 girls). According to the *Australian Bureau of Statistics* (ABS):

> The sex ratio for all births registered in Australia generally fluctuates around 105.5 male births per 100 female births.

This is close to the odds of 1.053 found above.

⚠ *Probabilities* describe the likelihood that an event will occur *before* the result is known. *Odds* and *proportions* can be used either *before* or *after* the result is known, provided the wording is correct.

For example, *proportions* describe how often an event has occurred *after* the result is known, and *expected proportions* describe the likelihood that an event will occur in many repetitions *before* the result is known.

The following example may help explain.

Example 18.10 (Probabilities, proportions and odds). *Before* a fair coin is tossed:

- the *probability* of throwing a Ⓗ is $1/2 = 0.5$ (or 50%).
- the *expected proportion* of Ⓗ for many coin tosses is 0.5 (or 50%).
- the *odds* of throwing a Ⓗ is $1/1 = 1$.

If we have *already* tossed a coin 100 times and found 47 heads:

- the *proportion* of Ⓗ in the sample is $47/100 = 0.47$ (or 47%).
- the odds that we *threw* a Ⓗ in the sample is $47/53 = 0.887$.

The 'probability that we just threw a Ⓗ' makes no sense, because the result is known.

18.3.3 Subjective approach

Many probabilities cannot be computed using the classical or relative frequency approach; for example, what is the probability that your sporting team wins their next game? It may depend on how important you deem the injuries to key players, whether you think recent form is crucial, or if you believe in a substantial home ground advantage. In this case, only a *subjective probability* can be given.

'Subjective' probabilities may be based on personal judgement or experience. They can also be given by considering some of the relevant issues that may impact the probability (and may, for example, be based on mathematical models that incorporate information from numerous inputs). Depending on how these other issues are considered and combined, different subjective probabilities may be given.

Weather forecasts are one example: they incorporate data from sea surface temperatures, local topography, air pressures, air temperatures and so on. Different models use different inputs, and may combine these inputs differently to produce different (subjective) forecast probabilities. Subjective probabilities are deductive probabilities (based on reasoning).

Definition 18.8 (Subjective approach to probability). In the *subjective approach to probability*, various factors are incorporated subjectively to determine the probability of an event occurring.

Example 18.11 (Subjective probability). During El Niño events, eastern Australia typically experiences drier-than-average winters and springs. The *Australian Broadcasting Corporation*'s news website reported (on 23 May 2023) that the Australian *Bureau of Meteorology* predicted a 50% probability of an El Niño event in 2023, while the American *National Oceanic and Atmospheric Administration* predicted a 90% chance of an El Niño event in 2023.

Despite this, '[both] agencies are looking at the same part of the Pacific Ocean' to make their predictions. However, 'the US and Australia base their probability on different criteria'. The probabilities are subjective probabilities, based on complex mathematical models.

18.4 Independence of events

One important concept in probability is *independence*. Two events are *independent* if the probability of one event happening is the same, whether or not the other event has happened. For example, the probability of getting a Ⓗ on a coin toss is the same whether you are sitting or not sitting: the result of the coin toss is *independent* of whether you are seated.

Definition 18.9 (Independence). Two events are *independent* if the probability of one event is the same, whether or not the other event has happened.

Example 18.12 (Independence). Consider drawing two cards from a well-shuffled, standard pack (of 52 cards), *without* returning the first card. For the *first* card, the sample space contains every card in the pack, and drawing any card is as equally likely as drawing any other. Since four cards are **Aces**, the probability of drawing an **Ace** on the first draw is 4/52 (using the classical approach).

If we drew an **Ace** for the first card, the probability of drawing an **Ace** for the *second* card is 3/51 (*three* **Aces** remain among the 51 remaining cards). Alternatively, if we *don't* draw an **Ace** for the first card, the probability of drawing an **Ace** second time is 4/51 (*four* **Aces** remain among the 51 remaining cards).

That is, the probability of drawing an **Ace** for the second card *depends* on whether an **Ace** was drawn for the first card. The two events 'Drawing an **Ace** for the first card' and 'Drawing an **Ace** for the second card' are *not independent* events.

ⓘ A 'standard' pack of cards has 52 cards, organised into four *suits*: spades ♠, clubs ♣ (both black), hearts ♡ and diamonds ♢ (both red). Each *suit* has 13 *denominations*: 2, 3, 4, 5, 6, 7, 8, 9, 10, Jack (J), Queen (Q), King (K), Ace (A). The Ace, King, Queen and Jack cards are called *picture cards*. (Most packs also contain two jokers, which are not considered part of a *standard* pack.)

⚠ Random samples produce *independent* units of analysis.

18.5 Conditional probability

Conditional probability refers to adjusting probabilities when extra information is known. For example, the probability of rolling a ⚀ is 1/6 using the classical approach, as the sample space has six equally-likely elements. However, if we are told that an *odd number* is rolled, only three elements in the sample space need now be considered (rolls of ⚀, ⚂, ⚄) rather than all six elements; other outcomes are impossible). So, the probability of rolling a ⚀ is 1/3. We say 'the probability of rolling a ⚀, *given that the roll is an odd number*, is 1/3'.

Example 18.13 (Conditional probability). Suppose someone draws a card from a pack of cards. The probability that the card is a ♣ is $13/52 = 1/4$, or 25%.

However, if you are told that the card is a *black* card, then the card must be either a ♣ or ♠. The probability that the card is a ♣, *given* that the card is black, is $13/26 = 1/2$, or 50%.

Example 18.14 (Wearing sunglasses). Dexter et al. [2019] recorded the number of people at the foot of the Goodwill Bridge, Brisbane, who wore sunglasses between 11:30am to 12:30pm (Table 18.1). The probability of an observed person wearing sunglasses is

$$\frac{126 + 123}{126 + 123 + 240 + 263} = 0.3311,$$

or about 33.1%.

Conditional probabilities can also be computed:

- *if the observed person is female*, the probability that she is wearing sunglasses is $126 \div (240 + 126) = 0.3443$, or about 34.4%.
- *if the observed person is male*, the probability that he is wearing sunglasses is $123 \div (263 + 123) = 0.3187$, or about 31.9%.

These probabilities are close, but not exactly equal.

If the two events were *independent*, then these two conditional probabilities would be the same: the probability of wearing sunglasses would be the same for females and males. In other words, the probability of wearing sunglasses did not depend on whether a female or a male was observed. We might say that wearing sunglasses is close to, but not exactly, independent of the sex of the person, in the *sample*. We cannot be sure if wearing sunglasses is independent of the sex of the person in the *population*.

TABLE 18.1: Females and males wearing sunglasses on the Goodwill Bridge, Brisbane.

	Female	**Male**
Not wearing sunglasses	240	263
Wearing sunglasses	126	123

18.6 Chapter summary

A *probability* is a number between 0 and 1 inclusive (or between 0% and 100% inclusive) that quantifies the likelihood of a certain event occurring. Three ways to think about probabilities are:

- the *classical approach*, which requires all outcomes to be *equally likely*.
- the *relative frequency* approach (which gives approximate probabilities).
- the *subjective approach* (deductive probabilities).

Two events are *independent* if the probability of one event is the same, whether the other event has happened or not. Conditional probability incorporates extra information when the probability is computed.

18.7 Quick review questions

Suppose *Event* A is defined as 'Rolling a ⚀ or a ⚁ on a fair die'. Also, suppose *Event* B is defined as 'Rolling an even number'.

Are the following statements *true* or *false*?

1. The best *approach* to computing the probability of Event A occurring is the *classical* approach.
2. The *probability* of Event A occurring is 2/6.
3. Rolling a ⚀ on the first roll is *independent* of rolling a ⚀ on a second roll.
4. The *probability* of 'A **and** B' occurring is 1/6.
5. The *probability* of 'A **or** B' occurring is 4/6.
6. The *probability* of '**not** B' occurring is 3/6.
7. The *odds* of '**not** B' occurring is 3/6.
8. The probability of Event B occurring, *if* Event A has already occurred, is 1/2.

18.8 Exercises

Answers to odd-numbered exercises are given at the end of the book.

Exercise 18.1. Which *approach* is best used to estimate a probability in these situations?

1. The probability that the stock market will rise next month.
2. The probability that a randomly-chosen person writes left-handed.

Exercise 18.2. Which *approach* is best used to estimate a probability in these situations?

1. The probability that a **King** will be chosen from a pack of cards.
2. The probability that Paris receives more than 50 mm of rain next May.

Exercise 18.3. Consider drawing cards from a fair pack. *Event* A is 'drawing a picture card', *Event* B is 'drawing a **King** or **Ace**' and *Event* C is 'drawing a ♠'.

1. What events are in 'A **and** B'?
2. Compute the probability of 'A **and** B'.

3. What events are in 'A **or** B'?
4. Compute the probability of 'A **or** B'.
5. What events are in 'A **and** C'?
6. Compute the probability of 'A **and** C'.
7. What events are in '**not** C'?
8. Compute the probability of '**not** C'.
9. Compute the probability of C, if A has already occurred.
10. Compute the probability of A, if C has already occurred.

Exercise 18.4. Consider rolling a fair die. *Event A* is 'rolling an *even* number', *Event B* is 'rolling an *odd* number' and *Event C* is 'rolling a ⚁.

1. What events are in 'A **and** B'?
2. Compute the probability of 'A **and** B'.
3. What events are in 'A **or** B'?
4. Compute the probability of 'A **or** B'.
5. What events are in 'A **and** C'?
6. Compute the probability of 'A **and** C'.
7. What events are in '**not** C'?
8. Compute the probability of '**not** C'.
9. Compute the probability of C, if A has already occurred.
10. Compute the probability of C, if B has already occurred.

Exercise 18.5. Consider these three events about tossing two fair coins, say Coin A and Coin B: *Event 1* is 'toss a Ⓗ on Coin A'; *Event 2* is 'toss a Ⓣ on Coin A'; and *Event 3* is 'toss a Ⓗ on Coin B'.

1. Are *Event 1* and *Event 2* independent events?
2. Are *Event 1* and *Event 3* independent events?
3. Compute the probability of *Event 3*.
4. What is the probability of *Event 3* occurring, if *Event 1* has already occurred?
5. List the sample space for the random procedure.

Exercise 18.6. Consider these three events about drawing one card from a fair pack: *Event 1* is 'draw a **Jack**'; *Event 2* is 'draw a ♡'; and *Event 3* is 'draw a ♣'.

1. Compute the probability of *Event 1*.
2. Compute the probability of *Event 1*, if *Event 2* has occurred.
3. Compute the probability of *Event 1*, if *Event 2* has *not* occurred.
4. Are *Event 1* and *Event 2* independent? Explain.
5. Compute the probability of *Event 3*.
6. Compute the probability of *Event 3*, if *Event 2* has occurred.
7. Compute the probability of *Event 3*, if *Event 2* has *not* occurred.
8. Are *Event 2* and *Event 3* independent? Explain.

Exercise 18.7. Suppose I roll a standard six-sided die.

1. What is the *probability* that I will roll a number larger than ⚁?
2. What are the *odds* of rolling a number smaller than ⚅?
3. Suppose I toss a coin after rolling the die. Is the result from the coin toss *independent* of what I rolled on the die?
4. What is the probability that I roll a number divisible by 2 on the die?
5. What is the probability that I roll a number divisible by 2 **and** divisible by 3 on the die?
6. What is the probability of rolling a ⚁, *given* that the number is smaller than ⚃?

Exercise 18.8. Suppose you have a well-shuffled, standard pack of 52 cards.

1. What is the *probability* that you will draw a **King**?
2. What are the *odds* that you will draw a **King**?
3. What is the *probability* that you will draw a picture card (**Ace**, **King**, **Queen** or **Jack**)?
4. What are the *odds* that you will draw a picture card (**Ace**, **King**, **Queen** or **Jack**)?
5. Suppose I draw two cards from the pack. Are the events 'Draw a **King** first' and 'Draw a **Queen** second' independent events?

6. Suppose I draw one card from the pack (drawing the second without replacing the first), then roll a six-sided die. Are the events 'Draw a **Jack** from the pack of cards' and 'Roll a ⚄ on the die' independent events? Explain.
7. If I draw a picture card, what is the probability the card is a **King**?

Exercise 18.9. Consider drawing a card from a standard, well-shuffled pack of cards. The first card is replaced, the pack reshuffled, and then a second card is drawn from the pack. The colour of the two cards is recorded (Black or Red).

1. Write down the sample space.
2. Define Event D as 'the total number of black cards drawn, minus the total number of red cards drawn'. Find the probability that D is zero.
3. Find the probability that D is zero **or** one.
4. Is the colour of the card drawn first *independent* of the colour of the card drawn second? Explain.

Exercise 18.10. Consider the random process 'tossing a coin'. Event H is of interest: 'the number of tosses until the first Ⓗ is thrown'.

1. What is the sample space?
2. Find the probability that H is one.
3. Find the probability that H is two.
4. Find the probability that H is one **or** two.

Exercise 18.11. Dunn [2023] tabulated information about Queensland school children (Table 18.2).

1. What is the probability that a randomly chosen student is a First Nations student?
2. What is the probability that a randomly chosen student is in a government school?
3. Is the sex of the student approximately independent of whether the student is a First Nations student, for students in government schools?
4. Is the sex of the student approximately independent of whether the student is a First Nations student, for students in non-government schools?
5. Is whether the student is a First Nations student approximately independent of the type of school, for female students?
6. Is whether the student is a First Nations student approximately independent of the type of school, for male students?
7. Based on the above, what can you conclude from the data?

TABLE 18.2: The number of First Nations and non-First Nations students in Queensland schools in 2019.

	Number of First Nations students	Number of non-First Nations students
Government schools		
Females	2 540	21 219
Males	2 734	22 574
Non-government schools		
Females	391	9 496
Males	362	9 963

Exercise 18.12. Kelishadi et al. [2017] recorded whether Iranian children aged 6–18 years ate breakfast (Table 18.3). Find the *probability* that a randomly chosen student is:

1. A female student.
2. A female student who skipped breakfast.
3. A female student, *if we already know* the child skipped breakfast.

Exercise 18.13. Are these pairs of events likely to be *independent* or *not independent*? Explain.

1. 'I walk to work tomorrow morning', and 'Rain is expected tomorrow morning'.
2. 'A person smokes more than 10 cigarettes per week' and 'A person gets lung cancer'.
3. 'It rains today' and 'I hose my garden today'.

TABLE 18.3: The number of Iranian children aged 6 to 18 who skip and do not skip breakfast.

	Skips breakfast	**Doesn't skip breakfast**	**Total**
Females	2 383	4 257	6 640
Males	1 944	4 902	6 846
Total	4 327	9 159	13 486

Exercise 18.14. In disease testing, two keys aspects of the test are:

- *sensitivity*: the probability of a *positive* test result among those *with* the disease; and
- *specificity*: the probability of a *negative* test result among those *without* the disease.

Both are important for understanding how well a test works in practice. Ideally, a test would have high sensitivity and high specificity.

A certain test has a *sensitivity* of 0.99 and a *specificity* of 0.98. Consider a group of 1 000 people, 100 of whom (unknowingly) have the disease and 900 who (unknowingly) do not have the disease. All the people are given the test.

1. Suppose the 100 people who *do* have a disease are tested. How many would be expected to return a positive test result?
2. Suppose the 900 people who *do not* have a disease are tested. How many would be expected to return a positive test result?
3. In total, how many positive tests would be expected from the 1 000 people?
4. Consider those people who return a positive test result. What is the probability that one of these people actually has the disease?

Exercise 18.15. Explain *why* the following argument is incorrect:

> When I toss two coins, there are only three outcomes: a **Head** and a **Head**, a **Tail** and a **Tail**, or one of each. So the probability of obtaining two **Tails** must be one-third.

Exercise 18.16. On 13 October, 1997, the American television programme *Nightline* interviewed Dr Richard Andrews, director of California's *Office of Emergency Services*, to discussed natural disasters being predicted. In the interview, Dr Andrews said (see *Chance News* 6.12):

> I listen to earth scientists talk about earthquake probabilities a lot and in my mind every probability is 50–50: either it will happen or it won't happen.

Explain why Dr Andrews is incorrect when he says that 'every probability is 50–50'. Give an example to show why he must be incorrect.

Answers to *Quick review* questions: 1. True. **2.** True. **3.** True. **4.** True. **5.** True. **6.** True. **7.** False: odds are 1. **8.** True.

19

Sampling variation

So far, you have learnt to ask an RQ, design a study, describe and summarise the data, and understand probability. **In this chapter**, you will learn to:

- explain what a sampling distribution describes.
- explain the difference between the variation between individuals and the variation in statistics.
- determine when a standard error is appropriate to use.
- explain the difference between standard errors and standard deviations.

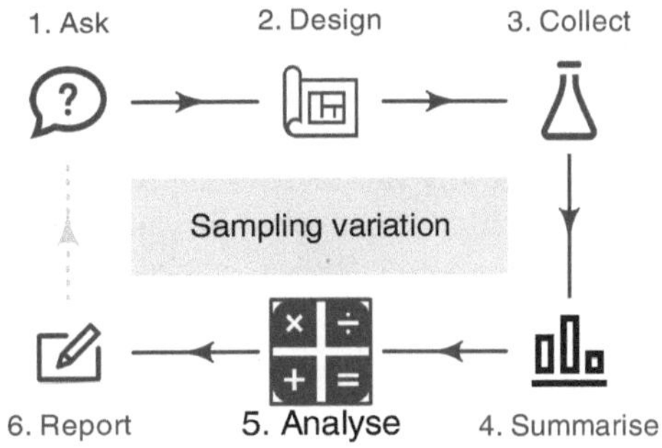

19.1 Introduction

One of the most important ideas in research and statistics is that the sample being studied is only one of countless possible samples that could have been selected to study.

⚠ Studying a sample leads to the following observations:

- every sample is likely to be different.
- we observe just one of the many possible samples.
- every sample is likely to yield a different value for the statistic (i.e., a different estimate for the parameter).
- we observe just one of the many possible values for the statistic.

Since many values for the statistic are possible, the values of the statistic vary and have a distribution.

In research, decisions need to be made about *populations* based on *samples*; that is, about *parameters* based on *statistics*. The challenge is that the decision must be made using only one of the many possible samples, and every sample is likely to be different. Each sample will produce a different value for the *statistic*. This is called *sampling variation*.

Definition 19.1 (Sampling variation). *Sampling variation* refers to how the sample estimates (statistics) vary from sample to sample, because every possible sample is different.

Any distribution that describes how a statistic varies for all possible samples is called a *sampling distribution*.

Definition 19.2 (Sampling distribution). A *sampling distribution* is the distribution of a statistic, showing how its value varies in all possible samples.

19.2 Sample proportions have a sampling distribution

Sample proportions, like all statistics, vary from sample to sample; that is, *sampling variation* exists, so sample proportions have a *sampling distribution.*

Consider a European roulette wheel (Fig. 19.1): a ball and wheel are spun, and the ball can land on any number on the wheel from 0 to 36 (inclusive). Using the classical approach to probability, the probability of the ball landing on an *odd* number (an '*odd spin*') is $p = 18/37 = 0.486$. This is the *population proportion* (the parameter).

If the wheel is spun (say) 15 times, the *sample* proportion of odd spins, denoted $\hat{p}$, will vary. But, *how* does $\hat{p}$ vary from one set of 15 spins to another set of 15 spins? Can we describe *how* the values of $\hat{p}$ vary across the possible samples?

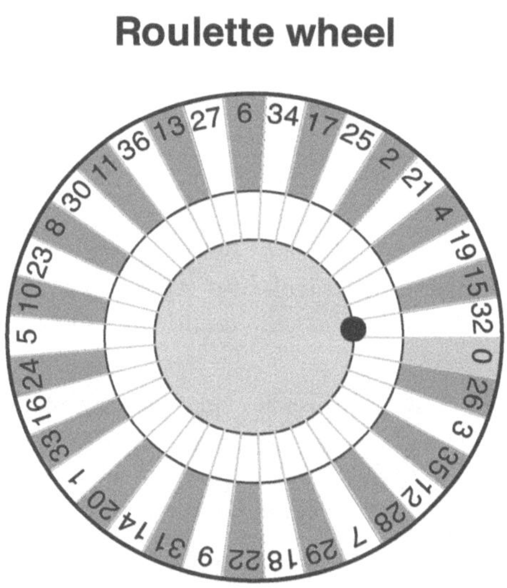

FIGURE 19.1: A European roulette wheel, with numbers 0 to 36. The ball landed on 32.

Computer simulation can be used to demonstrate what happens if the wheel was spun, over and over again, for $n = 15$ spins each time, and the proportion of odd-spins was recorded for each repetition. The proportion of odd spins $\hat{p}$ varies from sample to sample (sampling variation), as shown by the histogram (Fig. 19.2, top left panel). The *shape* of the distribution is approximately bell shaped. We can see that, for many repetitions of 15 spins, $\hat{p}$ is rarely smaller than 0.2, and rarely larger than 0.8. That is, reasonable values to expect for $\hat{p}$ are between about 0.2 and 0.8.

If the wheel was spun (say) $n = 25$ times (rather than 15 times), $\hat{p}$ again varies (Fig. 19.2, top right panel): the values of $\hat{p}$ vary from sample to sample. The same process can be repeated for many repetitions of (say) $n = 100$ and $n = 200$ spins (Fig. 19.2, bottom panels).

Notice that as the sample size n increases, the variation in the sampling distribution gets smaller. With 200 spins, for instance, observing a sample proportion smaller than 0.4 or greater than 0.6 seems highly unusual, but these are not uncommon at all for 15 spins.

The sampling distributions can be described by a mean (called the *sampling mean*) and a standard deviation (called the *standard error*).

Example 19.1 (Reasonable values for the sample proportion). Suppose a roulette wheel was spun 100 times, and 31 odd numbers were observed. The sample proportion is $\hat{p} = 31/100 = 0.31$. From Fig. 19.2 (bottom left panel), a sample proportion this low almost never occurs in a sample of 100 rolls.

This is very unlikely to occur from a fair roulette wheel. Hence, we either observed something highly unusual, or the wheel is not fair (e.g., a problem exists with the wheel).

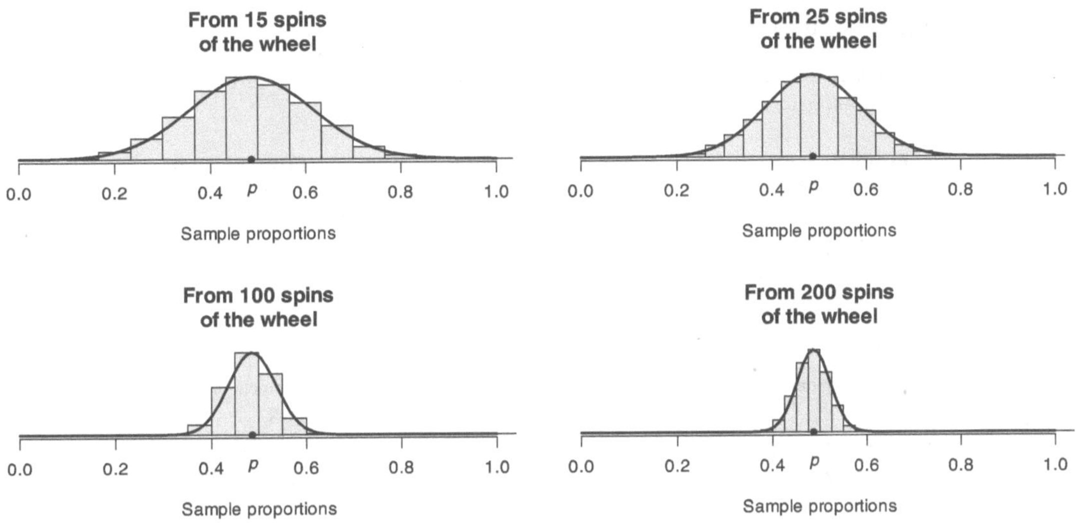

FIGURE 19.2: Sampling distributions for the proportion of roulette wheel spins that show an odd number, for sets of rolls of varying sizes.

⚠ The values of the sample proportion vary from sample to sample. The distribution of the possible values of the *sample* proportion is called a *sampling distribution.*

Under certain conditions, the sampling distribution of a sample proportion is described by an approximate bell-shaped distribution (formally called a *normal distribution*). In general, the approximation gets better as the sample size increases. In addition, the possible values of $\hat{p}$ vary less as the sample size increases.

The mean of the sampling distribution is called the *sampling mean*; the standard deviation of the sampling distribution is called the *standard error* (Fig. 19.4).

19.3 Sample means have a sampling distribution

The sample mean, like all statistics, varies from sample to sample; that is, *sampling variation* exists, so sample means have a *sampling distribution.*

Consider a European roulette wheel again (Sect. 19.2). Rather than recording the sample proportion of odd-spins, suppose the *sample mean* of the numbers was recorded. If the wheel is spun (say) 15 times, the *sample* mean of the spins $\bar{x}$ will vary from one set of 15 spins to another. But *how* does it vary?

Again, computer simulation can be used to demonstrate what could happen if the wheel was spun 15 times, over and over again, and the mean of the numbers was recorded for each repetition. Clearly, the sample mean spin $\bar{x}$ can vary from sample to sample (sampling variation) for $n = 15$ spins (Fig. 19.3, top left panel).

When $n = 15$, the sample mean $\bar{x}$ varies from sample to sample, and the *shape* of the distribution again is approximately bell-shaped. If the wheel was spun more than 15 times (say, $n = 50$ times) something similar occurs (Fig. 19.3, top right panel): the values of $\bar{x}$ vary from sample to sample, and the values have an approximate bell-shaped (normal) distribution. In fact, the values of $\bar{x}$ have a bell-shaped distribution for other numbers of spins also (Fig. 19.3, bottom panels).

The sampling distributions can be described by a mean (called the *sampling mean*) and a standard deviation (called the *standard error*).

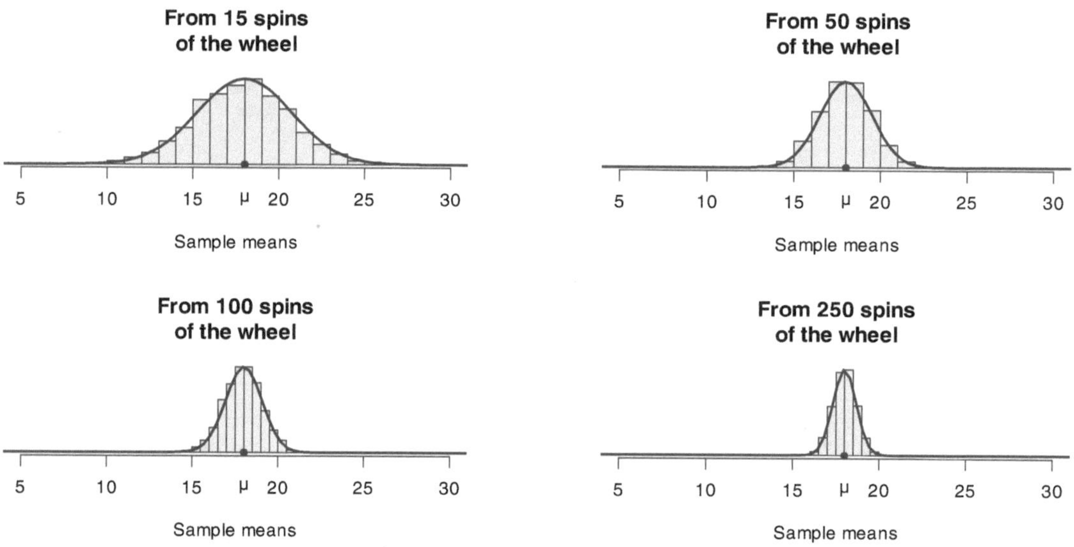

FIGURE 19.3: Sampling distributions for the mean of the numbers after a roulette wheel is spun a certain number of times.

The values of the sample mean vary from sample to sample. The distribution of the possible values of the *sample* mean is called a *sampling distribution.*

Under certain conditions, the sampling distribution of a sample mean is described by an approximate bell-shaped (or normal) distribution. In general, the approximation gets better as the sample size increases. In addition, the possible values of $\bar{x}$ vary less as the sample size increases.

The mean of the sampling distribution is called the *sampling mean*; the standard deviation of the sampling distribution is called the *standard error* (Fig. 19.4).

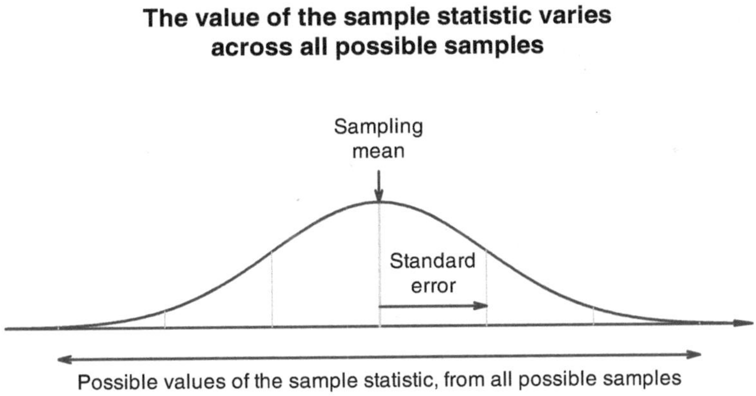

FIGURE 19.4: Describing how the value of the sample statistic varies across all possible samples, when the sampling distribution has a normal distribution.

Example 19.2 (Reasonable values for the sample mean). Suppose we spun a roulette wheel 100 times, and the mean of the observed numbers was $\bar{x} = 18.9$. From Fig. 19.3 (bottom left panel), a sample mean with this value does not look unusual at all; it would occur reasonably frequently. The evidence does not suggest a problem with the wheel.

As we have seen, each sample is likely to be different, so *any* statistic is likely to vary from sample to sample. (The value of the *population* parameter does not change.) This variation in the possible values of the observed sampling statistic is called *sampling variation.*

19.4 Sampling means and standard errors

As seen in the previous two sections, the value of a sample statistic varies from sample to sample. The value of the sample statistic that is observed depends on which one of the countless samples happens to be observed

The possible values of the statistic that we could potentially observe have a *distribution* (specifically, a *sampling distribution*); see Fig. 19.4. The *mean* of this sampling distribution is called the *sampling mean.* The sampling mean is the average value of all possible values of the statistic. Not all sampling distributions have a bell-shaped distribution.

Definition 19.3 (Sampling mean). The *sampling mean* is the mean of the sampling distribution of a statistic: the mean of the values of the statistic from all possible samples.

The *standard deviation* of this sampling distribution is called the *standard error.* The standard error measures how the value of the statistic varies across all possible values of the statistic; see Fig. 19.4. The standard error is a measure of how precisely the *sample* statistic estimates the *population* parameter. If every possible sample (of a given size) was found, and the statistic computed from each sample, the standard deviation of all these estimates is the *standard error.*

Definition 19.4 (Standard error). A *standard error* is the standard deviation of the sampling distribution of a statistic: the standard deviation of the values of the statistic from all possible samples.

Figures 19.2 and 19.3 show that the variation in the values of the statistic get smaller for larger sample sizes. That is, the standard error gets *smaller* as the sample sizes get *larger*: sample statistics show less variation for larger n. This makes sense: *larger* samples generally produce more precise estimates. After all, that's the advantage of larger samples: all else being equal, larger samples produce more precise estimates (Sect. 6.3).

Example 19.3 (Standard errors). In Fig. 19.3, a sample of 250 (i.e., 250 spins) is unlikely to produce a sample mean larger than 20, or smaller than 15. However, in a sample of size 15 (i.e., 15 spins) sample means near 15 and 20 are quite commonplace.

In samples of size 100, the variation in the mean spin is smaller than in samples of size 15. Hence, the *standard error* (the standard deviation of the sampling distributions) will be smaller for samples of size 250 than for samples of size 15.

For many sample statistics, *the variation from sample to sample can be approximately described by a bell-shaped (normal) distribution* (the *sampling distribution*) if certain conditions are met. Furthermore, the *standard deviation of this sampling distribution is called the standard error*. The standard error is a special name given to the *standard deviation* that describes the variation in the possible values of a statistic.

ⓘ 'Standard error' is an unfortunate label: it is not an *error*, or even *standard*. (For example, there is no such thing as a '*non*-standard error'.)

19.5 Standard deviation and standard error

Even experienced researchers confuse the meaning and the usage of the terms *standard deviation* and *standard error* [Ko et al., 2014]. Understanding the difference is important.

The *standard deviation*, in general, quantifies the amount of variation in any quantity that varies. The *standard error* only ever refers to the standard deviation that describes a sampling distribution.

Typically, in a research study, *standard deviations* describe the variation in the individuals in a sample: how observations vary from *individual to individual*. The *standard error* is only used to describe how *sample estimates* vary from sample to sample (i.e., to describe the precision of sample estimates).

The standard error *is* a standard deviation, but specifically describes the variation in sampling distributions. *Any* numerical quantity estimated from a sample (a *statistic*) can vary from sample to sample, and so has sampling variation, a sampling distribution, and hence a standard error. (Not all sampling distributions are *normal* distributions, however.)

⚠ *Any* quantity estimated from a sample (a statistic) has a standard error.

> ⓘ The *standard error* is often abbreviated to 'SE' or 's.e.'. For example, the 'standard error of the sample mean' is usually written s.e.($\bar{x}$), and the 'standard error of the sample proportion' is usually written s.e.($\hat{p}$).

Parameters do not vary from sample to sample, so *do not* have a sampling distribution (or a standard error).

19.6 Chapter summary

A *sampling distribution* describes how all possible values of a statistic vary from sample to sample. Under certain circumstances, the sampling distribution often can be described by a *bell-shaped (or normal) distribution*. The standard deviation of this normal distribution is called a *standard error*. The standard error is the name specifically given to the standard deviation that describes the variation in the statistic *across all possible samples*.

19.7 Quick review questions

Are the following statements *true* or *false*?

1. The phrase 'the standard error of the population proportion' is illogical.
2. The sample size *does not* have a standard error?
3. Sampling variation refers to how sample sizes vary.
4. Sampling distributions describe how parameters vary.
5. Statistics do not vary from sample to sample.
6. Populations are numerically summarised using parameters
7. The *standard deviation* is a type of *standard error* in a specific situation.
8. Sampling distributions are always *normal* distributions.
9. Sampling variation measures the amount of variation in the individuals in the sample.
10. The standard error measures the size of the error when we use a sample to estimate a population.
11. In general, a standard deviation measures the amount of variation.

19.8 Exercises

Answers to odd-numbered exercises are given at the end of the book.

Exercise 19.1. In the following scenarios, would a *standard deviation* or a *standard error* be the appropriate way to measure the amount of variation? Explain.

1. Researchers are studying the spending habits of customers. They would like to measure the variation in the amount spent by shoppers per transaction at a supermarket.
2. Researchers are studying the time it takes for inner-city office workers to travel to work each morning. They would like to determine the precision with which their estimate (a mean of 47 mins) has been measured.

Exercise 19.2. In the following scenarios, would a *standard deviation* or a *standard error* be the appropriate way to measure the amount of variation? Explain.

1. A study examined the effect of taking a pain-relieving drug on children. The researchers want to describe the sample they used in the study, including a description of how the ages of the children in the study vary.
2. A study estimated the proportion of children aged under 14 who owned a mobile (cell) phone. The researchers want to report this estimate, indicating the precision of that estimate.

Exercise 19.3. Which of the following have a *standard error*?

1. The population proportion.
2. The sample median.
3. The sample IQR.

Exercise 19.4. Which of the following have a *standard error*?

1. The sample standard deviation.
2. The population odds.
3. The sample odds ratio.

Exercise 19.5. Consider spinning a European roulette wheel.

1. Suppose the wheel was spun 15 times (Fig. 19.3, top left panel), and the mean spin was 22.1. What would you conclude about the wheel?
2. Suppose the wheel was spun 250 times (Fig. 19.3, bottom right panel), and the mean spin was 22.1. What would you conclude about the wheel?
3. Suppose the wheel was spun 50 times (Fig. 19.3, top right panel), and the mean spin was 22.1. What would you conclude about the wheel?
4. Suppose the wheel was spun 50 times (Fig. 19.3, top right panel), and the mean spin was 24.0. What would you conclude about the wheel?

Exercise 19.6. Consider spinning a European roulette wheel.

1. Suppose the wheel was spun 15 times (Fig. 19.2, top left panel), and the proportion of spins showing an odd number was 0.44. What would you conclude about the wheel?
2. Suppose the wheel was spun 15 times (Fig. 19.2, top left panel), and the proportion of spins showing an odd number was 0.13. What would you conclude about the wheel?
3. Suppose the wheel was spun 15 times (Fig. 19.2, top left panel), and the proportion of spins showing an odd number was 0.65. What would you conclude about the wheel?
4. Suppose the wheel was spun 200 times (Fig. 19.2, bottom right panel), and the proportion of spins showing an odd number was 0.65. What would you conclude about the wheel?

Exercise 19.7. A research article [Nagele, 2003] made this statement:

> ... authors often [incorrectly] use the standard error of the mean (SEM) to describe the variability of their sample...

What is wrong with using the standard error of the mean to describe the sample? How would you explain the difference between the *standard error* and the *standard deviation*?

Answers to *Quick review* questions: 1. True. **2.** True. **3.** False. **4.** False. **5.** False. **6.** True. **7.** False. **8.** False. **9.** False. **10.** False. **11.** True.

20

Models and normal distributions

So far, you have learnt to ask an RQ, design a study, describe and summarise the data, and understand sampling variation. **In this chapter**, you will learn to:

- describe and draw normal distributions.
- use z-scores to compute probabilities related to normal distributions.
- work 'backwards' from probabilities for normal distributions.

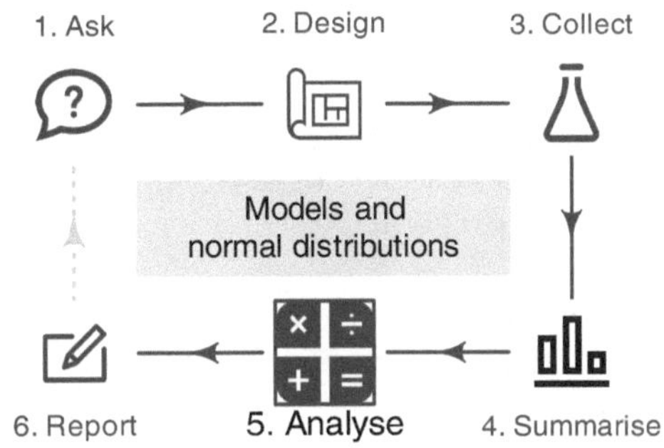

20.1 Introduction

As seen in Chap. 19, many different samples could be drawn from a population, and the value of the statistic varies from sample to sample. The challenge of research is that only one of these countless possible samples is observed. The distribution of possible values of the statistic that could be observed from all possible samples is a *sampling distribution.*

⚠ Remember: studying a sample leads to the following observations:

- every sample is likely to be different.
- we observe just one of the many possible samples.
- every sample is likely to yield a different value for the statistic.
- we observe just one of the many possible values for the statistic.

Since many values for the statistic are possible, the possible values of the statistic vary (called *sampling variation*) and have a *distribution* (called a *sampling distribution*).

As seen in Chap. 19, sampling distributions often have a *normal distribution* (or bell-shaped distribution). That is, the normal distribution is often used to describe the *sampling distribution.* We now study normal distributions, as they appear in many places in research.

20.2 Normal distributions: examples

In Chap. 19, we saw that the proportion of odd spins in 15 spins of a roulette wheel could vary; similarly, the mean spin from 15 spins could vary (Fig. 20.1). In both cases, these sampling distributions had a rough *normal distribution* shape. This is true for larger numbers of spins also (Figs. 19.2 and 19.3).

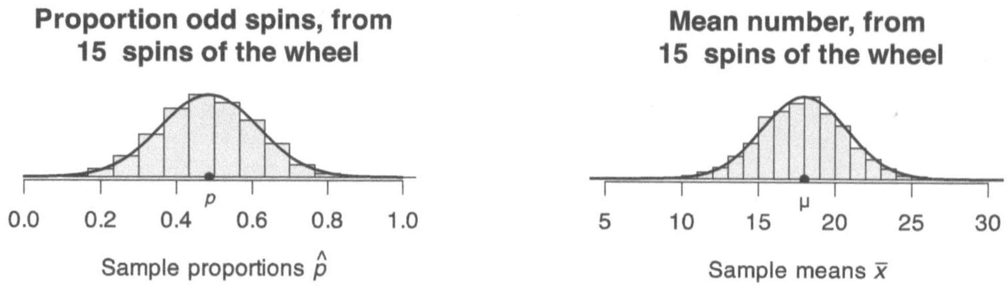

FIGURE 20.1: Sampling distributions for the proportion of odd spins (left), and the mean of the numbers after 15 roulette wheel spins (right) are approximate normal distributions. The solid lines are theoretical normal distributions.

The *histograms* in Fig. 20.1 are based on results from a limited number of simulations. The solid lines shown in Fig. 20.1 are actual *normal distributions*, and represent how the histogram might appear theoretically if we used an infinite number of simulations. The normal distributions are *models* for what might occur in the *population*, so normal distributions are also called *normal models*. Since the models represent *populations*, the mean of the model is denoted μ and the standard deviation is denoted σ.

A *model* is a theoretical or ideal concept. A model skeleton isn't 100% accurate and certainly not exactly like *your* skeleton; nonetheless, it suitably approximates reality. None of us probably have a skeleton *exactly* like the model, but the model is still useful and helpful. Likewise, a sampling distribution may not have *exactly* a normal shape, but the model is still useful and helpful. The model is a way of describing a *theoretical* distribution in the population. A model is a simple (but not overly simple) approximation to reality.

The histograms in Fig. 20.1 are not *exactly* normal distributions, but are very close to normal distributions, and certainly close enough for most purposes. Many, but not all, sampling distributions have approximate normal distributions.

Sampling distributions represent theoretical distributions of sample *statistics*, not the distribution of sample *data*. When the sampling distribution is a normal distribution, the mean of the distribution is called the *sampling mean* and the standard deviation is called the *standard error*.

Apart from their use in modelling theoretical sampling distributions, some quantitative variables have approximate normal distributions too, when the distribution of the data in the *population* can be approximately modelled by a normal distribution.

Example 20.1 (Normal distributions of data)**.** Some quantitative variables have approximate normal distributions. Figure 20.2 (left panel) shows the diastolic blood pressure of 398 Americans [Willems et al., 1997, Schorling et al., 1997]. Figure 20.2 (right panel) shows the weight of 83 male Leadbeater's possums [Williams et al., 2022].

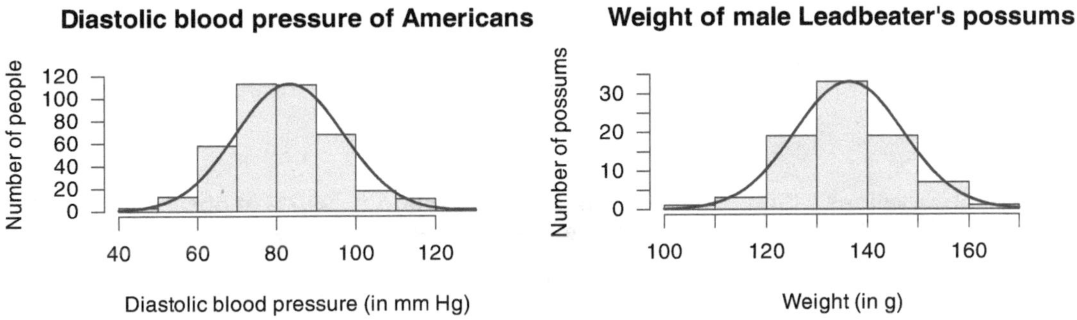

FIGURE 20.2: Two normal distributions. Left: diastolic blood pressure of a sample of 398 Americans. Right: the weight of a sample of 83 male Leadbeater's possums. The solid lines are the approximate normal model for the variable in the population.

20.3 Normal distributions and the 68–95–99.7 rule

Normal distributions have a shape that is symmetric about the mean, with a bell shape. Half the values are greater than the mean, and half the values are less than the mean. The total probability represented by a normal distribution is one (or 100%). For example, every sample will produce a sample proportion between 0 and 1 and so is represented somewhere in Fig. 20.1 (left panel); every American has a diastolic blood pressure and so is represented somewhere in Fig. 20.2 (left panel); every male Leadbeater's possum has a weight and so is represented somewhere in Fig. 20.2 (right panel).

In theory, no upper limit or lower limit exists for a variable modelled using a normal distribution. In practice, this is rarely true, but usually never presents a problem. Consider the normal distributions in Fig. 20.2, for example. The normal distribution shown for the diastolic blood pressure (left panel) has no lower or upper limit in theory, but all practical values of diastolic blood pressure are captured by that part of the normal distribution shown. The normal distribution implies almost no-one has a diastolic blood pressure below 40 mm Hg or above 130 mm Hg.

One of the most important properties of normal distributions is the *68–95–99.7 rule* (sometimes called the *empirical rule*).

Definition 20.1 (The 68–95–99.7 rule). For any quantity modelled by a normal distribution:

- *approximately* 68% of values lie within 1 standard deviation of the mean.
- *approximately* 95% of values lie within 2 standard deviations of the mean.
- *approximately* 99.7% of values lie within 3 standard deviations of the mean.

These properties are true for *all* normal distributions, whatever the quantity, whatever the value of the mean, and whatever the value of the standard deviation (Fig. 20.3).

Example 20.2 (Heights of females). Suppose the heights of Australian adult females in the population can be *modelled* with a normal distribution having a mean of $\mu = 162$ cm, and a standard deviation of $\sigma = 7$ cm, and follow a normal distribution (Fig. 20.4). Using the 68–95–99.7 rule, approximately 68% of Australian women will be between $162 - 7 = 155$ cm and $162 + 7 = 169$ cm tall using this model. Similarly, approximately 95% of Australian women will be between $162 - (2 \times 7) = 148$ cm and $162 + (2 \times 7) = 176$ cm tall.

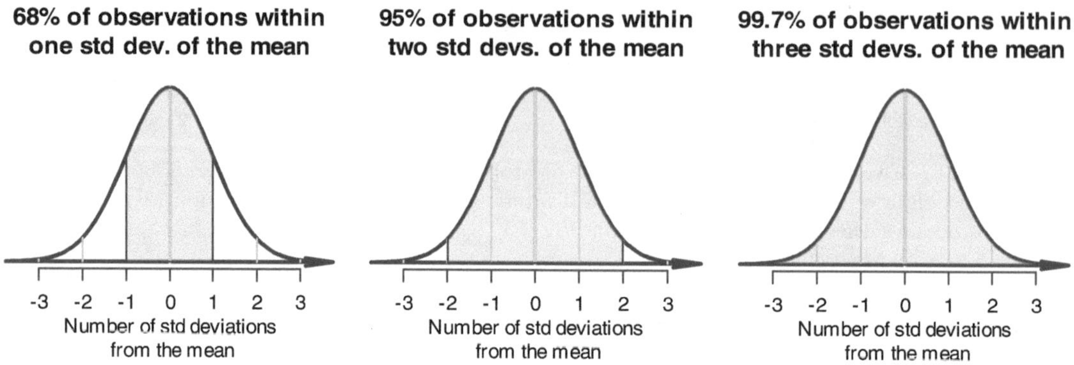

FIGURE 20.3: The 68–95–99.7 rule. The shaded regions correspond to the central 68%, 95% and 99.7%.

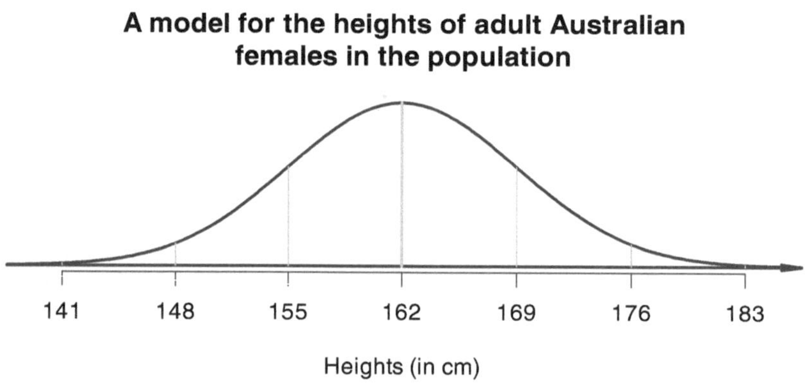

FIGURE 20.4: A model for the height of adult Australian females in the population.

These regions under the normal curve are probabilities, are often called areas, and are sometimes expressed as percentages.

20.4 Standardising (z-scores)

Since the 68–95–99.7 rule (Def. 20.1) applies for all normal distributions, the percentages in the rule only depend on how many standard deviations (σ) a value (x) is from the mean (μ). This information can be used to learn more about how values are distributed in a normal distribution.

For example, suppose heights of Australian adult females can be modelled with a normal distribution having a mean of $\mu = 162$ cm, and a standard deviation of $\sigma = 7$ cm (Example 20.2). Using this model, the proportion of Australian adult women *taller* than 169 cm can be determined.

From a picture (Fig. 20.5, left panel), $162 + 7 = 169$ cm is one standard deviation *above* the mean. Since 68% of values are within one standard deviation of the mean, 32% are outside that range (some shorter; some taller). Hence, 16% are taller than one standard deviation above the mean, so the answer is about 16%. (Another 16% are shorter than one standard deviation *below* the mean, or less than $162 - 7 = 155$ cm in height.)

Again, the percentages only depend on how many standard deviations (σ) the value (x) is from the mean (μ), and not the actual values of μ and σ.

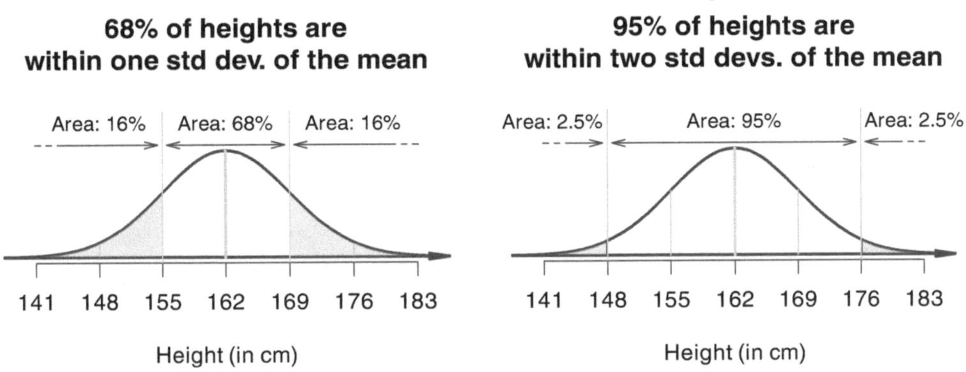

FIGURE 20.5: Left: what proportion of Australian adult females are taller than 169 cm? Right: what proportion of Australian adult females are shorter than 148 cm?

Example 20.3 (The 68–95–99.7 rule). Consider again the heights of Australian adult females. Using this model, what proportion are *shorter* than 148 cm?

Again, drawing a picture is helpful (Fig. 20.5, right panel). Since $162 - (2 \times 7) = 148$, 148 cm is two standard deviations *below* the mean. Since 95% of values are within two standard deviation of the mean, 5% are outside that range (half smaller, half larger; see Fig. 20.5, right panel), so that 2.5% are *shorter* than 148 cm. (Another 2.5% are *taller* than $162 + 14 = 176$ cm.)

Again, the percentages only depend on how many standard deviations (σ) the value (x) is from the mean (μ). The number of standard deviations that an observation is from the mean is called a *z-score*. A z-score is computed using

$$z = \frac{x - \mu}{\sigma},$$

where σ is the standard deviation quantifying the variation in the x-values. Converting values to z-scores is called *standardising*.

Definition 20.2 (z-score). A *z-score* measures how many standard deviations a value x is from the mean. In symbols:

$$z = \frac{x - \mu}{\sigma}, \tag{20.1}$$

where μ is the mean of the distribution, and σ is the standard deviation of the distribution (measuring the variation in the x-values).

The z-score is also called the *standardised value* or *standard score*. Note that:

- z-scores are negative for observations *below* the mean.
- z-scores are positive for observations *above* the mean.
- z-scores have no units (that is, not measured in kg, or cm, etc.).

Example 20.4 (z-scores). Consider the model for the heights of Australian adult females again. From earlier, the z-score for a height of 169 cm is

$$z = \frac{x - \mu}{\sigma} = \frac{169 - 162}{7} = 1,$$

one standard deviation *above* the mean. Similarly, the z-score for a height of 148 cm is

$$z = \frac{x - \mu}{\sigma} = \frac{148 - 162}{7} = -2,$$

two standard deviations *below* the mean.

Example 20.5 (The 68–95–99.7 rule). Consider the model for the heights of Australian adult females: a normal distribution, mean $\mu = 162$ cm, standard deviation $\sigma = 7$ cm (Fig. 20.6). Using this model:

- a height of 162 cm is zero standard deviations from the mean: $z = 0$.
- 155 cm is one standard deviation *below* the mean: $z = -1$.
- 169 cm is one standard deviation *above* the mean: $z = 1$.
- 148 cm and 176 cm correspond to $z = -2$ and $z = 2$ respectively.
- 141 cm and 183 cm correspond to $z = -3$ and $z = 3$ respectively.

20.5 Approximating areas (percentages) using the 68–95–99.7 rule

As seen above, the 68–95–99.7 rule can be used to approximate percentages under normal distributions. The rule can even be used for values that do not exactly align with 1, 2 or 3 standard deviations from the mean.

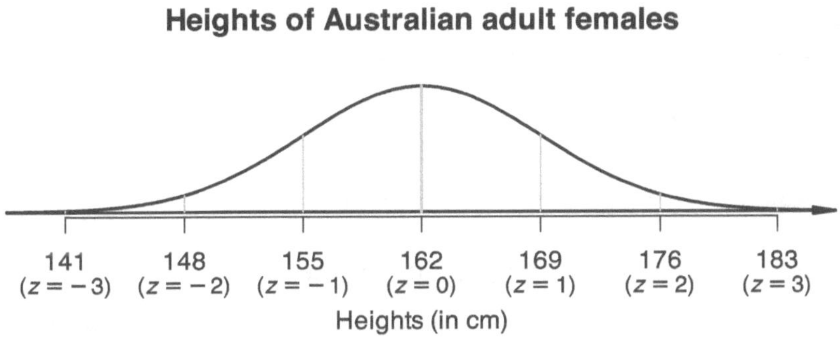

FIGURE 20.6: The 68–95–99.7 rule and the heights of Australian adult females.

Suppose again that heights of Australian adult females can be modelled with a normal distribution with a mean of $\mu = 162$ cm, and a standard deviation of $\sigma = 7$ cm (Fig. 20.6). To find the proportion of women *shorter* than 145 cm, first draw the situation (Fig. 20.7). Proceeding as before, we ask 'How many standard deviations from the mean is 145 cm?' Using Equation (20.1), 145 cm corresponds to a z-score of

$$z = \frac{145 - 162}{7} = -2.4285... \qquad (20.2)$$

which is about 2.43 standard deviations *below* the mean.

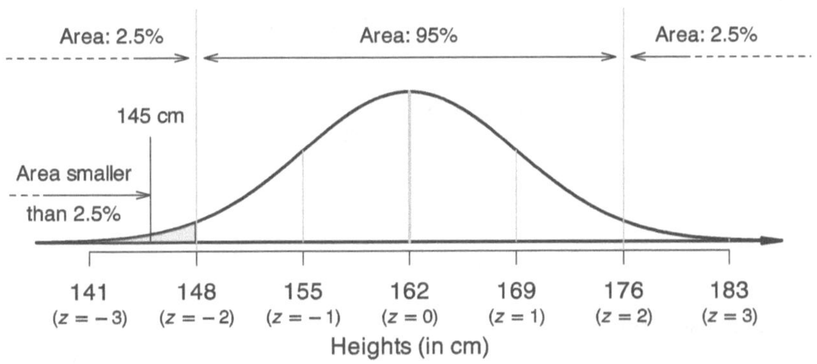

FIGURE 20.7: What proportion of Australian adult females are shorter than 145 cm?

What percentage of observations are less than this z-score? This case is not covered by the 68–95–99.7 rule, though the rule can be used to make *rough estimates*.

About 2.5% of observations are less than 2 standard deviations below the mean; that is, about 2.5% of women are shorter than 148 cm. So the percentage of females shorter than 145 cm (that is, even shorter than 148 cm, and so further into the tail of the distribution) will be *smaller* than 2.5%. While we don't know the probability exactly, it will be smaller than 2.5%.

Percentages found this way are very approximate, but often sufficient. However, more accurate percentages are found using tables compiled for this very purpose (Appendices B.1 and B.2). We now learn to use these tables.

20.6 Exact areas (percentages) using tables

Areas under normal distributions can be found using online tables, or hard copy tables, for any z-score. The online tables are easier to use, but only the hard-copy tables are explained in this book (see the online version of this book for the online tables, and instructions for using the online tables). The tables (Appendices B.1 and B.2) work with z-scores to two decimal places, so consider the z-score from Sect. 20.5 as $z = -2.43$.

Using Appendix B.1, find -2.4 in the *left* margin of the table (Fig. 20.8), and find the second decimal place (in this case, 3) in the *top* margin of the table: where these intersect is the area (or probability) *less than* the z-score of -2.43; that is, the probability of finding a z-score less than $z = -2.43$ is 0.0075, or about 0.75%.

	0.00	0.01	0.02	0.03	0.04
−3.5	0.0002	0.0002	0.0002	0.0002	0.0002
−3.4	0.0003	0.0003	0.0003	0.0003	0.0003
−3.3	0.0005	0.0005	0.0005	0.0004	0.0004
−3.2	0.0007	0.0007	0.0006	0.0006	0.0006
−3.1	0.0010	0.0009	0.0009	0.0009	0.0008
−3.0	0.0013	0.0013	0.0013	0.0012	0.0012
−2.9	0.0019	0.0018	0.0018	0.0017	0.0016
−2.8	0.0026	0.0025	0.0024	0.0023	0.0023
−2.7	0.0035	0.0034	0.0033	0.0032	0.0031
−2.6	0.0047	0.0045	0.0044	0.0043	0.0041
−2.5	0.0062	0.0060	0.0059	[illegible]	0.0055
−2.4	0.0082	0.0080	[illegible]	0.0075	0.0073
−2.3	0.0107	0.0104	0.0102	0.0099	0.0096

FIGURE 20.8: Using the z-score tables. When $z = -2.43$, the area to the left is 0.0075.

⚠ Our tables always give the area to the *left* of the z-score.

Either the hard-copy or online tables gives an answer of 0.75%. This is consistent with the rough answer using the 68–95–99.7 rule: a value less than 2.5%.

20.7 Examples using z-scores

The general approach to computing probabilities from normal distributions is:

- *draw a diagram*, and mark on the value(s) of interest.
- *shade* the required region of interest.
- *compute* the z-score(s) using Equation (20.1).
- *use* the tables in Appendices B.1 and B.2 to compute corresponding areas (percentages).
- *deduce* the answer.

This approach can be used to answer more complicated questions involving normal distributions.

Example 20.6 (Normal distributions). Mechanised forest harvesting systems were simulated by Aedo-Ortiz et al. [1997], and the diameters of a specific type of tree were modelled using:

- a normal distribution, with
- a mean of $\mu = 8.8$ inches, and
- a standard deviation of $\sigma = 2.7$ inches.

Using this model, what is the probability that a randomly-chosen tree has a diameter *greater* than 5 inches?

Following the steps identified earlier:

- *draw* the appropriate normal curve, and mark on 5 inches (Fig. 20.9, left panel).
- *shade* the region 'greater than 5 inches' (Fig. 20.9, centre panel).
- *compute* the z-score using Equation (20.1): $z = (5 - 8.8)/2.7 = -1.41$ to two decimal places.
- *use* tables: the probability of a tree diameter *shorter* than 5 inches is 0.0793. (Remember: the tables always give area *less* than the value of z.)
- *deduce* the answer (Fig. 20.9, right panel): since the *total* area under the normal distribution is one (or 100%), the probability of a tree diameter *greater* than 5 inches is $1 - 0.0793 = 0.9207$, or about 92%.

A randomly-chosen tree has a probability of 92% of having a diameter *greater* than 5 inches.

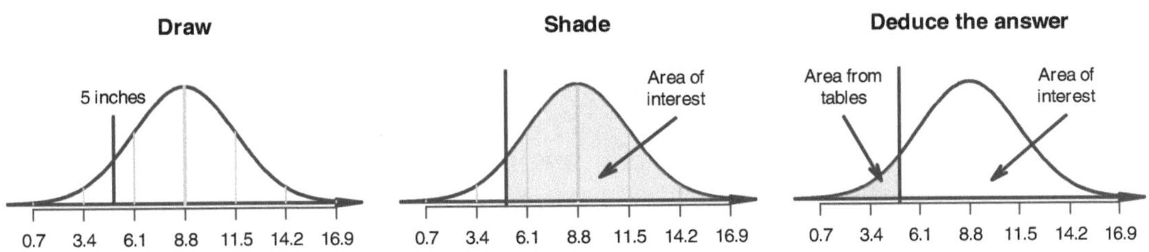

FIGURE 20.9: What proportion of tree diameters are greater than 5 inches?

⚠ Our normal-distribution tables *always* provide area to the *left* of the z-score. Drawing a picture of the situation is important: it helps visualise getting the area requested from the area the tables provide. Remember: the *total* area under the normal distribution is one (or 100%).

Example 20.7 (Normal distributions). These scenarios can be displayed on a diagram as shown in Fig. 20.10 (recall $\mu = 8.8$ inches).

1. Tree diameters between 3 and 5 inches: Diagram A.
2. Tree diameters greater than 11 inches: Diagram B.
3. Tree diameters between 5 and 11 inches: Diagram C.
4. Tree diameters less than 11 inches: Diagram D.

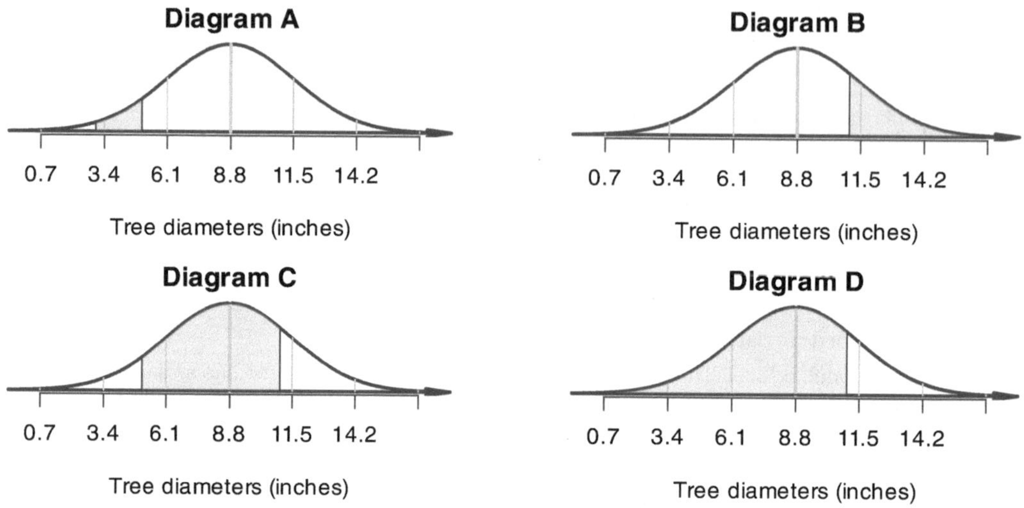

FIGURE 20.10: Scenarios with their corresponding diagrams.

Example 20.8 (Normal distributions). Using the model for tree diameters in Example 20.6, what is the probability that a tree has a diameter *between* 5 and 11 inches?

First, *draw* the situation, and *shade* 'between 5 and 10 inches' (Fig. 20.10, Diagram C). Then, *compute* the z-scores for *both* tree diameters:

- For 5 inches: $z = (5 - 8.8)/2.7 = -1.41$ (i.e., below the mean).
- For 11 inches: $z = (11 - 8.8)/2.7 = 0.81$ (i.e., above the mean).

The tables in Appendices B.1 and B.2 can then be used to find the area to the *left* of $z = -1.41$ (which is 0.0793), and also to find the area to the *left* of $z = 0.81$ (which is 0.791). However, neither of these provide the area *between* $z = -1.41$ and $z = 0.81$.

Looking carefully at the areas from the tables and the area sought, the required area is the *area* between the two z-scores (Fig. 20.11): $0.7910 - 0.0793 = 0.7117$. The probability that a tree has a diameter between 5 and 11 inches is about 0.7117, or about 71%.

20.8 Unstandardising: working backwards

Using the model for tree diameters in Example 20.6 again, different types of questions can be asked too. Suppose we needed to identify the diameters of the *smallest* 3% of trees.

This is a different type of problem than before; previously, the *tree diameter* was known, so a z-score could be computed, and hence a probability (Fig. 20.12). However, here the

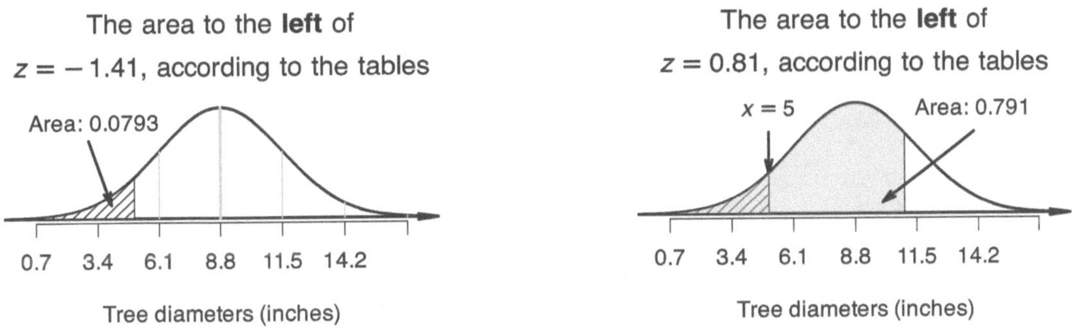

FIGURE 20.11: What proportion of tree diameters are between 5 and 11 inches? Left: the hatched area is the area to the left of $z = -1.41$. Right: the shaded area is the area to the left of $z = 0.81$. Neither give us the area we seek directly.

probability is known, and a tree diameter is sought. That is, working 'backwards' is necessary (Fig. 20.12), so the z-tables need to be used 'backwards' too.

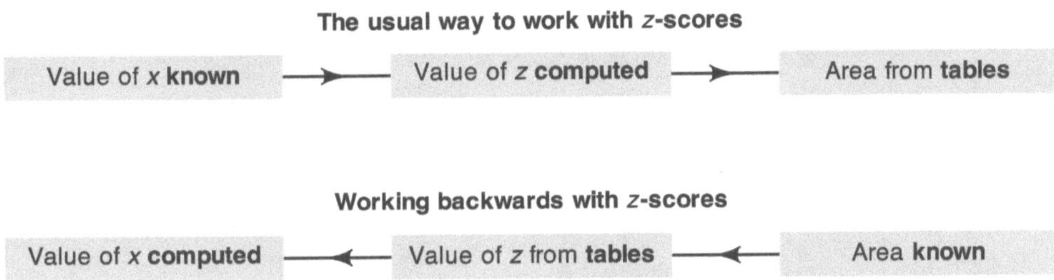

FIGURE 20.12: Working with z-scores. In the tables, the areas (probabilities) are in the body of the table, and the z-scores are in the margins of the table.

Drawing a rough diagram of the situation again is very helpful (Fig. 20.13). We can only mark the approximate location of the required score, but this is sufficient. Then, tables must be used to determine the corresponding z-score. Since the required value will be smaller than the mean, the z-score will be negative (to the *left* of the mean).

As before (Sect. 20.6), online tables or hard copy tables can be used (and again the online tables are easier to use). Only the hard-copy tables are explained in this book (see the online version of this book for the online tables, and instructions for their use).

When the z-scores (in the *margins* of the tables in Appendices B.1 and B.2) were known, the *areas* were found in the *body* of the table. If the area, or probability (in the *body* of the table) is known, the corresponding z-score can be found (in the *margins* of the table). Using the hard copy tables (Fig. 20.14), locate the area of 0.0300 (or as close as possible) in the *body* of the table, then read the z-score from the margins of the table.

Using hard copy tables, the closest value in the *body* of the table to 0.0300 (or 3%) is 0.0301. This corresponds to a z-score of $z = -1.88$ (from the *margins* of the table). Sometimes, the exact area can be found in the body of the table, but often the closest value in the body of the table must be used. (The online tables give a slightly more precise value of $z = -1.881$.)

⚠ Our tables always give the area to the *left* of the z-score.

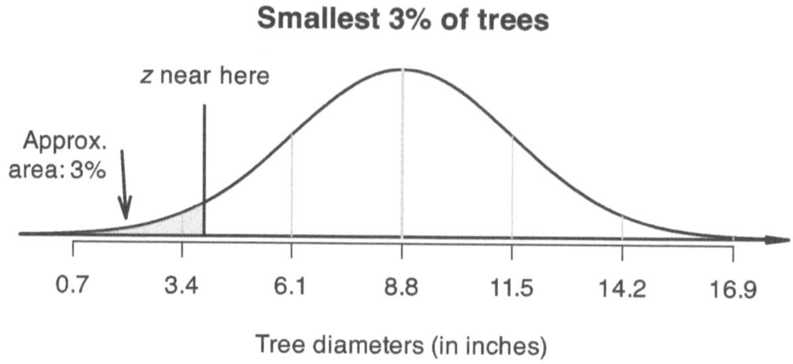

FIGURE 20.13: Tree diameters: the smallest 3% is shaded. The approximate location of the required z-score is drawn.

	0.00	0.01	0.02	0.03	0.04	0.05	0.06	0.07	0.08	0.09
−3.5	0.0002	0.0002	0.0002	0.0002	0.0002	0.0002	0.0002	0.0002	0.0002	0.0002
−3.4	0.0003	0.0003	0.0003	0.0003	0.0003	0.0003	0.0003	0.0003	0.0003	0.0002
−3.3	0.0005	0.0005	0.0005	0.0004	0.0004	0.0004	0.0004	0.0004	0.0004	0.0003
−3.2	0.0007	0.0007	0.0006	0.0006	0.0006	0.0006	0.0006	0.0005	0.0005	0.0005
−3.1	0.0010	0.0009	0.0009	0.0009	0.0008	0.0008	0.0008	0.0008	0.0007	0.0007
−3.0	0.0013	0.0013	0.0013	0.0012	0.0012	0.0011	0.0011	0.0011	0.0010	0.0010
−2.9	0.0019	0.0018	0.0018	0.0017	0.0016	0.0016	0.0015	0.0015	0.0014	0.0014
−2.8	0.0026	0.0025	0.0024	0.0023	0.0023	0.0022	0.0021	0.0021	0.0020	0.0019
−2.7	0.0035	0.0034	0.0033	0.0032	0.0031	0.0030	0.0029	0.0028	0.0027	0.0026
−2.6	0.0047	0.0045	0.0044	0.0043	0.0041	0.0040	0.0039	0.0038	0.0037	0.0036
−2.5	0.0062	0.0060	0.0059	0.0057	0.0055	0.0054	0.0052	0.0051	0.0049	0.0048
−2.4	0.0082	0.0080	0.0078	0.0075	0.0073	0.0071	0.0069	0.0068	0.0066	0.0064
−2.3	0.0107	0.0104	0.0102	0.0099	0.0096	0.0094	0.0091	0.0089	0.0087	0.0084
−2.2	0.0139	0.0136	0.0132	0.0129	0.0125	0.0122	0.0119	0.0116	0.0113	0.0110
−2.1	0.0179	0.0174	0.0170	0.0166	0.0162	0.0158	0.0154	0.0150	0.0146	0.0143
−2.0	0.0228	0.0222	0.0217	0.0212	0.0207	0.0202	0.0197	0.0192	0.0188	0.0183
−1.9	0.0287	0.0281	0.0274	0.0268	0.0262	0.0256	0.0250	0.0244	0.0239	0.0233
−1.8	0.0359	0.0351	0.0344	0.0336	0.0329	0.0322	0.0314	0.0307	0.0301	0.0294
−1.7	0.0446	0.0436	0.0427	0.0418	0.0409	0.0401	0.0392	0.0384	0.0375	0.0367
−1.6	0.0548	0.0537	0.0526	0.0516	0.0505	0.0495	0.0485	0.0475	0.0465	0.0455

FIGURE 20.14: Using the z-tables backwards. When $z = -1.88$, the area to the left is 0.0301, which is the closest we can get to 0.03 (or 3%).

Using either the hard-copy or online tables, the appropriate z-value is about -1.88 standard deviations *below* the mean; that is, $z = -1.88$ (Fig. 20.13). The z-score can be converted to an observation value x using the *unstandardising* formula:[1]

$$x = \mu + z\sigma.$$

Using this unstandardising formula:

$$\begin{aligned} x &= \mu + (z \times \sigma) \\ &= 8.8 + (-1.88 \times 2.7) = 3.724; \end{aligned}$$

that is, about 3% of trees have diameters less than about 3.72 inches.

Definition 20.3 (Unstandardising formula)**.** When the z-score is known, the corresponding value of the observation x is

$$x = \mu + z\sigma. \qquad (20.3)$$

This is called the *unstandardising formula.*

Example 20.9 (Normal distributions backwards)**.** Using the model for tree diameters in Example 20.6 again, suppose now the diameters of the *largest* 25% of trees needs to be identified.

The situation can be drawn (Fig. 20.15). Since an area is given, we need to work 'backwards', so the z-tables need to be used 'backwards' too. The *largest* 25% implies large trees, so required diameter is larger than the mean (so corresponds to a positive z-score).

The tables work with the area to the *left* of the value of interest, which is 75% (Fig. 20.15). Using either the hard-copy or online tables, the appropriate z-value is $z = 0.674$. Then, the z-score can be converted to an observation value x using the *unstandardising* formula:

$$\begin{aligned} x &= \mu + (z \times \sigma) \\ &= 8.8 + (0.674 \times 2.7) = 10.621. \end{aligned}$$

That is, about 25% of trees have diameters larger than about 10.6 inches.

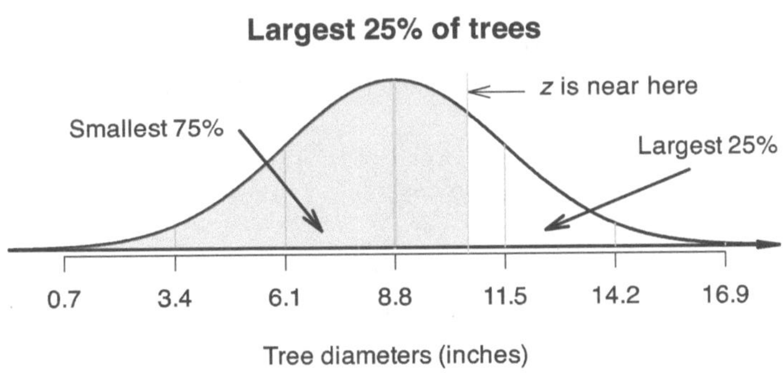

FIGURE 20.15: Tree diameters: the largest 25% is the same as the smallest 75%.

[1]This is found by re-arranging Equation (20.1).

20.9 Example: methane production

Huhtanen et al. [2016] modelled the retention time of food in sheep, using a normal distribution with the mean retention time as $\mu = 42.5$ h, and the standard deviation as $\sigma = 3.68$ h. We can draw this normal distribution (Fig. 20.16), and then apply the 68–95–99.7 rule:

- about 68% of retention times are between 38.82 and 46.18 h.
- about 95% of retention times are between 35.14 and 49.86 h.
- about 99.7% of retention times are between 31.46 and 53.54 h.

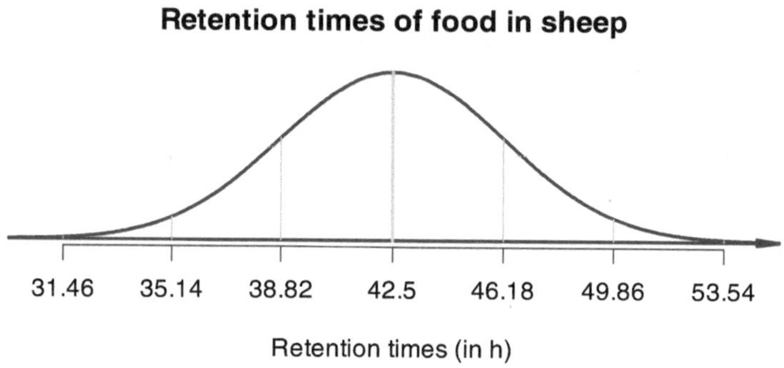

FIGURE 20.16: Retention times of food in sheep.

Example 20.10 (Working with the normal distribution). Using this model, what proportion of sheep have a retention time *less than* 40 h?

A retention time of 40 h corresponds to a z-score of (Fig. 20.17, top left panel):

$$z = \frac{40 - 42.5}{3.68} = -0.68.$$

This is a *negative* number, since 40 h is *below* the mean. Using the tables in Appendices B.1 and B.2 (that give the *area to the left* of the z-score), the area to the left of $z = -0.68$ is 0.2483, or about 24.8%. About 24.8% of sheep have a retention time *less* than 40 h.

Example 20.11 (Working with the normal distribution). What proportion of sheep have a retention time *greater than* 48 h (two days)?

A retention time of 48 h corresponds to a z-score of 1.49. Using the normal distribution tables, the area to the *left* of this z-score is 0.9319, so the area to the *right* of this z-score is 0.0681 (Fig. 20.17, top right panel).

Example 20.12 (Working with the normal distribution). What proportion of sheep have a retention time *between* 40 and 48 h?

A retention time of 40 h corresponds to $z = -0.68$ and, using the normal distribution tables, the area to the *left* of $z = -0.68$ is 0.2483 (Fig. 20.17, bottom left panel; hatched area). But this is not the area that we seek. From earlier, the area to the *left* of $z = 1.49$ is 0.9319 (Fig. 20.17, bottom left panel; shaded region). But this is not the area we seek either. From

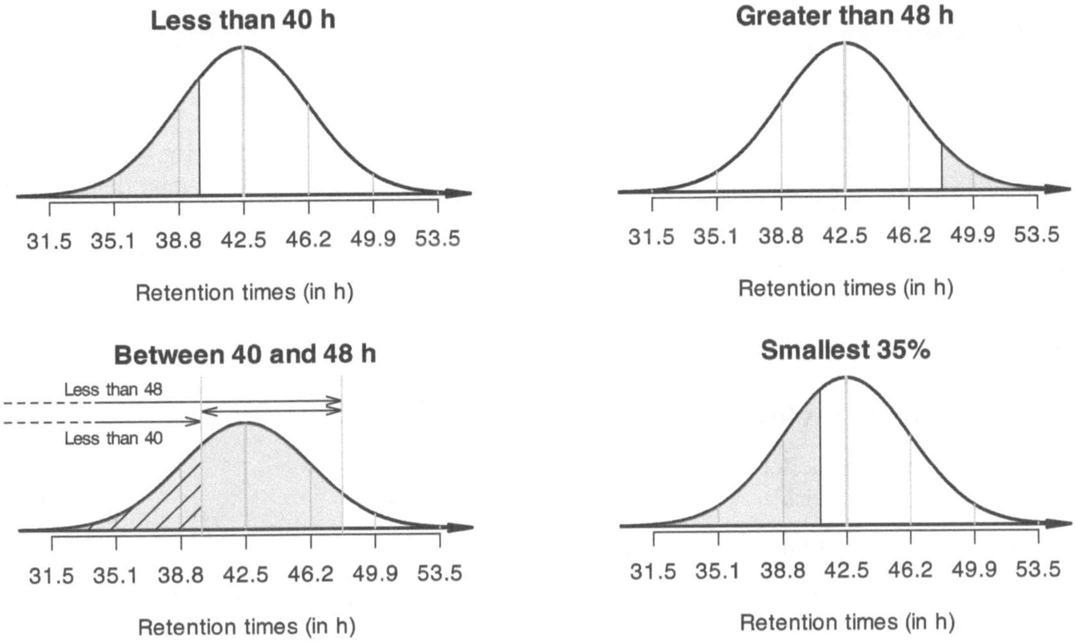

FIGURE 20.17: Plots for retention times in sheep.

the two areas that we know, we *can* find the area that we seek (Fig. 20.17, bottom left panel):

- 48 h corresponds to $z = 1.49$; the area to the *left* of this z-score is 0.9319.
- 40 h corresponds to $z = -0.68$; the area to the *left* of this z-score is 0.2483.
- the *difference* between these two *areas* is sought, which is $0.9319 - 0.2483 = 0.6836$.

So the proportion is about 0.684 (or 68.4%).

Example 20.13 (Working with the normal distribution)**.** Consider the 35% of sheep with the *shortest* retention times. What are these retention times?

The time we seek must be *smaller* than the mean if it defines the *shortest* 35% of retention times. We don't know *exactly* where to draw the retention time that this corresponds to on the diagram; it's just somewhere to the left of the mean (Fig. 20.17, bottom right panel).

This time, *we know the area to the left*, but we do not know the value (or z-score). This a 'backwards problem', and we need to find the z-score 'backwards' (Sect. 20.8). From the hard copy tables, a z-score of $z = -0.39$ has an area to the left of 0.3483, which is as close as we can get. (The online tables are more precise: $z = -0.385$.)

We know the z-score, so the retention value is found using the unstandardising formula:

$$x = \mu + (z \times \sigma) = 42.5 + (-0.385 \times 3.68) = 41.0832.$$

The retention time is about 41.1 h.

20.10 Chapter summary

A *model* is a way of describing the theoretical distribution of some quantitative quantity. One common model is a *normal model* or *normal distribution*, which is a bell-shaped distribution with a theoretical mean μ and a theoretical standard deviation σ. Probabilities can be computed from normal distributions using *z-scores*, the 68–95–99.7 rule, or tables.

20.11 Quick review questions

Consider again the model for tree diameters in Example 20.6 [Aedo-Ortiz et al., 1997]: a normal distribution with $\mu = 8.8$ inches, and $\sigma = 2.7$ inches.

Are the following statements *true* or *false*?

1. A tree diameter of 10.2 inches corresponds to a z-score of $(10.2 - 8.8)/2.7 = 0.519$.
2. The probability that a tree has a diameter *less* than 10.2 inches is about 0.70.
3. The probability that a tree has a diameter *greater* than 10.2 inches is about 0.70.
4. A tree diameter of 6 inches corresponds to a z-score of 1.04.
5. The probability that a tree has a diameter *less* than 6 inches is 0.15.
6. The probability that a tree has a diameter *greater* than 6 inches is 0.85.

20.12 Exercises

Answers to odd-numbered exercises are given at the end of the book.

Exercise 20.1. Are the following statements *true* or *false*?

1. The unstandardising formula can be used to compute probabilities.
2. About 68% of observations are within two standard deviations of the mean.
3. Positive z-scores correspond to values larger than the mean.
4. A z-score tells us how many standard deviations a value is away from the mean.

Exercise 20.2. Are the following statements *true* or *false*?

1. A z-score larger than 4 is impossible.
2. A z-score of zero is located at the mean value of the population.
3. About 5% of observations are less than two standard deviations below the mean.
4. A z-score of zero means a calculation error has been made.

Exercise 20.3. Determine the probability that an observation is *less* than the following z-scores.

1. $z = 1.84$.
2. $z = -2.09$.
3. $z = -5.34$.
4. $z = 4.25$

Exercise 20.4. Determine the probability that an observation is *greater* than the following z-scores.

1. $z = -0.48$.
2. $z = 1.03$.
3. $z = -4.00$.
4. $z = 0.00$

Exercise 20.5. Growth charts released by the *World Health Organisation* [WHO, 2006] showed that girls aged five-years-old with a height of 100 cm are said to have a z-score of $z = -2$. What does this mean?

Exercise 20.6. Growth charts released by the *World Health Organisation* [WHO, 2006] showed that girls aged five-years old with a height of 120 cm are said to have a z-score of $z = +2$. What does this mean?

Exercise 20.7. IQ scores are designed to have a mean of 100 and a standard deviation of 15. Match the diagram in Fig. 20.18 with the meaning.

1. IQs greater than 110.
2. IQs between 90 and 115.
3. IQs less than 110.
4. IQs greater than 85.

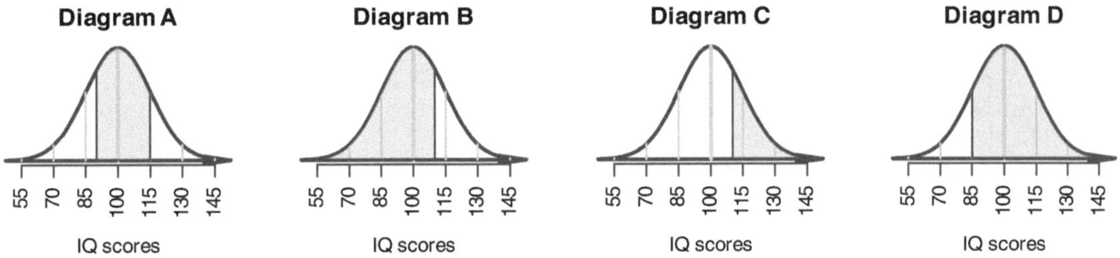

FIGURE 20.18: Match the diagram with the description.

Exercise 20.8. IQ scores are designed to have a mean of 100 and a standard deviation of 15. Match the diagram in Fig. 20.19 with the meaning.

1. The *largest* 25% of IQ scores.
2. The *smallest* 10% of IQ scores.
3. The *largest* 70% of IQ scores.
4. The *smallest* 60% of IQ scores.

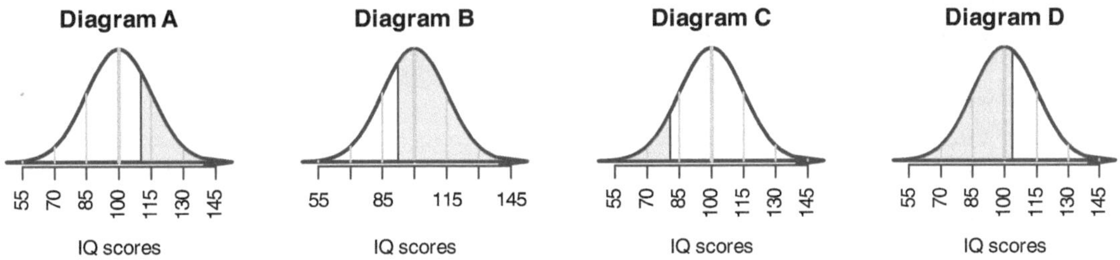

FIGURE 20.19: Match the diagram with the description.

Exercise 20.9. The 68–95–99.7 rule states that *approximately* 68% of observations are within one standard deviation of the mean. Use the tables in Appendices B.1 and B.2 to compute a more precise value for the percentage of observations within one standard deviation of the mean. Comment.

Exercise 20.10. The 68–95–99.7 rule states that *approximately* 95% of observations are within two standard deviations of the mean. Use the tables in Appendices B.1 and B.2 to compute a more precise value for the percentage of observations within two standard deviations of the mean. Comment.

Exercise 20.11. Consider again the study by Aedo-Ortiz et al. [1997] (Example 20.6), who studied the diameter of trees in certain forests. The tree diameters can be modelled as having a normal distribution, with a mean of $\mu = 8.8$ inches, and a standard deviation of $\sigma = 2.7$ inches. Using this model, answer these questions.

1. What is the probability that a tree will have a diameter *less than* 8 inches?
2. What is the probability that a tree will have a diameter *greater than* 9 inches?
3. What is the probability that a tree will have a diameter *between* 7 and 10 inches?

4. The largest 15% of trees have what diameters?
5. The smallest 25% of trees have what diameters?

Exercise 20.12. Pasha et al. [2016] simulated methods for coating corn seeds (with fertiliser and crop protection chemicals, etc.). The seed diameter was modelled with a normal distribution, with mean 7.5 mm and standard deviation of 0.225 mm. Using this model, answer these questions.

1. What is the probability that a seed has a diameter of more than 8 mm?
2. What is the probability that a seed has a diameter less than 7.1 mm?
3. What is the probability that a seed has a diameter between 7.5 and 8 mm?
4. What is the diameter of the smallest 30% of seeds?
5. What is the diameter of the largest 90% of the seeds?

Exercise 20.13. Snowden and Basso [2018] studied factors influencing preterm births. They modelled the gestation length of healthy babies with a normal distribution, having a mean of 40 weeks, and a standard deviation of 1.64 weeks. Using this model, answer these questions.

1. What proportion of births are *longer* than 39 weeks (that is, nine months)?
2. In Australia, a premature birth is defined as a birth occurring before 37 weeks. What proportion of births are expected to be premature?
3. According to *Health Direct*, 'Babies born between 32 and 37 weeks may need care in a special care nursery'. What proportion of healthy births would be expected to be born between 32 and 37 weeks gestation?
4. How long is the gestation length for the *longest* 5% of pregnancies?
5. How long is the gestation length for the *shortest* 10% of pregnancies?

Exercise 20.14. A new method for evaluating bridge loads [O'Brien et al., 2018] used a simulation to compare the new method to an existing method. For the simulation, they modelled the gross vehicle mass (GVM) of trucks as having a normal distribution, with a mean of 13 tonnes and a standard deviation of 1.3 tonnes.

The Isuzu F-Series trucks in 2025 were rated as having a GVM between 10.7 and 26.0 tonnes (depending on the configuration).

1. What is the z-score for the lower limit of 10.7 tonnes?
2. What is the z-score for the upper limit of 26.0 tonnes?
3. What does a negative z-score mean in this context?

Exercise 20.15. IQ scores are designed to have a mean of 100 and a standard deviation of 15. Mensa is a society for people with a high IQ; specifically, for people who have 'attained a score within the upper two percent of the general population' (Mensa webpage: https://www.mensa.org/). What IQ score is needed to join Mensa?

Exercise 20.16. IQ scores are designed to have a mean of 100 and a standard deviation of 15. Zagorsky [2016] reports that the US Military must 'reject all military recruits whose IQ is in the bottom 10% of the population' (Zagorsky [2016], p. 403). What IQs scores lead to a rejection from the US military?

Exercise 20.17. A study of the impact of charging electric vehicles (EVs) on electricity demands [Affonso and Kezunovic, 2018] modelled the *time* at which people began charging their EVs at home. Based on a survey [US Department of Transportation, 2011], they modelled the time at which EVs began charging as having a mean of 5:30pm, with a standard deviation of 2.28 h. For this model:

1. What is the probability that an EV will begin charging after 9pm?
2. What is the probability that an EV will begin charging before 5pm?
3. What is the probability that an EV will begin charging between 5pm and 6pm?
4. 30% of the EVs begin charging after what time?
5. The earliest 15% of charging begins when?

Hint: this question is easier if you convert times into 'minutes after 5:30'.

Answers to *Quick review* questions: 1. True. **2.** True. **3.** False: $1 - 0.70 = 0.30$. **4.** False: $z = -1.04$. **5.** True. **6.** True.

Part VI

Analysis

21

Introducing inference

So far, you have learnt to ask an RQ, design a study, describe and summarise the data, and model sampling variation. **In this chapter**, you will be introduced to the two big ideas in inference: *confidence intervals* and *hypothesis testing*. You will learn to:

- explain the purpose of a confidence interval (CI).
- explain the purpose of hypothesis testing.

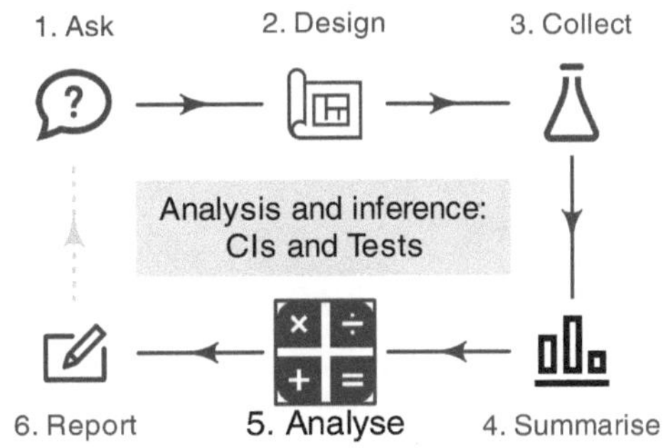

After posing an RQ (Chap. 2), a study is designed (Chaps. 4–8) to gather the evidence to answer the RQ (Chap. 9). Then the data are classified (Chap. 10) and summarised accordingly (Chaps. 12 to 17) in preparation for answering the RQ.

This part introduces *analysis*: where the data are used to answer the RQ about the population. Answering the RQ (which is about a *parameter*) is difficult, since we only study one of the countless possible samples, and hence observe only one of the countless possible values for the *statistic*. The variation in the values of the statistics from sample to sample is called *sampling variation* (Chap. 19).

Analysis provides the tools for learning about a population parameter, based on observing one of the numerous possible values of a sample statistic. The appropriate type of analysis depends upon the number and types of variables, and the purpose of the RQ (Sect. 2.8):

- *confidence intervals* answer estimation RQs, where the interest is in how precisely a *statistic* estimates a *parameter* (Chaps. 22 to 23; 29 to 31; Sect. 33.4.2).
- *hypothesis tests* answer decision-making RQs, where *decisions* are required about a *parameter* based on the value of the *statistic* (Chaps. 26 to 27; 29 to 31; Sects. 33.2.2 and 33.4.3.)

Different scenarios require different confidence intervals and hypothesis tests; those discussed in this book are shown in Table 21.1.

TABLE 21.1: Confidence intervals and hypothesis tests for different situations.

	Estimation RQs (for forming confidence intervals)	**Decision-making RQs (for conducting hypothesis tests)**
Descriptive RQs		
Single proportions	Chap. 22	Chap. 26
Single means	Chap. 23	Chap. 27
Repeated-measures RQs		
Mean differences (paired data)	Chap. 29	Chap. 29
Relational RQs		
Comparing two means	Chap. 30	Chap. 30
Comparing two odds or proportions	Chap. 31	Chap. 31
Correlational RQs		
Correlation	Sect. 33.2.1	Sect. 33.2.2
Regression	Sect. 33.4.2	Sect. 33.4.3

22

Confidence intervals: one proportion

So far, you have learnt to ask an RQ, design a study, classify and summarise the data, and model sampling variation. **In this chapter**, you will learn to:

- identify situations where estimating a proportion is appropriate.
- form confidence intervals for one proportion.
- determine whether the conditions for using the confidence intervals apply in a given situation.

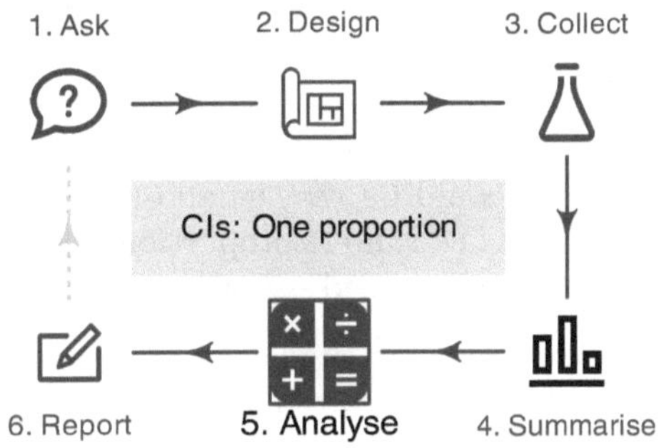

22.1 Introduction

Suppose a fair, six-sided die is rolled 25 times. What proportion of the rolls will produce an even number? That is, what will be the value of the *sample proportion* of numbers that are even? Of course, no-one knows, because the proportion of rolls that will be even will not be the same for every sample of 25 rolls. The value of the sample proportion (the statistic) *varies* from sample to sample: *sampling variation* exists.

22.2 Sampling distribution for $\hat{p}$: for p known

As seen in Chap. 19, sample statistics often vary with a normal distribution (whose standard deviation is called the *standard error*). However, being more specific about the details of the normal distribution (such as the values of its mean and standard deviation) is useful.

⚠ Remember: studying a sample leads to the following observations:

- every sample is likely to be different.
- we observe just one of the many possible samples.
- every sample is likely to yield a different value for the statistic.
- we observe just one of the many possible values for the statistic.

Since many values for the sample proportion are possible, the values of the sample proportion vary (called *sampling variation*) and have a *distribution* (called a *sampling distribution*).

To better understand the sampling distribution for the proportion of even numbers in 25 rolls of a die, statistical theory could be used, or thousands of repetitions of a sample of 25 rolls could be performed, or a computer could *simulate* many samples of 25 rolls (like we did for a roulette wheel in Sect. 19.2).

Here, the *population proportion* of even rolls is $p = 0.5$ (using the classical approach to probability: three of the six faces of the die are even). Each sample of $n = 25$ rolls produces a *sample* proportion, denoted by $\hat{p}$, which varies from sample to sample.

> ⓘ p refers to the *population* proportion, and $\hat{p}$ refers to a *sample* proportion.

The sample proportions would be expected to vary around $p = 0.5$ (the *population proportion*): some values of $\hat{p}$ larger than p, and some smaller than p. The value of the sample proportion in 25 rolls could be *very* small or *very* high by chance, but we wouldn't expect to see that very often. The sample proportions exhibit sampling variation, and the *amount* of sampling variation is quantified using a *standard error*.

Suppose a fair die was rolled 25 times, and this random procedure was repeated *thousands* of times, and the proportion of even rolls was recorded for every one of those thousands of sets of 25 rolls. These thousands of sample proportions $\hat{p}$ (one from every sample of $n = 25$ rolls) could be shown using a histogram (Fig. 22.1).

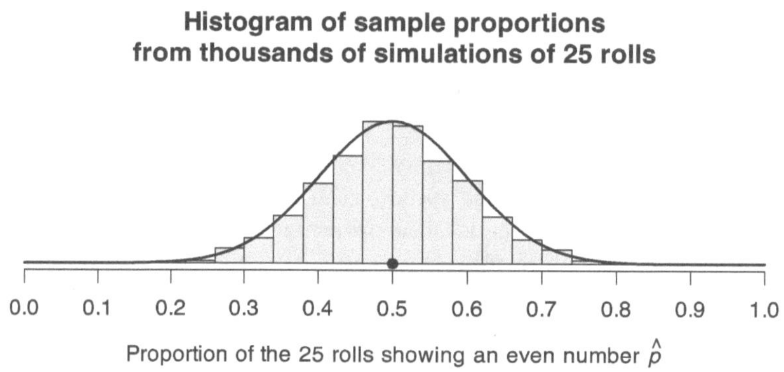

FIGURE 22.1: The proportion of rolls that are even, $\hat{p}$, is not the same for every sample of 25 rolls; it varies around a mean of $p = 0.5$. The solid line is the normal distribution used to model the sampling distribution.

The shape of the histogram is roughly a normal distribution. The sampling distribution for $\hat{p}$ will always have an approximately normal distribution when certain conditions are met: see Sect. 22.7. The mean of this distribution is called the *sampling mean*, and the standard deviation for this sampling distribution is called the *standard error*, denoted $\text{s.e.}(\hat{p})$ (see Fig. 22.2).

More specifically, the *values* of the mean and standard deviation of the normal distribution in Fig. 22.1 can be determined:

- the *sampling mean* has the value of $p = 0.5$ (i.e., the average value of $\hat{p}$ is 0.5).
- the standard deviation, called the *standard error* $\text{s.e.}(\hat{p})$, has the value 0.1. (The source of this number will be revealed soon, in Equation (22.2).)

This distribution is the *sampling distribution*, whose standard deviation is called a *standard error*. A picture of this normal distribution can be drawn (Fig. 22.2). While we still don't know *exactly* what the next roll will produce, we have some idea of *how* the sample proportion varies in samples of 25 rolls. For instance, values of $\hat{p}$ less than 0.2, or greater than 0.8 are unlikely to be observed from a fair die (with $p = 0.5$) in 25 rolls.

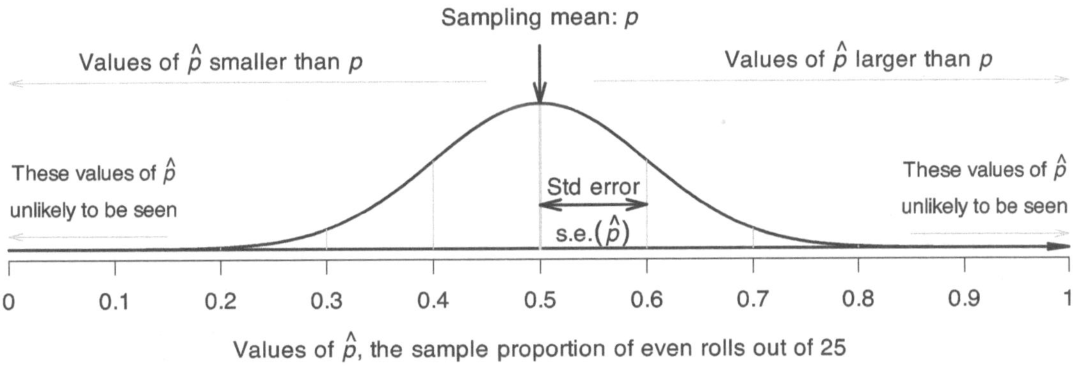

FIGURE 22.2: The sampling distribution is an approximate normal distribution with mean 0.5 and standard error 0.1; it is a model for how the proportion of even rolls varies when a die is rolled 25 times.

More generally, the sampling distribution of $\hat{p}$ is described as follows.

Definition 22.1 (Sampling distribution of a sample proportion with p known). When the value of p is *known*, the *sampling distribution of the sample proportion* is (when certain conditions are met; Sect. 22.7) described by

- an approximate normal distribution,
- centred around the sampling mean whose value is p,
- with a standard deviation (called the *standard error* of $\hat{p}$), whose value is

$$\text{s.e.}(\hat{p}) = \sqrt{\frac{p \times (1-p)}{n}}, \tag{22.1}$$

where n is the size of the sample used to compute $\hat{p}$, and p is the population proportion.

⚠ The parameter p and the statistic $\hat{p}$ are both *proportions*. However, the *average value* of the sample proportions from all possible samples can be described by a *sampling mean*, whose value is p. The sampling mean of the sampling distribution is the 'average' value of all possible sample proportions, $\hat{p}$.

For the die example, where $n = 25$ rolls and $p = 0.5$, using Equation (22.1) gives:

$$\text{s.e.}(\hat{p}) = \sqrt{\frac{0.5 \times (1-0.5)}{25}} = 0.1. \tag{22.2}$$

This standard error is the standard deviation of the normal distribution in Fig. 22.1.

In practice the value of p is almost always unknown. This situation is studied from Sect. 22.4 onwards.

22.3 Sampling intervals for $\hat{p}$

Since the possible values of the sample proportions $\hat{p}$ can be described by an approximate *normal distribution*, the 68–95–99.7 rule (Def. 20.1) applies.

For example, in Fig. 22.2 (where the sampling mean is 0.5 and the standard error is 0.1), about 68% of the time a sample of 25 rolls will have a value of $\hat{p}$ between 0.5 give-or-take *one* standard deviation (that is, give-or-take 0.1).

So, about 68% of the time, the proportion of even rolls in a sample of 25 rolls will lie between $0.5 - 0.1 = 0.4$ and $0.5 + 0.1 = 0.6$. Similarly, about 95% of the time, the proportion of even rolls will be between 0.5 give-or-take (2×0.1), or between 0.3 and 0.7.

These intervals tell us what values of $\hat{p}$ are likely to be observed in samples of size 25. Most of the time (i.e., approximately 95% of the time), the value of $\hat{p}$ is expected to be between 0.30 and 0.70 (Fig. 22.3).

Formally, the sample proportion $\hat{p}$ is likely to lie within the interval

$$p \pm \left(\text{multiplier} \times \text{s.e.}(\hat{p})\right),$$

where s.e.($\hat{p}$) is the *standard error of the sample proportion* (calculated using Equation (22.1)). The *multiplier* is a z-score, whose value depends on how confident we wish to be that the interval contains the value of $\hat{p}$. For a 95% interval, the multiplier is *approximately* 2, based on the 68–95–99.7 rule: approximately 95% of observations are within *two* standard deviations of the value of p (the mean of the normal distribution in Fig. 22.2). That is, the *approximate* 95% sampling interval is:

$$p \pm (2 \times \text{s.e.}(\hat{p})).$$

An exact value for the multiplier (i.e., a z-score) can be found using the tables in Appendices B.1 and B.2. *Any* level of confidence can be used (but different multipliers are then needed). This interval is called a *sampling interval.*

⚠ The symbol '±' means 'plus or minus', or (colloquially) 'give-or-take'.

22.4 Sampling distribution for $\hat{p}$: for p unknown

In the die example (Sects. 22.2 and 22.3), the value of p was known. However, usually the value of p (the *parameter*) is *unknown*; after all, the reason for taking a sample is to *estimate* the unknown value of p. When the value of p is unknown, the standard error is computed using the best available estimate of p, which is $\hat{p}$.

Definition 22.2 (Sampling distribution of a sample proportion with p unknown)**.** When the value of p is *unknown*, the *sampling distribution of the sample proportion* is (when certain conditions are met; Sect. 22.7) described by

- an approximate normal distribution,
- centred around the sampling mean, whose value is p,

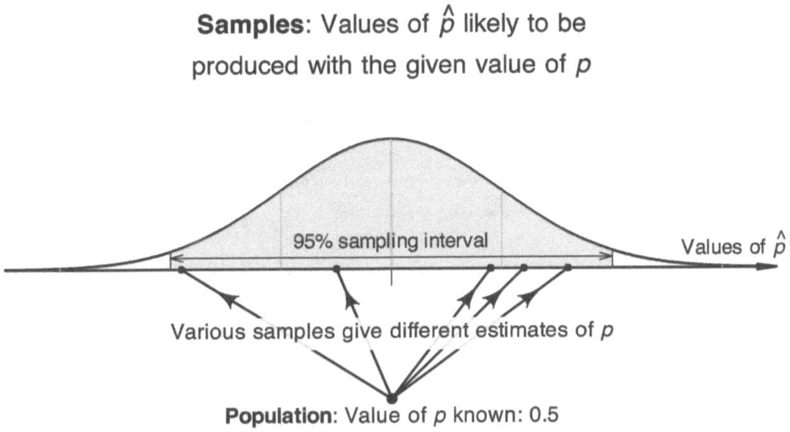

FIGURE 22.3: A known value of p produces a range of $\hat{p}$ values.

- with a standard deviation (called the *standard error* of $\hat{p}$) whose value is

$$\text{s.e.}(\hat{p}) = \sqrt{\frac{\hat{p} \times (1 - \hat{p})}{n}}, \tag{22.3}$$

where n is the size of the sample used to compute $\hat{p}$, and $\hat{p}$ is the sample proportion. In general, the approximation gets better as the sample size gets larger.

⚠ When computing the standard error for a proportion, take care! Make sure you use a *proportion* in the formula, not a *percentage* (i.e., 0.5 rather than 50%). Also: don't forget to take the square root.

22.5 Confidence intervals for p

Let's pretend for the moment that the population proportion of even rolls on a die is *unknown* (simply to demonstrate ideas). An *estimate* of the population proportion of even rolls could be found by rolling a die $n = 25$ times, and computing $\hat{p}$ (an estimate of p). Suppose 11 of the $n = 25$ rolls produce an even number, so $\hat{p} = 11/25 = 0.44$. The best estimate of p is therefore $\hat{p} = 0.44$. We might expect the (unknown) value of p to be a little smaller than this estimate $\hat{p}$, or a little larger.

Using Def. 22.2, the sample proportions vary with an approximate normal distribution around p (whose value is unknown), with a standard deviation (the standard error) of

$$\text{s.e.}(\hat{p}) = \sqrt{\frac{0.44 \times (1 - 0.44)}{25}} = 0.09927739.$$

Previously, the sampling distribution was used to construct a sampling interval that was likely to contain the unknown value of $\hat{p}$. However, here the value of $\hat{p}$ is known, so an interval is created that is likely to contain the unknown value of p that produced the observed value of $\hat{p}$ (Fig. 22.4).

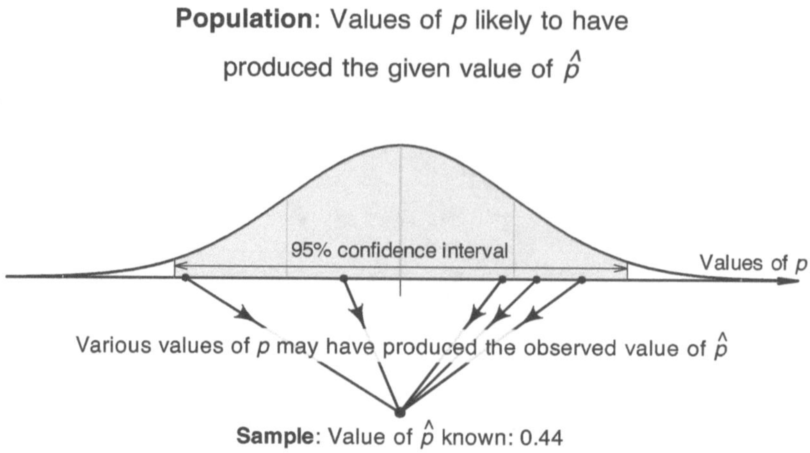

FIGURE 22.4: The sampling distribution for $\hat{p}$: many values of p may have produced the observed value of $\hat{p}$.

The unknown value of p could be a little smaller or a little larger than the value of $\hat{p}$; the interval is the value of $\hat{p}$, give-or-take a little. More formally:

$$\hat{p} \pm \big(\text{multiplier} \times \text{s.e.}(\hat{p})\big)$$

for a suitable multiplier. This interval for p is called a *confidence interval* (or a CI). The multiplier is a z-score, and the 68–95–99.7 rule gives approximate values for the multipliers. The give-or-take amount, called the *margin of error*, is (multiplier $\times$ s.e.($\hat{p}$)).

Using the approximate multiplier of 2 (from the 68–95–99.7 rule), the approximate 95% CI is

$$0.44 \pm (2 \times 0.099277), \quad \text{or } 0.44 \pm 0.1986;$$

that is, the margin of error is 0.1986. Computing the two values, the interval is from

$$\begin{array}{rll} & 0.44 - 0.1986 & \text{(which is 0.241)} \\ \text{to} & 0.44 + 0.1986 & \text{(which is 0.639).} \end{array}$$

The interval, from 0.241 to 0.639, is an interval containing values of p that could have reasonably produced the observed value of $\hat{p} = 0.44$ (Fig. 22.5). We can say the interval 0.241 to 0.639 has a 95% chance of straddling the unknown value of the population proportion p.

Definition 22.3 (Confidence interval for p). A *confidence interval* (CI) for the unknown value of the population proportion p is

$$\hat{p} \pm \big(\text{multiplier} \times \text{s.e.}(\hat{p})\big), \tag{22.4}$$

where $\big(\text{multiplier} \times \text{s.e.}(\hat{p})\big)$ is the *margin of error*, and s.e.($\hat{p}$) is the *standard error* of $\hat{p}$ (see Equation (22.3)), where $\hat{p}$ is the sample proportion. For an *approximate* 95% CI, the multiplier is 2.

In general, we do not know if the computed CI contains the actual value of p, since the value of p is usually unknown. However, in this contrived example, the CI *does* happen to straddle the value of $p = 0.5$.

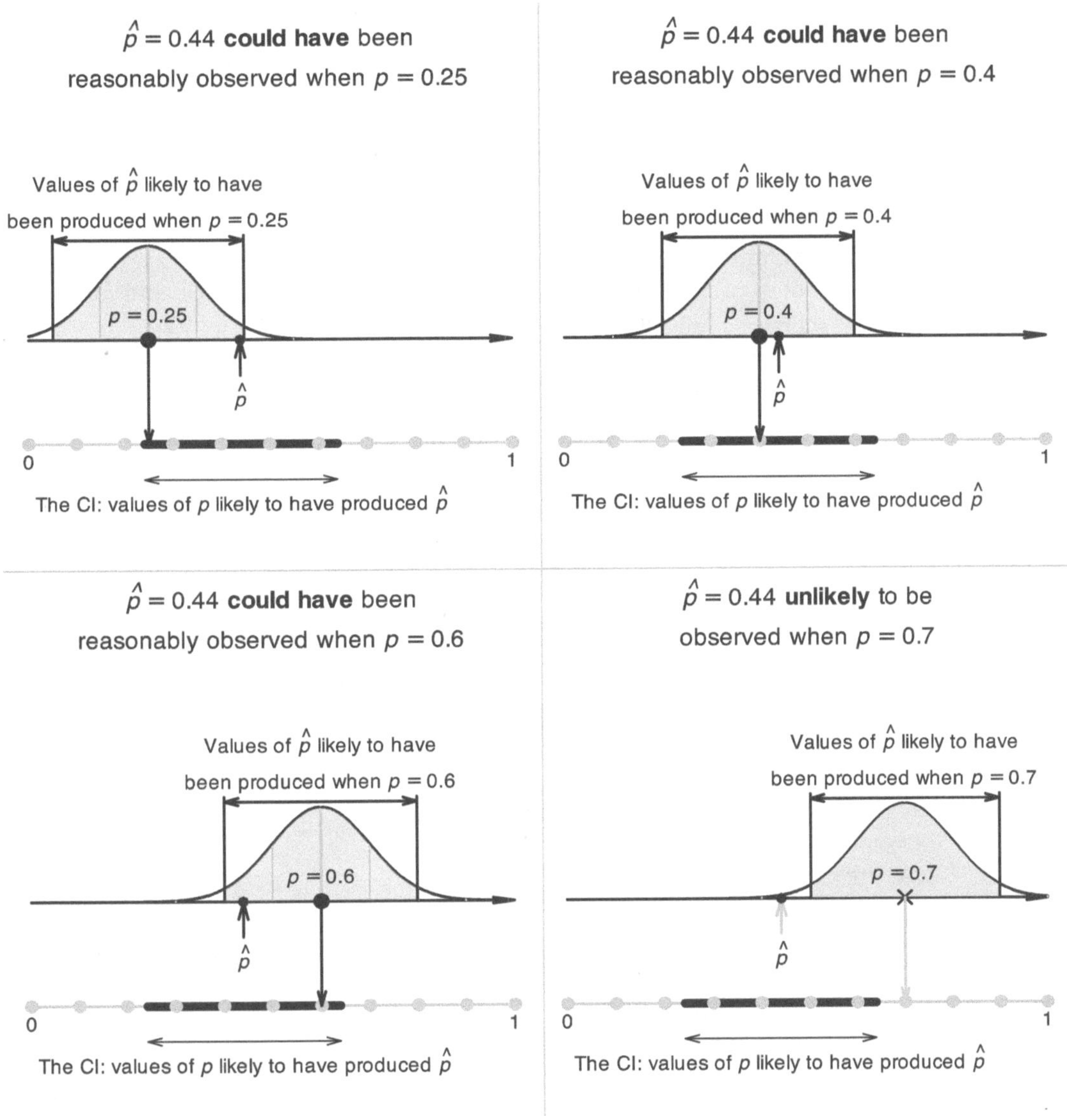

FIGURE 22.5: The CI gives an interval containing values of p that may have produced the observed value of $\hat{p}$. Here, the CI is 0.241 to 0.639, shown as the thick black horizontal line under the plots.

> In this case, we know the value of the population parameter: $p = 0.5$. Usually we do *not* know the value of the parameter. After all, that's why we take a sample: to *estimate* the unknown value of the population proportion.

Suppose *thousands* of people rolled a die 25 times, and *each* person found $\hat{p}$ for their sample, and hence computed the CI for their sample of 25 rolls. Every sample of 25 rolls could produce a different estimate $\hat{p}$, and so a different value for s.e.$(\hat{p})$, and hence a different 95% CI. However, *about 95% of these thousands of CIs from those thousands of samples would straddle the true proportion p.*

Since we usually don't know the value of p, and since we usually only have one sample (and hence one CI), in general *we never know whether the CI computed from the single sample we have straddles the value of p or not.*

Again, let's allow the computer to simulate the situation. Suppose the process of recording the sample proportion of even numbers in $n = 25$ rolls is repeated fifty times, and for each of those fifty sets of 25 rolls a CI is produced (Fig. 22.6). About 95% of those 95% CIs straddle the value $p = 0.5$ (shown as solid lines), but some do not (shown as dashed lines). Of course, value of p is rarely known, so we never know if the CI computed from our single sample contains p or not.

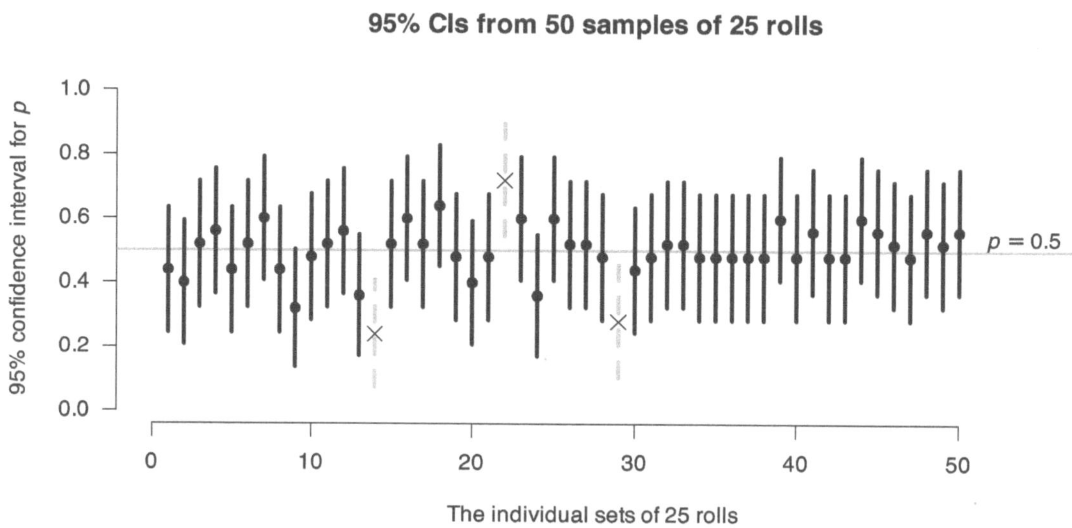

FIGURE 22.6: About 95% of CIs contain the population proportion. In the 50 samples, three produced a CI that did not straddle $p = 0.5$. In practice, we only have one sample.

> **Definition 22.4** (Confidence interval (CI)). A CI is an interval which contains the unknown value of the parameter a given percentage of the time (over repeated sampling).
>
> Informally: a *confidence interval* (CI) is an interval likely to contain the unknown value of the parameter.

In general, higher confidence means wider intervals (Fig. 22.7), since wider intervals are needed to be *more* certain that the interval contains the value of p that produced the observed value of $\hat{p}$.

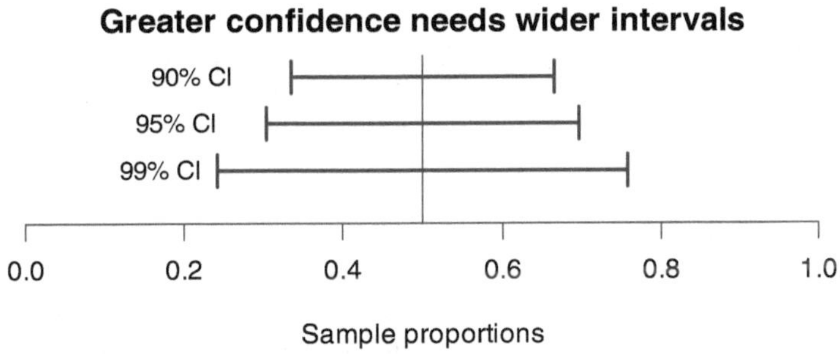

FIGURE 22.7: To have greater confidence that the interval straddles the value of the population proportion, the interval needs to be wider, for any given sample size.

Using the 68–95–99.7 rule produces *approximate* multipliers and hence *approximate* CIs. Exact multipliers (and hence exact CIs), which are z-scores, can be found using the tables in Appendices B.1 and B.2, or software. Except for small sample sizes, the approximate CIs are generally close to the exact CIs.

22.6 Interpretation of a CI

The *correct* interpretation (see Def. 22.4) of a 95% CI is the following:

If the same size samples were repeatedly taken many times, and the 95% CI computed for each sample, 95% of these CIs formed would contain the value of the parameter.

In Sect. 22.5, the CI was interpreted as giving a range of values of p that could reasonably be expected to produce the observed value of $\hat{p}$. The CI can also be seen as having a 95% chance of straddling the unknown value of the parameter. These are close to the correct interpretation.

Commonly, the CI is interpreted as having a 95% chance of containing the value of population parameter p (even though the CI either *does* or *does not* contain the value of p). This is like a convenience that captures the essence of the correct interpretation. More details on interpreting a CI are given in Sect. 24.4.

22.7 Statistical validity conditions

The CIs formed in this chapter assume the sampling distribution is approximately a normal distribution (and so, for example, the 68–95–99.7 rule can be applied). This is only true if certain conditions are met. If these conditions are met (so that the sampling distribution has an approximate normal distribution), the CI is called *statistically valid*. Whenever a CI is formed, the relevant statistical validity conditions need to be checked.

If the statistical validity conditions are not met, an alternative method [Conover, 2003] or resampling methods may be used [Efron and Hastie, 2021].

Definition 22.5 (Statistical validity). A result is *statistically valid* if the conditions for the underlying mathematical calculations to be approximately correct are met, such as the sampling distribution having an approximate normal distribution.

Example 22.1 (Statistical validity analogy). Suppose your doctor asks you to get a blood test, after fasting (refraining from eating) for 12 h before your test.

The next day, you have a big breakfast, lunch at a café, and then have your blood test. Your blood is analysed, and your doctor is sent the results of the blood test.

Since you did not fast, the results may or may not be valid. The doctor can learn *something*, but not as much as if you had followed instructions. Similarly, if the conditions for computing the CI are not met, the calculations still produce a CI, but the results may be slightly unreliable.

The CI for a single proportion is *statistically valid* if *both* of these are true:

- the number of individuals of interest exceeds 5.
- the number of individuals *not* of interest exceeds 5.

The value of 5 here is a rough figure; some books give other values (such as 10). The units of analysis are also assumed to be *independent* (e.g., ideally from a simple random sample).

These conditions ensure that the sampling distribution of $\hat{p}$ has an approximate normal distribution. If these conditions are not met, the normal model may not be a good approximation to the sampling distribution (so, for example, using the 68–95–99.7 rule may be inappropriate) and so the CI may also be slightly unreliable.

Example 22.2 (Statistical validity). For the die-throwing example in Sect. 22.5, 11 even rolls and 14 odd rolls were observed. Both exceed 5, so the CI is statistically valid.

Example 22.3 (Statistical validity conditions). Consider a situation where $p = 0.1$ is the population proportion for some result of interest.

A sample of size $n = 10$ is taken, with one positive result: $\hat{p} = 0.1$. The statistical validity conditions *are not* satisfied, and the sampling distribution is not well modelled by a normal distribution (Fig. 22.8, left panel). Using a normal distribution to model the sampling distribution would be silly.

In contrast, assume a sample of size $n = 150$ is taken, with 15 positive results: $\hat{p} = 0.1$ again. However, in this case, the statistical validity conditions *are* satisfied, and the sampling distribution is well modelled by a normal distribution (Fig. 22.8, right panel).

22.8 Example: female coffee drinkers

Kelpin et al. [2018] studied 360 female college students in the United States, and found that 61 drank coffee daily. The parameter is p, the unknown *population* proportion of female college students in the United States that drink coffee daily.

The sample size is $n = 360$, and the *sample* proportion of daily coffee drinkers is $\hat{p} = 61/360 = 0.16944$. Of course, the sample proportion varies from sample to sample, so the

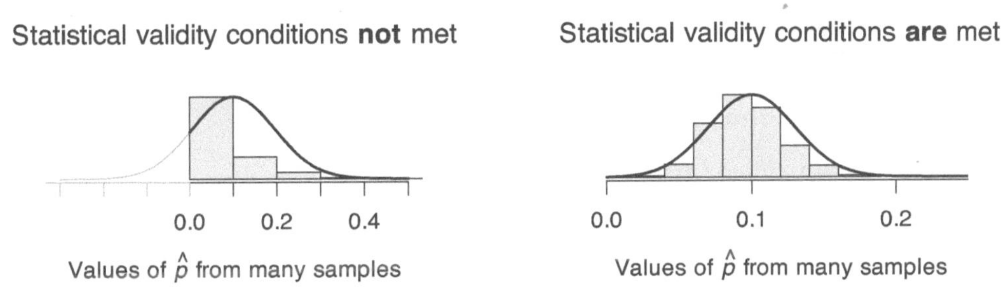

FIGURE 22.8: Two proposed sampling distributions. The sampling distribution from many simulated samples is shown in the histogram; the normal model is shown by the solid lines. Left: when the statistical validity conditions are not met. Right: when the statistical validity conditions are met.

sample proportion has *sampling variation*, measured by the *standard error*:

$$\text{s.e.}(\hat{p}) = \sqrt{\frac{0.16944 \times (1 - 0.16944)}{360}} = 0.01977.$$

An approximate 95% CI is $0.16944 \pm (2 \times 0.01977)$, or 0.16944 ± 0.03954 (i.e., the *margin of error* is 0.03954). Equivalently, the approximate 95% CI is from 0.130 to 0.209, after rounding appropriately. We write:

> The sample proportion of female US college students who drink coffee daily is $\hat{p} = 0.169$ ($n = 360$), with an approximate 95% CI from 0.130 to 0.209.

That is, values for p that may have led to this value of $\hat{p} = 0.1694$ are between 0.130 and 0.209 with 95% confidence. (This CI may or may not contain the true proportion p.) This CI is *statistically* valid, since 61 in the sample drink coffee, and 299 do not (and both exceed five).

⚠ Many decimal places are used in the working, but final answers are rounded.

22.9 Chapter summary

To compute a confidence interval (CI) for a proportion, compute the sample proportion, $\hat{p}$, and identify the sample size n used to compute $\hat{p}$. Then compute the standard error, which quantifies how much the value of $\hat{p}$ varies across all possible samples:

$$\text{s.e.}(\hat{p}) = \sqrt{\frac{\hat{p} \times (1 - \hat{p})}{n}}.$$

The *margin of error* is (multiplier × standard error), where the multiplier is 2 for an approximate 95% CI (from the 68–95–99.7 rule). Then the CI is:

$$\hat{p} \pm (\text{multiplier} \times \text{standard error}).$$

The statistical validity conditions should also be checked.

ⓘ You must use *proportions* in these formulas, **not** *percentages*; that is, use values between 0 and 1 (like 0.169 rather than 16.9%).

22.10 Quick review questions

Are the following statements *true* or *false*?

1. p is a *parameter.*
2. The value of p will vary from sample to sample.
3. The *standard error* refers to the sampling variation in p.
4. Suppose $n = 50$ and $\hat{p} = 0.4$; then the standard error of $\hat{p}$ is 0.06928.

22.11 Exercises

Answers to odd-numbered exercises are given at the end of the book.

Exercise 22.1. Won Lee et al. [2016] found that 708 of 864 patients examined with hiccups were male in their sample.

1. Compute the sample proportion of people with hiccups who are male.
2. Find an approximate 95% CI for the proportion of people with hiccups who are male.
3. Check if the statistical validity conditions are met or not.
4. Draw a sketch of how the sample proportion varies for samples of size 864.

Exercise 22.2. Lord et al. [2009] studied how paramedics administer pain medication, and found that 791 of patients reporting pain did *not* receive pain relief, out of 1 766 patients in the study who initially reported pain.

1. Compute the sample proportion of patients who did not receive pain medication.
2. Find an approximate 95% CI for the proportion of patients who did not receive pain medication.
3. Check if the statistical validity conditions are met or not.
4. Draw a sketch of how the sample proportion varies for samples of size 1 766.

Exercise 22.3. For an approximate 95% CI, the multiplier (from the 68–95–99.7 rule) is 2. Use Appendices B.1 and B.2 to find the *exact* value for the multiplier.

Exercise 22.4. Use Appendices B.1 and B.2 to find the *exact* value for the multiplier to create a 99% CI.

Exercise 22.5. Mann and Blotnicky [2017] studied the eating habits of university students in Canada. They found that 8 students out of 154 met the recommendation for eating a sufficient number of servings of grains each day.

1. Find an approximate 95% CI for the population proportion of Canadian students that meet the recommendation for eating a sufficient number of servings of grains each day.
2. Check if the statistical validity conditions are met or not.
3. Draw a sketch of how the sample proportion varies for samples of size 154.
4. Would these results be likely to apply to US university students? Explain.

Exercise 22.6. Dexter et al. [2018] found that 18 of the $n = 51$ koalas studied in a certain area over 30 months had crossed at least one road during that time. The parameter is p, the unknown *population* proportion of koalas that had crossed at least one road over the 30 months.

1. Find an approximate 95% CI for the proportion of koalas that had crossed the road at least once in the 30 months.
2. Check if the statistical validity conditions are met or not.
3. Draw a sketch of how the sample proportion varies for samples of size 51.

Exercise 22.7. Sutherland et al. [2012] studied salt intake in the United Kingdom, and found that 2 182 out of the 6 882 people sampled in 2007 'generally added salt at the table'. Find an approximate 95% CI for the population proportion of Britons that generally add salt at the table.

Exercise 22.8. A study of turbine failures [Myers et al., 2002, Nelson, 1982] ran 42 turbines for around 3 000 h, and found that nine developed fissures (small cracks). Find a 95% CI for the true proportion of turbines that would develop fissures after 3 000 h of use. Are the statistical validity conditions satisfied?

The study also ran 39 turbines for around 400 h, and found that zero developed fissures. Find a 95% CI for the true proportion of turbines that would develop fissures after 400 h of use. Are the statistical validity conditions satisfied?

Exercise 22.9. Hammond et al. [2018] studied young Canadians aged 12–24, and found 365 of the 1 516 respondents reported sleeping difficulties after consuming energy drinks. Find a 95% CI for the true proportion of young Canadians who experience sleeping difficulties after consuming energy drinks. Are the statistical validity conditions satisfied?

Exercise 22.10. McLaughlin [2010] studied the proportion of alcohol-associated calls to the ambulance service over four years in a midwestern American town. Of the 1 014 calls received over the four years, 500 were received on the weekend (Saturday and Sunday). Find an approximate 95% CI for the true proportion of alcohol-related calls that occur on the weekend.

Exercise 22.11. Oca et al. [2023] used three different AI chatbots to produce recommendations for ophthalmologist in the 20 largest cities in the USA. ChatGPT made 44 recommendations, of which 14 were female. Find an approximate 95% CI for the true proportion of female ophthalmologists recommended in those 20 cities.

Exercise 22.12. ChatGPT was launched in 2022. Lee et al. [2024] studied the impact on cheating for Californian high-school students in 2023. Students were asked to respond to this question (among many others):

> In the past month, how often have you used an Artificial Intelligence or digital device (e.g. ChatGPT, smart phone) as an unauthorised aid during an assessment, school assignment, or homework.

Options were 'Never', 'Once', '2 to 3 times' and '4 or more times'.

In private high schools, 13 of 202 students reported using AI in this manner at least once. Find an approximate 95% CI for the true proportion of students using ChatGPT in this manner in 2023.

Exercise 22.13. [*Dataset*: `HatSunglasses`] Dexter et al. [2019] recorded the number of people at the foot of the Goodwill Bridge, Brisbane, who wore hats between 11:30am to 12:30pm. Of the 752 people observed, 101 wore hats. Find an approximate 95% CI for the true proportion of people wearing hats at this time at the foot of the Goodwill Bridge.

Exercise 22.14. [*Dataset*: `HatSunglasses`] Dexter et al. [2019] recorded the number of people at the foot of the Goodwill Bridge, Brisbane, who wore sunglasses between 11:30am to 12:30pm. Of the 752 people observed, 249 wore sunglasses. Find an approximate 95% CI for the true proportion of people wearing sunglasses at this time at the foot of the Goodwill Bridge.

Answers to *Quick review* questions: 1. True. **2.** False. **3.** False. **4.** True.

23

Confidence intervals: one mean

So far, you have learnt to ask an RQ, design a study, classify and summarise the data, and construct a confidence interval for one proportion. **In this chapter**, you will learn to

- identify situations where estimating a mean is appropriate.
- form confidence intervals for one mean.
- determine whether the conditions for using the confidence intervals apply in a given situation.

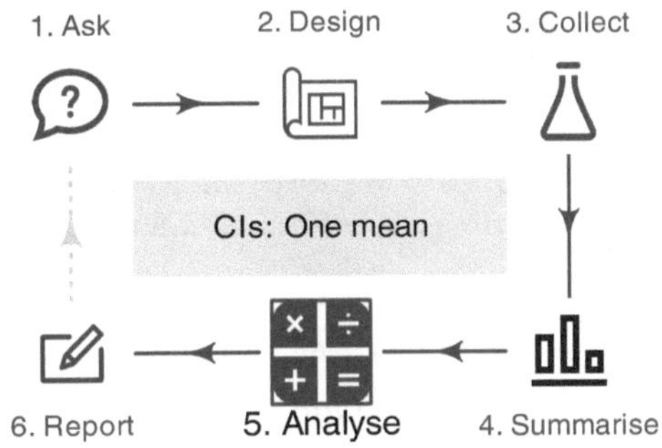

23.1 Introduction

Consider rolling a fair, six-sided die $n = 25$ times. Suppose we are interested in the *mean* of the numbers that are rolled. Since every face of the die is equally likely to appear on any one roll, the population mean of all possible rolls is $\mu = 3.5$ (in the middle of the numbers on the faces of the die, so this is also the *median*).

What will be the sample mean of the numbers in the 25 rolls? We don't know, as the sample mean varies from sample to sample (*sampling variation*).

> Remember: studying a sample leads to the following observations:
> - every sample is likely to be different.
> - we observe just one of the many possible samples.
> - every sample is likely to yield a different value for the statistic.
> - we observe just one of the many possible values for the statistic.
>
> Since many values for the sample mean are possible, the values of the sample mean vary (called *sampling variation*) and have a *distribution* (called a *sampling distribution*).

23.2 Sampling distribution for $\bar{x}$: for σ known

Suppose thousands of people made one set of 25 rolls each, and computed the mean for their sample. Then, every person would have a sample mean for their sample, and we could produce a histogram of all these sample means (Fig. 23.1). The mean for any single sample of $n = 25$ rolls will sometimes be higher than $\mu = 3.5$, and sometimes lower than $\mu = 3.5$, but often close to 3.5. Sample means larger than 4.5, or smaller than 2.5, would occur rarely.

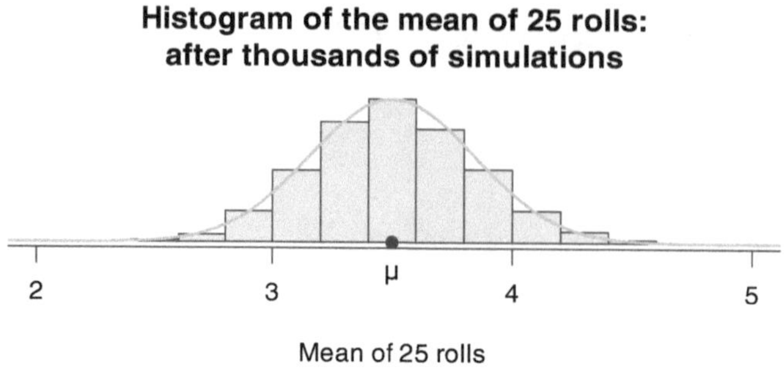

FIGURE 23.1: Rolling dice: the mean of 25 rolls, for thousands of repetitions. The solid line is the normal distribution used to model the sampling distribution.

From Fig. 23.1, the sample means vary with an approximate normal distribution (as with the sample proportions). This normal distribution does *not* describe the data; it describes how the *values of the sample means vary across all possible samples*. Under certain conditions (Sect. 23.5), the values of the sample mean vary with a normal distribution, and this normal distribution has a mean and a standard deviation.

The mean of this sampling distribution (the *sampling mean*) has the value μ. The standard deviation of this sampling distribution (the *standard error of the sample means*) is denoted $\text{s.e.}(\bar{x})$. When the *population* standard deviation σ is *known*, the value of the standard error happens to be

$$\text{s.e.}(\bar{x}) = \frac{\sigma}{\sqrt{n}}.$$

In summary, the values of the sample means have a *sampling distribution* described by:

- an approximate normal distribution,
- with a sampling mean whose value is μ, and
- a standard deviation, called the standard error, of $\text{s.e.}(\bar{x}) = \sigma/\sqrt{n}$.

However, since the *population* standard deviation is rarely ever known, we will focus on the case where the value of σ is unknown (and estimated by the *sample* standard deviation, s).

23.3 Sampling distribution for $\bar{x}$: for σ unknown

Since the value of the population standard deviation σ is almost never known, the sample standard deviation s is used to estimate of the standard error of the mean: $\text{s.e.}(\bar{x}) = s/\sqrt{n}$. With this information, the *sampling distribution of the sample mean* can be described.

Definition 23.1 (Sampling distribution of a sample mean for σ unknown). When the *population* standard deviation is unknown, the *sampling distribution of the sample mean* is (when certain conditions are met; Sect. 23.5) described by:

- an approximate normal distribution,
- centred around a sampling mean whose value is μ,

- with a standard deviation (called the *standard error of the mean*), denoted s.e.($\bar{x}$), whose value is

$$\text{s.e.}(\bar{x}) = \frac{s}{\sqrt{n}}, \qquad (23.1)$$

where n is the size of the sample, and s is the sample standard deviation of the observations.

⚠ A mean or a median may be appropriate for describing the *data*. However, the *sampling distribution* for the sample mean (under certain conditions) has a *normal distribution*. Hence, the mean is appropriate for describing the sampling distribution, even if not for describing the data.

23.4 Confidence intervals for μ

In practice, we do not know the value of μ. After all, that's why we take a sample: to *estimate* the value of the unknown population mean. Suppose, then, we did not know the value of μ (the parameter) for the die-rolling situation, but we have an *estimate*: the value of $\bar{x}$, the sample mean (the statistic). The value of $\bar{x}$ may be a bit smaller than μ, or a bit larger than μ (but we don't know which, since we do not know the value of μ). In other words, the values of μ that may have produced the observed value $\bar{x}$ may be less than the value of $\bar{x}$, or greater than the value of $\bar{x}$.

Since the values of $\bar{x}$ vary from sample to sample (*sampling variation*) with an approximate normal distribution (Def. 23.1), the 68–95–99.7 rule could be used to construct an approximate 95% interval for the plausible values of μ that may have produced the observed values of the sample mean. This is a *confidence interval* (or a CI).

A CI for the population mean is an interval surrounding a sample mean. In general, a CI for μ is

$$\bar{x} \pm \overbrace{\left(\text{multiplier} \times \text{s.e.}(\bar{x})\right)}^{\text{The `margin of error'}}.$$

For an approximate 95% CI, the multiplier is about 2 (since about 95% of values are within two standard deviations of the mean, from the 68–95–99.7 rule).

Definition 23.2 (Confidence interval for μ). A *confidence interval* (CI) for the unknown value of the population mean μ is

$$\bar{x} \pm \left(\text{multiplier} \times \text{s.e.}(\bar{x})\right), \qquad (23.2)$$

where $\left(\text{multiplier} \times \text{s.e.}(\hat{p})\right)$ is the *margin of error*, and s.e.($\bar{x}$) is the *standard error* of $\bar{x}$ (see Equation (23.1)), where $\bar{x}$ is the sample mean, and n is the sample size. For an *approximate* 95% CI, the multiplier is 2.

CIs are often 95% CIs, but any level of confidence can be used (with the appropriate multiplier). In this book, a multiplier of 2 is used when *approximate* 95% CIs are created manually, and otherwise software is used. Commonly, CIs are computed at 90%, 95% and 99% confidence levels.

In Chap. 22, the multiplier was a z-score, and approximate values for the multiplier were found using the 68–95–99.7 rule.

However, when computing the CI for a sample mean, the multiplier is *not* a z-score. The multiplier would be a z-score if the value of the *population* standard deviation was known (e.g., the situation in Sect. 23.2). When σ is unknown (almost always), and the *sample* standard deviation is used instead, the multiplier is a t-score (Sect. 27.4).

The values of t- and z-multipliers are *very* similar, and (except for small sample sizes) using an approximate multiplier of 2 is reasonable for computing *approximate* 95% CIs in either case.

Pretend for the moment that the value of μ was unknown, and we tossed a die 25 times, and found $\bar{x} = 3.2$ and $s = 2.5$. Then,

$$\text{s.e.}(\bar{x}) = \frac{s}{\sqrt{n}} = \frac{2.5}{\sqrt{25}} = 0.5.$$

Hence, the sample means vary with an approximate normal distribution, centred around the unknown value of μ, with a standard deviation of $\text{s.e.}(\bar{x}) = 0.5$ (Fig. 23.2).

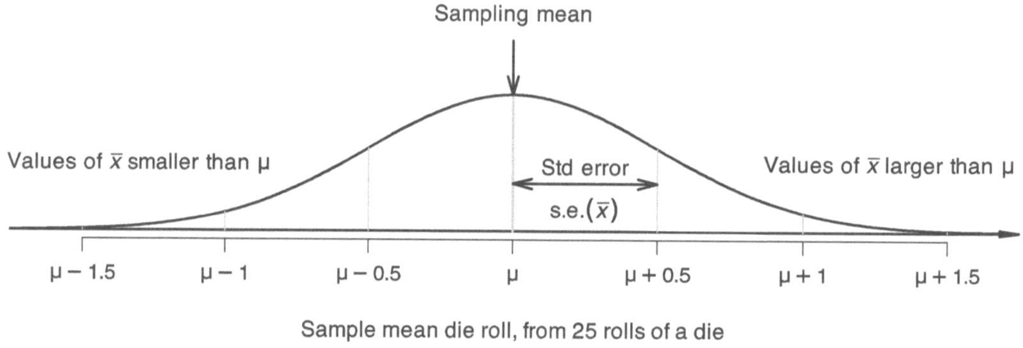

FIGURE 23.2: The sampling distribution is an approximate normal distribution with mean μ and standard error 0.5; it is a model of how the mean roll varies when a die is rolled 25 times.

Our estimate of $\bar{x} = 3.2$ may be a bit smaller than the value of μ, or a bit larger than the value of μ; that is, the value of μ is $\bar{x}$, give-or-take a bit. A range of μ values that are likely to straddle $\bar{x}$ is given by a CI. An *approximate* 95% CI is (using Equation (23.2)) from

$$\begin{aligned} & 3.2 - (2 \times 0.5) \qquad \text{(which is 2.2)} \\ \text{to}\quad & 3.2 + (2 \times 0.5) \qquad \text{(which is 4.2).} \end{aligned}$$

Hence, values of μ between 2.2 to 4.2 could reasonably have produced a sample mean of $\bar{x} = 3.2$. Using software, the exact 95% CI is from 2.17 to 4.23, the same as the approximate CI to one decimal place.

23.5 Statistical validity conditions

As with any CI, the underlying mathematics requires certain conditions to be met so that the results are statistically valid (i.e., the sampling distribution is sufficiently like a normal distribution).

The CI for a single mean is *statistically valid* if *either* of these is true:

- $n \geq 25$. (If the distribution of the data is highly skewed, the sample size may need to be larger.)
- $n < 25$, *and* the sample data come from a population with a normal distribution.

The sample size of 25 is a rough figure, and some books give other values (such as 30).

This condition ensures that the *sampling distribution of the sample means has an approximate normal distribution* (so that, for example, the 68–95–99.7 rule can be used). Provided the sample size is larger than about 25, this will be approximately true *even if* the distribution of the individuals in the population do not have a normal distribution. That is, when $n \geq 25$ the sample means generally have an approximate normal distribution, even if the data themselves do not follow a normal distribution. The units of analysis are also assumed to be *independent* (e.g., ideally from a simple random sample).

If the statistical validity conditions are not met, other methods (e.g., non-parametric methods [Bauer, 1972]; resampling methods [Efron and Hastie, 2021]) may be used.

⚠ When $n \geq 25$ approximately, the *data* do not have to have a normal distribution. The *sample means* need to have a normal distribution, which is approximately true if the statistical validity conditions are true.

Example 23.1 (Statistical validity). In the die example (Sect. 23.4), where $n = 25$, the CI is statistically valid.

The second statistical validity condition requires the *population* to have a normal distribution. Knowing this is obviously difficult; we do not have access to the whole population. All we can reasonably do is to identify (from the sample) whether the population is likely to be non-normal (when the CI would be not valid).

Example 23.2 (Statistical validity). Silverman et al. [1999] examine exposure to radiation for CT scans in the abdomen for $n = 17$ patients [Zou et al., 2003]. As the sample size is 'small' (less than 25), the *population data* must have a normal distribution for a CI for μ to be statistically valid.

A histogram of the total radiation dose received using the *sample* data (Fig. 23.3) suggests this is very unlikely. Even though the histogram is from *sample* data, it seems improbable that the data in the sample would have come from a *population* with a normal distribution.

A CI for the mean of these data will probably *not* be statistically valid. Other methods (such as resampling methods, which are beyond the scope of this book) are needed to compute a CI for the mean.

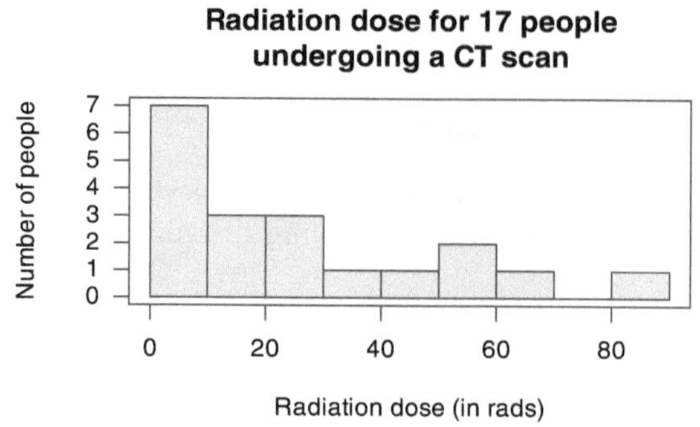

FIGURE 23.3: The radiation doses from CT scans for 17 people.

23.6 Example: cadmium in peanuts

Blair and Lamb [2017] studied peanuts gathered from a variety of regions in the United States over various times (perhaps a representative sample). They found the sample mean cadmium concentration was $\bar{x} = 0.0768$ ppm with a standard deviation of $s = 0.0460$ ppm, from a sample of 290 peanuts. The parameter is μ, the population mean cadmium concentration in peanuts.

Every sample of $n = 290$ peanuts is likely to produce a different sample mean, so *sampling variation* in $\bar{x}$ exists and can be measured using the standard error:

$$\text{s.e.}(\bar{x}) = \frac{s}{\sqrt{n}} = \frac{0.0460}{\sqrt{290}} = 0.002701\text{ ppm}.$$

The approximate 95% CI is $0.0768 \pm (2 \times 0.002701)$, or 0.0768 ± 0.00540, which is from 0.0714 to 0.0822 ppm. (The *margin of error* is 0.00540.) We write:

> The sample mean cadmium concentration of peanuts is 0.0768 ppm ($n = 290$), with an approximate 95% CI from 0.0714 to 0.0822 ppm.

If we repeatedly took samples of size 290 from this population, about 95% of the 95% CIs would contain the population mean (*our* CI may or may not contain the value of μ). The plausible values of μ that could have produced $\bar{x} = 0.0768$ are between 0.0714 and 0.0822 ppm. Alternatively, we are about 95% confident that the CI of 0.0714 to 0.0822 ppm straddles the population mean.

Since the sample size is larger than 25, the CI is statistically valid.

23.7 Chapter summary

To compute a confidence interval (CI) for a mean, compute the sample mean, $\bar{x}$, and identify the sample size n. Then compute the standard error, which quantifies how much the value of $\bar{x}$ varies across all possible samples:

$$\text{s.e.}(\bar{x}) = \frac{s}{\sqrt{n}},$$

where s is the sample standard deviation. The *margin of error* is (multiplier × standard error), where the multiplier is 2 for an approximate 95% CI (from the 68–95–99.7 rule). Then the CI is:

$$\bar{x} \pm (\text{multiplier} \times \text{standard error}).$$

The statistical validity conditions should also be checked.

23.8 Quick review questions

Are the following statements *true* or *false*?

1. The value of $\bar{x}$ varies from sample to sample.
2. A CI for μ is never statistically valid if the histogram of the *data* has a non-normal distribution.
3. A sample of data produces $s = 8$ and $n = 20$; the standard error of the mean is 1.7889.
4. When the sample size is less than 25, the standard error is not statistically valid.

23.9 Exercises

Answers to odd-numbered exercises are given at the end of the book.

Exercise 23.1. Bartareau [2017] studied American black bears, and found the mean weight of the $n = 185$ male bears was $\bar{x} = 84.9$ kg, with a standard deviation of $s = 51.1$ kg.

1. Define the *parameter* of interest.
2. Compute the standard error of the mean.
3. Draw a picture of the approximate sampling distribution for $\bar{x}$.
4. Compute the approximate 95% CI.
5. Write a conclusion.
6. Is the CI statistically valid?

Exercise 23.2. Dianat et al. [2014] studied the weight of the school bags of a sample of 586 children in Grades 6–8 in Tabriz, Iran. The mean weight was $\bar{x} = 2.8$ kg with a standard deviation of $s = 0.94$ kg.

1. Define the *parameter* of interest.
2. Compute the standard error of the mean.
3. Draw a picture of the approximate sampling distribution for $\bar{x}$.
4. Compute the approximate 95% CI.
5. Write a conclusion.
6. Is the CI statistically valid?

Exercise 23.3. [*Dataset*: **LungCap**] Tager et al. [1979] studied the lung capacity of children in East Boston. They measured the forced expiratory volume (FEV) of a sample of $n = 45$ eleven-year-old girls. For these children, the mean lung capacity was $\bar{x} = 2.85$ litres and the standard deviation was $s = 0.43$ litres [Kahn, 2005]. Find an approximate 95% CI for the population mean lung capacity of eleven-year-old females from East Boston.

Exercise 23.4. Taylor et al. [2013] studied lead smelter emissions near children's public playgrounds. They found the mean lead concentration at one playground (Memorial Park, Port Pirie, in South Australia) was 6 956.41 μg. m^{-2}, with a standard deviation of 7 571.74 μg. m^{-2}, from a sample of $n = 58$ wipes taken over a seven-day period. (As a reference, the Western Australian Government recommends a maximum of 400 μg. m^{-2}.)

Find an approximate 95% CI for the mean lead concentration at this playground. Would these results apply to playgrounds in other parts of Australia?

Exercise 23.5. Macgregor and Rugg-Gunn [1985] studied the brushing time for 60 young adults (aged 18–22 years old), and found the mean tooth brushing time was 33.0 s, with a standard deviation of 12.0 s. Find an approximate 95% CI for the mean brushing time for young adults.

Exercise 23.6. Williams and Boyle [2007] asked paramedics ($n = 199$) to estimate the amount of blood loss on four different surfaces. When the actual amount of blood spill on concrete was 1 000 mL, the mean guess was 846.4 mL (with a standard deviation of 651.1 mL).

1. What is the approximate 95% CI for the mean guess of blood loss?
2. Do you think the participants are good at estimating the amount of blood loss on concrete?
3. Is this CI statistically valid?

Exercise 23.7. [*Dataset*: **NHANES**] Using data from the NHANES study [Centers for Disease Control and Prevention (CDC), 1996], the approximate 95% CI for the mean direct HDL cholesterol is 1.356 to 1.374 mmol/L. Which (if any) of these interpretations are acceptable? Explain *why* are the other interpretations are incorrect.

1. In the *sample*, about 95% of individuals have a direct HDL concentration between 1.356 to 1.374 mmol/L.
2. In the *population*, about 95% of individuals have a direct HDL concentration between 1.356 to 1.374 mmol/L.
3. About 95% of the *samples* are between 1.356 to 1.374 mmol/L.
4. About 95% of the *populations* are between 1.356 to 1.374 mmol/L.
5. The *population* mean varies so that it is between 1.356 to 1.374 mmol/L about 95% of the time.
6. We are about 95% sure that *sample* mean is between 1.356 to 1.374 mmol/L.
7. It is plausible that the *sample* mean is between 1.356 to 1.374 mmol/L.

Exercise 23.8. Grabosky and Bassuk [2016] describe the diameter of *Quercus bicolor* trees planted in a lawn as having a mean of 25.8 cm, with a standard error of 0.64 cm, from a sample of 19 trees. Which (if either) of the following is correct?

1. About 95% of the trees in the *sample* will have a diameter between $25.8 - (2 \times 0.64)$ and $25.8 + (2 \times 0.64)$ cm (using the 68–95–99.7 rule).
2. About 95% of these types of trees in the *population* will have a diameter between $25.8 - (2 \times 0.64)$ and $25.8 + (2 \times 0.64)$ cm (using the 68–95–99.7 rule)?

Exercise 23.9. Watanabe et al. [1995] studied $n = 30$ five-year-old children, and found the mean time for the children to eat a cookie was 61.3 s, with a standard deviation of 29.4 s.

1. What is an approximate 95% CI for the population mean time for a five-year-old child to eat a cookie?
2. Is the CI statistically valid?

Exercise 23.10. [*Dataset*: **PizzaSize**] In 2011, *Eagle Boys Pizza* ran a campaign that claimed (among many other claims) that *Eagle Boys* pizzas were 'Real size 12-inch large pizzas' in an effort to out-market *Dominos Pizza*. *Eagle Boys* made the data behind the campaign publicly available [Dunn, 2012]. A summary of the diameters of a sample of 125 of *Eagle Boys* large pizzas is shown in Fig. 23.4.

1. What do μ and $\bar{x}$ represent in this context?
2. Write down the *values* of μ and $\bar{x}$.
3. Write down the *values* of σ and s.
4. Compute the value of the standard error of the mean s.e.$(\bar{x})$.
5. Explain the difference in *meaning* between s and s.e.$(\bar{x})$ here.

6. If someone else takes a sample of 125 *Eagle Boys* pizzas, will the sample mean be 11.486 inches again (as in this sample)? Why or why not?
7. Draw a picture of the approximate sampling distribution for $\bar{x}$.
8. Compute an approximate 95% CI for the mean pizza diameter.
9. Write a statement that communicates your 95% CI for the mean pizza diameter.
10. What are the *statistical* validity conditions? Is the computed CI statistically valid?
11. Do you think that, on average, the pizzas do have a mean diameter of 12 inches in the population, as Eagle Boys claim? Explain.

Exercise 23.11. Claire and Jake were wondering about the mean number of matches in a box. The boxes contain this statement:

> An average of 45 matches per box.

They purchased a carton containing $n = 25$ boxes of matches, and Jake counted the number of matches in *one* of those 25 boxes. The box contained 44 matches.

'Oh wow. Just wow.' said Jake. 'They lie. There's only 44 in this box.'

1. What is Jake's misunderstanding?
2. Then, they counted the number of matches in *each* of the $n = 25$ boxes, and found the mean number of matches per box was 44.9 matches, and the standard deviation was 0.124. Jake notes that the mean is 44.9 matches per box, and says: 'You can't have 0.9 of a match. That's dumb.' How would you respond?
3. 'Wow!' said Jake. 'The claim is 45 matches per box on average, but the mean really is 44.9! They're liars!' What is Jake's misunderstanding?
4. 'Come on, Jake,' said Claire. 'As if the mean will be *exactly* 45 in a sample every single time. Let's work out the confidence interval.' Why does Claire think a CI is needed? What will it tell them?
5. What is an approximate 95% CI for the mean for Claire's sample?
6. 'Aha,' said jake; 'I told you so! They *are* absolutely lying! Your confidence interval doesn't even include their mean of 45! 'The manufacturer *must* be lying!' Is Jake correct? Why or why not? What does the CI *mean*?
7. In this scenario, what does $\bar{x}$ represent? What is the *value* of $\bar{x}$?
8. In this scenario, what does μ represent? What is the *value* of μ?

Descriptives

	DiameterInches
N	125
Missing	0
Mean	11.486
Median	11.449
Standard deviation	0.247
Minimum	10.465
Maximum	12.228

FIGURE 23.4: Summary statistics for the diameter of *Eagle Boys* large pizzas.

Answers to *Quick review* questions: 1. True **2.** False. **3.** True. **4.** False.

24

More details about CIs

So far, you have learnt to ask an RQ, design a study, classify and summarise the data, and construct some confidence intervals. **In this chapter**, you will learn more about forming *confidence intervals.* You will learn to

- communicate confidence intervals.
- interpret confidence intervals.

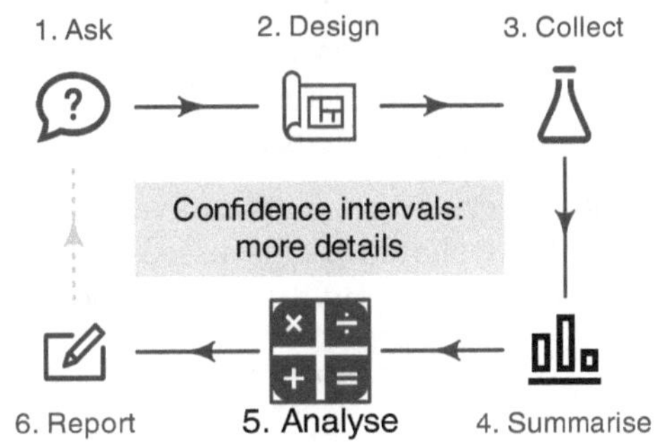

24.1 General comments

The previous chapters discussed forming confidence intervals (CIs) for estimating a population proportion, and for estimating a population mean. CIs will also be studied (Chaps. 29 to 31) in other contexts. This chapter discusses some principles that apply to CIs in general:

- statistical validity (Sect. 24.2).
- writing conclusions (Sect. 24.3).
- interpreting CIs (Sect. 24.4).

CIs are formed for an unknown value of a *population* parameter (such as the population proportion p), using the best estimate: the value of the *sample* statistic (such as the sample proportion $\hat{p}$). When the sampling distribution of the statistic has an approximate normal distribution (and not all sampling distribution do have a normal distribution), CIs have the form

$$\text{statistic} \pm (\text{multiplier} \times \text{standard error}),$$

where (multiplier × standard error) is called the *margin of error.*

When the sampling distribution has an approximate normal distribution, the *approximate* 95% CI uses a *multiplier* of 2 (from the 68–95–99.7 rule). To compute CIs other than 95% CIs (such as 99% CIs), and for *exact* 95% CIs, software is used.

⚠ *Confidence intervals (CIs)* tell us about the unknown value of the *population parameter*, based on what we learn from one of the countless possible sample statistics.

24.2 More details about statistical validity

When CIs are computed, *statistical validity conditions* must be true to ensure the mathematics behind the calculations are sound. For instance, many CIs assume the sampling distribution has a normal distribution (so that, for example, the 68–95–99.7 rule can be used); the statistical validity conditions state the conditions under which the sampling distribution has an approximate normal distribution. If these conditions are *not* met, the sampling distribution may not be close to an approximate normal distribution, so the 68–95–99.7 rule (on which the CI is based) may not be appropriate, and the CI itself may be inappropriate. Of course, if the statistical validity conditions are close to being satisfied, then the resulting CI will still be reasonably useful.

Besides checking the statistical validity conditions, the *internal validity* and *external validity* of the study should be discussed (Fig. 24.1). In addition, CIs also require that the sample size is less than about 10% of the population size; this is almost always the case.

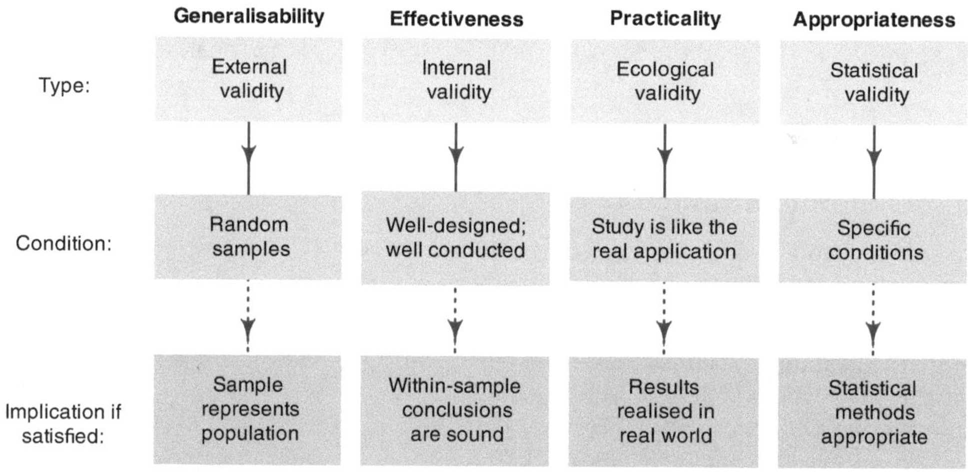

FIGURE 24.1: Four types of validities for studies.

24.3 More details about writing conclusions

When reporting a CI, include:

1. the CI (including units of measurement, if relevant).
2. the level of confidence for the CI (typically, a 95% CI).
3. the value of the statistic (the parameter estimate) and the sample size.

If the CI is an *approximate* CI (e.g., based on using an approximate multiplier of 2 from the 68–95–99.7 rule), this should also be clear.

Example 24.1 (Writing conclusions). In Sect. 23.6, the mean cadmium level of peanuts was estimated. The conclusion was:

> The sample mean cadmium concentration of peanuts is 0.0768 ppm ($n = 290$), with an approximate 95% CI from 0.0714 to 0.0822 ppm.

Each of the three elements above are given.

1. The CI: 0.0714 to 0.0822 ppm.
2. The level of confidence for the CI: 95%.
3. The value of the statistic: $\bar{x} = 0.0768$ ppm.

In addition, the CI is flagged as an *approximate* 95% CI.

24.4 More details about interpreting CIs

Interpreting CIs correctly takes care. The *correct* interpretation (Def. 22.4) of a 95% CI is:

> The CI is an interval which contains the unknown value of the parameter 95% of the time (over repeated sampling).

That is, if we *repeated* the process (of selecting a sample of a given size, then computing the CI) numerous times, 95% of those CIs formed would contain the value of the parameter. This is the idea shown in Fig. 22.6.

In practice, this definition is unsatisfying, since we only ever have *one* sample, not *many* samples. Furthermore, since the value of the parameter is unknown (after all, the reason for taking a sample was to *estimate* the value of the parameter), we don't know if the CI from *our* single sample straddles the population parameter or not.

Two reasonable alternative interpretations for a 95% CI are below.

- The 95% CI gives a range of values of the parameter that could reasonably (with 95% confidence) have produced the observed value of the statistic.
- There is a 95% chance that the 95% CI straddles the unknown value of the parameter.

These alternatives are adequate and common interpretations.

Frequently, the CI is described as having a 95% chance of containing the population parameter. This is not strictly correct (the CI either *does* or *does not* contain the value of the population parameter), but is a common and a brief paraphrase for the correct interpretation above.

I use this analogy: most people say the sun rises in the east. This is incorrect; the sun doesn't *rise* at all. People *say* the sun rises in the east as a convenient way to explain that we see the sun each morning in the east as the earth rotates. Similarly, most people say a CI is an interval with a certain chance of containing the value of the population parameter, as a convenient way to explain the CI.

Example 24.2 (Interpreting CIs). In Example 24.1, the approximate 95% CI for the cadmium concentration in peanuts was from 0.0714 to 0.0822 ppm. The correct interpretation is:

> If many samples of 290 peanuts were taken, and the approximate 95% CI computed for each one, about 95% of those CIs would contain the population mean.

Our CI may or may not include the value of μ, however. We might say:

> This 95% CI (from 0.0714 to 0.0822 ppm) has a 95% chance of straddling the actual value of μ.

Alternatively, we could say:

> The range of values of μ that could plausibly (with 95% confidence) have produced $\bar{x} = 0.0768$ is between 0.0714 to 0.0822 ppm.

In practice, the CI is usually interpreted as:

> There is a 95% chance that the population mean level of cadmium in peanuts is between 0.0714 to 0.0822 ppm.

This last statement is not strictly correct, but is commonly-used, and sufficient for our use.

24.5 Chapter summary

Confidence intervals (or CIs) tell us about the unknown *population parameter*, based on what we learn from one the countless possible sample statistics. CIs give an interval in which a parameter is likely to lie over repeated sampling. Since we only ever have one sample, two reasonable alternative interpretations for a 95% CI are:

- the 95% CI gives a range of values of the parameter that could reasonably (with 95% confidence) have produced our observed value of the statistic.
- there is a 95% chance that our 95% CI straddles the value of the parameter.

We never know if the CI from *our* single sample includes the population parameter or not. When reporting a CI, include:

1. the CI (including units of measurement, if relevant);
2. the level of confidence for the CI (typically, a 95% CI); and
3. the value of the statistic (the parameter estimate) and the sample size.

24.6 Quick review exercises

Are the following statements *true* or *false*?

1. CIs *always* have 95% confidence.
2. Statistical validity concerns the *generalisability* of the results.
3. CIs always include the value of the *population* parameter.

4. All else being equal, a 95% CI is *wider* than a 90% CI.
5. The 'multiplier times the standard error' is called the *margin of error*.
6. We are fairly sure (but *not certain*) that the CI includes the value of the statistic.

24.7 Exercises

Answers to odd-numbered exercises are given at the end of the book.

Exercise 24.1. Hirst and Stedman [1962] computed a 95% CI to estimate the proportion of trees with apple scab, and found $\hat{p} = 0.314$ and s.e.$(\hat{p}) = 0.091$. What would be *wrong* with the following conclusions?

1. An approximate 95% CI for the sample proportion is between 0.223 and 0.405.
2. This CI means we are 95% confident that between 22.3 and 40.5 trees are infected with apple scab.

Exercise 24.2. Fayet-Moore et al. [2017] studied the snacking habits of Australian children. In 2007 (for which $n = 3\,637$), the CI for the proportion of children snacking ('an eating occasion that occurred between meals based on time of day'; p. 1) was 0.981 ± 0.003 in 2007. What would be *wrong* with the following conclusion?

- An approximate 95% CI for the sample proportion of snacks (in 2007) is 0.981 ± 0.003.

Exercise 24.3. Guirao et al. [2017] studied how far amputees could walk in two minutes following a femoral (leg) implant. After 14 months, the sample of ten amputees walked a mean of 122.5 m; the 95% CI was computed as 96.4 m to 148.6 m. What would be *wrong* with the following conclusions?

1. Approximately 95% of the amputees walked between 96.4 and 1 488.6 m in two minutes.
2. The 95% CI for the sample mean distance walked in two minutes was between 96.4 and 1 488.6 m.

Exercise 24.4. A study of sodium intake in Thailand found the 95% CI for the mean daily sodium intake for subjects with a secondary school education was 3 565 to 3 903 mg. What would be wrong with the following conclusions?

1. This CI means that approximately 95% of the subjects had a daily sodium intake between 3 565 to 3 903 mg.
2. A 95% CI for the sample mean daily sodium intake is between 3 565 to 3 903 mg.

Exercise 24.5. In discussing the weight of adult male Leadbeater's possums, Williams et al. [2022] state (p. 170):

> The average adult male Leadbeater's possum weighed 137 g (95% CI = 135 g, 139 g), with 90% of weights between 122 and 153 g.

Figure 22.7 indicates that a *higher* value for the confidence level means *wider* CIs, since wider intervals are needed to be *more* certain that the interval contains the value of the parameter that produced the value of the statistic.

In light of this, explain why the 90% interval is *wider* than the 95% interval in the above quote.

Answers to *Quick review* questions: 1. False. **2.** False. **3.** False. **4.** True. **5.** True. **6.** False (CI *must* contain value of statistic).

25

Making decisions

So far, you have learnt to ask an RQ, design a study, classify and summarise the data, and form confidence intervals. **In this chapter**, you will learn to:

- state the two broad reasons that might explain the difference between the values of the statistic and parameter.
- explain how decisions are made in research.

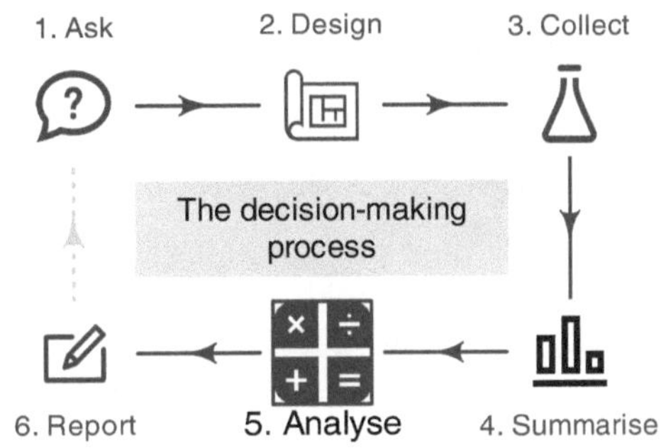

25.1 Introduction: drawing cards

Suppose I produce a pack of cards, and shuffle them well. The event of interest is 'the number of red cards when I draw 25 cards from the pack, with replacement'. ('With replacement' means that, after drawing a card, I place the card back into the pack, and reshuffle before drawing a new card; each draw is then from an identical, complete pack of 52 cards.) The pack of cards can be considered the *population*. In a standard pack, the proportion of red cards is $p = 0.5$ for each draw, because sampling is with replacement. $\hat{p}$ is the proportion of red cards in the *sample* of $n = 25$ cards.

Suppose the sample of 25 cards produces $\hat{p} = 1$; that is, *all* $n = 25$ cards in the sample are red cards (Fig. 25.1). What should you conclude? How likely is it that this would happen by chance from a fair pack? Is this evidence that the pack of cards is somehow unfair, poorly shuffled, or manipulated?

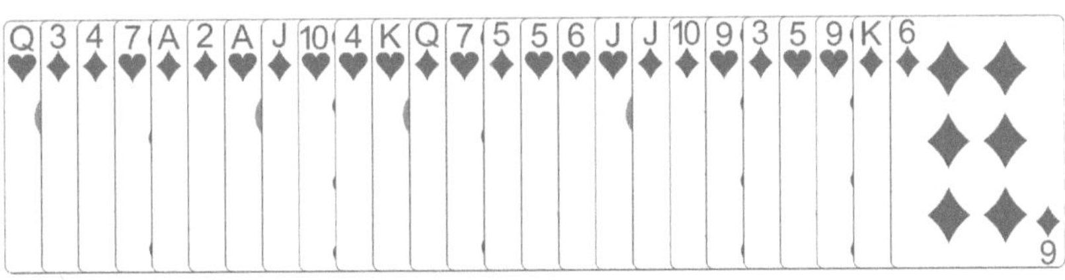

FIGURE 25.1: How likely is it to draw (with replacement) 25 red cards in a row from a fair pack?

Of course, the sample of 25 cards is just one of *countless* possible samples that could have been chosen to study. Different samples comprise different cards, and the sample proportion depends on which cards are drawn for the studied sample. This leads to one of the most important observations about sampling.

⚠ Studying a sample leads to the following observations:

- every sample is likely to be different.
- we observe just one of the many possible samples.
- every sample is likely to yield a different value for the sample statistic.
- we observe just one of the many possible values for the statistic.

Since many values for the sample are possible, the possible values of the statistic vary (this is called *sampling variation*) and have a *distribution* (this is called a *sampling distribution*).

In research, decisions need to be made about parameters, based on just one of many possible values of the statistic. Sensible decisions *can* be made (and *are* made) about parameters based on statistics. To do this though, the process of *how* decisions are made needs to be articulated, which is the purpose of this chapter.

In the cards example, obtaining 25 reds cards out of 25 (i.e., $\hat{p} = 1$) seems very unlikely from a fair pack; you would probably conclude that the pack is somehow unfair, or that I was cheating somehow. But importantly, *how* did you reach that decision? Your unconscious decision-making process may have followed these steps.

1. You *assumed*, quite reasonably, that I used a standard, well-shuffled pack of cards, where half the cards are red and half the cards are black. That is, you assumed the *population proportion* of red cards really was $p = 0.5$.
2. Based on that assumption, you *expected* about half the cards in the sample of 25 to be red (i.e., expected $\hat{p}$ to be about 0.5). You wouldn't necessarily have expected *exactly* half red cards (because of sampling variation), but you expected the value of $\hat{p}$ to be close to 0.5.
3. You then *observed* that *all* 25 cards were red. That is, you observed $\hat{p} = 1\dots$ which seems rather high.
4. You were expecting $\hat{p} = 0.5$ approximately, but instead observed $\hat{p} = 1$. What you observed ('all red cards') was not at all like what you were expecting ('about half red cards'); the sample *contradicts* what you were expecting (from a fair pack). This suggests your assumption of a fair pack is probably wrong.

Of course, finding 25 red cards in a row is *possible*... just *extremely unlikely*. For this reason, you would probably conclude that this is persuasive evidence that the pack is not fair.

25.2 Making decisions: hypotheses

Two reasons could explain why the value of the *sample* proportion of red cards in 25 cards ($\hat{p} = 1$) is not exactly equal to the value of the population proportion ($p = 0.5$).

1. *The* ***population*** *proportion of red cards really is* $p = 0.5$, and the value of the *sample* proportion $\hat{p}$ is not equal to 0.5 only due to *sampling variation*. That is, we just happen to have—by chance—one of those samples where the value of $\hat{p}$ is very large and not equal to p.
2. *The* ***population*** *proportion of red cards really isn't* $p = 0.5$, and this is simply reflected in the observed sample proportion.

These two possible explanations ('statistical hypotheses') have special names.

1. The first explanation is the *null hypothesis*, denoted H_0. This hypothesis proposes that *the population proportion is* 0.5; the value of the sample proportion is *not* 0.5 due to *sampling variation*.
2. The second explanation is the *alternative hypothesis*, denoted H_1. This hypothesis proposes that the population proportion is really not 0.5 at all, which is reflected in the value of the sample proportion.

How do we decide which of these explanations is supported by the data?

The usual approach to decision-making in science begins by assuming the null hypothesis (the sampling-variation explanation) is true. Then the data are examined to see if persuasive evidence exists to change our mind (and support the alternative hypothesis). Remember: conclusions drawn about the *population* from the *sample* can never be certain, since the sample studied is just one of many possible samples that could have been taken (and every sample is likely to be different).

⚠ The onus is on the data to refute the null hypothesis. That is, the null hypothesis is retained unless persuasive evidence suggests otherwise.

25.3 Making decisions: the process

The ideas in Sect. 25.1 suggest a formal process of decision-making in research (Fig. 25.2).

1. *Make an assumption about the value of the parameter*. By doing so, we assume that *sampling variation* explains any discrepancy between the value of the observed statistic and this assumed value of the parameter.
2. *Define the expectations of the statistic*. Based on the assumption made about the parameter, describe what values of the *statistic* might reasonably be observed from all the possible samples from the population (due to sampling variation).
3. *Evaluate the observations*. Take a sample (one of the many possible samples), and compute the observed sample statistic from these data; then compare this to what was expected.
4. *Make a decision*:
   - if the observed value of the *sample statistic* is *unlikely* to have been observed by chance, the statistic (i.e., the evidence) *contradicts* the assumption about the *parameter*, so the assumption is probably (but not certainly) *wrong*.
   - if the observed value of the *sample statistic* could easily be explained by chance, the statistic (i.e., the evidence) is *consistent with* the assumption about the *parameter*, so the assumption may be (but is not certainly) *correct*.

This approach is similar to how we unconsciously make decisions every day. For example, suppose I ask my son to brush his teeth [Budgett et al., 2013]. Later, I wish to decide if he really did. The decision-making process may proceed as follows.

1. *Assumption*: I *assume* my son brushed his teeth (because I asked him to).
2. *Expectation*: based on my assumption, I would *expect* to find his toothbrush is damp.
3. *Observation*: when I check later, I observe a *dry* toothbrush, which is unexpected.
4. *Decision*: the evidence *contradicts* what I expected to find based on my assumption, so my assumption is probably *false*: he probably *didn't* brush his teeth.

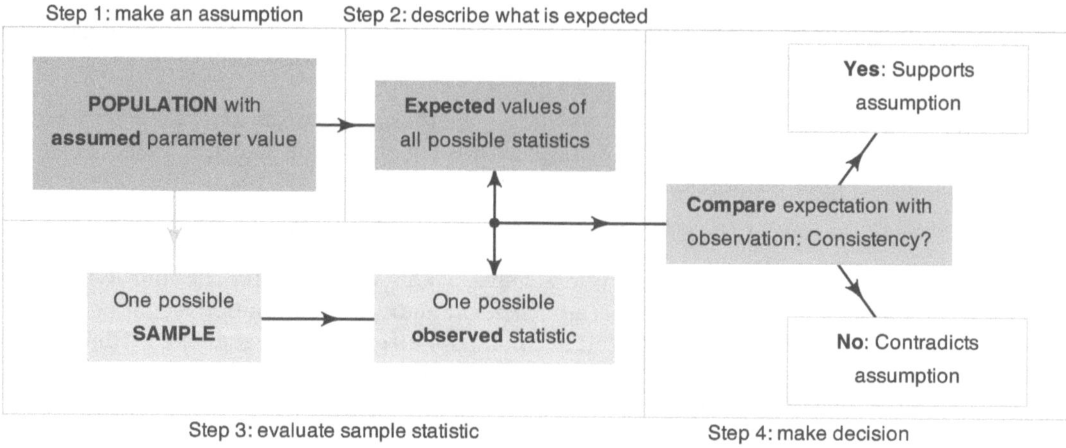

FIGURE 25.2: A way to make decisions.

I may have made the wrong decision: he may have brushed his teeth, then dried his toothbrush with a hair dryer. However, based on the evidence, he likely has not brushed his teeth.

The situation may have ended differently: when I check later, suppose I observe a *damp* toothbrush. Then, the evidence seems *consistent* with what I expected if he brushed his teeth (my assumption), so my assumption is probably *true*; he probably did brush his teeth. Again, I may be wrong: he may have rinsed his toothbrush under a tap. Nonetheless, I don't have evidence that he didn't brush his teeth.

Similar logic underlies most decision-making in science.[1]

25.4 Making decisions: the steps

Let's think about each step in the decision-making process (Fig. 25.2) individually:

- making an *assumption* about the parameter (Sect. 25.4.1).
- describing the *expectations* of the statistic (Sect. 25.4.2).
- taking the sample *observations* (Sect. 25.4.3).
- making a *decision* (Sect. 25.4.4).

25.4.1 Making an assumption about the parameter

The initial assumption is that sampling variation explains any discrepancy between the values of the parameter and the statistic. This assumption about the value of the parameter is called the *null hypothesis*, denoted H_0.

[1]Other ways exist to make decisions, such as using prior knowledge. For example, if my son had a reputation for wetting his toothbrush under the tap instead of brushing his teeth, that information can be incorporated into the decision-making. This is called a *Bayesian approach.*

The null hypothesis is always that sampling variation explains the difference between the observed value of the statistic and the assumed value of the parameter. Depending on the RQ and the context, this may mean:

- the parameter value has not changed (e.g., for descriptive or repeated-measures RQs). The value of the statistic might show a change, but only due to sampling variation.
- the value of some parameter is the same in all the groups being compared (e.g., for relational RQs). The values of the statistic are not exactly the same due to sampling variation.
- there is no relationship between the variables, as measured by some parameter (e.g., for correlational RQs). The value of the statistic is not exactly this value due to sampling variation.

In other words, the null hypothesis is the 'no change, no difference, no relationship' position.

Using this idea, a reasonable assumption can be made about the parameter. For example, when comparing the mean of two groups, we would initially assume *no difference* between the *population* means: any difference between the *sample* means would be attributed to sampling variation.

Example 25.1 (Assumptions about the population). Many dental associations recommend brushing teeth for two minutes. Macgregor and Rugg-Gunn [1979] recorded the toothbrushing time for 85 uninstructed schoolchildren from England to assess compliance with these guidelines.

We initially *assume* the *population* mean toothbrushing time is two minutes (H_0: $\mu = 2$). If the *sample* mean is not two minutes, the null hypothesis explains this discrepancy as sampling variation. A sample can then be obtained to determine if the sample mean is consistent with, or contradicts, this assumption.

25.4.2 Describing the expectations of the statistic

Based on the assumed value for the parameter, we then determine *what values to expect from the statistic* from all the possible samples we could select (of which we only select one). Since many samples are possible, and every sample is likely to be different (sampling variation), the observed value of the statistic depends on which one of the countless possible samples we obtain. To know what values of the statistic are expected, the *sampling distribution* needs to be described.

Think about the cards in Sect. 25.1. In a fair pack, *half* the cards are red in the *population* (the pack of cards), so the population proportion is assumed to be $p = 0.5$. In a *sample* of 25 cards, what values could be reasonably expected for the *sample* proportion $\hat{p}$ of red cards (the statistic)? If samples of size $n = 25$ were repeatedly taken from a fair pack with $p = 0.5$, the sample proportion of red cards would vary from hand to hand, of course. But *how* would $\hat{p}$ vary from sample to sample?

Suppose we simulated only ten hands of $n = 25$ cards each, using a computer; Fig. 25.3 shows the sample proportion of red cards from each repetition. Naturally, the value of $\hat{p}$ varies.

The distribution of the sample statistic is called the *sampling distribution* (Sect. 19.1). The sampling distribution for $\hat{p}$ is given in Sect. 22.2 when the value of p is known (as assumed here): an approximate normal distribution, with mean $p = 0.5$, and a standard deviation

FIGURE 25.3: Ten hands of 25 cards: the sample proportion of cards in each hand that is red (shown on the right-hand side) varies from hand to hand.

(the *standard error*) of

$$\text{s.e.}(\hat{p}) = \sqrt{\frac{p \times (1-p)}{n}} = \sqrt{\frac{0.5 \times (1-0.5)}{25}} = 0.1.$$

A picture of this sampling distribution can be drawn (Fig. 25.4). Values of $\hat{p}$ larger than 0.8 or smaller than 0.2 would appear very unlikely.

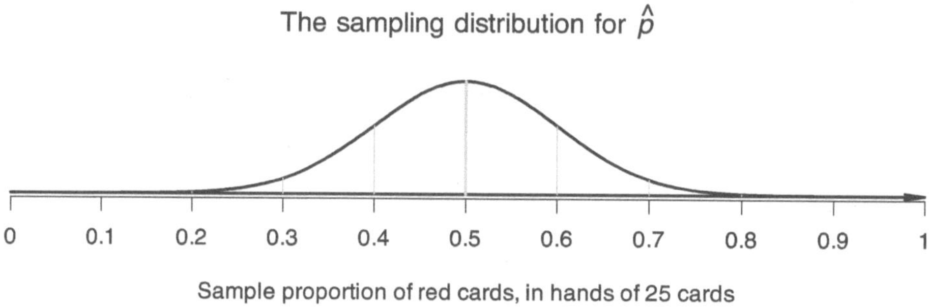

FIGURE 25.4: The sampling distribution for $\hat{p}$, the sample proportion of red cards in 25 cards.

25.4.3 Evaluating the sample observations

While many samples are possible, we only observe *one* of those countless possible samples. From our sample of 25 cards, all cards are red (Fig. 25.1), and so the value of the statistic is $\hat{p} = 25/25 = 1$. Assuming $p = 0.5$, is this value of $\hat{p}$ unusual, or not unusual? From Fig. 25.4, the value $\hat{p} = 1$ is *very* unusual: it would not be expected in a sample of 25 cards.

25.4.4 Making a conclusion

Observing 25 red cards out of 25 cards from a fair pack is highly unusual, so the chance that our specific, randomly-chosen sample produced $\hat{p} = 1$ is incredibly unlikely. So you could reasonably conclude that finding $\hat{p} = 1$ almost never occurs, *if the assumption of a fair pack was true.*

But since we *did* find $\hat{p} = 1$, the assumption of a fair pack is probably wrong. That is, there is persuasive evidence that the assumption is wrong, and that the pack of cards is not fair. The decision-making process is shown in Fig. 25.5.

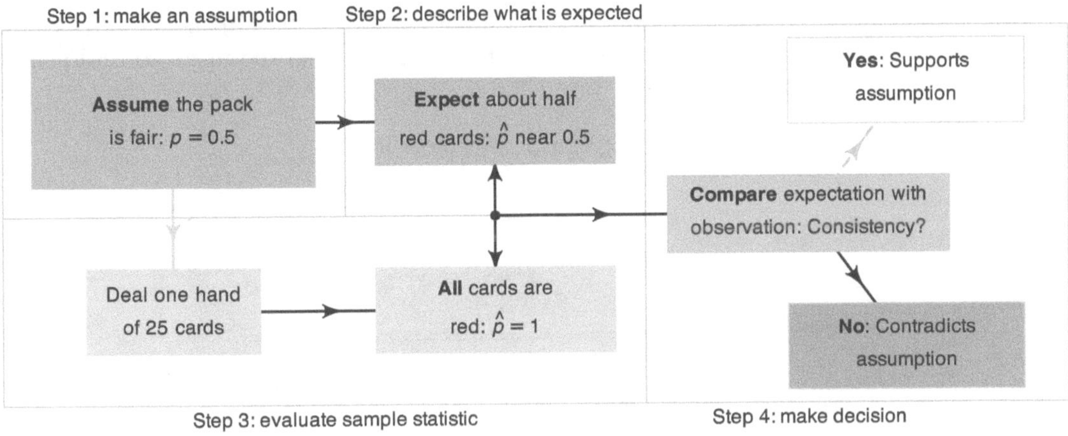

FIGURE 25.5: A way to make decisions for the cards example.

What if we had observed 18 red cards in a hand of 25 cards, a sample proportion of $\hat{p} = 18/25 = 0.72$? The conclusion is not quite so obvious then: Fig. 25.4 shows that $\hat{p} = 0.72$ is unlikely, but $\hat{p} = 0.72$ (and even larger values) certainly do occur.

What if 15 red cards were found in the 25 (i.e., $\hat{p} = 0.6$)? Figure 25.4 shows that $\hat{p} = 0.6$ could reasonably be observed, since there are many possible samples that lead to $\hat{p} = 0.6$, or even larger values. This would not seem unusual at all, and is certainly not persuasive evidence to change our mind. Many of the possible samples produce values of $\hat{p}$ near 0.6.

This process of decision-making is similar to the process used in research. This process will be studied in coming chapters.

25.5 Example: brushing teeth

Many dental associations recommend brushing teeth for two minutes (i.e., for 120 s). Macgregor and Rugg-Gunn [1979] recorded the toothbrushing time for 85 uninstructed schoolchildren from England to assess compliance with these guidelines. Of course, every possible sample of 85 children in England will include different children, and so produce various sample mean brushing times $\bar{x}$. Even if the *population* mean toothbrushing time really was 120 s (i.e., $\mu = 120$), the *sample* mean probably wouldn't be exactly 120 s, because of this sampling variation.

To begin, *assume* the population mean toothbrushing time is $\mu = 120$; that is, H_0 is $\mu = 120$. *If* this is true, we then could describe what values of the sample statistic $\bar{x}$ could be *expected* from all possible samples by describing the sampling distribution of $\bar{x}$: how sample means are likely to vary for samples of size 85 when $\mu = 120$.

The study found the mean time spent brushing was 60.3 s, with a standard deviation of 23.8 s. Using the ideas in Chap. 23, and in Sect. 23.3 specifically, the sampling distribution of $\bar{x}$ has an approximate normal distribution, with mean $\mu = 120$ (from H_0) and a standard deviation of $\text{s.e.}(\bar{x}) = 23.8/\sqrt{85} = 2.58$ s (shown in Fig. 25.6).

A sample mean of $\bar{x} = 60.3$ seems incredibly unlikely if $\mu = 120$. This suggests that the sample evidence *contradicts* the assumption that $\mu = 120$, and so the mean toothbrushing time in the population is very unlikely to be 120 s.

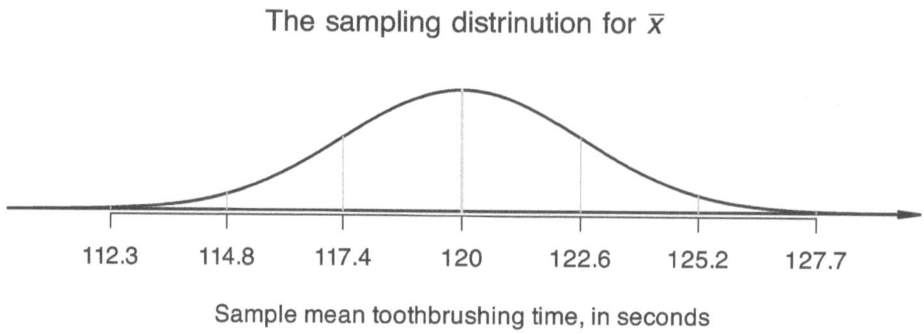

FIGURE 25.6: The sampling distribution for $\bar{x}$, the mean toothbrushing time in schoolchildren from England. A sample mean of 60.3 s seems very unlikely.

25.6 Chapter summary

Making decisions about parameters based on a statistic is challenging: only one of the many possible samples is observed. Since every sample is likely to be different, different values of the sample statistic are possible. In this chapter, though, a process for making decisions has been studied (Fig. 25.2).

Decisions are often made by making an *assumption* about the parameter, which leads to an *expectation* of what values of the statistic are reasonably possible. We can then make *observations* about our sample, and then make a *decision* about whether the sample data support or contradict the initial assumption.

25.7 Quick review questions

Are the following statements *true* or *false*?

1. Parameters describe *populations*.
2. Both $\bar{x}$ and μ are *statistics*.
3. The value of a statistic is likely to be *same* in every sample.
4. *Sampling variation* describes how the value of a *statistic* varies from sample to sample.
5. An initial assumption is made about the *sample statistic*.
6. If the sample results seem inconsistent with what was expected, then the assumption about the population is probably true.
7. In the sample, we know exactly what value of $\hat{p}$ to expect.
8. Hypotheses are made about the population.

25.8 Exercises

Answers to odd-numbered exercises are given at the end of the book.

Exercise 25.1. While playing a die-based game, your opponent rolls a ⚅ ten times in a row.

1. Do you think there is a problem with the die?
2. Explain how you came to this decision.

Exercise 25.2. A friend tosses a coin, and obtains Ⓗ two times in a row.

1. Do you think there is a problem with the coin?
2. Explain how you came to this decision.

Exercise 25.3. Since my wife and I have been married, I have been called to jury service four times. The latest notice reads: 'Your name has been selected at random from the electoral roll'.

In the same time, my wife has *never* been called to jury service. Do you think the selection process really is 'at random'? Explain.

Exercise 25.4. In a 2012 advertisement, an Australian pizza company claimed that their 12-inch pizzas were 'real 12-inch pizzas' [Dunn, 2012].

1. What is a reasonable assumption to make to test this claim?
2. The claim is supported by a sample of 125 pizzas, which gave the sample mean pizza diameter as $\bar{x} = 11.48$ inches. What are the two reasons why the sample mean is not exactly 12-inches?
3. Does the claim appear to be supported by, or contradicted by, the data? Explain.
4. Would your conclusion change if the sample mean was $\bar{x} = 11.25$ inches? Explain.
5. Does your answer depend on the sample size? For example, is observing a sample mean of 11.25 inches from a sample of size $n = 10$ equivalent to observing a sample mean of 11.25 inches from a sample of size $n = 125$? Explain.

Exercise 25.5. Suppose that 36% of all students at a certain large university are aged over 30. A student takes a sample of $n = 40$ students from the School of Arts to determine if students in that school are somehow different from the general university population in terms of age.

1. What is the null hypothesis?
2. If the student researcher finds 13 students in the sample aged over 30, does this present persuasive evidence to change your mind? Explain.
3. If the student researcher finds three students in the sample aged over 30, does this present persuasive evidence to change your mind? Explain.

Answers to *Quick review* questions: 1. True. **2.** False. **3.** False. **4.** True. **5.** False. **6.** False. **7.** False. **8.** True.

26

Hypothesis tests: one proportion

So far, you have learnt to ask an RQ, design a study, classify and summarise the data, and form confidence intervals. **In this chapter,** you will learn to:

- identify situations where conducting a test for a proportion is appropriate.
- conduct hypothesis tests for one sample proportion, using a z-test.
- determine whether the conditions for using these methods apply in a given situation.

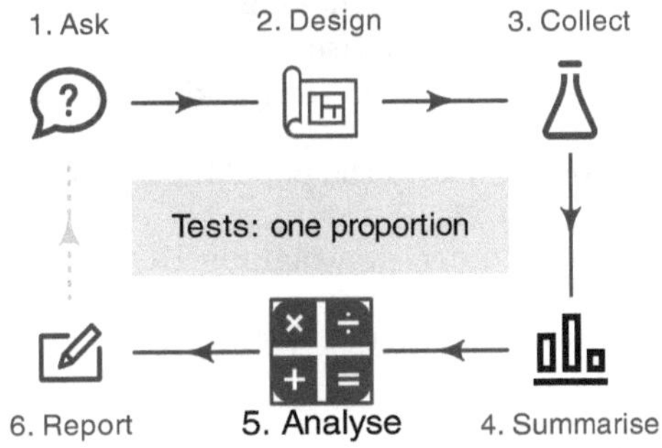

26.1 Introduction: rolling dice

FIGURE 26.1: The packaging (Photo: P. K. Dunn).

When in a toy store one day (for my children, of course), I saw 'loaded dice' for sale (Fig. 26.1). The packaging claimed ONE LOADED & ONE NORMAL. I bought two sets! However, there was no indication as to *which* die was the loaded die. How could I determine which of the dice was loaded? That is, how could I make a *decision* about which die was loaded?

For a die that is *not* loaded, the population proportion of rolling any face of the die is $p = 1/6$. So, for example, the population proportion of rolls that show a ⚀ is $p = 1/6$, using the classical approach to probability. In any *sample* of rolls, however, the proportion of rolls showing a ⚀ would vary due to sampling variation, but would be approximately $\hat{p} = 1/6$ with a fair die.

Suppose I rolled one die a certain number of times (say, $n = 50$ times), then determined the value of the sample proportion $\hat{p}$, the sample proportion of rolls that show a ⚀. It is unlikely that the value of $\hat{p}$ will be *exactly* $1/6$ (the population proportion). If the observed value of $\hat{p}$ was not exactly $1/6$, two possible reasons could explain this discrepancy between the value of the statistic and the assumed value of the parameter (Chap. 25):

- I was rolling the *fair* die (with $p = 1/6$), and the discrepancy between the values of the *population* and *sample* proportions was simply due to sampling variation.
- I was rolling the *loaded* die (with $p \neq 1/6$), and the discrepancy between the values of the *population* and *sample* proportion simply reflected this.

If I observed an unusually small or unusually large sample proportion of rolls that showed a ⚀, I would suspect that I had the loaded die: I was observing something unusual if I had rolled the fair die. This is exactly the decision-making process seen in Chap. 25.

More formally then, the decision-making process (Chap. 25) could proceed as follows.

- Make an *assumption* about the parameter: assume I have a fair die, so that $p = 1/6$, where p is the population proportion of rolls that show a ⚀.
- Describe the *expectations* of the statistic: describe what value of the *sample* proportion $\hat{p}$ could reasonably be expected from a fair die in 50 rolls.
- Evaluate the sample *observations*: roll the die 50 times to find a value of $\hat{p}$ and compare to what was expected.
- Make a *decision* based on what is observed in the sample.

Using this decision-making process (Fig. 26.2), I could decide if the die I had rolled seemed to be the fair die (based on rolling a ⚀; the die may be loaded in a different way, of course). For one specific die, I am asking the decision-making RQ:

> For this die, is the population proportion of rolls that show a ⚀ equal to 1/6?

Answering a decision-making RQ such as this requires a *hypothesis test*.

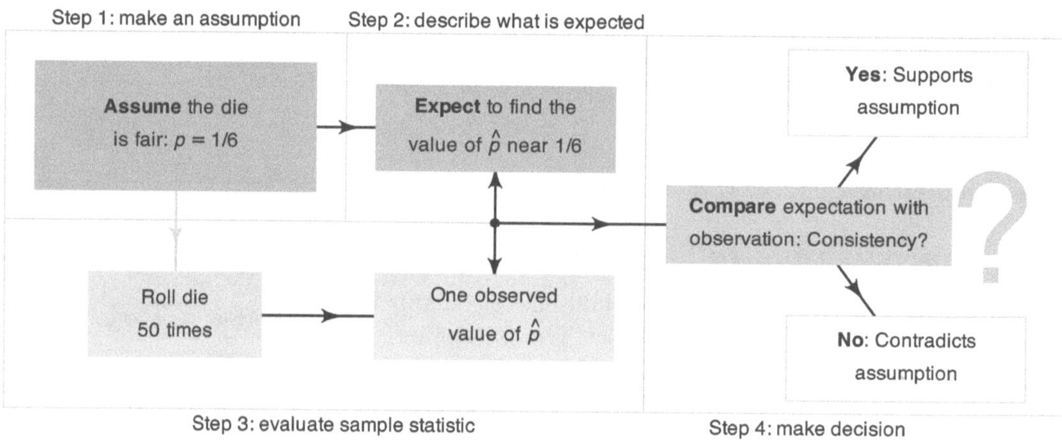

FIGURE 26.2: A way to make decisions for the dice example.

> p refers to the *population* proportion, and $\hat{p}$ refers to a *sample* proportion.

26.2 Rolling dice: the sampling distribution of $\hat{p}$

When a fair, six-sided die is rolled 50 times, what proportion of the rolls will produce a ⚀? That is, what will be the value of the *sample proportion* $\hat{p}$? Of course, no-one knows, because the sample proportion will not be the same for every sample of 50 rolls. The sample proportion *varies* from sample to sample: *sampling variation* exists and is described by the *sampling distribution*.

> Remember: studying a sample leads to the following observations:
> - every sample is likely to be different.
> - we observe just one of the many possible samples.
> - every sample is likely to yield a different value for the statistic.
> - we observe just one of the many possible values for the statistic.
>
> Since many values for the sample proportion are possible, the values of the sample proportion vary (called *sampling variation*) and have a *distribution* (called a *sampling distribution*).

The sampling distribution of $\hat{p}$ was described in Def. 22.1 (and repeated in Def. 26.1 below). The sample proportions are described by

- an approximate normal distribution,
- centred around the *sampling mean*, with a value of $p = 1/6$ (assumed, from H_0),
- with a standard deviation, called the *standard error* s.e.$(\hat{p})$, of

$$\text{s.e.}(\hat{p}) = \sqrt{\frac{p \times (1-p)}{n}} = \sqrt{\frac{\frac{1}{6} \times (1 - \frac{1}{6})}{50}} = 0.0527. \tag{26.1}$$

> **Definition 26.1** (Sampling distribution of a sample proportion with p known). For a known value of p, the *sampling distribution of the sample proportion* is (when certain conditions are met; Sect. 22.7) described by
>
> - an approximate normal distribution,
> - centred around the sampling mean whose value is p,
> - with a standard deviation (called the *standard error* of $\hat{p}$), denoted s.e.$(\hat{p})$, whose value is
>
> $$\text{s.e.}(\hat{p}) = \sqrt{\frac{p \times (1-p)}{n}}, \tag{26.2}$$
>
> where n is the size of the sample used to compute $\hat{p}$, and p is the population proportion.

A picture of this normal distribution can be drawn (Fig. 26.3); the standard error is the standard deviation of the normal distribution in Fig. 26.3. While we still don't know *exactly* what values of $\hat{p}$ the next set of $n = 50$ rolls will produce, we have some idea of *how* the sample proportion varies in samples of 50 rolls. For instance, values of $\hat{p}$ greater than about 0.35 are unlikely to be observed from a fair die (with $p = 1/6$).

26.3 Rolling dice: making a decision

Figure 26.3 shows what values of the sample proportion $\hat{p}$ are expected when a fair die is rolled. Step 3 of the decision-making process (Fig. 26.2) is to roll the die.

When I rolled the die, a ⚀ appeared 19 times in my 50 rolls, a sample proportion of

$$\hat{p} = \frac{19}{50} = 0.38.$$

In this unusual or unexpected? Locating this value of $\hat{p}$ on the sampling distribution in Fig. 26.3 shows that a sample proportion of $\hat{p} = 0.38$ is *highly* unusual from a fair die with

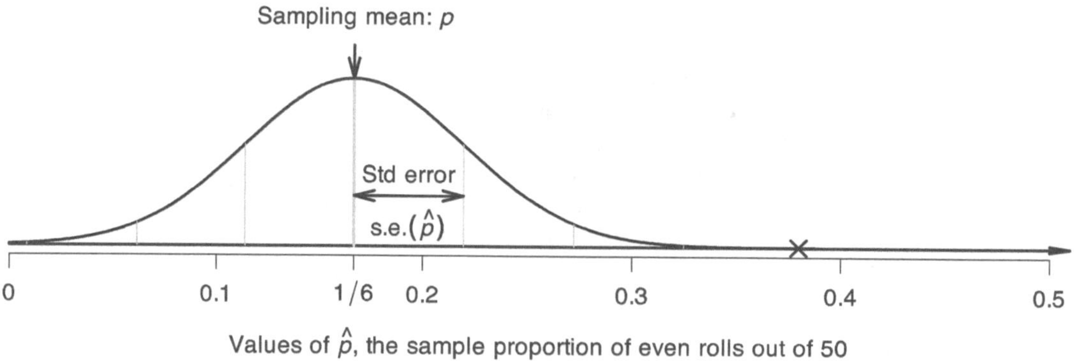

FIGURE 26.3: The sampling distribution is an approximate normal distribution; it shows a model of how the proportion of rolls showing a ⚀ varies, when a die is rolled 50 times. The cross represents the observed sample proportion, $\hat{p} = 0.38$.

$p = 1/6$. More specifically, since the sampling distribution has a normal distribution, a z-score can be computed:

$$z = \frac{\text{statistic} - \text{mean of the distribution}}{\text{std dev. of the distribution}} = \frac{0.38 - (1/6)}{0.05270} = 4.05,$$

which is a *very* large z-score (based on the 68–95–99.7 rule). Using a fair die, observing $\hat{p} = 0.38$ would almost never occur. But I *did* observe $\hat{p} = 0.38$, which suggests that the die I was rolling was *not* the fair die.

I concluded that the die I was rolling was loaded (that is, $p \neq 1/6$). I may be incorrect (after all, it is not *impossible* to observe $\hat{p} = 0.38$), but the evidence is certainly persuasive. Using the decision-making process, a decision has been made about the die.

The process described above is called *hypothesis testing*. Hypothesis testing is used to make decisions about a population after observing just one of the countless possible samples. Formally, the hypothesis test above proceeds as described in the following sections.

26.4 Assumptions: hypotheses

Step 1 in the decision-making process is to make an assumption about the parameter. For the die example, the parameter is p, the population proportion of rolls that show a ⚀. The assumption is that $p = 1/6$. This is called the *null hypothesis*, denoted by H_0:

$$H_0\colon p = 1/6.$$

The null hypothesis states the value of p is $1/6$; in other words, if the sample proportion $\hat{p}$ is not equal to $1/6$, the discrepancy is explained by sampling variation. The null hypothesis is always the 'sampling variation' explanation for the discrepancy between the values of the statistic and the parameter (Sect. 25.4.1).

The other explanation for why the value of the sample proportion $\hat{p}$ is not equal to $1/6$ is called the *alternative hypothesis* (denoted H_1), that the population proportion is *not* $1/6$, and this is the cause of the discrepancy between the values of the statistic and the parameter:

$$H_1\colon p \neq 1/6.$$

These two hypotheses offer different explanations for the discrepancy between the values of the population proportion (the parameter) and the sample proportion (the statistic). The null hypothesis H_0 states that $p = 1/6$ and the discrepancy is due to sampling variation. The alternative hypothesis H_1 states that $p \neq 1/6$, which explains the discrepancy.

Here, the RQ here is open to the value of p being smaller *or* larger than 1/6; that is, two possibilities are considered. Hence, we write $p \neq 1/6$, which is called a *two-tailed* alternative hypothesis. Alternative hypotheses like $p > 1/6$ (the population proportion is *larger* than 1/6) or $p < 1/6$ (the population proportion is *smaller* than 1/6) are *one-tailed* hypothesis.

⚠ The form of the alternative hypothesis (either one- or two-tailed) depends on what the research question asks, *not the data.*

26.5 Expectations: sampling distribution for $\hat{p}$

Step 2 in the decision-making process is to describe what values of the statistic (i.e., $\hat{p}$) could be expected under the assumption about the parameter (i.e., *when the null hypothesis is true*). Hypothesis testing *always* begins by assuming the null hypothesis is true.

⚠ The decision-making process begins by assuming the *null hypothesis* is true. Thus, *the onus is on the data to refute the null hypothesis, the initial assumption.*

That is, the null hypothesis is retained unless persuasive evidence emerges to change our mind.

Effectively, this step requires describing the sampling distribution of the statistic. For the die example, the sampling distribution for $\hat{p}$ is (see Def. 26.1):

- an approximate normal distribution,
- centred around the sampling mean whose value is $p = 1/6$,
- with a standard deviation, whose value is $\text{s.e.}(\hat{p}) = 0.05270\ldots$

Drawing the picture of the sampling distribution (like Fig. 26.3) using this information is not necessary, but may be helpful.

26.6 Observations: z-score

Step 3 in the decision-making process is to evaluate the observations. As noted above, a ⚀ was observed in 19 of the 50 rolls, so $\hat{p} = 0.38$. Since the sampling distribution has a normal distribution, the corresponding z-score was computed as $z = 4.05$, which very large.

In hypothesis testing, the z-score is called the *test statistic*. The test statistic measures how far, in relative terms, the sample proportion is from the assumed value of the parameter.

26.7 Decision: P-value

Step 4 of the decision-making process is to use the information to make a decision: is the sample statistic *consistent* with what was expected under the assumption that $p = 1/6$, or does it *contradict* what was expected?

For the die example, the decision is reasonably easy: $z = 4.05$ is *very* large and *very* unlikely to be observed if $p = 1/6$. This means the sample evidence *contradicts* what was expected if the assumption was true: persuasive evidence exists that the die is loaded.

More generally, evidence is evaluated using a P-value. P-values refer to the area *more extreme* than the calculated test statistic in the sampling distribution.

For this situation, where the sampling distribution has a normal distribution, P-values refer to the area *more extreme* than the calculated z-score (the statistic) in the normal distribution; that is, the area in the *tails* of the distribution (see Fig. 26.4). This is a way to measure how unusual the calculated z-score is.

For *two-tailed* alternative hypotheses, the P-value is the combined area in the lower and upper tails that correspond to the positive *and* negative values of the test statistic. For *one-tailed* alternative hypotheses, the P-value is the area in one tail only. Clearly, since the P-value is a probability, its value is always between 0 and 1.

Since the sampling distribution has a normal distribution in this example, P-values can be approximated using the 68–95–99.7 rule and a diagram (Sect. 20.5; Sect. 26.7.1), or more precisely using the z-tables in Appendices B.1 and B.2 (Sect. 20.7; Sect. 26.7.2). P-values are also reported by software for most statistical tests.

26.7.1 Approximating P-values using the 68–95–99.7 rule

The 68–95–99.7 rule can be used to determine *approximate* P-values. To demonstrate, suppose the computed z-score was $z = 1$. Then, the two-tailed P-value is the shaded tail-area in Fig. 26.4 (top left panel): about 32%, based on the 68–95–99.7 rule. The two-tailed P-value would be the same if $z = -1$ (as both tails are of interest). The *one-tailed* P-value would be the area in one-tail (Fig. 26.4, bottom left panel): about 16%, based on the 68–95–99.7 rule.

As another example, suppose the calculated z-score was $z = -2$. Then, the two-tailed P-value is the shaded area shown in Fig. 26.4 (top right panel): about 5%, based on the 68–95–99.7 rule. The two-tailed P-value would be the same if $z = 2$. The *one-tailed* P-value would be the area in one tail only (Fig. 26.4, bottom right panel): about 2.5%, based on the 68–95–99.7 rule.

Of course, calculated z-scores are unlikely to be exactly $z = 1$ or $z = -2$. Suppose the z-score is a little *larger* than $z = 1$; say $z = 1.2$. Then, the two-tailed area will be a little *smaller* than the tail area when $z = 1$ (Fig. 26.5, left panel). The two-tailed P-value is a little *smaller* than 0.32.

Similarly, suppose the z-score is not quite equal to $z = -2$; say $z = -1.9$. Then, the two-tailed area will be a little *larger* than the tail area when $z = -2$ (Fig. 26.5, right panel). The two-tailed P-value is a little *larger* than 0.05.

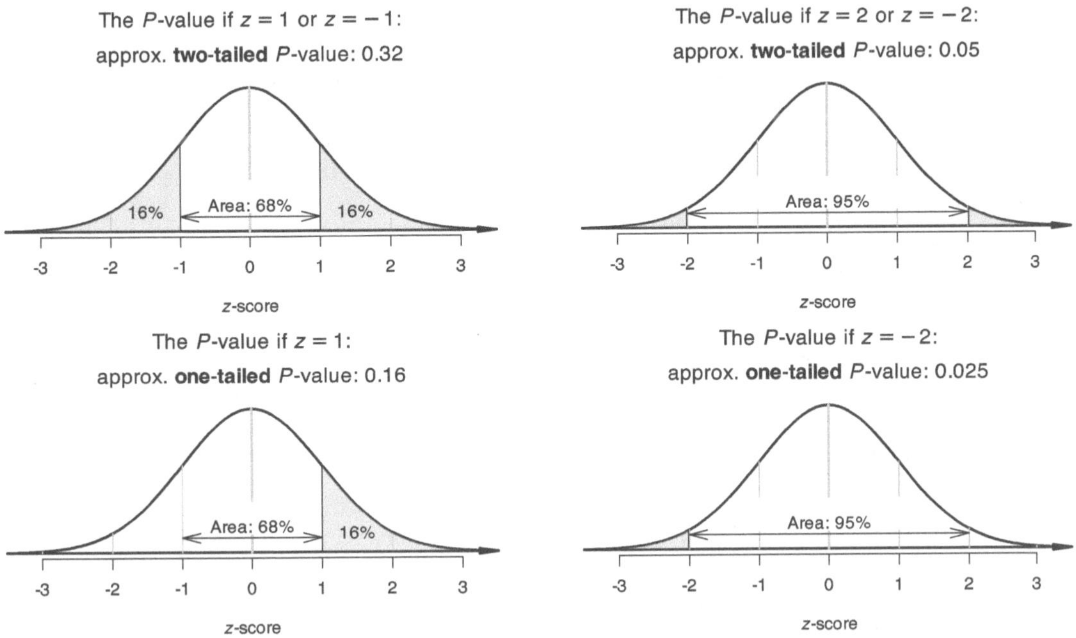

FIGURE 26.4: The two-tailed P-value is the combined area in the two tails of the distribution; the one-tailed P-value is the area in one tail only. Top left panel: if $z = 1$ (or $z = -1$), the two-tailed P-value is approximately 0.16. Top right panel: if $z = 2$ (or $z = -2$), the two-tailed P-value is approximately 0.05. The corresponding one-tailed P-values are half the two-tailed P-values, and are shown in the bottom panels.

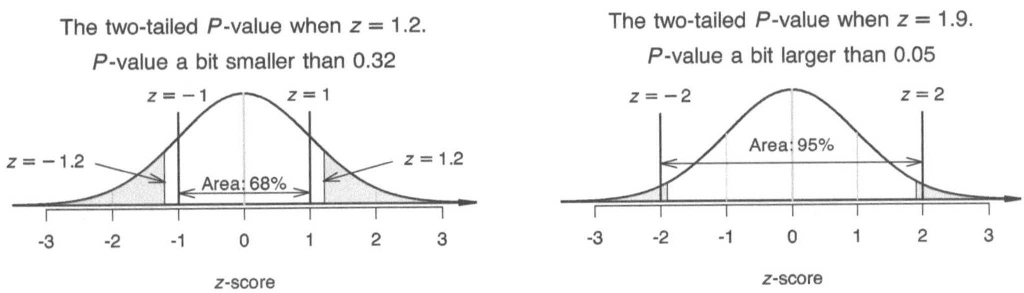

FIGURE 26.5: The two-tailed P-value for z-scores not aligned with the 68–95–99.7 rule. Left panel: when $z = 1.2$ (or $z = -1.2$). Right panel: when $z = 1.9$ (or $z = -1.9$).

26.7.2 More precise P-values using tables

Using the tables of areas under normal distributions (Appendices B.1 and B.2), more precise P-values can be found using the ideas from Sect. 20.6. For instance (see Fig. 26.5):

- for $z = 1.2$: the area to the *left* of $z = -1.2$ is 0.1151, and the area to the *right* of $z = 1.2$ is 0.1151, so the *two-tailed* P-value is $0.1151 + 0.1151 = 0.2302$. This is a little smaller than 0.32, as estimated above.
- for $z = 1.9$: the area to the *left* of $z = -1.9$ is 0.0287, and the area to the *right* of $z = 1.9$ is 0.0287, so the *two-tailed* P-value is $0.0287 + 0.0287 = 0.0574$. This is a little larger than 0.05, as estimated above.

In this die-rolling example, where $z = 4.05$, the tail area is *very* small (using Appendices B.1 and B.2), and zero to four decimal places. P-values are never exactly zero, so we write $P < 0.0001$ (that is, the P-value is *less than* 0.0001).

P-values tells us the probability of observing the sample statistic (or a value even more extreme), assuming the null hypothesis is true. In the die-rolling example, the P-value is the probability of observing the value of $\hat{p} = 0.38$ (or a more extreme value), just through sampling variation if $p = 1/6$. Then (see Fig. 26.6):

- 'big' P-values mean the sample statistic (i.e., $\hat{p}$) could reasonably have occurred through sampling variation in one of the many possible samples, if the assumption made about the parameter (stated in H_0) was true; the data *do not* contradict the assumption in H_0. There *is no* persuasive evidence to support the alternative hypothesis.
- 'small' P-values mean the sample statistic (i.e., $\hat{p}$) is *unlikely* to have occurred through sampling variation in one of the many possible samples, if the assumption made about the parameter (stated in H_0) was true; the data *do* contradict the assumption in H_0. There *is* persuasive evidence to support the alternative hypothesis.

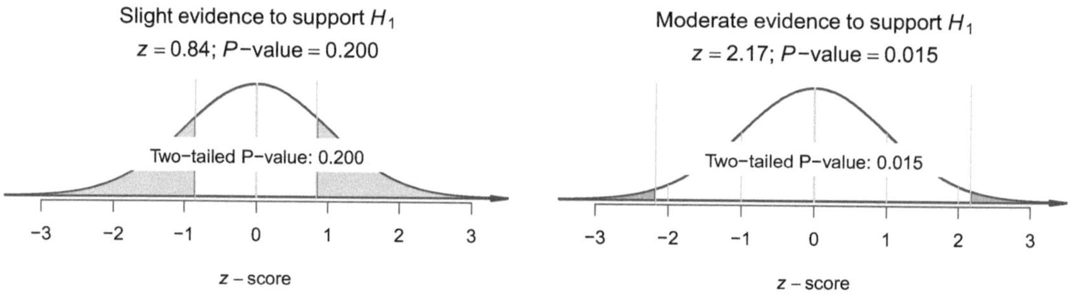

FIGURE 26.6: The strength of evidence: P-values. As the z-score becomes larger, the P-value becomes smaller, and it is more likely that the evidence contradicts the null hypothesis.

What is meant by 'small' and 'big' in this context? What represents persuasive evidence to support the alternative hypothesis? A P-value smaller than 5% (or 0.05) is usually considered 'small', and persuasive evidence to support the alternative hypothesis. In contrast, a P-value larger than 5% (or 0.05) is usually considered 'big', and *not* persuasive evidence to support the alternative hypothesis.

> ⚠ The value of 0.05 given here is *arbitrary*, and in some disciplines the distinction is made when $P = 0.01$ or $P = 0.10$ instead.

Rather than having an arbitrary boundary between 'big' and 'small', a more sensible approach is to qualify the strength of the evidence that supports the alternative hypotheses (discussed further in Sect. 28.6).

In this die-rolling example, where the P-value is *very* small, the data contradict the null hypothesis that $p = 1/6$: the evidence supports the alternative hypothesis that $p \neq 1/6$. This suggests that the die is very likely *not* fair.

⚠ *Be careful interpreting the results!* We cannot be *sure* that the die is unfair. *A small P-value is not proof that the die is loaded.* The die may be fair but, due to sampling variation, the sample we observed may simply have produced an unusually high proportion of rolls that show a ⚀ by chance.

The result is interpreted as 'there is strong evidence that the die is unfair'. Remember: *the onus is on the data to refute the null hypothesis, the initial assumption.*

Example 26.1 (Interpreting P-values). In the die example, suppose we found the two-tailed P-value as 0.26. This is 'large' (i.e., much larger than 0.05). Then the observed value of $\hat{p}$ could easily be explained by chance, and is *not* persuasive evidence to support the alternative hypothesis (that the die is unfair). There is no persuasive evidence that p is not 1/6.

Note that a different value for s.e.$(\hat{p})$ is required to produce the CI (see Def. 22.2).

26.8 Writing conclusions

In general, communicating the results of any hypothesis test requires:

- an answer to the RQ, worded in terms of how much evidence exists to support the *alternative* hypothesis.
- a summary of the evidence used to reach that conclusion (such as the z-score and P-value, including if the P-value is one- or two-tailed).
- sample summary information (see Chap. 22), summarising the data used to make the decision (which usually includes a CI for the parameter).

So for the die-rolling example, write:

> The sample provides strong evidence ($z = 4.05$; two-tailed $P < 0.001$) that the proportion of rolls that show a ⚀ is not 1/6 ($\hat{p} = 0.38$; approx. 95% CI: 0.243 to 0.517; $n = 50$ rolls) in the population.

This statement includes the three necessary components:

- an answer to the RQ: 'The sample provides very strong evidence... that the population proportion is not 1/6'.
- the evidence used to reach the conclusion: '$z = 4.05$; two-tailed $P < 0.001$'.
- sample summary information (including a CI).

⚠ Since the *null* hypothesis is initially assumed to be true, *the onus is on the evidence to refute the null hypothesis.* That is, we retain the null hypothesis unless there is persuasive evidence to stop doing so. Hence, conclusions are worded in terms of how strongly the evidence (i.e., sample data) supports the alternative hypothesis.

The alternative hypothesis *may* or *may not* be true, but we report how strongly the evidence (data) supports the alternative hypothesis. Conclusions are *not* worded in terms of how much evidence supports the null hypothesis.

26.9 Process overview

Let's recap the decision-making process, in this context of rolling a ⚀ (Fig. 26.7):

1. *Assumption.* Write the *null hypothesis* and *alternative hypothesis* about the *parameter* (based on the RQ), where p is the population proportion of rolls that are a ⚀:
   - H_0: $p = 1/6$ (i.e., sampling variation explains the discrepancy between p and $\hat{p}$).
   - H_1: $p \neq 1/6$ (this is a two-tailed alternative hypothesis).
2. *Expectation.* The sampling distribution describes what values to reasonably expect from the sample statistic across all possible samples, *if* the null hypothesis is true. In this situation, the sampling distribution has an approximate normal distribution.
3. *Observation.* Compute the z-score ($z = 4.05$), a measure of the discrepancy between the assumed population value, and the observed sample value. This is a very large value.
4. *Decision.* Determine if the data are consistent with the assumption, by computing the P-value. Here, the two-tailed P-value is (much) less than 0.0001, so strong evidence exists that p is *not* 1/6.

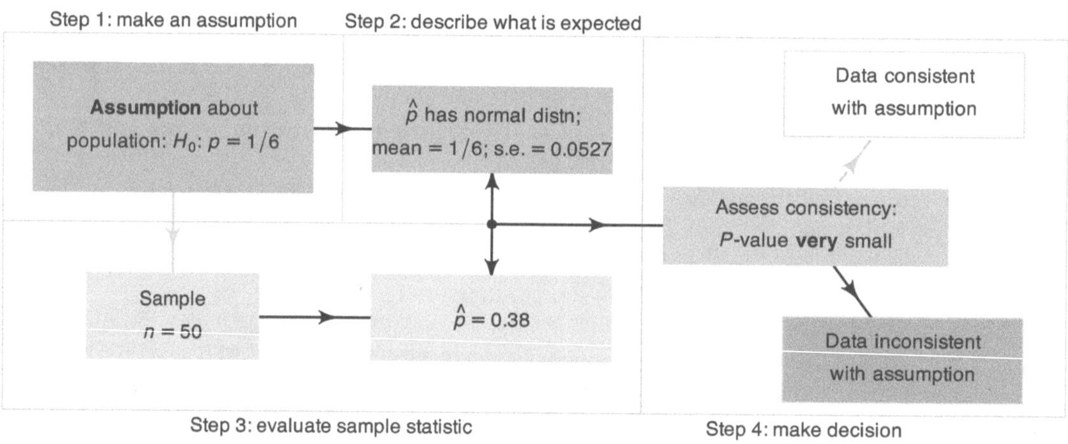

FIGURE 26.7: The decison-making process for the die-rolling data.

26.10 Statistical validity conditions

The hypothesis test conducted in this chapter assumes the sampling distribution is approximately a normal distribution (and so, for example, the 68–95–99.7 rule can be applied). This is only true if certain conditions are met.

The *statistical validity conditions* for a test for a single proportion is that the *expected* number of individuals in the group of interest (i.e, $n \times p$) and in the group *not* of interest (i.e., $n \times (1 - p)$) both exceed five; that is:

- both $n \times p > 5$ *and* $n \times (1 - p) > 5$.

The value of 5 here is a rough figure; some books give other values (such as 10). This condition ensures that the *sampling distribution of the sample proportions has an approximate normal distribution* (so that, for example, the 68–95–99.7 rule can be used). The units of analysis are also assumed to be *independent* (e.g., from a simple random sample). For a test for one proportions, these conditions are similar to those for the CI for one proportion (Sect. 22).

If the statistical validity conditions are not met, other similar options include using a binomial test [Conover, 2003].

Example 26.2 (Statistical validity). The hypothesis test regarding the dice is statistically valid. Firstly, $n \times p = 50 \times (1/6) = 8.666\ldots$ (i.e., expect about 8.7 rolls to show a ⚀), and $n \times (1 - p) = 41.666\ldots$ (i.e., expect about 41.7 rolls to *not* show a ⚀). *Both* comfortably exceed five, so the normal distribution will be a good approximation for the sampling distribution. This is what we observe from a computer simulation (Fig. 26.8, left panel).

Example 26.3 (Statistical validity). Suppose the die was rolled 10 times rather than 50 times. Then, $n \times p = 10 \times (1/6) = 1.666\ldots$ and $n \times (1 - p) = 10 \times (1 - 1/6) = 8.333\ldots$. These do not *both* exceed five, so the normal distribution may be a poor approximation for the sampling distribution.

This is what we observe from simulating the situation (Fig. 26.8, right panel). The normal model is poor: the simulation shows that the sample proportions are not even symmetrically distributed.

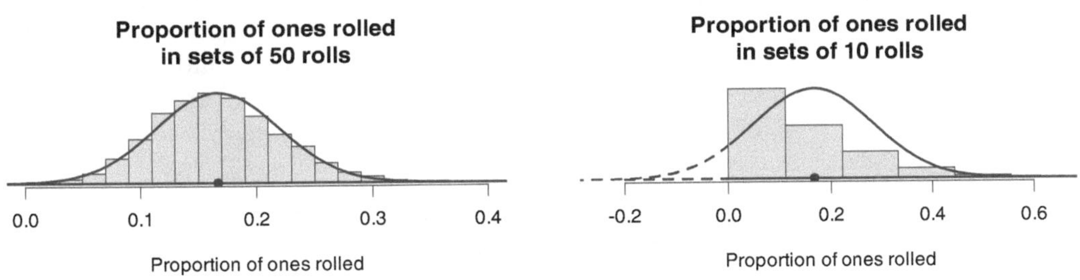

FIGURE 26.8: The sampling distributions for two situations for rolling a die. Left: for sets of 50 rolls, the sampling distribution does have an approximate normal distribution. Right: for sets of 10 rolls, the sampling distribution does not have a normal distribution. The solid lines show the approximate normal distributions, and the histograms show the simulated distribution of the sample proportions over many sets of rolls. The solid dots are the value $p = 1/6$, the population proportion of rolls that show a ⚀.

26.11 Example: rolling the other die

In 50 rolls of the *other* die, I found a ⚀ on 7 rolls, so that $\hat{p} = 7/50 = 0.14$. To determine if this die appears loaded, the hypotheses are the same as before:

$$H_0\colon p = 1/6 \qquad \text{and} \qquad H_1\colon p \neq 1/6.$$

Following the procedures above (check!) and using the same hypotheses, $z = -0.506$ and (using tables) the two-tailed P-value is $2 \times 0.3061 = 0.6122$. This means that the sample result was not unusual if $p = 1/6$, and is certainly not persuasive evidence to support the alternative hypothesis. There is *no evidence* to suggest the second die is loaded.

This all implies the first die was the loaded die. Now I need to decide how to distinguish the two dice so I can tell which is which...

A large P-value does not prove that the die is fair! It only means that the proportions of rolls that produce a ⚀ is not unusual... but perhaps the die is loaded in some other way (i.e., to produce more-than-expected rolls of a ⚄).

A large P-value does not necessarily mean that the die is fair! The die may indeed be loaded to produce a larger-than-expected numbers of rolls that show a ⚀, but (due to sampling variation) the sample we observed simply did not provide evidence to make that conclusion.

The result is interpreted in terms of how much evidence exists to support the alternative hypothesis. The onus is on the data (i.e., evidence) to refute the assumption made in the null hypothesis.

26.12 Example: dominance of birds

Barve and Dhondt [2017] compared two types of birds (male green-backed tits; male cinereous tits) to see which was more behaviourally dominant over winter. If the species were equally-dominant, then about 50% of the interactions would be won by each species. If we define p as the proportion of interactions won by green-backed tits, then we would expect $p = 0.50$. However, in the 45 interactions observed between the two species, green-backed tits won 37 interactions (i.e., $\hat{p} = 37/45 = 0.82222$). A discrepancy exists between the sample proportion ($\hat{p} = 0.8222$) and the expected population proportion $p = 0.50$.

Of course, different sample of 45 interactions would produce different values of $\hat{p}$. To test if the population proportion of interaction wins could be equally shared, the hypotheses are:

$$H_0\colon p = 0.5 \quad \text{and} \quad H_1\colon p \neq 0.5 \text{ (two-tailed)}.$$

The test is statistically valid, since both $n \times p = 45 \times 0.5 = 22.5$ and $n \times (1 - p) = 22.5$ exceed five. The *standard error* is

$$\text{s.e.}(\hat{p}) = \sqrt{\frac{p \times (1-p)}{n}} = \sqrt{\frac{0.50 \times (1 - 0.50)}{45}} = 0.0745356...$$

Then, the value of the *test statistic* is:

$$z = \frac{\hat{p} - p}{\text{s.e.}(\hat{p})} = \frac{0.82222 - 0.50}{0.0745356} = 4.322.$$

This is a *very* large z-score, so the P-value will be very small, using the 68–95–99.7 rule, or using tables. This is persuasive evidence to support the alternative hypothesis. We write:

> *Very* strong evidence exists in the sample ($P < 0.0001$; $z = 4.325$) that the interactions were not won equally by each species ($\hat{p} = 0.8222$ won by green-backed tits; approx. 95% CI: 0.708 to 0.936; $n = 45$) in the population.

Note that a different value for s.e.($\hat{p}$) is required to produce the CI (see Def. 22.2).

26.13 Chapter summary

These steps are used to test a hypothesis about a population proportion p.

- Write the null hypothesis (H_0; the sampling variation explanation) and the alternative hypothesis (H_1); initially *assume* the value of p in the null hypothesis to be true.
- Describe the *sampling distribution*, which describes what to *expect* from the sample statistic across all possible samples, based on this assumption: under certain statistical validity conditions, the sample mean varies with:
  - an approximate normal distribution,
  - with sampling mean, whose value is the value of p,
  - with a standard deviation of $\text{s.e.}(\hat{p}) = \sqrt{\frac{p \times (1-p)}{n}}$, where p is the hypothesised value given in the null hypothesis, and n is the sample size.
- Compute the value of the *test statistic*:

$$z = \frac{\hat{p} - p}{\text{s.e.}(\hat{p})}.$$

- Compute an approximate P-*value* using the 68–95–99.7 rule, or using tables. Use the P-value to make a decision, and write a conclusion.
- Check the statistical validity conditions.

26.14 Quick review questions

A study of diseases in Native Americans [Kizer et al., 2006] found 381 obese or overweight patients in 449 patients. Across all the US population, the percentage obese or overweight was 65%. The researchers wanted to determine if the percentage of obesity/overweight Native Americans was *greater* than that of the general population.

Are the following statements *true* or *false*?

1. The sample size is $n = 381$.
2. The value of the *sample* proportion is $\hat{p} = 381$.
3. The *null* hypothesis is H_0: $p = 0.65$.

4. The *alternative* hypothesis is H_0: $p = 0.8486$.
5. We initially assume the *population* proportion of overweight/obese Native Americans is 0.65.
6. The *alternative* hypothesis is *one*-tailed.
7. In a one-sample test of proportion, the z-score is always large.
8. The value of the z-score for this example is 8.82.
9. We have evidence to support the alternative hypothesis in this example.
10. We always accept the *null* hypothesis.

26.15 Exercises

Answers to odd-numbered exercises are given at the end of the book.

Exercise 26.1. Explain *why* the standard error is computed using p for hypothesis testing, but using $\hat{p}$ for CIs.

Exercise 26.2. Explain why describing the sampling distribution is difficult if we *assume* $p \neq 1/6$.

Exercise 26.3. In the die example, the observed proportion is 0.38. Could we simply state that the proportion clearly is not $1/6 = 0.1666$? Explain.

Exercise 26.4. Explain why we compute s.e.$(\hat{p})$ and not s.e.(p).

Exercise 26.5. What is wrong with the following statement, after testing H_0: $p = 0.25$:

> There is very strong evidence that the sample proportion is greater than 0.25.

Exercise 26.6. Explain what is wrong with this statement from Davis et al. [2024], that appears under their Table 2:

> One proportion z-test with $H_0 = 0.076$, the proportion of UDT in our sample...

Exercise 26.7. The study of herbal medicines is complicated, as *blinding* subjects is difficult: placebos are often easily identifiable by eye, by taste, or by smell.

Loyeung et al. [2018] studied if subjects could identify potential placebos at a *better* rate than just guessing. The 81 subjects were each presented with a choice of five different supplements, four of which were placebos. Subjects were asked to select which one was the legitimate herbal supplement based on the *taste*; 50 subjects correctly selected the true herbal supplement.

1. If the subjects were selecting the true herbal supplement randomly, what proportion of subjects would be expected to select the correct supplement as the true herbal medicine?
2. Write the hypotheses for addressing the aims of the study.
3. Is this a one- or two-tailed test? Explain.
4. Sketch the *sampling distribution* of the sample proportion, assuming H_0 is correct, for $n = 81$.
5. Is there evidence that people can identify the true supplement by taste?
6. Are the statistical validity conditions satisfied?

Exercise 26.8. Kim et al. [2004] studied the measles-rubella vaccination-rates in Korea, comparing the proportion of children susceptible to measles with the *World Health Organization* target proportion (for children aged 5 to 9 years old: 10%).

The aim was to test if the proportion of Korean children susceptible to measles in the *population* was 10% or *lower* (i.e., better). In the study, 55 children out of 972 were susceptible to measles.

1. Compute the sample proportion $\hat{p}$ of children susceptible to measles.
2. Write the hypotheses for the test. Is the test one- or two-tailed?

3. Compute the standard error for the test.
4. Compute the z-score and determine the P-value.
5. Write a conclusion.
6. Are the statistical validity conditions satisfied?

Exercise 26.9. Streeting et al. [2022] studied western saw-shelled turtles. When eggs were incubated at 27°C, they observed that 29 males and 44 females hatched. Are the proportions of male and female turtles that hatch at this temperature equal?

Exercise 26.10. [*Dataset*: `PremierL`] In the 2019/2020 English Premier League (EPL), the home team won 91 games, and the away team won 67 games. (Another 50 games were draws.)

Use the $n = 158$ games with a result to determine if there is evidence that the home team wins more often than 50% (i.e., that there is a home-side advantage).

Exercise 26.11. Maeda [2013] introduced pedal machines on the first floor of the Joyner Library for use by students at East Carolina University (ECU) to increase activity in library users. At ECU, 60.2% of all students were females (i.e., in the population). Students were observed using the machine on 589 occasions, of which 295 times were by females

Is there evidence that the proportion of female users of the machines was *lower* than the overall female proportion at the university? What would you conclude?

Exercise 26.12. Koenen [1995] found that 88 of the 357 visitors to Las Vegas casinos in 1995 were smokers. At the time, 25.5% of the general US population were smokers (based on data from the US *National Center for Health Statistics*). Is the proportion of smokers among casino-goers the same as for the general US population?

Exercise 26.13. Nochera and Ragone [2019] developed gluten-free pasta made from breadfruit. In the study sample, 57 of the 71 participants stated that they liked the pasta. Do the researchers have sufficient evidence to claim that the 'majority of people like breadfruit pasta'?

Exercise 26.14. Carpal Tunnel Syndrome (CTS) is a painful condition in the wrists. Boltuch et al. [2020] were interested in whether 'a relationship exists between the palmaris tendon [and] carpal tunnel syndrome (CTS)' (p. 493). The palmaris longus (PL) tendon is visually absent in about 15% of the population. The researchers found PL was visually absent in 33 of 516 CTS wrists in their sample. Is there evidence to suggest that rate of PL absence is *different* in CTS cases, compared to the general population?

Exercise 26.15. Siegfried et al. [2014] studied resistance of some commercial corn varieties to the European corn borer. Borers were collected from corn in Iowa and Nebraska.

Researchers aimed to estimate the frequency of resistance to the toxin in the corn. By mating borers collected from the field with various resistant laboratory individuals, they could determine what proportion of resistant individuals to expect in the second generation offspring. In one study of $n = 172$ second-generation individuals, 24 were found to be resistant. The theoretical expectation was that 1-in-16 of the second-generation borers would be resistant if the field borers were resistant. Perform a hypothesis test to determine if the data suggest that the field borers were resistant (that is, if the population proportion is 1/16) as expected.

Exercise 26.16. Davidovic et al. [2019] studied street-light preferences of drivers. Drivers were asked to conduct a series of manoeuvres under 3 000K LED light and then under 4 000K LED lights. They were then asked to decide which street light they preferred. Out of the 52 subjects, 29 preferred the 3 000K LED lights. Is there evidence that the choice between the two street lights is random, or is there evidence of a preference for one over the other?

Exercise 26.17. The euro was introduced as a currency on 01 January 1999. According to a report by the *New Scientist*, students in Poland spun a Belgian one-euro coin 250 times, and found 140 heads (as reported by Gelman and Nolan [2002]). This resulted in an 'accusation of bias' in the *New Scientist* article. However, every set of 250 spins can produce a different proportion of heads, so perhaps the results is just due to randomness. Does this sample of 250 spins suggest that the one-euro Belgian coin is biased?

Exercise 26.18. As noted in Sect. 18.3.2, the *Australian Bureau of Statistics* (ABS) stated that:

> The sex ratio for all births registered in Australia generally fluctuates around 105.5 male births per 100 female births.

(This statistic does not use births registered as 'other' or 'not stated'.)

1. The value of 105.5 is effectively a population odds ratio of male-to-female births. Show that this is equivalent to the population proportion of male births as 0.51338 (not including 'other' or 'not stated').
2. In 2021, there were 148 636 male births and 140 944 female births. Compute the *sample* proportion of male births in 2021 (to five decimal places).
3. Conduct a test to determine if the 2021 data appear different to the long-term proportion.

Answers to *Quick review* questions: **1.** False. **2.** False; $\hat{p} = 381/449 = 0.84855$. **3.** True. **4.** False. **5.** True. **6.** True. **7.** False. **8.** True. **9.** True. **10.** False.

27

Hypothesis tests: one mean

You have learnt to ask an RQ, design a study, classify and summarise the data, construct confidence intervals, and perform a hypothesis test for one proportion. **In this chapter**, you will learn to:

- identify situations where conducting a test for a mean is appropriate.
- conduct hypothesis tests for one sample mean, using a t-test.
- determine whether the conditions for using these methods apply in a given situation.

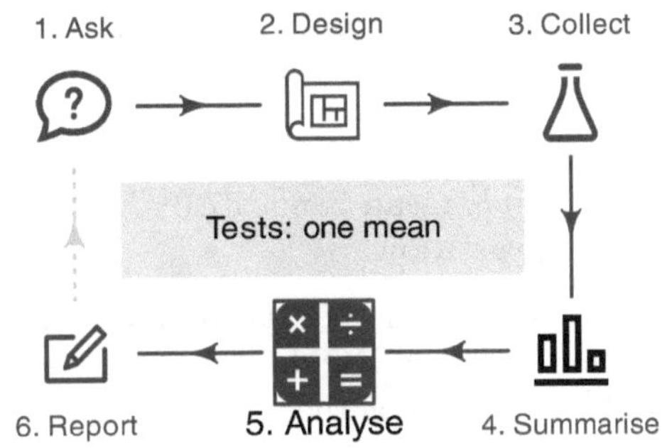

27.1 Introduction: body temperatures

The average internal body temperature is commonly believed to be 37.0°C (98.6°F). This value is based on data over 150 years old [Wunderlich, 1868]. Since then, the methods for measuring internal body temperature have changed substantially:

> Thermometers used by Wunderlich were cumbersome, had to be read in situ, and, when used for axillary measurements [i.e., under the armpit]... required 15 to 20 mins to equilibrate. Today's thermometers are smaller and more reliable and equilibrate more rapidly. In addition, the mouth and rectum have replaced the axilla [armpit] as the preferred sites for monitoring body temperature.
>
> — Mackowiak et al. [1992], p. 1579

For this reason, the reported internal body temperature (recorded by newer instruments, in different locations) may have changed since the 1860s. Therefore, we could ask:

> Is the *population* mean internal body temperature equal to 37.0°C?

A *decision* is sought about the value of the *population* mean body temperature. Presumably, the intended population is all people, though the population, in practice, may depend on what population is represented by the available data.

The population mean internal body temperature will never be known: the internal body temperature of every person alive would need to be measured, and even those not yet born. A *sample* must be studied.

Define the parameter as μ, the population mean internal body temperature (in °C). A *sample* of people can be used to determine if evidence exists that the *population* mean internal body temperature is not 37.0°C, using the decision-making process (Sect. 25.3).

27.2 Assumptions: hypotheses

Step 1 of the decision-making process is to *assume* a value for the parameter. The established claim is that the population mean internal body temperature is 37.0°C, so we assume this value. This assumption becomes the null hypothesis:

$$H_0: \mu = 37.0.$$

If the *sample* mean is not 37.0°C, this hypothesis proposes that the discrepancy is due to sampling variation.

The RQ asks if the *population* mean internal body temperate μ is *equal* to 37.0°C, or if it has *changed*. The RQ does not specifically ask if μ is smaller than 37.0°C, or larger than 37.0°C. This means the alternative hypothesis is two-tailed:

$$H_1: \mu \neq 37.0.$$

27.3 Expectations: sampling distribution for $\bar{x}$

Step 2 of the decision-making process is to describe what values of the statistic (in this case, the sample mean $\bar{x}$) can be expected if the value of μ is assumed to be 37.0 (the value specified in H_0). In other words, the *sampling distribution* of $\bar{x}$ needs to be described.

The sample mean *varies* from sample to sample, and varies with a normal distribution (whose standard deviation is called the *standard error*) under certain conditions (given in Sect. 27.8). The sampling distribution of $\bar{x}$ was described in Sect. 23.3, and repeated below.

Definition 27.1 (Sampling distribution of a sample mean). When the *population* standard deviation is unknown, the *sampling distribution of the sample mean* is (when certain conditions are met; Sect. 23.5) described by:

- an approximate normal distribution,
- centred around a sampling mean whose value is μ,
- with a standard deviation (called the *standard error of the mean*), denoted s.e.$(\bar{x})$, whose value is

$$\text{s.e.}(\bar{x}) = \frac{s}{\sqrt{n}}, \qquad (27.1)$$

where n is the size of the sample, and s is the sample standard deviation of the observations.

The mean of this sampling distribution—the *sampling mean*—has the value μ. The standard deviation of this sampling distribution is called the *standard error of the sample means*, denoted s.e.$(\bar{x})$. When the *population* standard deviation σ is *unknown*, the value of the standard error happens to be (see Equation (27.1))

$$\text{s.e.}(\bar{x}) = \frac{s}{\sqrt{n}}.$$

Mackowiak et al. [1992] gathered body-temperature data for $n = 130$ people, collated by Shoemaker [1996] (Table 27.1; Fig. 27.1). The data all come from

> ... volunteers participating in Shigella vaccine trials conducted at the University of Maryland Center for Vaccine Development, Baltimore...
>
> — Mackowiak et al. [1992], p. 1578

Hence, the population for the study (and RQ) should be redefined accordingly. From software output (Fig. 27.2), the *sample* mean is $\bar{x} = 36.8052$°C and the *sample* standard deviation is $s = 0.4073$°C. Using this value of s, the sampling distribution of $\bar{x}$ can be described, if μ really was 37.0:

- an approximate normal distribution,
- with a sampling mean whose value is $\mu = 37.0$ (from H_0),
- with a standard deviation of $\text{s.e.}(\bar{x}) = s/\sqrt{n} = 0.4073/\sqrt{130} = 0.0357$ (as in the output).

TABLE 27.1: The body temperature data: the first nine and last nine of the 130 ordered observations.

Body temperature (in °C)			
35.72	36.17	⋮	37.39
35.94	36.17	37.28	37.44
36.06	36.22	37.28	37.72
36.11	36.28	37.33	37.78
36.17	⋮	37.33	38.22

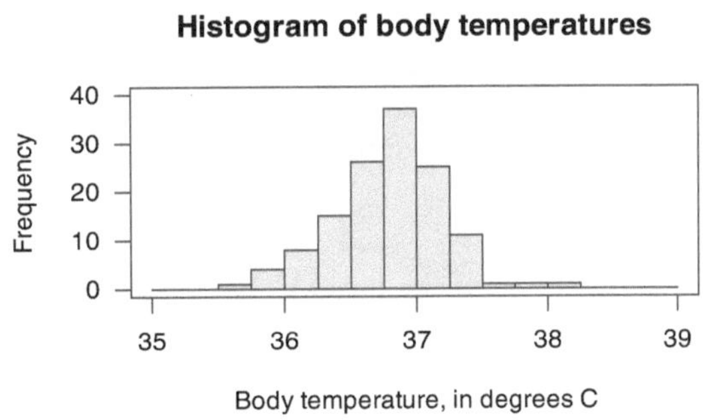

FIGURE 27.1: The histogram of the body temperature data.

Descriptives

	N	Mean	Median	SD	SE
BodyTempC	130	36.8052	36.8330	0.4073	0.0357

FIGURE 27.2: The software output summary for the body temperature data.

A picture of this sampling distribution (Fig. 27.3) shows how the sample mean varies when $n = 130$, for all possible samples when $\mu = 37.0$. For example, the value of $\bar{x}$ will be *larger* than 37.0357°C about 16% of the time (using the 68–95–99.7 rule) if μ really is 37.0.

27.4 Observations: t-score

Step 3 of the decision-making process is to evaluate the observations. Locating $\bar{x} = 36.8052$ on the sampling distribution (Fig. 27.4) shows that this observed sample mean is, relatively speaking, *extremely* low: a sample mean this low is very unlikely to occur in any sample of $n = 130$ when $\mu = 37.0$. How many standard deviations is $\bar{x}$ away from $\mu = 37.0$? Compute:

$$\frac{\text{statistic} - \text{mean of the distribution}}{\text{std dev. of the distribution}} = \frac{36.8052 - 37.0}{0.035724} = -5.453.$$

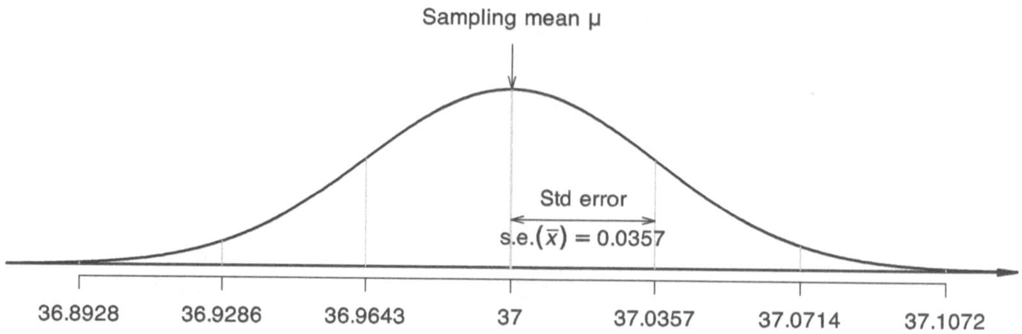

FIGURE 27.3: The distribution of sample mean body temperatures, if the population mean is 37.0°C and $n = 130$. The grey vertical lines are 1, 2 and 3 standard deviations from the mean.

This is like a z-score: it measures the number of standard deviations that the value is from the mean. However, it is not a z-score; it is a t-score. Both t- and z-scores measure *the number of standard deviations that a value is from the mean.* Here the value is a t-score, because the *population* standard deviation σ is unknown, and the *sample* standard deviation is used instead to compute s.e.($\bar{x}$).

ⓘ Like z-scores, t-scores measure the number of standard deviations that a value is from the mean of the distribution.

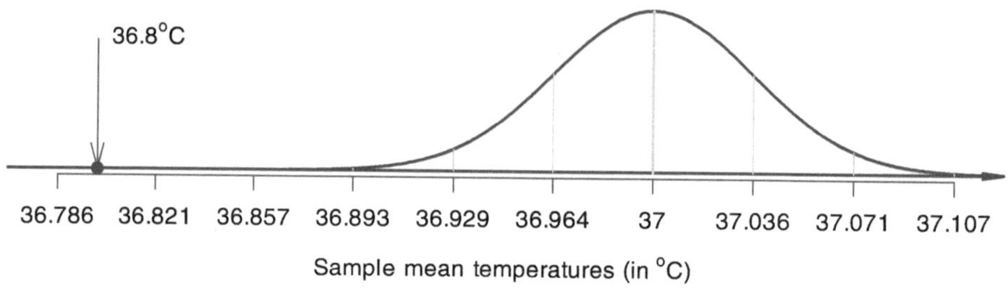

FIGURE 27.4: The sample mean of $\bar{x} = 36.8041$°C is very unlikely to be observed in any sample of size $n = 130$, if $\mu = 37.0$°C. The standard deviation of the distribution is s.e.($\bar{x}$) = 0.035724.

The calculation is therefore:

$$t = \frac{36.8052 - 37.0}{0.035724} = -5.453.$$

The observed sample mean is *more than five standard deviations below the population mean*, which is highly unusual based on the 68–95–99.7 rule (Fig. 27.4). This is very persuasive evidence that μ is not 37.0.

In general, when the sampling distribution has an approximate normal distribution and the *sample* standard deviation is used to compute the standard error, the t-score is

$$t = \frac{\text{sample statistic} - \text{mean of the sampling distribution}}{\text{standard error of the sampling distribution}} = \frac{\bar{x} - \mu}{\text{s.e.}(\bar{x})}. \qquad (27.2)$$

27.5 Decision: P-value

As seen in Sect. 26.7, a P-value quantifies how unusual the observed sample statistic is, after assuming H_0 is true. Since t-scores and z-scores are very similar, the P-value can be *approximated* using the 68–95–99.7 rule and a diagram, or *approximated* using z-tables (Appendices B.1 and B.2). Usually, however, software is used to compute the P-value. t-scores and z-scores with the same value produce almost the same P-values, except for small sample sizes.

Both methods produce approximate P-values only, since the approximations are based on using z-scores rather than t-scores. Usually, software is used to determine precise P-values for t-scores (Fig. 27.5). The output (under the heading `p`) shows that the P-value is indeed very small: less than 0.001 (written $P < 0.001$).

> Some software reports a P-value of `0.000`, which really means that the P-value is zero to three decimal places. Since P-values can never be exactly zero, we should write $P < 0.001$: that is, the P-value is *smaller* than 0.001.

This P-value means that, if $\mu = 37.0$, a sample mean as low as 36.8052 would be *very* unusual to observe (from a sample size of $n = 130$). And yet, we did. Using the decision-making process, this implies that the initial assumption (i.e., H_0) is contradicted by the data: we observed something extremely unlikely if $\mu = 37.0$. That is, there is very persuasive evidence that the *population* mean body temperature is *not* 37.0°C.

One Sample T-Test

		statistic	df	p
BodyTempC	Student's t	-5.45	129	< .001

Note. H_a population mean ≠ 37

FIGURE 27.5: Software output for conducting the t-test for the body temperature data.

> For *one-tailed tests*, the P-value is *half* the value of the two-tailed P-value.

As seen in Sect. 26.7, P-values measure the probability of observing the sample statistic (or something more extreme), assuming the population parameter is the value given in H_0. For the body-temperature data then, where $P < 0.001$, the P-value is *very* small, so *very strong evidence* exists that the population mean body temperature is not 37.0°C.

27.6 Writing conclusions

Communicating the results of any hypothesis test requires an *answer to the RQ*, a summary of the *evidence* used to reach that conclusion (such as the t-score and P-value, stating if the

P-value is one- or two-tailed), and some *sample summary information* (including a CI). For the body-temperature example, write:

> The sample provides very strong evidence ($t = -5.45$; two-tailed $P < 0.001$) that the population mean body temperature is *not* 37.0°C ($\bar{x} = 36.81$; 95% CI: 36.73 to 36.88°C; $n = 130$).

This statement contains the three components.

1. The *answer to the RQ*: the sample provides very strong evidence that the population mean body temperature is not 37.0°C. The alternative hypothesis is two-tailed, so the conclusion is that the population mean body temperature is *not equal to* 37.0°C.
2. The *evidence* used to reach the conclusion: $t = -5.45$; two-tailed $P < 0.001$.
3. Some *sample summary information*: the sample mean (with the CI) and the sample size.

The test is about the *mean* internal body temperature; *individuals* have internal body temperatures ranging from 35.722°C to 38.222°C.

The difference between the value of 37.0°C and the sample mean of 36.81°C is small in absolute terms, and is probably of little practical importance for most applications. Notice that the CI does *not* include the value of $\mu = 37.0$.

27.7 Process overview

Let's recap the decision-making process for this body temperatures (Fig. 27.6) example:

1. *Assumption*. Write the *null hypothesis* about the parameter (based on the RQ): H_0: $\mu = 37.0$. In addition, write the *alternative hypothesis*: H_1: $\mu \neq 37.0$. (This alternative hypothesis is two-tailed.)
2. *Expectation*. The *sampling distribution* describes what to expect from the statistic *if* the null hypothesis is true. The sampling distribution is an approximate normal distribution.
3. *Observation*. Compute the t-score: $t = -5.45$. The t-score can be computed by software, or using the general equation in Equation (27.2).
4. *Decision*. Determine if the data are *consistent* with the assumption, by computing the P-value. Here, the P-value is much smaller than 0.001. The P-value can be computed by software, or approximated using the 68–95–99.7 rule. The *conclusion* is that there is very strong evidence that μ is not 37.0.

27.8 Statistical validity conditions

All hypothesis tests have underlying conditions to be met so that the results are statistically valid. For a test of one mean, this means that the sampling distribution must have an approximate normal distribution so that P-values can be found.

The test for a single mean is *statistically valid* if *either* of these is true:

- when $n \geq 25$. (If the distribution of the data is highly skewed, the sample size may need to be larger.)
- when $n < 25$, *and* the sample data come from a *population* with a normal distribution.

The sample size of 25 is a rough figure; some books give other values (such as 30).

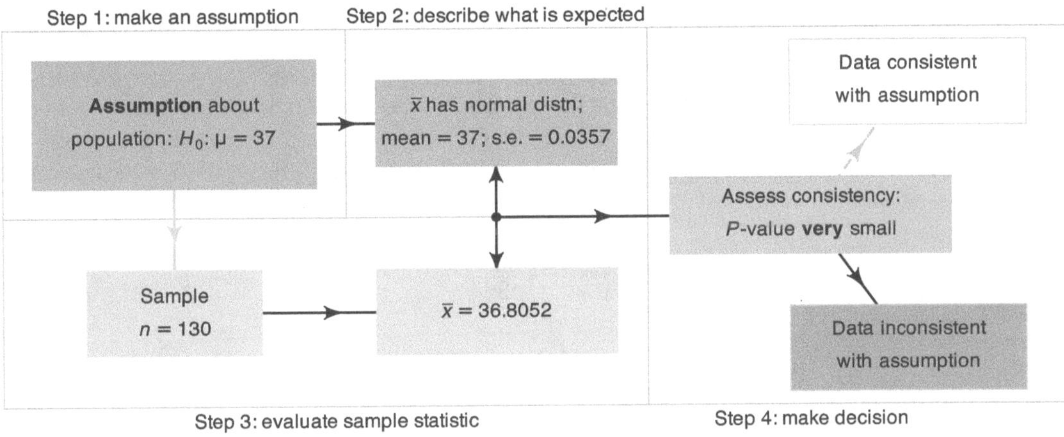

FIGURE 27.6: The decison-making process for the body-temperature data.

This condition ensures that the *distribution of the sample means has an approximate normal distribution* (so that, for example, the 68–95–99.7 rule can be used). Provided the sample size is larger than about 25, this will be approximately true *even if* the distribution of the individuals in the population does not have a normal distribution. That is, when $n \geq 25$ the sample means generally have an approximate normal distribution, even if the data themselves do not have a normal distribution. The units of analysis are also assumed to be *independent* (e.g., from a simple random sample).

If the statistical validity conditions are not met, other similar options include a sign test or a Wilcoxon signed-rank test [Conover, 2003], or using resampling methods [Efron and Hastie, 2021].

Example 27.1 (Statistical validity). The hypothesis test regarding body temperature is statistically valid since the sample size is larger than 25 ($n = 130$). (The data, as displayed in Fig. 27.1, do *not* need to come from a population with a normal distribution.)

27.9 Example: student IQs

Standard IQ scores are designed to have a mean in the general population of $\mu = 100$. Researchers at Griffith University (GU) asked:

For students at Griffith University, is the mean IQ higher than 100?

The parameter is μ, the population mean IQ for students at GU.

To answer this RQ, Reilly et al. [2022] studied $n = 224$ students at Griffith University (GU), finding a sample mean IQ of 111.19 and a standard deviation of 14.21. Is this evidence that GU students have a *higher* mean IQ than the general population? The hypotheses are:

$$H_0\colon \mu = 100 \qquad \text{and} \qquad H_1\colon \mu > 100.$$

This test is *one-tailed*, since the RQ asks if the mean IQ of GU students is *greater* than 100, the one-tailed P-value will be in the tail corresponding to *larger* IQ scores (i.e., to the right of the mean). (Writing H_0: $\mu \leq 100$ is also correct (and equivalent), though the test still proceeds as though $\mu = 100$, the largest option permitted by $\mu \leq 100$.)

We do not have the original data, but the summary information is sufficient: $\bar{x} = 111.19$ with $s = 14.21$ from a sample of size $n = 224$. The *sample* mean is higher than 100, but since sample means vary, the difference may be just due to sampling variation. The sample means vary with a normal distribution, with mean 100 and a standard deviation of

$$\text{s.e.}(\bar{x}) = \frac{s}{\sqrt{n}} = \frac{14.21}{\sqrt{224}} = 0.94945.$$

The t-score is

$$t = \frac{\bar{x} - \mu}{\text{s.e.}(\bar{x})} = \frac{111.19 - 100}{0.94945} = 11.786.$$

This t-score is *huge*: a sample mean as large as 111.19 would be highly unlikely to occur in any sample of size $n = 224$, simply by sampling variation, if the population mean really was 100. Since the alternative hypothesis is *one-tailed*, and $\mu > 100$ specifically, the P-value is the area in the right-side tail of the distribution only (Fig. 27.7); it will be extremely small. This is very persuasive evidence to support the alternative hypothesis.

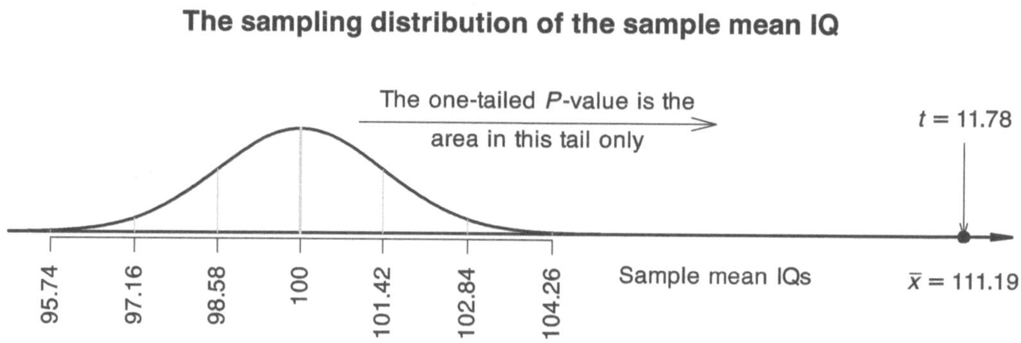

FIGURE 27.7: The sampling distribution for the IQ data. The RQ is one-tailed so the P-value is the area in one tail.

We conclude:

> Very strong evidence exists in the sample ($t = 11.78$; one-tailed $P < 0.001$) that the population mean IQ in students at Griffith University is greater than 100 (mean 111.19; 95% CI: 109.29 to 113.09; $n = 224$).

The test is about the *mean* IQ; *individual* students may have IQs less than 100.

Since the sample size is much larger than 25, this conclusion is *statistically valid*. The sample is not a random sample from the population of all GU students (the students are mostly first-year, undergraduate psychological science students). However, these students may be somewhat representative of all GU students. In any case, the results probably apply to first-year, undergraduate psychological science students at GU.

The difference between the general population IQ of 100 and the sample mean IQ of GU students is only small: about 11 IQ units (less than one standard deviation). Possibly, this difference has very little practical importance, even though the statistical evidence suggests that the difference cannot be explained by chance.

IQ scores are designed to have a standard deviation of $\sigma = 15$ in the general population. If this applies for university students too (and we do not know if it does), the standard error is $\text{s.e.}(\bar{x}) = \sigma/\sqrt{n} = 15/\sqrt{130} = 1.0022$, and the test-statistic is then a z-score:

$$z = \frac{\bar{x} - \mu}{\text{s.e.}(\bar{x})} = \frac{111.19 - 100}{1.0022} = 11.87.$$

The conclusions do not change: the P-value is still extremely small.

27.10 Chapter summary

These steps are used to test a hypothesis about a population mean μ.

- Write the null hypothesis (H_0) and the alternative hypothesis (H_1); initially *assume* the value of μ in the null hypothesis to be true.
- Describe the *sampling distribution*, which describes what to *expect* from the sample mean based on this assumption: under certain statistical validity conditions, the sample mean varies with:
  - an approximate normal distribution,
  - with sampling mean whose value is the value of μ (from H_0), and
  - having a standard deviation of $\text{s.e.}(\bar{x}) = \dfrac{s}{\sqrt{n}}$.
- Compute the value of the *test statistic*:

  $$t = \frac{\bar{x} - \mu}{\text{s.e.}(\bar{x})},$$

  where μ is the hypothesised value given in the null hypothesis.
- The t-value is like a z-score, and so an approximate *P-value* can be estimated using the 68–95–99.7 rule or tables, or found using software. Use the P-value to make a decision, and write a conclusion.
- Check the statistical validity conditions.

27.11 Quick review questions

The usual recommendation for a safe gap between travelling vehicles in traffic (a 'headway') is *at least* 1.9 s (often rounded to 2 s for the public). Majeed et al. [2014] studied $n = 28$ streams of traffic in Birmingham, Alabama found the mean headway was 1.1915 s, with a standard deviation of 0.231 s. The researchers wanted to test if the mean headway in Birmingham was *less than* the recommended 1.9 s.

Are the following statements *true* or *false*?

1. The standard error of the mean is 0.231 s.
2. The null hypothesis is 'The sample mean headway is 1.9 s'.
3. The alternative hypothesis 'The population mean is less than 1.9 s'.
4. The test is *one-tailed*.
5. The value of the test statistic is $t = -16.23$.

6. The one-tailed P-value is very small.
7. There is no evidence to support the *alternative* hypothesis.

27.12 Exercises

Answers to odd-numbered exercises are given at the end of the book.

Exercise 27.1. Azwari and Hamsa [2021] studied driving speeds in Malaysia, and recorded the speeds of vehicles on various roads. One RQ was whether the mean speed of cars on one particular road was the posted speed limit of 90 km.h^{-1}, or whether it was *higher*.

The researchers recorded the speed of $n = 400$ vehicles on this road, and found the mean and standard deviation of the speeds of individual vehicles were $\bar{x} = 96.56$ and $s = 13.874$ km.h^{-1}.

1. Define the parameter of interest.
2. Write the statistical hypotheses.
3. Compute the standard error of the sample mean.
4. Sketch the sampling distribution of the sample mean for $n = 400$.
5. Compute the test statistic, a t-score.
6. Determine the P-value, and write a conclusion.
7. Is the test statistically valid?

Exercise 27.2. A competitive slalom competitor completed $n = 30$ attempts on a 38.8 m kayak slalom course to assess the accuracy of a GPS tracking system [Macdermid et al., 2022]. The trials produced a mean distance, recorded by the GPS, as 36.54 m with a standard deviation of 2.07 m.

1. Define the parameter of interest.
2. Write the statistical hypotheses.
3. Compute the standard error of the sample mean.
4. Sketch the sampling distribution of the sample mean for $n = 30$.
5. Compute the test statistic, a t-score.
6. Determine the P-value, and write a conclusion.
7. Is the test statistically valid?

Exercise 27.3. Greenlee et al. [2018] conducted a study of human–automation interaction with automated vehicles. They were interested in whether the average mental demand of 'drivers' of automated vehicles was *higher* than the average mental demand for ordinary tasks.

In the study, the $n = 22$ participants 'drove' (in a simulator) an automated vehicle for 40 mins. While driving, the drivers monitored the road for hazards. The researchers assessed the 'mental demand' placed on these drivers, where scores over 50 'typically indicate substantial levels of workload' (p. 471). For the sample, the mean score was 84.00 with a standard deviation of 22.05.

Is there evidence of a 'substantial workload' associated with monitoring roadways while 'driving' automated vehicles?

Exercise 27.4. Health departments recommend that hot water be stored at 60°C or higher, to kill *legionella* bacteria. Alary and Joly [1991] studied $n = 178$ Quebec homes with electric water heaters to see if the mean water temperature was less than 60°C (i.e., at risk).

The mean temperature was 56.6°C, with a standard error of 0.4°C. Is there evidence the mean water temperature in Quebec is too low to kill *legionella* bacteria?

Exercise 27.5. [*Dataset*: `CherryRipe`] A *Cherry Ripe* is a popular Australian chocolate bar. In 2017, 2018 and 2019, I sampled some *Cherry Ripe* Fun Size bars. The packaging claimed that the Fun Size bars weigh 14 g (on average).

1. Use the software output (Fig. 27.8) to determine if the mean weight is 14 g or not.
2. Explain the difference in the meaning of `SD` and `SE` in this context.

Descriptives

	N	Mean	Median	SD	SE
BarWt	67	14.9033	14.9900	0.5496	0.0671

FIGURE 27.8: Software output for the *Cherry Ripes* data.

Exercise 27.6. (This study was also seen in Exercise 23.6.) Williams and Boyle [2007] asked $n = 199$ paramedics to estimate the amount of blood on four different surfaces. When the actual amount of blood spilt on concrete was 1000 mL, the mean guess was 846.4 mL (with $s = 651.1$ mL).

Is there evidence that the mean guess is 1 000 mL (the true amount)? Is this test statistically valid?

Exercise 27.7. Lin et al. [2021] compared the average sleep times of Taiwanese pre-school children to the recommendation (of *at least* 10 h per night). Using the summary of the data for weekend sleep-times (Table 27.2), do girls get *less than* 10 h of sleep per night, on average? Do boys?

TABLE 27.2: Summary information for the Taiwanese pre-schoolers sleep times (in h).

	Sample size	Sample mean	Sample std dev.
Boys	47	8.50	0.48
Girls	39	8.64	0.37

Exercise 27.8. [*Dataset*: `LHconc`] Feng et al. [2017] assessed the accuracy of two instruments from a clinical laboratory, by comparing the reported luteotropichormone (LH) concentrations to known, pre-determined values using $n = 36$ samples. Use hypothesis tests to determine how the instruments perform, for both high- and mid-level LH concentrations (using the information in Table 27.3).

TABLE 27.3: The quality-control data: LH levels (in mIU.mL^{-1}) for two instruments (only the first four of 36 observations shown).

	Instrument 1	
	High level	Mid level
	61.63	18.36
	63.11	18.77
	66.88	18.98
	62.56	17.97
	⋮	⋮
Mean	64.31	19.24
Std deviation	1.70	0.59
Target	64.22	19.01

	Instrument 2	
	High level	Mid level
	62.64	19.12
	64.36	19.07
	66.06	19.58
	65.39	19.35
	⋮	⋮
Mean	64.97	19.40
Std deviation	1.03	0.41
Target	65.05	19.45

Exercise 27.9. [*Dataset*: `PizzaSize`] (This study was also seen in Exercise 23.10.) In 2011, *Eagle Boys Pizza* ran a campaign that claimed that *Eagle Boys* pizzas were 'Real size 12-inch large pizzas' [Dunn, 2012]. *Eagle Boys* made the data from the campaign publicly available. Using the summary of the diameters of a sample of 125 of their large pizzas (Fig. 27.9), test the company's claim:

For *Eagle Boys*' pizzas, is mean diameter actually 12 inches, or not?

1. What is the parameter of interest?
2. Write down the values of $\bar{x}$ and s.
3. Determine the value of the standard error of the mean.
4. Write the hypotheses to test if the mean pizza diameter is 12 inches.
5. Is the alternative hypothesis one- or two-tailed? Why?

6. Draw the normal distribution that shows how the *sample mean pizza diameter* would vary by chance, *even if* the population mean diameter was 12 inches.
7. Compute the t-score for testing the hypotheses.
8. What is the approximate P-value using the 68–95–99.7 rule?
9. Write a conclusion: do pizzas have a mean diameter of 12 inches, as claimed?
10. Is it reasonable to assume the *statistical* validity conditions are satisfied?

Descriptives

	DiameterInches
N	125
Missing	0
Mean	11.486
Median	11.449
Standard deviation	0.247
Minimum	10.465
Maximum	12.228

FIGURE 27.9: Summary statistics for the diameter of *Eagle Boys* large pizzas.

Exercise 27.10. Saxvig et al. [2021] studied the length of sleep each night for a 'large and representative sample of Norwegian adolescents' (p. 1) aged 16 and 17 years of age. The recommendation is for adolescents to have at least 8 h of sleep each night.

In the sample of $n = 3\,972$ individuals, the mean amount of sleep on schools days was 6 h 43 mins (i.e., 403 mins), with a standard deviation of 87 mins. On non-school days, the mean amount of sleep was 8 h 38 mins (i.e., 518 mins), with a standard deviation of 98 mins.

Do Norwegian adolescents appear to meet the guidelines of having '*at least* 8 h' sleep each night on school days? On non-school days?

Answers to *Quick review* questions: 1. False: 0.0436. **2.** False: *population* mean. **3.** True. **4.** True. **5.** True. **6.** True. **7.** False.

28

More details about hypothesis testing

You have learnt to ask an RQ, design a study, classify and summarise the data, construct confidence intervals, and conduct hypothesis tests. **In this chapter**, you will learn more about *hypothesis tests*. You will learn to:

- understand the process of hypothesis testing.
- communicate the results of hypothesis tests.
- interpret P-values.

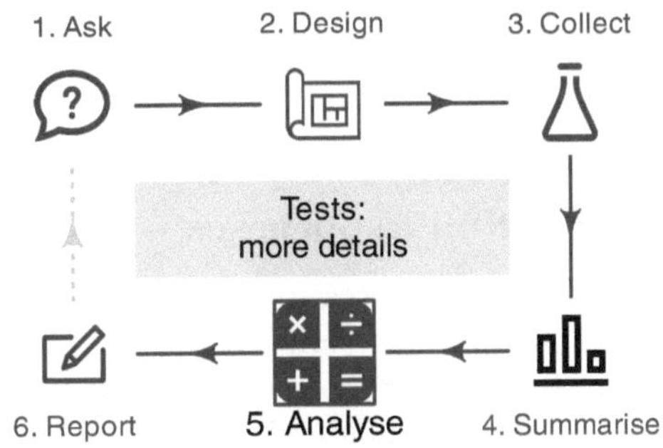

28.1 Introduction

In Chaps. 26 and 27, hypothesis tests for one proportion and one mean were studied. Later chapters discuss hypothesis tests in other contexts, too. However, the general approach to hypothesis testing is the same for *any* hypothesis test. This chapter discusses some general ideas in hypothesis testing:

- stating the *assumptions* and forming hypotheses (Sect. 28.2).
- describing the *expectations* of the statistic using the sampling distribution (Sect. 28.3).
- evaluating *observations* and the test statistic (Sect. 28.4).
- quantifying the *consistency* between the values of the statistic and parameter using P-values (Sect. 28.5).
- interpreting P-values (Sect. 28.6).
- how conclusions can go wrong (Sect. 28.7).
- wording *conclusions* (Sect. 28.8).
- practical importance and statistical significance (Sect. 28.9).
- statistical validity in hypothesis testing (Sect. 28.10).

28.2 More details about hypotheses and assumptions

Two *statistical* hypotheses are stated about the population parameter: the null hypothesis H_0, and the alternative hypothesis H_1. The null hypothesis is assumed to be true, and retained unless persuasive evidence exists to change our mind.

ⓘ The word *hypothesis* means 'a possible explanation'.

Scientific hypotheses refer to potential *scientific* explanations that can be tested by collecting data. For example, an engineer may hypothesise that replacing sand with glass in the manufacture of concrete will produce desirable characteristics [Devaraj et al., 2021]. Scientific hypotheses lead to research questions.

Statistical hypotheses refer to statements made about a parameter that may explain the value of a sample statistic. The statistical hypotheses are the foundation of the logic of hypothesis testing. One of the statistical hypotheses usually align with the scientific hypothesis.

This book discusses forming *statistical hypotheses*.

28.2.1 Null hypotheses

Statistical hypotheses *are always about a parameter*. Hypothesising, for example, that the *sample* mean body temperature (in Chap. 27) is equal to 37.0°C is silly: the *sample* mean clearly is 36.8052°C for the sample taken, and its value will vary from sample to sample anyway. The RQ is about the unknown *population*: the **P** in **P**OCI stands for **P**opulation.

The *null hypothesis* H_0 proposes that *sampling variation* is why the value of the statistic (such as the sample mean) is not the same as the assumed value of the parameter (such as the population mean). Every sample is different, and the observed data is from just one of the many possible samples. The value of the *statistic* will vary from sample to sample; the statistic may not be equal to the *parameter*, just because of the random sample obtained and sampling variation.

Definition 28.1 (Null hypothesis). The *null hypothesis* proposes that *sampling variation* explains the discrepancy between the proposed value of the parameter, and the observed value of the statistic.

Null hypotheses always contain an 'equals', because (as part of the decision-making process) a specific value must be assumed for the parameter, so we can describe what we might expect from the sample. For example: the population mean *equals* 100, is *less than or equal to* 100 ($\mu \leq 100$), or is *more than or equal to* 100 ($\mu \geq 100$).

The null hypothesis always assumes the discrepancy between the statistic and the assumed value of the parameter is due to sampling variation. This may mean, for example:

- there is *no change* in the value of the parameter compared to an established or accepted value (for descriptive RQs), such as in the body-temperature example in Chap. 27.
- there is *no change* in the value of the parameter for the units of analysis (i.e., for repeated-measures RQs).
- there is *no difference* between the value of the parameter in two (or more) groups (i.e., for relational RQs).
- there is *no relationship* between the variables, as measured by some parameter (for correlational RQs).

⚠ The *null hypothesis* always has the form 'no difference, no change, no relationship' regarding the population parameter. It is the 'sampling variation' explanation for the discrepancy between the value of the parameter and the value of the statistic.

⚠ Defining the parameter carefully is important!

28.2.2 Alternative hypotheses

The alternative hypothesis H_1 (or H_a) offers another possible reason why the value of the statistic (such as the sample proportion) is not the same as the proposed value of the parameter (such as the population proportion): the value of the parameter really is not the value claimed in the null hypothesis.

Definition 28.2 (Alternative hypothesis). The *alternative hypothesis* proposes that the discrepancy between the proposed value of the parameter and the observed value of the statistic cannot be explained by *sampling variation.* It proposes that the value of the parameter is not the value claimed in the null hypothesis.

Alternative hypotheses can be *one-tailed* or *two-tailed.* A *two*-tailed alternative hypothesis means, for example, that the population mean could be either smaller *or* larger than what is claimed. A *one*-tailed alternative hypothesis admits only one of those two possibilities. Most (but certainly not all) hypothesis tests are two-tailed.

The decision about whether the alternative hypothesis is one- or two-tailed depends on what the RQ asks (*not* by looking at the data). *The RQ and hypotheses should (in principle) be formed before the data are obtained*, or at least before looking at the data if the data are already collected.

The idea of hypothesis testing is the same whether the alternative hypothesis is one- or two-tailed: based on the data and the statistic, a decision is to be made about whether the data provides persuasive evidence to support the alternative hypothesis.

Example 28.1 (Alternative hypotheses). For the body-temperature study (Chap. 27), the alternative hypothesis is *two-tailed* (i.e., H_1: $\mu \neq 37.0$): the RQ asks if the population mean is 37.0°C or *not.* Two possibilities are considered: that μ could be either larger *or* smaller than 37.0.

A *one-tailed alternative hypothesis* would be appropriate if the RQ asked 'Is the *population* mean internal body temperature *greater* than 37.0°C?' (i.e., H_1: $\mu > 37.0$), or 'Is the *population* mean internal body temperature *smaller* than 37.0°C?' (i.e., H_1: $\mu < 37.0$). One-tailed RQs such as these would only be asked if there were good scientific reasons to suspect a difference in one direction specifically.

⚠ Important points about forming hypotheses:

- hypotheses always concern a *population* parameter.
- hypotheses emerge from the RQ (not the data).
- null hypothesis always have the form 'no difference, no change, no relationship' (i.e., sampling variation explains the discrepancy between the values of the parameter and statistic).
- null hypotheses always contain an 'equals'.
- alternative hypotheses may be one- or two-tailed, depending on the RQ.

28.3 More details about sampling distributions and expectations

The *sampling distribution* describes, approximately, how the value of the statistic (such as $\hat{p}$ or $\bar{x}$) varies across all possible samples, when H_0 is true; it describes the *sampling distribution*. Some sampling distributions have an approximate normal distribution.

When the sampling distribution is described by a normal distribution, the *mean* of the normal distribution (the sampling mean) is the parameter value given in the *assumption* (H_0), and the *standard deviation* of the normal distribution is called the *standard error*. However, *not all sampling distributions are normal distributions.*

The variation in the sampling distribution (as measured by the standard error) depends on the sample size. For example, suppose p is defined as the probability of rolling a ⊡ on a die. In one roll, finding a sample proportion of $\hat{p} = 1$, is not unreasonable. However, in 20 000 rolls, a sample proportion of $\hat{p} = 1$ would be *incredibly* unlikely for a fair die.

28.4 More details about observations and the test statistic

The sampling distribution describes what values the statistic can take over all possible samples of a given size. When the sampling distribution has an approximate normal distribution, the observed value of the *test statistic* is

$$\text{test statistic} = \frac{\text{value of sample statistic} - \text{centre of the sampling distribution}}{\text{standard deviation of the sampling distribution (i.e., standard error)}}.$$

The 'standard deviation of the sampling distribution' is called the standard error of the statistic. This is called a '*test statistic*', since the calculation is based on sample data (so it is a *statistic*) and used in a hypothesis *test*. This test statistic may be a z-score or a t-score. Other test statistics, when the sampling distribution is not described by a normal distribution, are used too (as in Chap. 31).

ⓘ For sampling distributions with an approximate normal distribution, a t-score and z-score both measure the number of standard deviations that a value is from the mean:

$$\frac{\text{a value that varies} - \text{mean of the distribution}}{\text{standard deviation of the distribution}}.$$

Then:

- if the quantity that varies is an *individual* observation x, the measure of variation is the standard deviation of the individual observations.
- if the quantity that varies is a *sample statistic*, the measure of variation is a *standard error*, which measures the variation in a sample statistic.

When conducting hypothesis tests about means, the test statistic is a t-score if the measure of variation uses a *sample* standard deviation.

28.5 More details about finding P-values

When the sampling distribution has an approximate normal distribution, P-values can be *approximated* (using the 68–95–99.7 rule or tables), as demonstrated in Sect. 26.7. The P-value is the area *more extreme* than the calculated z- or t-score (i.e., in the *tails* of the distribution). The 68–95–99.7 rule can be used to approximate this tail area (when the sampling distribution has an approximate normal distribution).

⚠ A lower-case p or upper-case P can be used to denote a P-value. We use an upper-case P, since we use p to denote a population proportion.

For *two-tailed* tests, the P-value is the *combined* area in the left and right tails. For *one-tailed* tests, the P-value is the area in just the left or right tail (as appropriate, according to the alternative hypothesis; see Sect. 27.9).

⚠ If the sampling distribution has an approximate normal distribution, the one-tailed P-value is half the value of the two-tailed P-value.

💻 Some software always reports two-tailed P-values.

More accurate approximations of the P-value can be found using tables. Precise P-values are found using the P-values from software output.

28.6 More details about interpreting P-values

Understanding P-values requires care.

Definition 28.3 (P-value). A P-value is the likelihood of observing the sample statistic (or something more extreme) over repeated sampling, under the assumption that the null hypothesis about the population parameter is true.

Since the null hypothesis is initially assumed true, *the onus is on the data to present evidence to contradict the null hypothesis.* That is, the null hypothesis is retained unless persuasive evidence suggests otherwise.

⚠ Conclusions are *always* about the parameters. P-values tell us about the unknown *parameters*, based on the data from one of the many possible values of the *statistic*.

A 'big' P-value means that the sample statistic (such as $\hat{p}$) could reasonably have occurred through sampling variation in one of the many possible samples, if the assumption made about the parameter (stated in H_0) was true. A 'small' P-value means that the sample statistic (such as $\hat{p}$) is unlikely to have occurred through sampling variation in one of the many possible samples, if the assumption made about the parameter (stated in H_0) was true. 'Small' P-values provide persuasive evidence to support the alternative hypothesis.

Commonly, a P-value smaller than 5% (or 0.05) is considered 'small' but this is *arbitrary*, and sometimes the threshold is discipline-dependent. More reasonably, P-values should be interpreted as giving varying degrees of evidence in support of the alternative hypothesis (Table 28.1), but these too are only guidelines.

> ⚠ The threshold for a 'small' P-value is very commonly 0.05, but this is arbitrary and not universal. There is nothing special about the value 0.05, and there is very little difference in the meaning of a P-value of 0.051 and a P-value of 0.049.

TABLE 28.1: A guideline for interpreting P-values. P-values should be interpreted in context, and indicate the strength of evidence to support the alternative hypothesis.

If the P-value is...	**Write the conclusion as...**
Larger than 0.10	*Insufficient* evidence to support H_1
Between 0.05 and 0.10	*Slight* evidence to support H_1
Between 0.01 and 0.05	*Moderate* evidence to support H_1
Between 0.001 and 0.01	*Strong* evidence to support H_1
Smaller than 0.001	*Very strong* evidence to support H_1

Identifying a P-value of 0.05 as 'small' (and hence providing 'persuasive evidence' to support H_1) is arbitrary; it means that, if H_0 is true, there is a 1-in-20 chance that the value of the statistic (or a value more extreme) would be observed due to sampling variation. In many situations, the evidence must be more persuasive than this.

To appreciate the concept of a 0.05 (or a 1-in-20) chance:

- the probability of throwing 5 or more Ⓗ in a row using a fair coin is about 0.063.
- the probability of drawing a black Ace from a pack of cards is about 0.038.
- the probability of rolling two or more consecutive throws of a ⚅ is about 0.033.

These events are improbable, without being essentially impossible.

P-values are commonly used in research, but must be used and interpreted correctly [Greenland et al., 2016]. Specifically:

- a P-value *is not* the probability that the null hypothesis is true.
- a P-value *does not prove* anything (only one possible sample was studied).
- a big P-value *does not* mean the null hypothesis H_0 is true, or that H_1 is false.
- a small P-value *does not* mean the null hypothesis H_0 is false, or that H_1 is true.
- a small P-value *does not* mean the results are practically important (Sect. 28.9).
- a small P-value does not necessarily mean a large difference between the statistic and parameter; it means that the difference (whether large or small) could not reasonably be attributed to *sampling variation* (chance).

> ⚠ P-values are never *exactly* zero. Some software reports very small P-values as '$P < 0.001$' (i.e., the P-value is smaller than 0.001). Some software reports very small P-values as '$P = 0.000$' (i.e., zero to three decimal places). In either case, we should still write $P < 0.001$.
>
> Some software only reports two-tailed P-values.

⚠ Sometimes the results of a study are reported as being *statistically significant*. This usually means that the P-value is less than 0.05, though a different P-value is sometimes used as the 'threshold', so check!

To avoid confusion, the word 'significant' should be avoided in writing about research unless 'statistical significance' is actually meant. In other situations, consider using words like 'substantial'.

28.7 More details about how conclusions can go wrong

In hypothesis testing, a decision is made about a *population* using *sample* information. Since the observed sample is just one of countless possible samples that could have been observed, making an incorrect conclusion is always a possibility.

Two mistakes can be made when making a conclusion:

- *incorrectly* concluding that evidence supports the alternative hypothesis. Of course, the researchers *do not know they are incorrect*, but the possibility of making this mistake is always present. This is a *false positive*, or a *Type I error*.
- *incorrectly* concluding there is *no* evidence to support the alternative hypothesis. Of course, the researchers *do not know they are incorrect*, but the possibility of making this mistake is always present. This is a *false negative*, or a *Type II error*.

Ideally, neither of these errors would be made; however, sampling variation means that neither can ever be completely eliminated. In practice, hypothesis testing begins by assuming the null hypothesis is true, and hence places the onus on the data to provide persuasive evidence in favour of the alternative hypothesis. This means researchers usually prioritise minimising the chance of a Type I error.

A Type I error is like declaring an innocent person guilty (recall: innocence is presumed in the judicial system). Similarly, a Type II error is like declaring a guilty person innocent. The law generally sees a Type I error as more grievous than a Type II error, just as in research. In general, larger sample sizes reduce the probability of making Type I and Type II errors.

In medical contexts, the similar concepts of *sensitivity* and *specificity* are often used rather than the terms *Type I* and *Type II errors*. *Sensitivity* is the probability of a *positive* test result among those *with* the disease, and *specificity* is the probability of a *negative* test result among those *without* the disease. High sensitivity is associated with a low chance of Type II error, and higher specificity is associated with a low chance of a Type I.

Example 28.2 (Type I errors). For the body-temperature example (Chap. 27), the conclusion was that the sample provided very strong evidence that the population mean body temperature was *not* 37.0°C. However, in truth, the mean internal body may not have changed, and is still 37.0°C; that is, the null hypothesis actually is true, but we incorrectly decided it was probably not true.

This would be a Type I error: we *incorrectly* concluded that the evidence supported the alternative hypothesis. Of course, since the value of μ is unknown, we do not know if we have made a Type I error or not.

28.8 More details about writing conclusions

In general, communicating the result of a hypothesis test requires stating:

1. the *answer* to the RQ.
2. the *evidence* used to reach that conclusion (such as the t-score and P-value, clarifying if the P-value is *one-tailed* or *two-tailed*).
3. *sample summary statistics* (such as sample means, with CIs and sample sizes).

Since we initially assume the null hypothesis is true, conclusions are worded (in context) in terms of how strongly the evidence supports the alternative hypothesis.

⚠ Since the null hypothesis is initially assumed to be true, the onus is on the data to provide evidence in support of the alternative hypothesis: the null hypothesis is retained unless persuasive evidence suggests otherwise. Hence, conclusions are always worded in terms of how much evidence supports the *alternative* hypothesis.

We *do not* say whether the evidence supports the null hypothesis; the null hypothesis is already assumed to be true. Even if the current sample presents no evidence to contradict the assumption, future evidence may emerge. That is:

'No evidence of a difference' is *not* the same as 'evidence of no difference'.

Example 28.3 (No evidence of a difference)**.** Suppose, when we tested if the mean internal body temperature remained 37.0°C (Chap. 27), that we found *no evidence* that the temperature had changed. This *does not* provide evidence that the mean internal body temperature is 37.0°C. It just means that the sample provided no evidence to change our initial *assumption* that the mean internal body temperature is 37.0°C.

28.9 More details about practical importance, statistical significance

Hypothesis tests assess *statistical significance*, which answers the question: 'Can sampling variation reasonably explain the discrepancy between the value of the statistic and the assumed value of the parameter?' Even very small discrepancies between the statistic and the parameter can be *statistically* different if the sample size is sufficiently large.

In contrast, *practical importance* answers the question: 'Is the discrepancy between the values of the statistic and the parameter of any importance *in practice*?' Whether a result is of practical importance depends upon the context: what the data are being used for. 'Practical importance' and 'statistical significance' are separate issues.

Example 28.4 (Practical importance)**.** In the body-temperature study (Sect. 27.1), very strong evidence exists that the mean body temperature had changed ('statistical significance'). But the change was so small that, for most purposes, it has no practical importance. In other (e.g., medical) situations, it *may* have practical importance.

Example 28.5 (Practical importance). Maunder et al. [2020] studied the use of herbal medicines for weight loss, and found that the intervention (p. 891)

> ... resulted in a statistically significant weight loss compared to placebo, although this was not considered clinically significant.

This means that the difference in mean weight loss between the placebo and intervention groups was unlikely to be explained by chance ($P < 0.001$; i.e., 'statistical significant'), but the difference was so small that it was unlikely to be of any use in practice ('practical importance'). In this context, the researchers decided that a weight loss of at least 2.5 kg was of practical importance. However, in the study, the sample mean weight loss was 1.61 kg.

28.10 More details about statistical validity

When performing hypothesis tests, *statistical validity conditions* must be true to ensure that the mathematics behind computing the P-value is sound. For instance, the statistical validity conditions may ensure that the sampling distribution is sufficiently like a normal distribution for the 68–95–99.7 rule to apply.

If the statistical validity conditions are *not* met, the P-values (and hence conclusions) may be inappropriate or only approximately correct.

28.11 Chapter summary

Hypothesis testing formalises the decision-making process. Starting with an *assumption* about a parameter of interest, a description of what values the statistic might take is produced (the sampling distribution): this describes what values the statistic is *expected* to take over all possible samples. This sampling distribution is often a normal distribution.

The statistic (the *sample estimate*) is then *observed*, and a *test statistic* is computed to quantify the discrepancy between the values of the parameter (given in H_0) and statistic. Using a P-value, a decision is made about whether the sample evidence supports or contradicts the initial assumption, and hence a *conclusion* is made. When the sampling distribution is an approximate normal distribution, the test statistic is a t-score or z-score, and P-values can often be approximated using the 68–95–99.7 rule.

28.12 Quick review questions

Are the following statements *true* or *false*?

1. When a P-value is very small, a very large difference *must* exist between the statistic and parameter.

2. The alternative hypothesis is one-tailed if the sample statistic is larger than the hypothesised population parameter.
3. When the sampling distribution has an approximate normal distribution, the standard deviation of this normal distribution is called the *standard error.*
4. Both z-scores and t-scores can be test statistics.
5. P-values can never be exactly zero.
6. A P-value is the probability that the null hypothesis is true.

Select the correct answer:

7. What is wrong (if anything) with this null hypothesis: $H_0 = 37$?
    a. There is nothing wrong.
    b. The value of 37 is probably a sample value.
    c. No parameter is given.
    d. This is the alternative (not the null) hypothesis.

28.13 Exercises

Answers to odd-numbered exercises are given at the end of the book.

Exercise 28.1. Assuming the statistical validity conditions are satisfied, use the 68–95–99.7 rule to approximate the *two*-tailed P-value if:

1. the t-score is 3.4.
2. the t-score is -2.9.
3. the z-score is -2.1.
4. the t-score is -6.7.

Exercise 28.2. Assuming the statistical validity conditions are satisfied, use the 68–95–99.7 rule to approximate the *two*-tailed P-value if:

1. the z-score is 1.05.
2. the t-score is -1.3.
3. the t-score is 6.7.
4. the t-score is 0.1.

Exercise 28.3. Consider the test statistics in Exercise 28.1. Use the 68–95–99.7 rule to approximate the *one*-tailed P-values in each case.

Exercise 28.4. Consider the test statistics in Exercise 28.2. Use the 68–95–99.7 rule to approximate the *one*-tailed P-values in each case.

Exercise 28.5. Suppose a hypothesis test results in a P-value of 0.0501. What would we conclude? What if the P-value was 0.0499? Comment.

Exercise 28.6. Suppose a hypothesis test results in a P-value of 0.011. What would we conclude? What if the P-value was 0.009? Comment.

Exercise 28.7. Consider the study to determine if the mean body temperature (Chap. 27) was 37.0°C, where $\bar{x} = 36.8052$°C. Explain *why* each of these sets of hypotheses are incorrect.

1. H_0: $\bar{x} = 37.0$; H_1: $\bar{x} \neq 37.0$.
2. H_0: $\mu = 37$; H_1: $\mu > 37$.
3. H_0: $\mu = 37$; H_1: $\mu = 36.8052$.
4. H_0: $\bar{x} = 36.8052$; H_1: $\bar{x} > 36.8052$.
5. H_0: $\mu = 36.8052$; H_1: $\mu \neq 36.8052$.
6. H_0: $\mu > 37.0$; H_1: $\bar{x} > 37.0$.

Exercise 28.8. Consider the study to determine if a die was loaded (Chap. 26) by studying the proportion of rolls that showed a ⚀, and where $\hat{p} = 0.41$. Explain *why* each of these sets of hypotheses are incorrect.

1. H_0: $\hat{p} = 1/6$; H_1: $\hat{p} \neq 1/6$.
2. $H_0 = 1/6$; $H_1 \neq 1/6$.
3. H_0: $p = 1/6$; H_1: $\hat{p} = 0.41$.
4. H_0: $\hat{p} = 1/6$; H_1: $\hat{p} = 0.41$.
5. H_0: $p = 1/6$; H_1: $p > 1/6$.
6. H_0: $p = 1/6$; H_1: $p = 0.41$.

Exercise 28.9. The recommended daily energy intake for women is 7 725 kJ (for a particular cohort, in a particular country; Altman [1991]). The daily energy intake for 11 women was measured to see if this is being adhered to. The RQ was 'Is the population mean daily energy intake 7 725 kJ?'

The test produced $P = 0.018$. What, if anything, is wrong with these conclusions after completing the hypothesis test?

1. There is moderate evidence ($P = 0.018$) that the energy intake is not meeting the recommended daily energy intake.
2. There is moderate evidence ($P = 0.018$) that the sample mean energy intake is not meeting the recommended daily energy intake.
3. There is moderate evidence ($P = 0.018$) that the population energy intake is not meeting the recommended daily energy intake.
4. The study proves that the population energy intake is not meeting the recommended daily energy intake ($P = 0.018$).
5. There is some evidence that the population energy intake is not meeting the recommended daily energy intake ($P < 0.018$).

Exercise 28.10. [*Dataset*: `Battery`] A study compared ALDI batteries to another brand of battery. In one test (comparing the time taken for 1.5 V AA batteries to reach 1.1 V), the ALDI brand battery took 5.73 h, and the other brand (Energizer) took 5.44 h [Dunn, 2013].

1. What is the null hypothesis for the test?
2. The P-value for comparing these two means is about $P = 0.70$. What does this mean?
3. Is this difference likely to be of any practical importance? Explain.
4. What would be a correct conclusion for ALDI to report from the study? Explain.
5. What else would be useful to know when comparing the two brands of batteries?

Exercise 28.11. An ecologist was compared the proportion of female and male dingoes kept in zoos that showed signs of mange (a skin disease). She finds 'no statistically significant' difference between the proportions of female and male dingoes with evidence of mange.

Which of these statements is *consistent* with this conclusion?

1. The difference in proportions is 0.27 and $P = 0.36$.
2. The difference in proportions is 0.27 and $P = 0.0001$.
3. The difference in proportions is 0.04 and $P = 0.36$.
4. The difference in proportions is 0.04 and $P = 0.0001$.

How would the other statements be interpreted then?

Exercise 28.12. The study of body temperatures (Chap. 27) also compared the mean internal body temperatures for females and males [Mackowiak et al., 1992]. The study concludes that there is moderate evidence of a difference between the mean temperatures of females and males.

Which of these statements is *consistent* with this conclusion?

1. The difference between the mean temperatures is 0.289°C and $P = 0.024$.
2. The difference between the mean temperatures is 2.89°C and $P = 0.024$.
3. The difference between the mean temperatures is 0.289°C and $P = 0.39$.
4. The difference between the mean temperatures is 2.89°C and $P = 0.39$.

How would the other statements be interpreted then?

Answers to *Quick review* questions: 1. False. **2.** False. **3.** True. **4.** True. **5.** True. **6.** False. **7.** c. No parameter is given; perhaps H_0: $\mu = 37$.

29

Mean differences (paired data): CIs and tests

You have learnt to ask an RQ, design a study, classify and summarise the data, construct confidence intervals, and conduct hypothesis tests. **In this chapter**, you will learn to:

- identify situations where mean differences are appropriate.
- construct confidence intervals for a mean difference.
- conduct hypothesis tests for the mean difference with paired data.
- determine whether the conditions for using these methods apply in a given situation.

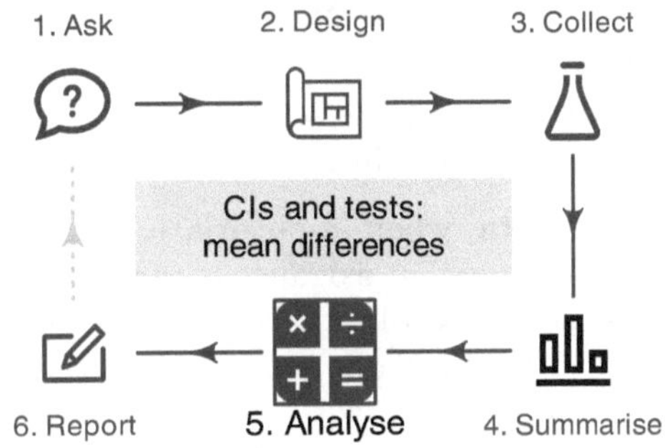

29.1 Introduction: six-minute walk test

The six-minute walk test (6MWT) measures how far subjects can walk in six minutes, and is used as a simple, low-cost evaluation of fitness and other health-related measures. The recommended setting for the test is usually a walkway of at least 30 m. Saiphoklang et al. [2022] measured the 6MWT distance when the same subjects used *both* 20 m and 30 m walkways.

The comparison is *within* individuals (Sect. 2.4); this is a *repeated-measures* study. Each subject has a *pair* of 6MWT measurements, and the study produced *paired data* (Table 29.1), the topic of this chapter.

TABLE 29.1: The six-minute walk test (6MWT) distance, for walkways of 20 m and 30 m length. These are the first five and the last five of the 50 total observations. (A negative difference means the 20 m distance is greater than the 30 m distance.)

	Distance walked (in m)		
Person	**20 m w'way**	**30 m w'way**	**Diff.**
1	272.1	281.6	9.5
2	425.3	454.4	29.0
3	338.2	330.0	−8.2
4	240.0	271.0	31.0
5	518.3	555.3	37.0
⋮	⋮	⋮	⋮

	Distance walked (in m)		
Person	**20 m w'way**	**30 m w'way**	**Diff.**
⋮	⋮	⋮	⋮
46	245.2	245.2	53.2
47	184.4	184.4	34.0
48	400.0	400.0	13.9
49	344.8	344.8	39.4
50	285.3	285.3	12.6

⚠ Some differences are *negative*. This does *not* mean a negative distance. Since the differences are computed as the 30 m distance minus the 20 m distance, a negative difference means the 20 m distance is a larger value than the 30 m distance.

29.2 Paired data

The data in Table 29.11 are *paired*. Computing the *differences* or *changes* between the pairs of observations makes sense, since the values for each pair belong to the same unit of analysis (the same person, in this case).

Pairing data, when appropriate, is useful because individuals can vary substantially, and pairing means that extraneous variables (potentially, *confounding* variables) are held constant for those paired observations. For example, each pair of measurements in Table 29.11 are recorded for the same person, so both measurements are recorded for someone of the same age, same sex, and with the same physical attributes.

Pairing is a form of blocking (Sect. 7.2). Pairing is a good design strategy when the individuals in the pair are the same, or are very similar, for many extraneous variables. (For example, the pair may comprise two different people, of the same sex, with similar age, height and weight.) Pairing often involves taking two measurements from the *same* individuals, as in Table 29.11.

Definition 29.1 (Paired data). *Paired data* occurs when the outcome is compared for two different, distinct situations for each unit of analysis.

Paired studies appear in many situations; for example, when:

- heart rate is measured for each twin in a pair (the twin-pair is the 'individual'), one of whom exercises regularly and one who does not. Pairing the twins is reasonable, given the shared genetics (and probably childhood environments also). The *difference* between the hearts rates of the twins can be recorded for each pair.
- the body temperature of dogs (the 'individuals') is measured using *both* rectal and ear thermometers for each dog. The *difference* between the two recorded temperatures from the thermometers for each dog is recorded.
- blood pressure is recorded from some individuals (Group A) after receiving Drug A, and from another group of individuals (Group B) after receiving Drug B. Each person in Group A is matched with someone in Group B of the same sex, similar age and similar weight (e.g., in one of the pairs, both individuals are male, about 30 years-of-age, and weighing about 95 kg). The *difference* between the blood pressure measurements for the individual in Group A and the matched person in Group B is recorded for each pair.
- the number of campers is recorded at many national parks (the 'individuals') on the first weekend in summer, and on the first weekend in winter. The *difference* in camper numbers for each national park between these time points is recorded.

Many of these examples can be extended to beyond two measurements. For instance, temperatures can be compared on each dog using three different types of thermometers. In this chapter, only *pairs* of measurements are studied, and only for quantitative variables.

29.3 Summarising the data

For the 6MWT study, the distance is measured for the same subjects for two different walkway distances. Each subject receives two measurements, and the *difference* between the distances walked for each individual is computed.

Since the data are paired, an appropriate graph is a histogram of the differences (Sect. 13.3.1); specifically, 30 m distance minus the 20 m distance. A boxplot comparing 6MWT distance for both walkway lengths (that is, *not* pairing the data) shows the distribution of distances, and the median distances, are very similar (Fig. 29.1, left panel). Any difference in individuals' 6MWT distances is difficult to see and detect. In addition, linking the 20 m and 30 m distances that belong together for each individual patient is not possible.

However, using a histogram of the differences makes the individuals' differences easier to see (Fig. 29.1, right panel). The histogram also makes it easy to see that some subjects walked further with a 20 m walkway, and some further for a 30 m walkway. Individually graphing the distances for both walkway distances may also be useful too (e.g., using two histograms), but a graph of the differences is *crucial*, as the RQ is about those differences. A case-profile plot (Sect. 13.3.2) is also appropriate, but is difficult to read for these data because sample size is large (a line is needed for each of the 50 units of analysis).

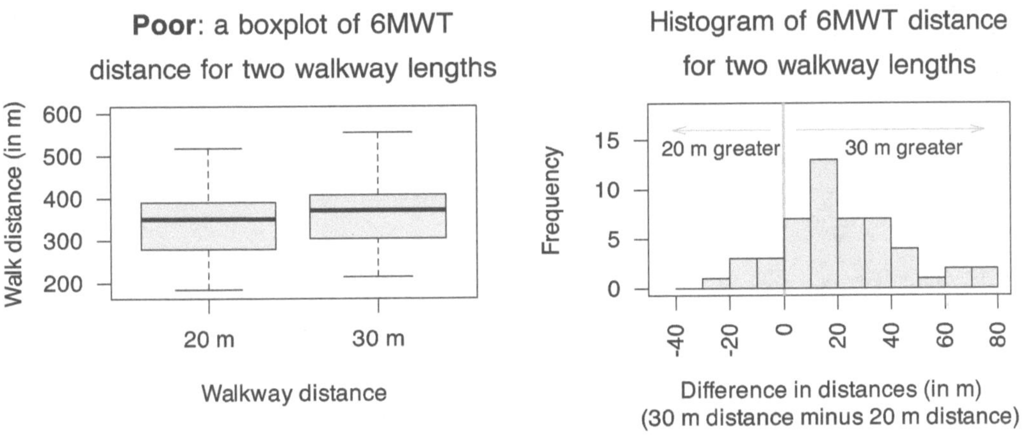

FIGURE 29.1: Plots of the 6MWT data. Left: graphing the data *incorrectly* as unpaired. Right: a histogram of 6MWT distances changes (30 m walkway distance *minus* 20 m walkway distance; the vertical grey line represents no change in distance).

The 6MWT distances for each walkway length can be summarised individually (the first two rows of Table 29.2) using the methods of Chap. 23, using software (Fig. 29.2). All statistics are slightly different for the two walkway distances; in particular, the mean 30 m walkway distance is slightly larger. However, since the RQ is about the difference between the distances, a numerical summary of the *differences* is essential (third row of Table 29.2, based on Fig. 29.2). Notice that the third row of information is computed from the values in the **Diff.** column in Table 29.11, not by (for instance) finding the difference between the standard deviations in the first two rows.

The differences (i.e., the **Diff.** column in Table 29.1) can be treated like a single sample of data (Table 29.3), with the notation adapted accordingly:

TABLE 29.2: The numerical summary of the 6MWT data. (The differences are the 30 m distances minus the 20 m differences.)

	Mean	Median	Standard deviation	Standard error	Sample size
20m walkway distance (in m)	337.82	351.0	71.801	10.154	50
30m walkway distance (in m)	359.85	371.4	77.250	10.925	50
Difference (in m)	22.03	17.0	22.039	3.117	50

Descriptives

	N	Mean	Median	SD	SE
Dist30	50	359.8	371.4	77.25	10.92
Dist20	50	337.8	351.0	71.80	10.15

Paired Samples T-Test

								95% Confidence Interval	
			statistic	df	p	Mean difference	SE difference	Lower	Upper
Dist30	Dist20	Student's t	7.067	49.00	< .001	22.03	3.117	15.76	28.29

Note. Hₐ $\mu_{\text{Measure 1 - Measure 2}} \neq 0$

FIGURE 29.2: The 6MWT data: numerical summary software output for each group (top), and the CI and test results (bottom).

- μ_d: the mean *difference* in the *population* (in m).
- $\bar{d}$: the mean *difference* in the *sample* (in m).
- s_d: the *sample* standard deviation of the *differences* (in m).
- n: the number of *differences*.

TABLE 29.3: The notation used for mean differences (paired data) compared to the notation used for one sample mean.

	One sample mean	Mean difference
The observations:	Values: x	Differences: d
Population mean:	μ	μ_d
Sample mean:	$\bar{x}$	$\bar{d}$
Standard deviation:	s	s_d
Standard error of $\bar{x}$:	$\text{s.e.}(\bar{x}) = \frac{s}{\sqrt{n}}$	$\text{s.e.}(\bar{d}) = \frac{s_d}{\sqrt{n}}$
Sample size:	Number of *observations*: n	Number of *differences*: n

29.4 Confidence intervals for μ_d

The data in Table 29.1 can be used to answer this repeated-measures, estimation RQ:

For Thai patients with chronic obstructive pulmonary disease, what is the mean difference between the 6MWT distance when subjects use a 20 m walkway and a 30 m walkway?

Every possible sample of $n = 50$ subjects comprises different people, and hence produces different 6MWT distances for 20 m and 30 m walkways. For this reason, the 6MWT distance summaries in Table 29.2 include standard errors. Since the 6MWT distance varies from sample to sample for each person, the *differences* between the distances for each person varies from sample to sample too, and also have a *sampling distribution.*

Definition 29.2 (Sampling distribution of a sample mean difference). The *sampling distribution of a sample mean difference* is (when certain conditions are met; Sect. 29.6) described by:

- an approximate normal distribution,
- centred around the *sampling mean* whose value is the population mean *difference* μ_d,
- with a standard deviation, called the standard error of the difference, of $\text{s.e.}(\bar{d}) = \dfrac{s_d}{\sqrt{n}}$,

where n is the number of differences, and s_d is the standard deviation of the individual differences in the sample.

For the 6MWT data, the sample mean differences $\bar{d}$ are described by (Fig. 29.3):

- approximate normal distribution,
- with a sampling mean whose value is μ_d,
- with a *standard error* of

$$\text{s.e.}(\bar{d}) = \frac{22.039}{\sqrt{50}} = 3.117. \tag{29.1}$$

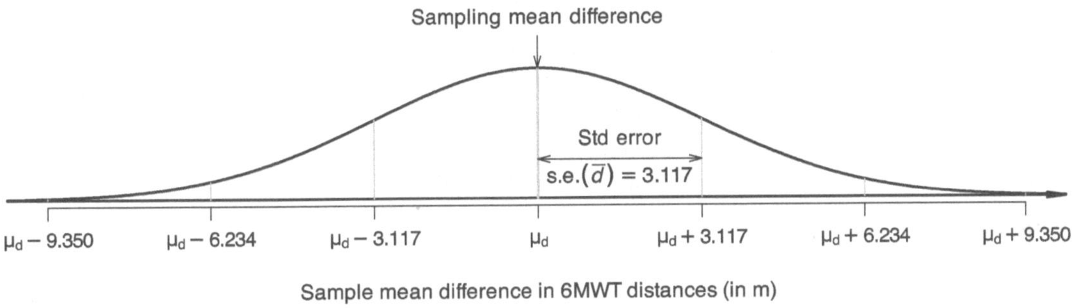

FIGURE 29.3: The sampling distribution is a normal distribution; it describes how the sample mean difference between the 6MWT distances varies in samples of size $n = 50$.

The CI for the mean difference has the same form as for a single mean (Chap. 23). The 95% confidence interval (CI) for μ_d is

$$\bar{d} \pm \left(\text{multiplier} \times \text{s.e.}(\bar{d})\right).$$

As usual when the sampling distribution has an approximate normal distribution, an approximate 95% CI uses the approximate multiplier of 2 (from the 68–95–99.7 rule). This is the same as the CI for $\bar{x}$ if the differences are treated as the data.

For the 6MWT data, the approximate 95% CI is:

$$22.03 \pm (2 \times 3.117),$$

or 22.03 ± 6.234 m (so the *margin of error* is 6.234 m). Equivalently, the CI is from $22.03 - 6.234 = 15.796$ m, up to $22.03 + 6.234 = 28.264$ m. We write:

The mean difference in the 6MWT distances when using a 20 m and 30 m walkway is 22.03 m (s.e. = 3.117; $n = 50$), with an approximate 95% CI from 15.80 m to 28.26 m, further for a 30 m walkway.

The CI means that the reasonable values for the population mean difference in 6MTW distances are between 15.80 m and 28.26 m. Alternatively, we are 95% confident that the population mean difference between the 6MWT distances is between 15.80 m and 28.26 m (further for 30 m walkway). A difference of this magnitude probably has practical importance. Also notice that the *direction* of the difference is given: 'further for 30 m walkway'.

Statistical software produces *exact* 95% CIs, which may be slightly different from the *approximate* 95% CI (recall: the 68–95–99.7 rule gives *approximate* multipliers). For the 6MWT data, the *approximate* and *exact* 95% CIs are the same to one decimal place (Fig. 29.2). We write:

The mean difference in the 6MWT distances when using a 20 m and 30 m walkway is 22.03 m (s.e. = 3.117; $n = 50$), with a 95% CI from 15.76 m to 28.29 m further for a 30 m walkway.

29.5 Hypothesis tests for μ_d: *t*-test

The data in Table 29.1 can be used to answer this repeated-measures, decision-making RQ:

For Thai patients with chronic obstructive pulmonary disease, is there a mean increase in 6MWT distance using a 30 m walkway compared to a 20 m walkway?

In Sect. 29.1, the differences were defined as the 30 m distance minus the 20 m distance, which is consistent with the wording in this RQ. This RQ asks if the mean walking distance is, in general, a smaller value when subjects use a 20 m walkway compared to a 30 m walkway (that is how *positive* differences eventuate). The *parameter* is the *population mean difference* in 6MWT, μ_d. Note that the RQ is worded as one-tailed.

The *null* hypothesis is that 'there is *no mean change* in 6MWT, in the population':

- H_0: $\mu_d = 0$.

This hypothesis, which we initially *assume* to be true, postulates that the mean reduction may not be zero in the sample, due to sampling variation.

Since the RQ asks specifically if the mean distance is *smaller* for a 20 m walkway, the alternative hypothesis is *one-tailed* (Sect. 28.2). According to how the differences have been defined, the alternative hypothesis is:

- H_1: $\mu_d > 0$ (i.e., one-tailed).

This hypothesis says that the mean change in the population is *greater than* zero, because of the wording of the RQ, and because of how the differences were defined. If the differences were defined in the opposite way (as 'the 20 m distance minus the 30 m distance') then the alternative hypothesis would be $\mu_d < 0$, which has the same *meaning*.

The sampling distribution, as described in Sect. 29.2, still applies, where μ_d is assumed to be the value given in H_0 (see Fig. 29.4):

- an approximate normal distribution,

- centred around the *sampling mean* whose value is the population mean *difference* $\mu_d = 0$ (from H_0),
- with a standard deviation of $\text{s.e.}(\bar{d}) = 3.117$ (from Equation (29.1)).

The sample mean difference can be located on the sampling distribution by computing the t-score:

$$t = \frac{\bar{d} - \mu_d}{\text{s.e.}(\bar{d})} = \frac{22.026 - 0}{3.117} = 7.07,$$

following the ideas in Equation (27.2). Software displays the same t-score (Fig. 29.2). This is a *huge* t-score.

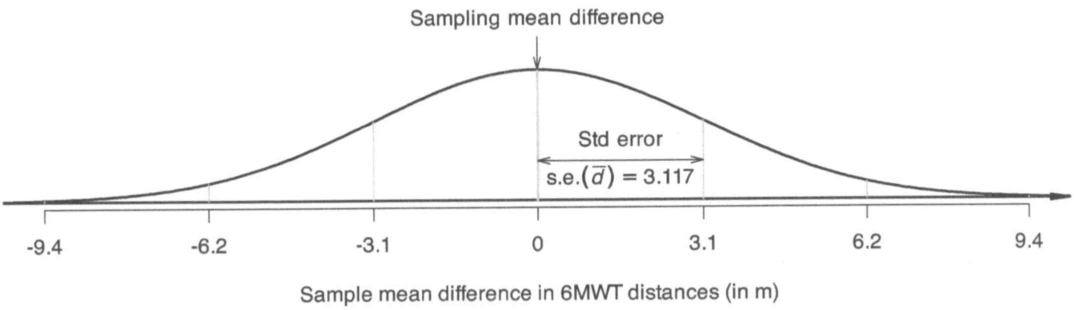

FIGURE 29.4: The sampling distribution is a normal distribution; it describes how the sample mean difference between the 6MWT distances varies in samples of size $n = 50$.

A P-value determines if the sample data are consistent with the assumption (Table 28.1). Since $t = 7.07$, and since t-scores are like z-scores, the *one*-tailed P-value will be very small (based on the 68–95–99.7 rule). Software (Fig. 29.2) reports that the *two*-tailed P-value is less than 0.0001. Hence, the *one*-tailed P-value is less than $0.0001/2 = 0.00005$.

⚠ The software clarifies *how* the differences have been computed. At the left of the output (Fig. 29.2), the order implies the differences are found as `Dist30` (the 30 m walk distance) minus `Dist20` (the 20 m walk distance), the same as our definition.

The one-tailed P-value is less than 0.00005, suggesting very strong evidence (Table 28.1) to support H_1. A conclusion requires an *answer to the RQ*, a summary of the *evidence* leading to that conclusion, and some *summary statistics*:

> Very strong evidence exists in the sample (paired $t = 7.07$; one-tailed $P < 0.0005$) of a mean reduction in 6MWT for a 20 m walkway compared to a 30 m walkway (mean reduction: 22.03 m; 95% CI: 15.76 m to 28.29 m; $n = 50$).

Note that the direction of the difference is provided.

⚠ Saying 'there is evidence of a difference' is insufficient. You must state *which* measurement is, on average, higher (that is, what the differences *mean*).

29.6 Statistical validity conditions

As with any CI and hypothesis test, these results apply under certain conditions. The conditions under which the results are statistically valid for paired data are similar to those for one sample mean, rephrased for differences.

The CI and test for a mean difference is *statistically valid* if *either* of these is true:

- when $n \geq 25$. (If the distribution of the differences is highly skewed, the sample size may need to be larger.)
- when $n < 25$, *and* the sample data come from a *population* with a normal distribution.

The sample size of 25 is a rough figure; some books give other values (such as 30).

This condition ensures that the *distribution of the sample mean differences has an approximate normal distribution* (so that, for example, the 68–95–99.7 rule can be used). Provided the sample size is larger than about 25, this will be approximately true *even if* the distribution of the differences in the population does not have a normal distribution. That is, when $n \geq 25$ the sample mean differences generally have an approximate normal distribution, even if the differences themselves don't have a normal distribution. The units of analysis are also assumed to be *independent* (e.g., from a simple random sample).

If the statistical validity conditions are not met, other methods (e.g., non-parametric methods [Conover, 2003]; resampling methods [Efron and Hastie, 2021]) may be used. For paired qualitative data, McNemar's test can be used [Conover, 2003].

Example 29.1 (Statistical validity)**.** For the 6MWT data, the sample size is $n = 50$, so the results are statistically valid. Neither the differences *in the population*, nor the distances *in the population* for the individual walkway lengths, need to follow a normal distribution.

29.7 Example: invasive plants

Skypilot is an alpine wildflower native to the Colorado Rocky Mountains (USA). In recent years, a willow shrub (*Salix*) has been encroaching on skypilot territory and, because willow often flowers early, Kettenbach et al. [2017] studied whether the willow may 'negatively affect pollination regimes of resident alpine wildflower species' (p. 6 965).

Data for both species was collected at $n = 25$ different sites, so the data are *paired* by site (Sect. 29.1). The data are shown in Table 13.3. The parameter is μ_d, the population mean *difference* in day of first flowering for skypilot, less the day of first flowering for willow. A *positive* value for the difference means that the skypilot values are larger, and hence that willow flowered first. The RQ is:

In the Colorado Rocky Mountains, is there a mean difference between first-flowering day for the native skypilot and encroaching willow?

The hypotheses are

$$H_0\colon \mu_d = 0 \quad \text{and} \quad H_1\colon \mu_d \neq 0,$$

where the alternative hypothesis is two-tailed, and μ_d is the mean difference between first-flowering day for the native skypilot and encroaching willow.

> ⓘ Explaining *how* the differences are computed is important. The differences here are skypilot minus willow first-flowering days.
>
> However, the differences could be computed as willow minus skypilot first-flowering days. *Either is fine*, as long as you remain consistent. The *meaning* of any conclusions will be the same.

The data are summarised graphically in Fig. 13.4 and numerically (Table 29.4, after rounding) using software output (Fig. 29.5).

Paired Samples T-Test

								95% Confidence Interval	
			statistic	df	p	Mean difference	SE difference	Lower	Upper
Skypilot	Willow	Student's t	1.4474	24.0000	0.16073	1.3600	0.9396	–0.5793	3.2993

Note. H_a $\mu_{\text{Measure 1 - Measure 2}} \neq 0$

Descriptives

	N	Mean	Median	SD	SE
Skypilot	25	190.7600	190	13.0616	2.6123
Willow	25	189.4000	189	12.1997	2.4399

FIGURE 29.5: Software output for the flowering-day data.

TABLE 29.4: The day of first flowering for encroaching willow and native skypilot.

	Mean	**Standard deviation**	**Standard error**	**Sample size**
Willow (encroaching)	189.40	12.200	2.440	25
Skypilot (native)	190.76	13.062	2.612	25
Differences	1.36	4.698	0.940	25

The standard error of the mean difference is $\text{s.e.}(\bar{d}) = 0.9396$ (Fig. 29.5 or Table 29.4). The sampling distribution for $\bar{d}$ has a normal distribution, centred around μ_d with a standard deviation of $\text{s.e.}(\bar{d}) = 0.9396$.

The approximate 95% CI for the mean difference is

$$1.36 \pm (2 \times 0.9396),$$

or from -0.52 to 3.24 days. The exact 95% CI (Fig. 29.5) is -0.58 to 3.30 days; the difference is because the approximate CI uses the *approximate* multiplier of 2 from the 68–95–99.7 rule.

The value of the test statistic (i.e., the t-score) is

$$t = \frac{\bar{d} - \mu_d}{\text{s.e.}(\bar{d})} = \frac{1.36 - 0}{0.9396} = 1.45,$$

as in the output. This is a relatively small value of t, so a large P-value is expected using the 68–95–99.7 rule. Indeed, the output shows that $P = 0.161$: there is *no evidence* of a mean difference in first-flowering day (i.e., the sample mean difference could reasonably be explained by sampling variation if $\mu_d = 0$).

Since *positive* differences mean that willow flowers earlier, we write (using the exact CI):

> No evidence exists ($t = 1.45$; two-tailed $P = 0.161$) that the day of first-flowering is different for the encroaching willow and the native skypilot (mean difference: 1.36 days earlier for willow; approximate 95% CI between 0.52 days earlier for skypilot to 3.24 days earlier for willow; $n = 25$).

The CI is statistically valid since $n = 25$.

⚠ Be clear in your conclusion about *how* the differences are computed. Make sure to interpret the test and CI consistently with how the differences are defined.

⚠ We *do not* say whether the evidence supports the null hypothesis. We assume the null hypothesis is true, so we state how strong the evidence is to support the alternative hypothesis. The current sample presents no evidence to contradict the assumption (but future evidence may emerge).

29.8 Example: chamomile tea

Rafraf et al. [2015] studied patients with Type 2 diabetes mellitus (T2DM). They randomly allocated 32 patients into a control group (who drank hot water), and another 32 patients to receive chamomile tea (p. 164):

> The study was blinded so that the allocation of the intervention or control group was concealed from the researchers and statistician [...] The intervention group ($n = 32$) consumed one cup of chamomile tea [...] three times a day immediately after meals (breakfast, lunch, and dinner) for 8 weeks. The control group ($n = 32$) consumed an equivalent volume of warm water during the 8-week period...

The total glucose (TG) was measured for each individual both *before* the intervention and *after* eight weeks on the intervention (a within-individuals comparison), in both the control and treatment groups (a between-individuals comparison). The data are not available, so no graphical summary of the data can be produced; however, the article gives a data summary (motivating Table 29.5).

Is there a mean reduction in TG in either group? Estimates of the mean reduction in each group can be found by constructing a CI for each group. First, the standard errors for each reduction are needed:

- Tea-drinking group: $\text{s.e.}(\bar{d}) = 30.37/\sqrt{32} = 5.37$.
- Control group: $\text{s.e.}(\bar{d}) = 36.66/\sqrt{32} = 6.48$.

Then the approximate 95% CIs are:

- Tea-drinking group: $38.62 \pm (2 \times 5.37)$, or from 27.88 to 49.36 mg.dL^{-1}.
- Control group: $-7.12 \pm (2 \times 6.48)$, or from -20.08 to 5.84 mg.dL^{-1}.

TABLE 29.5: The total glucose (TG; in mg.dL^{-1}) for two groups: those who drank chamomile tea, and those who drank hot water (the control group). The **Reduction** columns summarise the reduction in TG for each group.

		Baseline		After 8 weeks		Reduction		
	n	**Mean**	**Std dev.**	**Mean**	**Std dev.**	**Mean**	**Std dev.**	**Std error**
Chamomile tea	32	203.00	54.96	164.37	50.70	38.62	30.37	5.37
Control	32	178.25	53.06	185.37	52.59	−7.12	36.66	6.48
Difference		24.75		21.00		45.74		

(A *negative reduction* in TG means an *increase* in TG.) The first CI suggests that the population mean difference is almost certainly larger than zero; the second suggests that a population mean difference of zero could reasonably have produced the sample data.

Of course, the sample mean differences in TG may be non-zero due to sampling variation. So, the following repeated-measures RQs can be asked:

- For patients with T2DM, is there a mean *change* in TG after eight weeks drinking *chamomile tea*?
- For patients with T2DM, is there a mean *change* in TG after eight weeks drinking *hot water*?

Then, the hypotheses are (where μ_d represent the mean change in TG (in mg.dL^{-1}) after eight weeks):

- Tea-drinking group: H_0: $\mu_d = 0$ vs H_1: $\mu_d \neq 0$.
- Control group: H_0: $\mu_d = 0$ vs H_1: $\mu_d \neq 0$.

The two test statistics are:

$$t_T = \frac{38.62 - 0}{5.37} = 7.19 \qquad \text{and} \qquad t_W = \frac{-7.12 - 0}{6.48} = -1.10,$$

where the subscripts T and W refer to the tea and hot-water groups respectively. The t-score for the tea-drinking group is *huge*, so the two-tailed P-value will be *very small* using the 68–95–99.7 rule, and certainly smaller than 0.001. This means that there is evidence that chamomile tea had an impact on the mean change in TG.

In contrast, the t-score for the water-drinking group is *small*, so the two-tailed P-value will be *large* using the 68–95–99.7 rule, and certainly larger than 0.10. This means there is no evidence that placebo treatment (hot water) had any impact on mean change in TG (as one might expect for a placebo).

We write:

> There is very strong evidence ($t = 7.19$; two-tailed $P < 0.001$) of a mean change in TG for the chamomile-drinking groups (mean reduction: 38.62 mg.dL^{-1}; approx. 95% CI: 27.88 to 49.36 mg.dL^{-1}; $n = 32$), but *no* evidence ($t = -1.10$; two-tailed $P > 0.10$) of a mean change in the hot-water drinking group (mean reduction: −7.12 mg.dL^{-1}; approx. 95% CI: −20.08 and 5.84 mg.dL^{-1}; $n = 32$).

The intervals have a 95% chance of straddling the population mean reduction in TG. The sample sizes are larger than 25, so the results are statistically valid.

These hypothesis tests have allowed decisions to be made about each group individually. However, the two groups ultimately need to be *compared*; this is considered in Sect. 30.8.

29.9 Chapter summary

To compute a confidence interval (CI) for a mean difference, compute the sample mean difference, $\bar{d}$, and identify the sample size n. Then compute the standard error, which quantifies how much the value of $\bar{d}$ varies across all possible samples:

$$\text{s.e.}(\bar{d}) = \frac{s_d}{\sqrt{n}},$$

where s_d is the sample standard deviation. The *margin of error* is (multiplier × standard error), where the multiplier is 2 for an approximate 95% CI (using the 68–95–99.7 rule). Then the CI is:

$$\bar{d} \pm (\text{multiplier} \times \text{standard error})\,.$$

The statistical validity conditions should also be checked.

These steps are used to test a hypothesis about a population mean difference μ_d.

- Write the null hypothesis (H_0) and the alternative hypothesis (H_1); initially *assume* the value of μ_d in the null hypothesis to be true.
- Describe the *sampling distribution*, which describes what to *expect* from the sample mean difference based on this assumption: under certain statistical validity conditions, the sample mean difference varies with:
  - an approximate normal distribution,
  - with sampling mean whose value is the value of μ_d (from H_0), and
  - having a standard deviation of $\text{s.e.}(\bar{d}) = \dfrac{s_d}{\sqrt{n}}$.
- Compute the value of the *test statistic*:

$$t = \frac{\bar{d} - \mu}{\text{s.e.}(\bar{d})},$$

  where μ_d is the hypothesised value given in the null hypothesis.
- The t-value is like a z-score, and so an approximate *P-value* can be estimated using the 68–95–99.7 rule, or found using software. Use the P-value to make a decision, and write a conclusion.
- Check the statistical validity conditions.

29.10 Quick review questions

Bacho et al. [2019] compared joint pain in stroke patients receiving a supervised exercise treatment. The same participants ($n = 34$) were assessed *before* and *after* treatment. The mean *improvement* in joint pain after 13 weeks was 1.27 (with a standard error of 0.57) measured using a standardised tool.

Are the following statements *true* or *false*?

1. For paired data, the mean of the *differences* is treated like the mean of a single variable.
2. An appropriate graph for displaying these data is a histogram of the differences.
3. The *population* mean difference is denoted μ_d.

4. The standard error of the sample mean difference is denoted s_d.
5. Only 'before and after' studies can be paired.
6. The null hypothesis is about the *population* mean difference.
7. The value of the test statistic is 2.23.
8. The approximate value of the two-tailed P-value is very small.
9. The 'test statistic' for this test is a t-score.

29.11 Exercises

Answers to odd-numbered exercises are given at the end of the book.

Exercise 29.1. Which (if any) of these scenarios are *paired*?

1. Heart rate is measured for each individual when sitting and when standing. (Some individuals have their heart rate recorded first while sitting, and some first while standing.) Each person receives two measurements, and the *difference* in heart rate between sitting and standing is recorded.
2. The mean protein concentrations were compared in sea turtles before and after being rehabilitated [March et al., 2018].

Exercise 29.2. Which (if any) of these scenarios are *paired*?

1. The mean HDL cholesterol concentration is recorded for a group of males and a group of females, and the means compared.
2. Heart rate was recorded for 36 people, both before and after exercise, to determine how much the average heart rate increases.

Exercise 29.3. A group of primary school children was asked to complete a certain task on both a personal computer (PC) and using a tablet computer.

If the differences were defined as the time to complete the task on the PC, minus the time to complete the same task on a tablet (one difference for each child), what do the differences *mean*?

Exercise 29.4. Suppose water quality was recorded 500 m upstream and 500 m downstream of 28 different copper mines.

If the differences were defined as the pH downstream minus the water pH upstream for each river, what do the differences *mean*?

Exercise 29.5. Suppose, in the example of Sect. 29.7, the differences were defined as the day of first flowering for willow, less the day of first flowering for skypilot. Write down, and interpret the meaning of, the approximate 95% CI for the mean difference in first-flowering times.

Exercise 29.6. Suppose, in the example of Sect. 29.8, the differences were defined as *increase* in total glucose (TG). Write down, and interpret the meaning of the approximate 95% CI for the mean increase in TG for the tea-drinking group.

Exercise 29.7. [*Dataset*: **Fruit**] Mukherjee et al. [2019] studied the effect of rainfall on growing Chayote squash (*Sechium edule*). They compared the size of the fruit in a year with normal rainfall (2015) compared to a dry year (2014) on 24 farms:

> For Chayote squash grown in Bangalore, what is the mean difference in fruit weight between a normal and dry year?

Ten fruits were gathered from each farm in both years, and the average (mean) weight of the fruit recorded for the farm. Since the same farms are used in both years, the data are *paired* (Table 29.6). Data is missing for Farm 20 in the dry year (2014), so there are $n = 23$ differences.

1. What is the *unit of analysis*? What is the *units of observation*?
2. What is the advantage of using the same 24 farms twice each?

TABLE 29.6: The average weight of fruits (in g) in two different years, from 24 farms. One observation is missing for Farm 20. The change is computed as the normal year (2015) minus dry year (2014).

	Average fruit weight (kg)		
Farm	**Dry**	**Normal**	**Change**
1	367.75	371.05	3.30
2	238.25	218.85	−19.40
3	271.25	217.55	−53.70
4	286.27	221.70	−64.57
5	259.20	268.95	9.75
⋮	⋮	⋮	⋮
20	—	223.70	—
21	257.50	258.75	1.25
22	230.70	248.95	18.25
23	260.50	155.95	−104.55
24	231.85	219.30	−12.55

Paired Samples T-Test

			statistic	df	p	Mean difference	SE difference
FWeight2015	FWeight2014	Student's t	0.2050	22.00	0.839	2.230	10.88

Note. H_a $\mu_{\text{Measure 1 - Measure 2}} \neq 0$

Descriptives

	N	Mean	Median	SD	SE
FWeight2015	23	247.4	225.9	49.70	10.363
FWeight2014	23	245.1	248.1	42.22	8.803

FIGURE 29.6: Software output for the fruit data.

3. Construct a suitable graph to display the differences.
4. Create a numerical summary table for the data (use Fig. 29.6).
5. What is the parameter? Carefully describe what it means.
6. Write down the hypotheses.
7. Sketch the sampling distribution.
8. Compute the t-score.
9. Determine the P-value.
10. Construct an approximate 95% CI for the mean difference in fruit weight.
11. Are the test and the CI statistically valid?
12. Write a conclusion.

Exercise 29.8. [*Dataset*: **Captopril**] In a study of hypertension [Hand et al., 1996, MacGregor et al., 1979], $n = 15$ patients were given a drug (Captopril) and their systolic blood pressure measured (in mm Hg) immediately before and two hours after being given the drug.

The aim is to see if there is evidence of a *reduction* in blood pressure after taking Captopril. Use the data (Table 13.7) and the software output (Fig. 29.7) to answer these questions.

1. Explain why it is probably more sensible to compute differences as the *Before* minus the *After* measurements. What do the differences *mean* when computed this way?
2. What is the advantage of using the same patients for both the before and after measurements, rather than one group for before measurements and a different group of people for after measurements?
3. What is the parameter? Carefully describe what it means.
4. Construct a suitable graph to display the differences.
5. Write down the hypotheses.
6. Sketch the sampling distribution.
7. Write down the t-score.

8. Write down the P-value.
9. Write down the *exact* 95% CI using the computer output (Fig. 29.7).
10. Compute an *approximate* 95% CI for the mean difference.
11. Why are the two CIs different?
12. Write a conclusion.
13. Are the CI and test statistically valid?

Paired Samples T-Test

						95% Confidence Interval	
			statistic	df	p	Lower	Upper
Before	After	Student's t	8.12	14.0	<.001	13.9	23.9

FIGURE 29.7: Software output for the Captopril data.

Exercise 29.9. People often struggle to eat the recommended intake of vegetables. Fritts et al. [2018] explored ways to increase vegetable intake in teens. Teens rated the taste of raw broccoli, and raw broccoli served with a specially-made dip.

Each teen ($n = 100$) had a *pair* of measurements: the taste rating of the broccoli *with* and *without* dip. Taste was assessed using a '100 mm visual analogue scale', where a *higher* score means a *better* taste. In summary:

- for raw broccoli, the mean taste rating was 56.0 (with a standard deviation of 26.6);
- for raw broccoli served with dip, the mean taste rating was 61.2 (with a standard deviation of 28.7).

Because the data are paired, the *differences* are the best way to describe the data. The mean difference in the ratings was 5.2, with standard error of 3.06.

1. Construct a suitable numerical summary table.
2. What does a positive difference mean?
3. Perform a hypothesis test to see if the use of dip *increases* the mean taste rating.
4. Compute the approximate 95% CI for the mean difference in taste ratings.
5. Are the CI and test statistically valid?

Exercise 29.10. Allen et al. [2018] examined the effect of exercise on smoking. Men and women were assessed on their 'intention to smoke', both before and after exercise for each subject (using two quantitative questionnaires). Smokers ('smoking at least five cigarettes per day') aged 18 to 40 were enrolled for the study. For the 23 women in the study, the mean intention to smoke after exercise *reduced* by 0.66 (with a standard error of 0.37). (Larger values for 'intention to smoke' mean a greater intent to smoke.)

1. Perform a hypothesis test to determine if there is evidence of a population mean *reduction* in intention-to-smoke for women after exercising.
2. Find an approximate 95% CI for the population mean reduction in intention to smoke for women after exercising.
3. Are the CI and test statistically valid?

Exercise 29.11. [*Dataset*: `Ferritin`] In a study [Cressie et al., 1984] conducted at the Adelaide Children's Hospital (p. 107; emphasis added):

> ... a group of beta thalassemia patients [...] were treated by a continuous infusion of desferrioxamine, in order to *reduce* their ferritin content...

Using the data in Table 29.7, conduct a hypothesis test to determine if there is evidence that the treatment reduces the ferritin content, as intended. Make sure to include a 95% CI in the conclusion.

Exercise 29.12. [*Dataset*: `Stress`] The concentration of beta-endorphins in the blood is a sign of stress. Hoaglin et al. [2011] measured the beta-endorphin concentration for 19 patients about to

TABLE 29.7: The ferritin content (in μg.L) for 20 thalassemia patients at the Adelaide Children's Hospital.

Sept.	March	Reduction	Sept.	March	Reduction
6630	5100	1530	5360	6780	−1420
4590	3510	1080	6110	7250	−1140
3510	6600	−3090	5300	6000	−700
6375	8000	−1625	3120	4300	−1180
2500	2800	−300	3300	4680	−1380
1400	2860	−1460	11400	8500	2900
4580	3640	940	3100	3735	−635
6885	9030	−2145	2800	2730	70
4200	4420	−220	3500	6600	−3100
5600	7910	−2310	12700	7000	5700

undergo surgery [Hand et al., 1996]. Each patient had their beta-endorphin concentrations measured 12–14 h before surgery, and also 10 mins before surgery (in fmol.mL^{-1}).

A numerical summary (Table 29.8) was produced from output.

1. Use the output to test the RQ.
2. Use the software output in Fig. 29.8 to construct an *approximate* 95% CI for the *increase* in beta-endorphin concentrations as surgery gets closer.
3. Use the software output in Fig. 29.8 to write down the *exact* 95% CI for the *increase* in beta-endorphin concentrations as surgery gets closer.
4. Why is there a difference between the two CIs?
5. Are the CI and test statistically valid?

TABLE 29.8: The surgery-stress data (in fmol.mL^{-1}).

	Sample mean	Standard deviation	Standard error	n
12–14 h before surgery	8.35	4.40	1.01	19
10 min before surgery	16.05	12.51	2.87	19
Increase	7.70	13.52	3.10	19

Descriptives

	BeforeHours	BeforeMins	Increase
N	19	19	19
Missing	0	0	0
Mean	8.35	16.05	7.70
Std. error mean	1.01	2.87	3.10
95% CI mean lower bound	6.23	10.02	1.18
95% CI mean upper bound	10.47	22.08	14.22
Median	7.50	14.00	4.00
Standard deviation	4.40	12.51	13.52
Minimum	2.00	2.00	-3.50
Maximum	17.00	52.00	45.00

FIGURE 29.8: Software output for the surgery-stress data.

Exercise 29.13. A study of $n = 213$ Spanish health students [Romero-Blanco et al., 2020] measured (among other variables) the number of minutes of vigorous physical activity (PA) performed by students weekly *before* and *during* the COVID-19 lockdown (from March to April 2020 in Spain). Since the *before* and *during* lockdown were both measured on *each* participant, the data are *paired.* The data are summarised in Table 29.9.

1. Explain what the differences *mean.*
2. Compute the standard error of the differences.
3. Perform a hypothesis test to determine if mean minutes of vigorous PA changed from before to during the lockdowns.

TABLE 29.9: Summary information for the COVID-lockdown exercise data for $n = 214$ Spanish students: weekly minutes of vigorous physical activity.

	Mean (mins)	**Std dev. (mins)**
Before lockdown	28.47	54.13
During lockdown	30.66	30.04
Increase	2.68	51.30

Exercise 29.14. What happens when students start university? Many students will be responsible for their own meals for the first time, so some may forgo healthy foods for convenient, but less healthy, foods. Alternatively, some may not be able to afford sufficient or healthy food. Exercise regimes may also change.

Levitsky et al. [2004] recorded some students' weights as they began university, and then *the same* students' weight some later time. They asked the RQ:

> For Cornell University students, what is the *mean weight change* in students after 12 weeks at university?

The data collected to answer this RQ are shown in Table 29.10 [Levitsky].

1. Use the software output (Fig. 29.9) to compute an *approximate* 95% CI for the weight *gain* from Weeks 1 to 12.
2. Use the software output to write down an *exact* 95% CI for the weight *gain* from Weeks 1 to 12.
3. Comment on the two CIs.
4. Are the CIs statistically valid?
5. Conduct a hypothesis tests to determine if there is a change in mean weight from Weeks 1 to 12.
6. Do you think the weight gain would be of practical importance?

TABLE 29.10: The student weight-change data, showing the weight of students in Week 1 at university, in Week 12, and the weight gain (all in kg). These are the first five and the last five of the 68 total observations. (A negative weight gain means a weight loss.)

	Weight (in kg)				**Weight (in kg)**		
Student	**Week 1**	**Week 12**	**Weight gain**	**Student**	**Week 1**	**Week 12**	**Weight gain**
1	77.0	75.6	−1.4	⋮	⋮	⋮	⋮
2	49.5	50.0	0.5	64	69.8	71.1	1.3
3	60.3	61.2	0.9	65	72.0	72.4	0.4
4	51.8	53.6	1.8	66	51.8	53.6	1.8
5	67.5	69.8	2.3	67	75.2	76.5	1.3
⋮	⋮	⋮	⋮	68	59.0	59.0	0.0

Exercise 29.15. [*Dataset*: `Anorexia`] Young girls with anorexia ($n = 29$) received cognitive behavioural treatment (Hand et al. [1996]), and their weight before and after treatment were recorded. In summary:

Descriptives

	N	Mean	Median	SD	SE
Week12	68	62.099	60.300	11.073	1.343
Week1	68	61.237	60.300	10.970	1.330

Paired Samples T-Test

			statistic	df	p	Mean difference	SE difference	95% Confidence Interval Lower	95% Confidence Interval Upper
Week12	Week1	Student's t	7.431	67.000	<.0001	0.862	0.116	0.630	1.093

Note. H_a $\mu_{\text{Measure 1 - Measure 2}} \neq 0$

FIGURE 29.9: The weight-gain data: software output.

- Before the treatment, the mean weight was 82.69 pounds ($s = 4.845$ pounds);
- After the treatment, the mean weight was 85.70 pounds ($s = 8.352$ pounds).

The mean weight gain per girls was 3.01 pounds, with a standard deviation of 7.31 pounds. Find an approximate 95% CI for the population mean weight gain. Do you think the treatment had any meaningful impact on the mean weight gain of the girls, based solely on these data?

Exercise 29.16. [*Dataset*: `SoilCN`] Lambie et al. [2021] compared the percentage nitrogen (%N) in soils from intensively-grazed irrigated and non-irrigated pastures. The researchers *paired* similar irrigated and non-irrigated sites (p. 338):

> The irrigated and non-irrigated pairs within each site were within 100 m of each other and were on the same soil, landform and usually the same farm with the same farm management...

One RQ in the study was:

> For intensively grazed pastures sites, is there a mean reduction in percentage soil nitrogen (%N) when sites are irrigated, compared to non-irrigated?

The data are shown in Table 29.11. Use the data to answer the RQ.

TABLE 29.11: The percentage total soil nitrogen (%N) in irrigated and non-irrigated soils in 28 sites.

%N: irrigated	%N: not irrigated	%N: reduction when irrigated	%N: irrigated	%N: not irrigated	%N: reduction when irrigated
0.35	0.38	0.03	0.27	0.33	0.06
0.42	0.43	0.01	0.29	0.31	0.02
0.27	0.23	−0.04	0.40	0.43	0.03
0.18	0.24	0.06	0.26	0.26	0.00
0.56	0.58	0.02	0.52	0.53	0.01
0.34	0.26	−0.08	0.30	0.41	0.11
0.26	0.25	−0.01	0.20	0.32	0.12
0.58	0.44	−0.14	0.30	0.30	0.00
0.50	0.49	−0.01	0.24	0.26	0.02
0.47	0.55	0.08	0.49	0.67	0.18
0.55	0.55	0.00	0.27	0.29	0.02
0.41	0.45	0.04	0.44	0.47	0.03
0.51	0.54	0.03	0.27	0.28	0.01
0.47	0.56	0.09	0.40	0.50	0.10

Paired Samples T-Test

			statistic	df	p	Mean difference	SE difference	95% Confidence Interval	
								Lower	Upper
NonirrigatedN	IrrigatedN	Student's t	2.4147	27.0000	0.02279	0.0282	0.0117	0.0042	0.0522

Note. H_a $\mu_{\text{Measure 1 - Measure 2}} \neq 0$

Descriptives

	N	Mean	Median	SD	SE
NonirrigatedN	28	0.4039	0.4200	0.1266	0.0239
IrrigatedN	28	0.3757	0.3750	0.1186	0.0224

FIGURE 29.10: Software output for the nitrogen data. In the top table, the difference is implied as non-irrigated minus irrigated.

Exercise 29.17. [*Dataset*: `Jumping`] Hébert-Losier et al. [2023] recorded double-legged jumping distance for 80 healthy people, when they wore both shoes and were barefoot (Exercise 13.7). Use the data to form a 95% CI to estimate the mean distance people can jump further when barefoot.

Exercise 29.18. [*Dataset*: `WCTennis`] Alberca et al. [2022] recorded the push time (the time between a shot and resetting) for French wheelchair tennis players, while holding a racquet and not holding a racquet (Table 13.10; Alberca [2022]). Use the data to form a 95% CI to estimate the mean difference between push times with and without a racquet.

Answers to *Quick review* questions: 1. True. **2.** True. **3.** True. **4.** False. **5.** False. **6.** True. **7.** True. **8.** True. **9.** True.

30

Comparing two means: CIs and tests

You have learnt to ask an RQ, design a study, classify and summarise the data, construct confidence intervals, and conduct hypothesis tests. **In this chapter**, you will learn to:

- identify situations where comparing two means is appropriate.
- construct confidence intervals for the difference between two means.
- conduct hypothesis tests for comparing two means.
- determine whether the conditions for using these methods apply in a given situation.

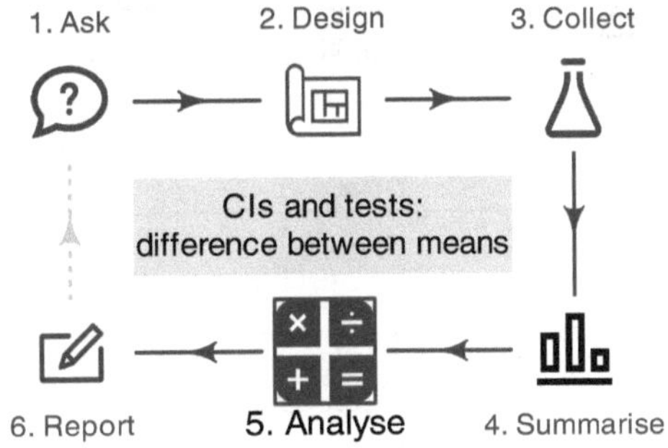

30.1 Introduction: garter snakes

Some Mexican garter snakes (*Thamnophis melanogaster*) live in habitats with no crayfish, while some live in habitats with crayfish and hence use crayfish as a food source. Manjarrez et al. [2017] were interested in whether the snakes in these two regions were different:

> For female Mexican garter snakes, is the mean snout–vent length (SVL) different for those in regions with crayfish and without crayfish?

Two different groups of snakes are studied, so this is a relational RQ (the study uses a between-individuals comparison) with no intervention, and the data are shown in Table 30.1.

TABLE 30.1: Snout–vent length (in cm) for female Mexican garter snakes living in crayfish ($n = 35$) and non-crayfish ($n = 41$) regions.

Non-crayfish region								
52	50	51	52	44	34	39	38	44
48	43	35	48	43	54	48	26	
29	44	33	48	48	43	45	50	
40	36	26	47	48	24	50	38	
48	44	40	38	40	36	26	26	

Crayfish region						
51	26	19	46	49	46	18
16	21	22	32	34	17	32
38	34	20	39	40	49	20
44	46	56	26	38	18	52
40	46	47	24	20	24	45

30.2 Summarising the data and error bar charts

A numerical summary *must* summarise the difference between the means, because the RQ is about this difference. Both groups should be summarised too. The information can be found using software (Fig. 30.1), and compiled into a table (Table 30.2). The appropriate summary for graphically summarising the *data* is (for example) a boxplot (Fig. 30.2, left panel).

⚠ No sample size or standard deviation is provided for the differences in Table 30.2; these make no sense in the context of comparing two means.

Independent Samples T-Test

		Statistic	df	p	Mean difference	SE difference	95% Confidence Interval	
							Lower	Upper
SVL	Student's t	3.569 [a]	74.000	0.0006	8.394	2.352	3.707	13.082
	Welch's t	3.445	55.135	0.0011	8.394	2.437	3.511	13.278

Note. H_a $\mu_{NoCfish} \neq \mu_{Cfish}$

[a] Levene's test is significant ($p < .05$), suggesting a violation of the assumption of equal variances

Group Descriptives

	Group	N	Mean	Median	SD	SE
SVL	NoCfish	41	42.566	43.500	7.786	1.216
	Cfish	35	34.171	34.000	12.493	2.112

FIGURE 30.1: Software output for the garter-snakes data.

TABLE 30.2: Numerical summaries of SVL (in cm) for female garter snakes in two regions.

	Mean	Standard deviation	Sample size	Standard error
Non-crayfish region	42.57	7.79	41	1.216
Crayfish region	34.17	12.49	35	2.112
Difference	8.39			2.437

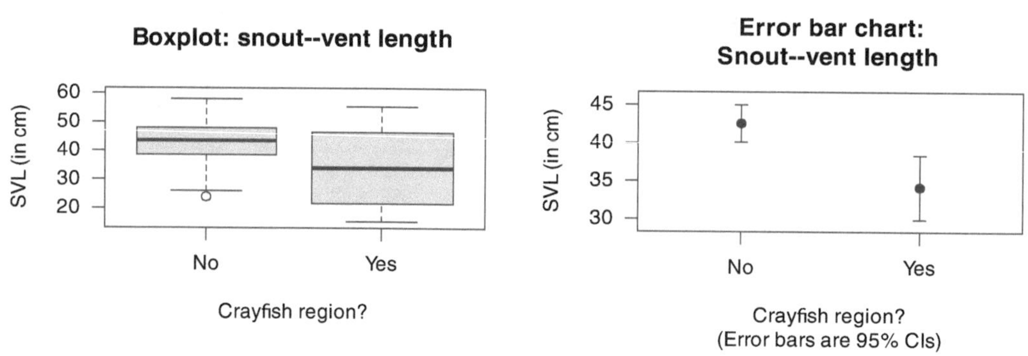

FIGURE 30.2: Boxplot (left) and error bar chart (right) of SVL for female snakes in two regions.

Since two groups are being compared, subscripts are used to distinguish between the statistics for the two groups; say, Groups 1 and 2 in general (Table 30.3). Using this notation, the *parameter* in the RQ is the difference between population means: $\mu_1 - \mu_2$. As usual, the population values are unknown, so this is estimated using the statistic $\bar{x}_1 - \bar{x}_2$.

TABLE 30.3: Notation used to distinguish the two independent groups.

	Group 1	**Group 2**	**Comparing groups**
Sample sizes:	n_1	n_2	
Population means:	μ_1	μ_2	$\mu_1 - \mu_2$
Sample means:	$\bar{x}_1$	$\bar{x}_2$	$\bar{x}_1 - \bar{x}_2$
Standard deviations:	s_1	s_2	
Standard errors:	$\text{s.e.}(\bar{x}_1) = \dfrac{s_1}{\sqrt{n_1}}$	$\text{s.e.}(\bar{x}_2) = \dfrac{s_2}{\sqrt{n_2}}$	$\text{s.e.}(\bar{x}_1 - \bar{x}_2)$

For the garter-snakes data, define the differences as the mean for females snakes living in non-crayfish regions (N), *minus* the mean for female snakes in crayfish regions (C): $\mu_N - \mu_C$. This is the *parameter*. By this definition, the differences refer to how much larger (on average) the SVL is for snakes living in non-crayfish regions.

⚠ Here the difference is computed as the mean SVL for snakes living in non-crayfish regions, *minus* the mean SVL for snakes living in crayfish regions. Computing the difference as the mean SVL for snakes in crayfish regions, *minus* non-crayfish regions is also correct.

You need to be clear about how the difference is computed, and be consistent throughout. The *meaning* of the conclusions will be the same whichever direction is used.

A useful way to compare the means of two (or more) groups is to display the CIs for the means of the groups being compared in an *error bar chart*. Error bars charts display the expected variation *in the sample means* from sample to sample, while boxplots display the variation *in the individual observations*. For the garter-snakes data, the error bar chart (Fig. 30.2, right panel) shows the 95% CI for each group; the mean appears as a dot.

The two CIs for the SVL are (using information from the bottom table in Fig. 30.1):

- Crayfish regions: $34.171 \pm (2 \times 2.112)$, or from 29.94 to 38.40 cm.
- Non-crayfish regions: $42.566 \pm (2 \times 1.216)$, or from 40.13 to 45.00 cm.

However, the error bar chart, and these CIs, do not give a CI for the *difference* between the two means, as relevant to the RQ.

Example 30.1 (Error bar charts)**.** Schepaschenko et al. [2017] studied the foliage biomass of small-leaved lime trees from three sources: coppices; natural; planted. Three graphical summaries are shown in Fig. 30.3: a boxplot (showing the variation in *individual* trees; left), an error bar chart (showing the variation in the *sample means*; centre) on the same vertical scale as the boxplot, and the same error bar chart using a more appropriate scale for the error bar plot (right).

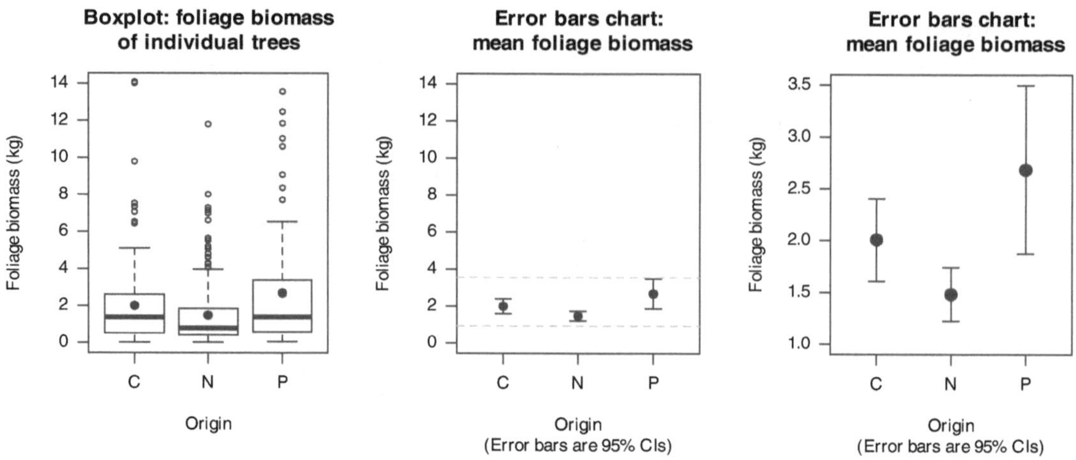

FIGURE 30.3: Boxplot (left) and error bar charts (centre; right) comparing the mean foliage biomass for small-leaved lime trees from three sources (C: Coppice; N: Natural; P: Planted). The centre panel shows an error bar chart using the same vertical scale as the boxplot; the dashed horizontal lines are the limits of the error bar chart on the right. The right error bar chart uses a more appropriate scale on the vertical axis. The solid dots show the mean of the distributions.

30.3 Confidence intervals for $\mu_1 - \mu_2$

Each sample will comprise different snakes, and give different SVLs. The sample means for each group will differ from sample to sample, and the *difference* between the sample means will be different for each sample also. The *difference* between the sample means varies from sample to sample, and so has a sampling distribution and a standard error.

Definition 30.1 (Sampling distribution for the difference between two sample means). The *sampling distribution of the difference between two sample means* $\bar{x}_1$ and $\bar{x}_2$ is (when the appropriate conditions are met; Sect. 30.5) described by:

- an approximate normal distribution,
- centred around a sampling mean whose value is $\mu_1 - \mu_2$, the difference between the *population* means,
- with a standard deviation, called the standard error of the difference between the means, of $\text{s.e.}(\bar{x}_1 - \bar{x}_2)$.

The standard error for the difference between the means is found using

$$\text{s.e.}(\bar{x}_1 - \bar{x}_2) = \sqrt{\text{s.e.}(\bar{x}_1)^2 + \text{s.e.}(\bar{x}_2)^2},$$

though this value will often be *given* (e.g., on computer output) rather than needing to be computed.

For the garter-snakes data, the differences between the sample means will have:

- an approximate normal distribution,
- centred around the sampling mean whose value is $\mu_N - \mu_C$,
- with a standard deviation, called the *standard error* of the difference, of $\text{s.e.}(\bar{x}_P - \bar{x}_C) = 2.437$.

The standard error of the difference between the means was computed using

$$\text{s.e.}(\bar{x}_N - \bar{x}_C) = \sqrt{\text{s.e.}(\bar{x}_N)^2 + \text{s.e.}(\bar{x}_C)^2} = \sqrt{1.216^2 + 2.1112^2} = 2.437,$$

the same value shown in the *second row* of the software output (Fig. 30.1).

The sampling distribution describes how the values of $\bar{x}_N - \bar{x}_C$ vary from sample to sample. Then, finding a 95% CI for the difference between the mean SVLs is similar to the process used in Chap. 23, since the sampling distribution has an approximate normal distribution:

$$\text{statistic} \pm \big(\text{multiplier} \times \text{s.e.(statistic)}\big).$$

When the statistic is $\bar{x}_P - \bar{x}_C$, the approximate 95% CI is

$$(\bar{x}_N - \bar{x}_C) \pm \big(2 \times \text{s.e.}(\bar{x}_N - \bar{x}_C)\big).$$

So, in this case, the approximate 95% CI is

$$8.394 \pm (2 \times 2.437)$$

or 8.394 ± 4.874, after rounding appropriately. We write:

> The difference between mean SVLs is 8.39 cm, shorter for those living in a crayfish region (mean: 34.17 cm; s.e.: 2.112; $n = 35$) compared to those *not* living in a crayfish region (mean: 42.57 cm; s.e.: 1.216; $n = 41$), with an *approximate* 95% CI for the difference between mean SVLs from 3.52 to 13.27 cm.

The plausible values for the difference between the two population means SVLs are between 3.52 to 13.27 cm (shorter for those living in crayfish regions).

> ⚠ Giving the CI alone is insufficient; the *direction* in which the differences were calculated must be given, so readers know which group had the higher mean.

Output from software often shows two CIs for the difference between the two means (Fig. 30.1). *We will use the results from Welch's test (the second row)*, as this row of output is more general, and makes fewer assumptions, than the results in the first row. The information in the second row makes fewer assumptions, and is more widely applicable.

> ⚠ Most software gives *two* CIs: one assuming the standard deviations in the two groups are the same (Student's), and another *not* assuming the standard deviations in the two groups are the same (Welch's).
>
> We will use the information that does *not* assume the standard deviations in the two groups are the same. In the software output in Fig. 30.1, this is the second row of the top table (labelled 'Welch's t'). (The information in both rows are often similar anyway.)

From the output, the 95% CI for the difference is from 3.51 to 13.28 cm. The *approximate* CI and the *exact* (from software) CIs are only slightly different, as the samples sizes are not too small. (Recall: the t-multiplier of 2 is an approximation, based on the 68–95–99.7 rule.)

30.4 Hypothesis tests for $\mu_1 - \mu_2$: t-test

A hypothesis test can be used to decide if the SVL is different for the two regions. The parameter for the test is $\mu_N - \mu_C$.

As always, the null hypothesis is the default 'no difference, no change, no relationship' position; any difference between the parameter and statistic is due to sampling variation (Sect. 28.2). Hence, the null hypothesis is 'no difference' between the population mean SVL of the two groups:

- H_0: $\mu_N - \mu_C = 0$ (equivalent to $\mu_N = \mu_C$).

From the RQ, the alternative hypothesis is *two*-tailed:

- H_1: $\mu_N - \mu_C \neq 0$ (equivalent to $\mu_N \neq \mu_C$).

The alternative hypothesis proposes that any difference between the *sample* means is because a difference really exists between the *population means.* The alternative hypothesis is two-tailed, based on the RQ.

The difference between the sample mean SVLs in the two groups depends on which one of the many possible samples is randomly obtained, *even if* the difference between the means in the population is zero. The difference between the sample means is 8.394 cm, but this difference will vary from sample to sample; that is, *sampling variation* exists.

For the SVL data, the sampling distribution of $\bar{x}_N - \bar{x}_C$ can be described as (see Def. 30.1):

- an approximate normal distribution,
- centred around the sampling mean whose value is $\mu_N - \mu_C = 0$, the difference between the population means (from H_0),
- with a standard deviation of $\text{s.e.}(\bar{x}_N - \bar{x}_C) = 2.4368$.

> Most software gives *two* hypothesis test results: one assuming the standard deviations in the two groups are the same, and another *not* assuming the standard deviations in the two groups are the same.
>
> We will use the information that does *not* assume the standard deviations in the two groups are the same. In the software output in Fig. 30.1, this is the second row of the bottom table (labelled 'Welch's t'). (The information in both rows are often similar anyway.)

The observed difference between sample means, relative to what was expected, is found by computing the test statistic; in this case, a t-score. The software output (Fig. 30.1) gives the t-score, but the t-score can also be computed using the information in Table 30.2:

$$\begin{aligned} t &= \frac{\text{sample statistic} - \text{mean of sampling distribution (from } H_0)}{\text{standard deviation of sampling distribution}} \\ &= \frac{(\bar{x}_P - \bar{x}_C) - (\mu_P - \mu_C)}{\text{s.e.}(\bar{x}_P - \bar{x}_C)} = \frac{8.39 - 0}{2.4368} = 3.44, \end{aligned}$$

as in the software output.

A P-value determines if the sample statistic is consistent with the assumption (i.e., H_0). Since the t-score is large, the P-value will be small using the 68–95–99.7 rule (and less than 0.003). This is confirmed by the software (Fig. 30.1): the two-tailed P-value is 0.0011.

A small P-value suggests the observations are *inconsistent* with the assumption of no difference (Table 28.1), and the difference between the sample means could *not* be reasonably explained by sampling variation, if $\mu_N - \mu_C = 0$.

In conclusion, write:

> Strong evidence exists in the sample (two independent samples $t = 3.445$; two-tailed $P = 0.0011$) that the population mean SVL is different for female snakes living in crayfish regions (mean: 34.17 cm; $n = 35$) and non-crayfish regions (mean: 42.57 cm; $n = 41$; 95% CI for the difference: 3.51 to 13.28 cm longer for those in non-crayfish regions).

The conclusion contains an *answer to the RQ*, the *evidence* leading to this conclusion ($t = 3.44$; two-tailed $P = 0.0011$), and *sample summary statistics*, including a CI.

30.5 Statistical validity conditions

As usual, these results apply under certain conditions. The CI and test for comparing two means is *statistically valid* if *either* of these is true:

- when *both* samples have $n \geq 25$. (If the distribution of a sample is highly skewed, the sample size for that sample may need to be larger.)
- when one or both groups have 25 or fewer observations, *and* the *populations* corresponding to the groups with samples sizes under 25 have an approximate normal distribution.

The sample size of 25 is a rough figure; some books give other values (such as 30).

This condition ensures that the *distribution of the difference between sample means has an approximate normal distribution* (so that, for example, the 68–95–99.7 rule can be used). The histograms of the *sample data* can be used to determine if normality of the *populations* seems reasonable. The units of analysis are also assumed to be *independent* (e.g., from a simple random sample).

If the statistical validity conditions are not met, other similar options include using a Mann-Whitney test [Conover, 2003] or using resampling methods [Efron and Hastie, 2021].

> **Example 30.2** (Statistical validity). For the garter-snakes data, both samples sizes exceed 25 (41 and 35), so the test is statistically valid. The data in each group do not need to be normally distributed, since both sample sizes are larger than 25, and the data are not severely skewed (Fig. 30.2, left panel).

30.6 Tests for comparing more than two means: ANOVA

Often, more than two means need to be compared. This requires a different method, called *analysis of variance* (or ANOVA). The details are beyond the scope of this book. In this section, a very brief overview of using a one-way ANOVA is given, using an example. Importantly, this example shows that the basic principles of hypothesis testing from Chap. 28 still apply.

ANOVA is a general tool that can be extended beyond just comparing more than two means, and used in many and varied context for the analysis of data.

Example 30.3 (ANOVA). [*Dataset:* `Lime`] Schepaschenko et al. [2017] studied the foliage biomass of small-leaved lime trees from three sources: coppices (C); natural (N); planted P); see Example 30.1. A boxplot and error bar chart are shown in Fig. 30.3. A numerical summary is shown in Table 30.4 (based on the output in Fig. 30.4).

To compare the mean foliage biomass of trees from the three sources, the null hypothesis is 'no difference' between the population means:

$$H_0\colon \mu_C = \mu_N = \mu_P.$$

The alternative hypothesis is that the three means are not all equal. This hypothesis encompasses many possibilities: for example, that the three means are *all* different from each other, or that the first is different from the other two (which are the same). Because the alternative hypothesis encompasses many possibilities, we write:

$$H_1\colon \text{Not all means are equal.}$$

⚠ For comparing more than two means, the alternative hypothesis *is always two-tailed.*

Performing an ANOVA using software (Fig. 30.4) gives $P = 0.005$. (The *test statistic* here is an F-score; we don't discuss these further, but the F-score measures the overall difference between the three means.) The small P-value in this context means the same as usual (Sect. 28.6): there is persuasive evidence to support the alternative hypothesis (that the three means are *not* all equal).

While we know the means are not all the same, we do not know *which* group means are different from which other group means. One option might be to compare all possible combinations of two groups (i.e., the means of groups C and N; the means of groups C and P; the means of groups N and P) using three separate two-sample t-tests. While this approach is possible, it increases the probability of declaring a false positive (i.e., of making a Type I error; Sect. 28.7): *incorrectly* declaring that a difference exists between two sets of means. The correct approach requires methods beyond this book.

One-Way ANOVA (Welch's)

	F	df1	df2	p
Foliage	5.536	2	149.8	0.005

Group Descriptives

	Origin	N	Mean	SD	SE
Foliage	Coppice	133	2.007	2.294	0.1989
	Natural	185	1.482	1.758	0.1293
	Planted	67	2.684	3.321	0.4058

FIGURE 30.4: Software output for testing hypotheses for the lime-trees data.

TABLE 30.4: Foliage biomass of lime trees (in kg) from different origins.

	Mean	Standard deviation	Standard error	n
Coppice	2.01	2.29	0.199	133
Natural	1.48	1.76	0.129	185
Planted	2.68	3.32	0.406	67

30.7 Example: speed signage

To reduce vehicle speeds on freeway exit ramps, Ma et al. [2019] studied the impact of additional signage. At one site (Ningxuan Freeway), speeds were recorded for 38 vehicles *before* the extra signage was added, and then for 41 different vehicles *after* the extra signage was added.

The researchers are hoping that the addition of extra signage will *reduce* the mean speed of the vehicles. The RQ is:

At this freeway exit, does the mean vehicle speed *reduce* after extra signage is added?

The data are *not* paired: different vehicles are measured before (B) and after (A) the extra signage is added. Define μ as the mean speed (in km.h^{-1}) on the exit ramp, and the parameter as $\mu_B - \mu_A$, the *reduction* in the mean speed.

The data can be summarised (Table 30.5) using the software output (Fig. 30.5), where

$$\text{s.e.}(\bar{x}_B - \bar{x}_A) = \sqrt{\text{s.e.}(\bar{x}_B)^2 + \text{s.e.}(\bar{x}_A)^2} = \sqrt{2.140^2 + 2.051^2} = 2.965,$$

as in the output table (Row 2). A boxplot of the data is shown in Fig. 30.6 (left panel), and an error bar chart in Fig. 30.6 (right panel).

Independent Samples T-Test

		Statistic	df	p	Mean difference	SE difference	95% Confidence Interval Lower	95% Confidence Interval Upper
Speed	Student's t	1.914	77.000	0.0593	5.674	2.964	−0.228	11.576
	Welch's t	1.914	76.492	0.0594	5.674	2.965	−0.229	11.578

Note. H_a $\mu_{\text{Before}} \neq \mu_{\text{After}}$

Group Descriptives

	Group	N	Mean	Median	SD	SE
Speed	Before	38	98.016	98.200	13.194	2.140
	After	41	92.341	93.900	13.134	2.051

FIGURE 30.5: Software output for the speed data.

An approximate 95% CI for the difference between the mean speeds is

$$5.674 \pm (2 \times 2.9642),$$

TABLE 30.5: The signage data summary (in km.h^{-1}).

	Mean	Median	Standard deviation	Standard error	Sample size
Before	98.02	98.2	13.194	2.140	38
After	92.34	93.9	13.134	2.051	41
Speed reduction	5.68			2.965	

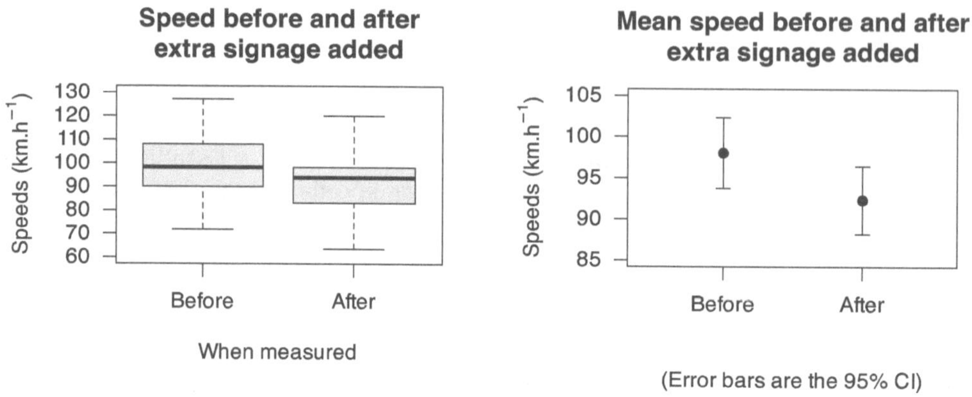

FIGURE 30.6: Boxplot (left) and error bar chart (right) showing the mean speed before and after the addition of extra signage, and the 95% CIs. The vertical scales on the two graphs are different.

or from -0.25 to $11.60\,\text{km.h}^{-1}$. (This is very similar to the 95% CI shown in Fig. 30.5.) The negative value is not a negative speed. Since the difference between the means is defined as a *reduction*, this CI means that the *reduction* in the populations mean speed is likely between -0.25 to $11.64\,\text{km.h}^{-1}$. Since a negative reduction is an increase, this is more easily understood as the difference being located between a $0.25\,\text{km.h}^{-1}$ *increase* before the signage was added to an $11.64\,\text{km.h}^{-1}$ *reduction* after the signage was added.

The hypotheses are:

- H_0: $\mu_B - \mu_A = 0$: there is no difference in the population mean speeds.
- H_1: $\mu_B - \mu_A > 0$: the population mean speed has *reduced* after the addition of signage.

The best estimate of the difference in *population* means is the difference between the *sample* means: $(\bar{x}_B - \bar{x}_A) = 5.68$. Since $\text{s.e.}(\bar{x}_B - \bar{x}_A) = 2.965$, the t-score is

$$t = \frac{(\bar{x}_B - \bar{x}_A) - (\mu_B - \mu_A)}{\text{s.e.}(\bar{x}_B - \bar{x}_A)} = \frac{5.674 - 0}{2.9642} = 1.91.$$

using Equation (27.2). (Recall that $\mu_B - \mu_A = 0$ is initially assumed, from the null hypothesis.)

Remembering that the alternative hypothesis is *one-tailed*, the P-value (using the 68–95–99.7 rule) is larger than 0.025, but smaller than 0.32, so making a clear decision is difficult without using software. However, since the t-score is *just* less than 2, we suspect that the P-value is likely to be closer to 0.025 than to 0.32.

From software, $P = 0.0297$ (you cannot be this precise just using the 68–95–99.7 rule). Using Table 28.1, this P-value provides moderate evidence of a reduction in mean speeds. We conclude:

Moderate evidence exists in the sample ($t = 1.91$; one-tailed $P = 0.030$) that mean speeds have reduced after the addition of extra signage (mean reduction: 5.67 km.h^{-1}; 95% CI for the difference: -0.23 to 11.6 km.h^{-1}; s.e.: 2.96 km.h^{-1}). The before mean speed was 98.02 km.h^{-1} ($n = 38$; standard deviation: 13.19 km.h^{-1}); the after mean speed was 92.34 km.h^{-1} ($n = 41$; standard deviation: 13.13 km.h^{-1}).

Whether the mean speed reduction of 5.67 km.h^{-1} has *practical importance* is a separate issue. Using the validity conditions, the CI and the test are statistically valid.

ⓘ Remember: the conclusion must make clear *which* mean is larger!

30.8 Example: chamomile tea

(This study was seen in Sect. 29.8.) Rafraf et al. [2015] studied patients with Type 2 diabetes mellitus (T2DM). They randomly allocated 32 patients into a control group (who drank hot water), and 32 to receive chamomile tea (Rafraf et al. [2015]).

The total glucose (TG) was measured for each individual in both groups, both before the intervention and after eight weeks on the intervention. Summary data are given in Table 29.5. Evidence suggests that the chamomile tea group shows a mean reduction in TG (Sect. 29.8), while the hot-water group shows no evidence of a reduction. That is, there appears to be a difference between the two groups regarding the *change* in TG. However, the differences between the chamomile-tea and the hot-water groups may be due to the samples selected (i.e., sampling variation), so comparing the changes between the two groups is helpful.

The following relational RQ can be asked:

For patients with T2DM, is the mean reduction in TG *greater* for the chamomile tea group compared to the hot water group?

Notice the RQ is one-tailed; the aim of the study is to determine if the chamomile-tea drinking group performs *better* (i.e., reduces the mean TG) than the control group.

This RQ is comparing two separate groups; specifically, comparing the *differences* between the two groups. This study contains both *within*-individuals comparisons (see Sect. 29.8) and a *between*-individuals comparison (this section); see Fig. 30.7. This is equivalent to treating the *differences* for both groups as the two separate sets of data in the two-sample analysis.

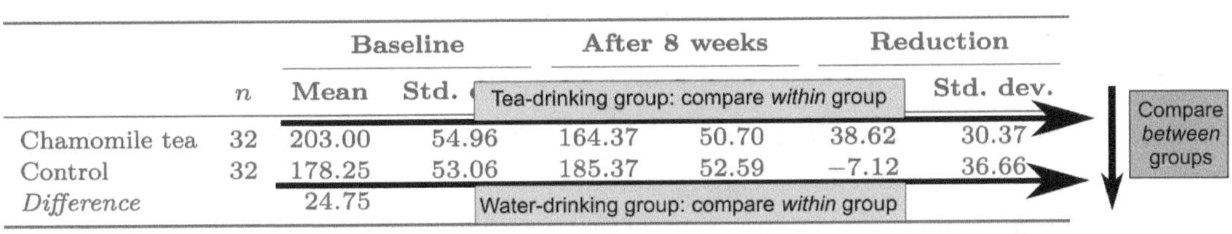

FIGURE 30.7: The chamomile-tea study has two within-individuals comparisons, and a between-individuals comparison (comparing the differences in each group).

The corresponding hypotheses are:

$$H_0\colon \mu_T - \mu_W = 0 \qquad \text{and} \qquad H_1\colon \mu_T - \mu_W > 0$$

where μ refers to the mean *reduction* in TG, T refers to the tea-drinking group, and W to the hot-water drinking group.

The parameter $\mu_T - \mu_W$ is estimated by the statistic $\bar{x}_T - \bar{x}_W = 45.74\,\text{mg.dL}^{-1}$. The standard error for the statistic was found as s.e.$(\bar{x}_T - \bar{x}_W) = 8.42$ (using the information in Table 29.5). Hence, the test statistic is:

$$t = \frac{(\bar{x}_T - \bar{x}_W) - (\mu_T - \mu_W)}{\text{s.e.}(\bar{x}_T - \bar{x}_W)} = \frac{45.75 - 0}{8.42} = 5.43,$$

which is very large, so the P value will be very small (using the 68–95–99.7 rule), and certainly smaller than 0.001.

We write:

> There is very strong evidence ($t = 5.43$; one-tailed $P < 0.001$) that the mean reduction in TG for the chamomile-tea drinking group (mean reduction: $36.62\,\text{mg.dL}^{-1}$) is greater than the mean reduction in TG for the hot-water drinking group (mean reduction: $-7.12\,\text{mg.dL}^{-1}$; difference between means: $45.74\,\text{mg.dL}^{-1}$; approx. 95% CI: 28.64 to $62.84\,\text{mg.dL}^{-1}$).

Again, the sample sizes are larger than 25, so the results are statistically valid.

30.9 Chapter summary

To compute a confidence interval (CI) for the difference between two means, compute the difference between the two sample means, $\bar{x}_1 - \bar{x}_2$, and identify the sample sizes n_1 and n_2. Then compute the standard error, which quantifies how much the value of $\bar{x}_1 - \bar{x}_2$ varies across all possible samples:

$$\text{s.e.}(\bar{x}_1 - \bar{x}_2) = \sqrt{\text{s.e}(\bar{x}_1) + \text{s.e.}(\bar{x}_2)},$$

where s.e.$(\bar{x}_1)$ and s.e.$(\bar{x}_2)$ are the standard errors of Groups 1 and 2. The *margin of error* is (multiplier $\times$ standard error), where the multiplier is 2 for an approximate 95% CI (using the 68–95–99.7 rule). Then the CI is:

$$(\bar{x}_1 - \bar{x}_2) \pm (\text{multiplier} \times \text{standard error})\,.$$

The statistical validity conditions should also be checked.

These steps are used to test a hypothesis about a difference between two population means $\mu_1 - \mu_2$.

- Write the null hypothesis (H_0) and the alternative hypothesis (H_1); initially *assume* the value of $(\mu_1 - \mu_2)$ in the null hypothesis to be true.

- Describe the *sampling distribution*, which describes what to *expect* from the difference between the sample means based on this assumption: under certain statistical validity conditions, the difference between the sample means vary with:
  - an approximate normal distribution,
  - with sampling mean whose value is the value of $(\mu_1 - \mu_2)$ (from H_0), and
  - having a standard deviation of $\text{s.e.}(\bar{x}_1 - \bar{x}_2)$.
- Compute the value of the *test statistic*:

$$t = \frac{(\bar{x}_1 - \bar{x}_2) - (\mu_1 - \mu_2)}{\text{s.e.}(\bar{x}_1 - \bar{x}_2)},$$

  where $\mu_1 - \mu_2$ is the hypothesised difference given in the null hypothesis.
- The t-value is like a z-score, and so an approximate *P-value* can be estimated using the 68–95–99.7 rule, or found using software. Use the P-value to make a decision, and write a conclusion.
- Check the statistical validity conditions.

ANOVA is used to compare means for more than two groups.

30.10 Quick review questions

Lee et al. [2016] studied iron levels in Koreans with Type 2 diabetes, comparing people on a vegan ($n = 46$) and a conventional ($n = 47$) diet for 12 weeks. A summary of the data for iron levels are shown in Table 30.6.

Are the following statements *true* or *false*?

1. An appropriate graph for displaying the *data* is a boxplot.
2. The difference between the means in the population is denoted $\mu_V - \mu_C$, where V represent the vegan diet, and C represents the conventional diet.
3. The standard error of the difference between the sample means is denoted $\text{s.e.}(\bar{x}_V) - \text{s.e.}(\bar{x}_C)$.
4. An error bar chart displays the variation in the *data*.
5. The sample size is missing from the *Difference* row, but the value is $47 - 46 = 1$.
6. The standard deviation is missing from the *Difference* row, but the value is 0.4.
7. The standard error for the difference cannot be computed, as not enough information is given.
8. The two-tailed P-value for the comparison is $P = 0.046$. This means that *no evidence* that the population means are different.

TABLE 30.6: Comparing the iron levels (mg) for subjects using a vegan or conventional diet for 12 weeks.

	Mean	**Standard deviation**	n
Vegan diet	13.9	2.3	46
Conventional diet	15.0	2.7	47
Difference	1.1		

30.11 Exercises

Answers to odd-numbered exercises are given at the end of the book.

Exercise 30.1. Suppose researchers are comparing the cell diameter of lymphocytes (a type of white blood cell) and tumour cells. Define the mean diameter of lymphocytes as μ_L, and the mean diameter of tumour cells as μ_T.

If the difference between the means were defined as $\mu_L - \mu_T$, what does this *mean*?

Exercise 30.2. Suppose researchers are comparing the braking distance of cars using two different types of brake pads (Type A and Type B). Define the mean breaking distance for cars with Type A brake pads as μ_A, and mean breaking distance for cars with Type B brake pads as μ_B.

If the difference between the means were defined as $\mu_B - \mu_A$, what does this *mean*?

Exercise 30.3. Sketch the sampling distribution for the difference between the mean speeds before and after adding extra signage (Sect. 30.7).

Exercise 30.4. Sketch the sampling distribution for the difference between reduction in mean TG for the tea-drinking and the hot-water drinking group (Sect. 30.8).

Exercise 30.5. Agbayani et al. [2020] measured (among other variables) the length of gray whales (*Eschrichtius robustus*) at birth. Are female gray whales longer than males, on average, in the population at birth? Summary information is shown in Table 30.7.

TABLE 30.7: Numerical summary of length of whales at birth (in m).

	Mean	**Standard deviation**	**Sample size**
Female	4.66	0.38	26
Male	4.60	0.30	30

1. Define the parameter, and write down its estimate. Carefully describe what it means.
2. Sketch an error bar chart.
3. Compute the standard error of the difference between the two means.
4. Compile a numerical summary table.
5. Compute the approximate 95% CI.
6. Write the hypotheses to answer the RQ.
7. Compute the t-score, and approximate the P-value using the 68–95–99.7 rule.
8. Write a conclusion.
9. Are the CI and test statistically valid?

Exercise 30.6. [*Dataset*: **NHANES**] Earlier, the NHANES study (Exercise 14.7) was used to summarise the data used to answer this RQ:

> Among Americans, is the mean direct HDL cholesterol (in mmol.L^{-1}) different for current smokers and non-smokers?

Use the software output in Fig. 30.8 to answer these questions.

1. Define the parameter of interest, and write down its estimate. Carefully describe what it means.
2. Sketch an error bar chart.
3. Compile a numerical summary table.
4. Compute the approximate 95% CI, and write a conclusion.
5. Write down the exact 95% CI, and write a conclusion.
6. Write the hypotheses to answer the RQ.
7. Write down the standard error of the difference.
8. Write down the t-score and the P-value.
9. Write a conclusion.

10. Are the CI and test statistically valid?
11. Is the difference between the means likely to be of practical importance?

Independent Samples T-Test

		Statistic	df	p	Mean difference	SE difference	95% Confidence Interval	
							Lower	Upper
DirectChol	Student's t	5.473	3054.000	<.001	0.085	0.015	0.054	0.115
	Welch's t	5.478	2964.437	<.001	0.085	0.015	0.054	0.115

Note. H_a $\mu_{No} \neq \mu_{Yes}$

Group Descriptives

	Group	N	Mean	Median	SD	SE
DirectChol	No	1668	1.392	1.320	0.428	0.010
	Yes	1388	1.308	1.240	0.424	0.011

FIGURE 30.8: Software output for the NHANES data.

Exercise 30.7. Barrett et al. [2010] studied the effectiveness of echinacea to treat the common cold, and compared the mean duration of the cold for participants treated with echinacea or a placebo to determine if using echinacea *reduced* the mean duration of symptoms. Participants were blinded to the treatment, and allocated to the groups randomly. A summary of the data is given in Table 30.8.

1. What is the parameter? Carefully describe what it means.
2. Compute the standard error for the mean duration of symptoms for each group.
3. Compute the standard error for the difference between the means.
4. Sketch an error bar chart.
5. Compute an approximate 95% CI for the *difference* between the mean durations for the two groups.
6. Compute an approximate 95% CI for the population mean duration of symptoms for those treated with echinacea.
7. Write the hypotheses to answer the RQ.
8. Compute the standard error of the difference.
9. Compute the t-score, and approximate the P-value using the normal distribution tables.
10. Write a conclusion.
11. Are the CI and test statistically valid?
12. Are the results likely to be of practical importance?

TABLE 30.8: Numerical summary of duration (in days) of common cold symptoms, for blinded patients taking echinacea or a placebo.

	Mean	**Standard deviation**	**Standard error**	**Sample size**
Placebo	6.87	3.62		176
Echinacea	6.34	3.31		183
Difference	0.53			

Exercise 30.8. Carpal tunnel syndrome (CTS) is pain experienced in the wrists. Schmid et al. [2012] compared two different treatments: night splinting, or gliding exercises.

Participants were *randomly allocated* to one of the two groups. Pain intensity (measured using a quantitative visual analogue scale; *larger* values mean *greater* pain) were recorded after one week of treatment. The data are summarised in Table 30.9.

1. What is the parameter? Carefully describe what it means.
2. Compute the standard error for the mean pain intensity for each group.
3. Compute the standard error for the difference between the mean of the two groups.

4. Sketch an error bar chart.
5. Compute an approximate 95% CI for the *difference* in mean pain intensity for the treatments.
6. Compute an approximate 95% CI for the population mean pain intensity for those treated with splinting.
7. Write the hypotheses to answer the RQ.
8. Compute the t-score, and approximate the P-value using the 68–95–99.7 rule.
9. Write a conclusion.
10. Are the CI and test statistically valid?

TABLE 30.9: Numerical summary of pain intensity for two different treatments of carpal tunnel syndrome.

	Mean	Standard deviation	Standard error	Sample size
Exercise	0.8	1.4		10
Splinting	1.1	1.1		10
Difference	0.3			

Exercise 30.9. [*Dataset*: **Dental**] Woodward and Walker [1994] recorded the sugar consumption in industrialised (mean: 41.8 kg/person/y) and non-industrialised (mean: 24.6 kg/person/y) countries. The software output is shown in Fig. 30.9.

1. What is the parameter? Carefully describe what it means.
2. Write the hypotheses.
3. Using the software output (Fig. 30.9), write down and interpret the CI.
4. Write a conclusion for the hypothesis test.
5. Is the test statistically valid?

Independent Samples T-Test

							95% Confidence Interval	
		statistic	df	p	Mean difference	SE difference	Lower	Upper
Sugar	Student's t	-5.25 [a]	88.0	< .001	-17.2	3.29	-23.8	-10.7
	Welch's t	-6.47	87.2	< .001	-17.2	2.66	-22.5	-11.9

[a] Levene's test is significant (p < .05), suggesting a violation of the assumption of equal variances

Group Descriptives

	Group	N	Mean	Median	SD	SE
Sugar	No	61	24.6	24.2	16.6	2.13
	Yes	29	41.8	44.0	8.63	1.60

FIGURE 30.9: Software output for the sugar-consumption data; the Groups refer to whether the country is industrialised (Yes) or not (No).

Exercise 30.10. [*Dataset*: **Deceleration**] To reduce vehicle speeds on freeway exit ramps, Ma et al. [2019] studied using additional signage. At one site studied (Ningxuan Freeway), speeds were recorded at various points on the freeway exit for vehicles *before* the extra signage was added, and then for different vehicles *after* the extra signage was added.

In addition, the *deceleration* of each vehicle was determined (Table 14.8) as the vehicle left the 120 km.h^{-1} speed zone and approached the 80 km.h^{-1} speed zone. Use the data, and the summary in Table 30.10, to test the RQ:

At this freeway exit, is the mean vehicle deceleration the same before extra signage is added and after extra signage is added?

Identify clearly the parameter of interest to understand how much the deceleration *increased* after adding the extra signage. Remember to compute and interpret the CI for this parameter.

TABLE 30.10: The signage deceleration data summary (in m.s^{-2}).

	Mean	**Standard deviation**	**Standard error**	**Sample size**
Before	0.0745	0.0494	0.00802	38
After	0.0765	0.0521	0.00814	41
Change	−0.0020		0.01143	

Exercise 30.11. [*Dataset*: **ForwardFall**] A study [Wojcik et al., 1999] compared the lean-forward angle in younger and older women (Table 14.6). An elaborate set-up was constructed to measure this lean-forward angle, using harnesses. Consider this RQ:

> Among healthy women, is the mean lean-forward angle *greater* for younger women compared to older women?

Use the software output (Fig. 30.10) to answer these questions:

1. What is the parameter? Carefully describe what it means.
2. What is an appropriate graph to display the *data*?
3. Construct an appropriate numerical summary from the software output (Fig. 14.11).
4. Construct *approximate* and *exact* 95% CIs. Explain any differences.
5. Is the test one- or two-tailed?
6. Write the statistical hypothesis.
7. Use the software output to conduct the hypothesis test.
8. Write a conclusion.
9. Are the CI and test statistically valid?

Independent Samples T-Test

							95% Confidence Interval	
		Statistic	df	p	Mean difference	SE difference	Lower	Upper
LeanAngle	Student's t	7.875	13.000	<.0001	14.500	1.841	10.522	18.478
	Welch's t	6.691	5.592	0.0007	14.500	2.167	9.102	19.898

Note. H_a $\mu_{\text{Younger}} \neq \mu_{\text{Older}}$

Group Descriptives

	Group	N	Mean	Median	SD	SE
LeanAngle	Younger	10	30.700	31.500	2.751	0.870
	Older	5	16.200	15.000	4.438	1.985

FIGURE 30.10: Software output for the face-plant data.

Exercise 30.12. Becker et al. [1991] compared the access to health promotion (HP) services for people with and without a disability in southwestern of the USA. 'Access' was measured using the quantitative *Barriers to Health Promoting Activities for Disabled Persons* (BHADP) scale. *Higher* scores mean *greater* barriers to health promotion services. The RQ is:

> Is there a difference between the mean BHADP scores, for people with and without a disability, in southwestern USA?

1. What is the parameter? Carefully describe what it means.
2. Sketch an error bar chart.
3. Compute the standard error of the difference.

4. Compile a numerical summary table.
5. Compute the approximate 95% CI, and write a conclusion.
6. Write down the hypotheses.
7. Compute the t-score.
8. Determine the P-value.
9. Write a conclusion.
10. Are the CI and test statistically valid?

TABLE 30.11: The data summary for BHADP scores (no measurement units).

	Sample mean	Standard deviation	Sample size	Standard error
Disability	31.83	7.73	132	0.67280
No disability	25.07	4.80	137	0.41010
Difference	6.76			

Exercise 30.13. [*Dataset*: **BodyTemp**] Consider again the body temperature data from Sect. 27.1. The researchers also recorded the gender of the patients, as they also wanted to compare the mean internal body temperatures for females and males.

Use the software output in Fig. 30.11 to perform this test and to construct an approximate 95% CI appropriate for answering the RQ. Comment on the practical significance of your results.

Independent Samples T-Test

		statistic	df	p	Mean difference	SE difference	95% Confidence Interval	
							Lower	Upper
BodyTempC	Student's t	-2.29	128	0.024	-0.161	0.0703	-0.300	-0.0216
	Welch's t	-2.29	128	0.024	-0.161	0.0703	-0.300	-0.0216

Group Descriptives

	Group	N	Mean	Median	SD	SE
BodyTempC	Male	65	36.7	36.7	0.388	0.0481
	Female	65	36.9	36.9	0.413	0.0512

FIGURE 30.11: Software output for the body-temperature data.

Exercise 30.14. Chapman et al. [2007] compared 'conventional' male paramedics in Western Australia with male 'special-operations' paramedics. Some information comparing their physical profiles is shown in Table 30.12.

1. Compute the missing standard errors.
2. Compare the mean grip strength for the two groups of paramedics. (The *standard error for the difference between the means* is 3.30.)
3. Compare the mean number of push-ups completed in one minute for the two groups of paramedics. (The *standard error for the difference between the means* is 4.0689.)

Exercise 30.15. [*Dataset*: **Anorexia**] Young girls ($n = 29$) with anorexia received cognitive behavioural treatment (Hand et al. [1996]), while another $n = 26$ young girls received a control treatment (the 'standard' treatment). All girls had their weight recorded before and after treatment.

1. Determine the mean *gain* for individual girls using software.
2. Compute a CI for the mean weight gain for the girls in each group.
3. Compute a CI for the difference between the mean weight gains for the two treatment groups.
4. Conduct a test to determine if there is a difference between the mean weight gains for the two treatment groups.

TABLE 30.12: The physical profile of conventional ($n = 18$) and special operation ($n = 11$) paramedics in Western Australia.

	Conventional	Special Operations
Grip strength (in kg)		
Mean	51	56
Standard deviation	8	9
Standard error		
Push-ups (per minutes)		
Mean	36	47
Standard deviation	10	11
Standard error		

Exercise 30.16. Researchers studied the impact of a gluten-free diet on dental cavities [Khalaf et al., 2020]. Some summary information regarding the number decayed, missing and filled teeth (DMFT) is shown in Table 30.13. An *exact* 95% CI is given as for the difference is −2.32 to 2.76.

1. Using the 68–95–99.7 rule gives a slightly different CI. Why?
2. True or false: the difference is computed as the number of DMFT for coeliacs minus non-coeliacs.
3. True or false: one of the values for the CI is a negative value, which must be an error (as a negative number of DMFT is impossible).
4. We are 95% confident that the difference between the population means is:
    - smaller for coeliacs;
    - between 2.32 higher for non-coeliacs to 2.76 higher for coeliacs.
    - between 2.76 higher for non-coeliacs to 2.32 higher for coeliacs.

TABLE 30.13: The summary of the number of DMFT for coeliacs and non-coeliacs.

	Sample size	Mean	Standard deviation	Standard error
Coeliacs	23	8.39	4.4	0.92
Non-coeliacs	23	8.17	4.1	0.86
Difference		0.22		1.30

Exercise 30.17. [*Dataset*: `ReactionTime`] Strayer and Johnston [2001] examined the reaction times, while driving, for students from the University of Utah [Agresti and Franklin, 2007]. In one study, students were randomly allocated to one of two groups: one group *used* a mobile phone while driving in a driving simulator, and one group *did not use* a mobile phone while driving in a driving simulator. The reaction time for each student was measured. The data are shown in Table 30.14.

Use the data to answer this RQ:

> For students, what is the difference between the mean reaction time while driving when using a mobile phone and when *not* using a mobile phone?

TABLE 30.14: Reaction times (in milliseconds) for students using, and not using, mobile phones while driving.

Reaction time: using phone						
636	600	609	554	578	688	527
623	542	559	626	560	679	536
615	554	595	501	525	960	
672	543	565	574	647	558	
601	520	573	468	456	482	

Reaction time: not using phone						
557	506	626	436	617	539	512
572	648	626	642	528	523	449
457	485	426	476	578	479	
489	610	585	586	472	535	
532	444	487	565	485	603	

Exercise 30.18. [*Dataset*: `BMI`] Johnson et al. [2021] collected data from hospital outpatients at an Irish hospital. One RQ in the study concerns comparing the mean number of days per week that patients exercise for more than 30 mins (say, μ) according to their smoking status: daily (D), occasionally (O) or not at all (N).

Use the output (Fig. 30.12) to answer the questions that follow.

1. Construct an error bar chart to summarise the data.
2. Construct a numerical summary table.
3. Perform a suitable hypothesis test, and answer the RQ,

One-Way ANOVA (Welch's)

	F	df1	df2	p
exercise	13.64	2	27.98	< .001

Group Descriptives

	smoke	N	Mean	SD	SE
exercise	daily	11	1.273	0.7862	0.2371
	not at all	46	3.152	1.9318	0.2848
	occasionally	13	2.769	1.6408	0.4551

FIGURE 30.12: Software output for testing hypotheses for the BMI data.

Answers to *Quick review* questions: **1.** True. **2.** True. **3.** False: s.e.$(\bar{x}_C - \bar{x}_V)$. **4.** False: variation for the sample means. **5.** False: sample size makes no sense. **6.** False: standard deviation makes no sense **7.** False: 0.5197. **8.** False: *slight* evidence population means are different.

31

Comparing two odds or proportions: CIs and tests

You have learnt to ask an RQ, design a study, classify and summarise the data, construct confidence intervals, and conduct hypothesis tests. **In this chapter**, you will learn to:

- identify situations where comparing proportions or odds is appropriate.
- form confidence intervals for the difference between two proportions.
- form confidence intervals for odds ratios.
- conduct hypothesis tests for comparing two proportions.
- conduct hypothesis tests for comparing two odds.
- determine whether the conditions for using these methods apply in a given situation.

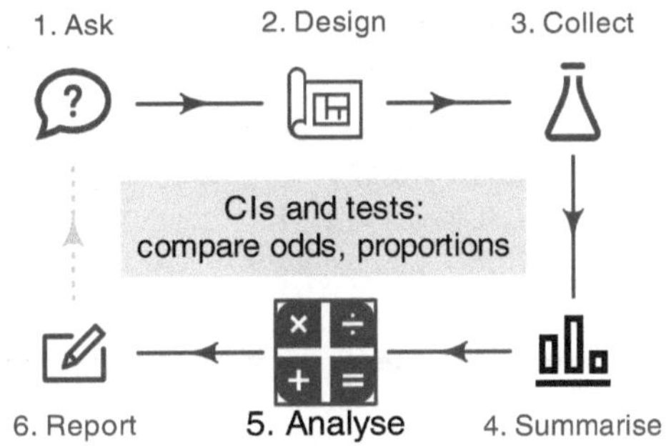

31.1 Introduction: meals on-campus

Mann and Blotnicky [2017] examined the relationship between where university students usually ate, and where the student lived, for students from two Canadian universities. The researchers cross-classified the $n = 183$ students (the units of analysis) according to two *qualitative* variables:

- where the student lived, *with* their parents or *not with* their parents.
- where the student ate most meals, *off*-campus or *on*-campus.

Both variables are qualitative, so means are not appropriate for summarising the data. The data can be compiled into a two-way table of counts (Table 31.1), also called a *contingency table*. Both qualitative variables have two levels, so this is a 2×2 table. Every cell in the 2×2 table contains different students, so the comparison is *between* individuals.

> ⚠ The study has one sample of students, classified according to two variables (i.e., each student is placed into one of the four cells in the 2×2 table).

The *proportion* of students who eat most meals off-campus can be compared between those who live with their parents and those who do *not* live with their parents. Then, the parameter is the difference between the population proportions in each group.

Alternatively, the *odds* of students who eat most meals off-campus can be compared between those who live with their parents and those who do *not* live with their parents. Then, the

TABLE 31.1: Where university students live and eat.

	Has most meals off-campus	Has most meals on-campus
Living with parents	52	2
Not living with parents	105	24

parameter is the comparison of the odds in both groups, the *odds ratio* (OR); specifically, the OR of eating most meals off-campus, comparing those living with parents to those not living with parents.

The table can be constructed with either variable as the rows. However, software commonly compares *rows*, so it makes sense to place the groups to be compared (i.e., the levels of the explanatory variable) in the rows of the table.

31.2 Summarising data

Since two groups are being compared, subscripts are used to distinguish between the two groups; say, Groups 1 and 2 in general (Table 31.2). For this example, we use N to refer to students *not* living with their parents, and L for students living with their parents.

TABLE 31.2: Notation used to distinguish the two independent groups.

	Group 1	Group 2	Comparing groups
Sample sizes:	n_1	n_2	
Sample odds:	Odds_1	Odds_2	Odds ratio (OR) $= \text{Odds}_1/\text{Odds}_2$
Sample proportions:	$\hat{p}_1$	$\hat{p}_2$	$\hat{p}_1 - \hat{p}_2$
Standard errors:	$\text{s.e.}(\hat{p}_1)$	$\text{s.e.}(\hat{p}_2)$	$\text{s.e.}(\hat{p}_1 - \hat{p}_2)$

The parameter is either a difference between two population proportions, or a population OR. For example, the parameter could be the difference between population proportion of students eating most meals *off*-campus, comparing students living with their parents, to students *not* living with their parents. Alternatively (and equivalently), the parameter could be the population OR of eating most meals *off*-campus, comparing students living with their parents, to students *not* living with their parents.

Since software commonly compares *rows* (for example, see the text under the bottom table in Fig. 31.1), it makes sense to place the groups to be compared (i.e., the explanatory variable) in the rows of the table.

Then, the difference between the two proportions are usually calculated as the Row 1 proportion minus the Row 2 proportion. Similarly, the odds then can be interpreted as comparing Column 1 counts to Column 2 counts, and the *odds ratio* as comparing the Row 1 odds to the Row 2 odds.

The RQ and the hypotheses can be written as comparing *proportions* (Sect. 31.4), comparing *odds* (Sect. 31.6), or about *ORs*. With two qualitative variables, an appropriate numerical

χ^2 Tests

	Value	df	p
χ^2	6.9341	1	0.00846
z test difference in 2 proportions	2.6333		0.00846
N	183		

Comparative Measures

		95% Confidence Intervals	
	Value	Lower	Upper
Difference in 2 proportions	0.1490 [a]	0.0651	0.2330
Odds ratio	5.9429	1.3524	26.1141

[a] Rows compared

FIGURE 31.1: Software output for comparing the odds and proportion of students eating most meals at home, for students living with and not with their parents

summary includes the odds and proportions (or percentages) for the outcome for both comparison groups, and the sample sizes (Table 31.3).

To compare the *proportions*, define the sample proportion of students eating most meals off-campus as $\hat{p}$, and write $\hat{p}_L$ for the proportion living with parents and $\hat{p}_N$ for the proportion *not* living with parents. Then,

$$\hat{p}_L = \frac{52}{52+2} = 0.96296 \quad \text{and} \quad \hat{p}_N = \frac{105}{105+24} = 0.813953.$$

The *difference* between the two proportions is

$$\hat{p}_L - \hat{p}_N = 0.9630 - 0.8140 = 0.1490,$$

(as in the software output: Fig. 31.1). By this definition, the difference is how much greater the proportion eating most meals off-campus is for students *living* with their parents, compared to students *not living* with their parents.

> ⓘ Be clear about how differences are defined! Differences could be computed as:
>
> - the proportion eating most meals off-campus for those living with their parents, *minus* the proportion *not* living with their parents. This measures how much greater the proportion is for those living with their parents; or
> - the proportion eating most meals off-campus for those *not* living with their parents, *minus* the proportion living with their parents. This measures how much greater the proportion is for those *not* living with their parents.
>
> Either is fine, provided you are *consistent*, and *clear* about how the differences are computed. The *meaning* of any conclusions will be the same.

To compare the *odds*, first see that the odds of eating most meals *off-campus* is:

- $52 \div 2 = 26$ for students *living with their parents* (Row 1 of Table 31.1).
- $105 \div 24 = 4.375$ for students *not living with their parents* (Row 2 of Table 31.1).

(Notice the numbers in the *second* column are always on the bottom of the fraction.) So the OR of eating most meals *off-campus* (the *first* column), comparing students living with parents to students *not* living with parents (*second* column), is $26 \div 4.375 = 5.943$ (as in the software output: Fig. 31.2).

The numerical summary (Table 31.3) shows the proportion and odds of eating most meals off-campus, comparing students living at home and those not living at home.

> ⚠ The OR can be interpreted in *either* of these ways (i.e., both are correct):
>
> - the *odds* compare Row 1 counts to Row 2 counts, for both columns. The OR then compares the Column 1 odds to the Column 2 odds.
> - the *odds* compare Column 1 counts to Column 2 counts. The OR then compares the Row 1 odds to the Row 2 odds.
>
> Odds and ORs are computed with the *first row* and *first column* values on the *top* of the fraction. Since the explanatory variable is usually in the rows, the second is usually the most useful. In this case, both of the above approaches produces an OR of 5.943.

An appropriate graph is a side-by-side bar chart or a stacked bar chart (Fig. 31.2). The side-by-side bar is useful for comparing odds. For instance, in the two left-most bars in Fig. 31.2 (left panel), the first bar is 26 times as high as the second bar (and 26 is the odds); in the two right-most bars, the first bar is 4.375 times as high as the second bar (and 4.375 is the odds). A stacked bar chart is useful for comparing proportions.

TABLE 31.3: The odds and proportion of university students eating most meals off-campus.

	Odds having most meals off-campus	**Proportion having most meals off-campus**	**Sample size**
Living with parents	26.000	0.963	54
Not living with parents	4.375	0.814	129
	OR: 5.943	*Difference:* 0.149	

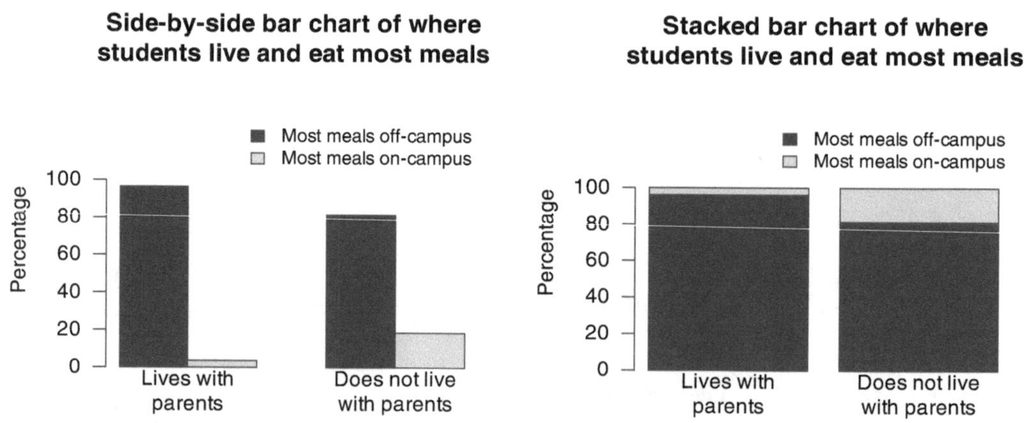

FIGURE 31.2: The student-eating data. Left: a side-by-side bar chart. Right: a stacked bar chart.

31.3 Confidence intervals for comparing proportions

The sample proportions for each group will vary from sample to sample, and the *difference* between the sample proportions will be different for each sample. Hence, the *difference* between the two sample proportions has a sampling distribution and *standard error*. Under certain conditions (Sect. 31.7), this sampling distribution has a normal distribution.

Definition 31.1 (Sampling distribution for the difference between two sample proportions for a CI). When constructing a CI, the *sampling distribution of the difference between two sample proportions* $\hat{p}_1$ and $\hat{p}_2$ is (when the appropriate conditions are met; Sect. 31.7) described by:

- an approximate normal distribution,
- centred around a sampling mean whose value is $p_1 - p_2$, the difference between the *population* proportions,
- with a standard deviation, called the standard error of the difference between the proportions, of $\text{s.e.}(\hat{p}_1 - \hat{p}_2)$.

The standard error for the difference between the proportions is found using

$$\text{s.e.}(\hat{p}_1 - \hat{p}_2) = \sqrt{\text{s.e.}(\hat{p}_1)^2 + \text{s.e.}(\hat{p}_2)^2}, \tag{31.1}$$

though this value will often be *given* (e.g., on computer output).

For the student-eating data, the standard errors of the sample proportions for each group are computed using Equation (22.3) as

$$\text{s.e.}(\hat{p}_L) = \sqrt{\frac{0.962963 \times (1 - 0.962963)}{54}} = 0.025700, \text{and}$$

$$\text{s.e.}(\hat{p}_N) = \sqrt{\frac{0.8139535 \times (1 - 0.8139535)}{129}} = 0.034262.$$

The standard error of the difference between the proportions is then

$$\text{s.e.}(\hat{p}_L - \hat{p}_N) = \sqrt{\text{s.e.}(\hat{p}_L)^2 + \text{s.e.}(\hat{p}_N)^2} = \sqrt{0.025700^2 + 0.034262^2} = 0.042830.$$

Thus, the differences between the sample proportions will have:

- an approximate normal distribution,
- centred around the sampling mean whose value is $p_L - p_N$,
- with a standard deviation of $\text{s.e.}(\hat{p}_L - \hat{p}_N) = 0.0428295$.

The sampling distribution describes how the values of $\hat{p}_L - \hat{p}_N$ vary from sample to sample. Then, finding a 95% CI for the difference between the proportions is similar to the process used previously, since the sampling distribution has an approximate normal distribution:

$$\text{statistic} \pm \big(\text{multiplier} \times \text{s.e.(statistic)}\big).$$

When the statistic is $\hat{p}_L - \hat{p}_N$, the approximate 95% CI is

$$(\hat{p}_L - \hat{p}_N) \pm \big(2 \times \text{s.e.}(\hat{p}_L - \hat{p}_N)\big).$$

So, in this case, the approximate 95% CI is

$$0.1490 \pm (2 \times 0.042830),$$

or 0.149 ± 0.0857 after rounding (i.e., from 0.0633 to 0.235). This approximate CI is very similar to the (exact) CI from software (Fig. 31.2). We write:

> The difference between the proportions of students eating most meals at home is 0.1490, higher for those living with their parents (0.963; $n = 52$) than those not living with their parents (0.814; $n = 129$), with the approximate 95% CI from 0.0633 to 0.235.

The plausible values for the difference between the two population proportions are between 0.063 to 0.235, larger for those living with parents.

> ⚠ Giving the CI alone is insufficient; the *direction* in which the differences were calculated must be given, so readers know which group had the higher proportion.

31.4 Hypothesis tests for comparing proportions: z-test

To compare the two proportions using a hypothesis test, the two-tailed RQ is:

> Is the *population* proportion of students eating most meals off-campus the same for students *living with* their parents and for students *not living with* their parents?

As usual, the population values are unknown, so the parameter $p_L - p_N$ is estimated using the statistic $\hat{p}_L - \hat{p}_N$.

Hypothesis testing always begins by assuming that the null hypothesis is true (Sect. 28.2.1). In this context, that means assuming that the population proportion of eating most meals off-campus is the same in both groups:

- H_0: $p_L - p_N = 0$ (equivalent to $p_L = p_N$).

From the RQ, the alternative hypothesis is *two*-tailed:

- H_1: $p_L - p_N \neq 0$ (equivalent to $p_L \neq p_N$).

Because the null hypothesis is assumed to be true, the proportions are assumed to have the same value for both groups. Hence, the data from the two groups can be combined to determine an overall (or common) proportion of students eating most meals off-campus:

$$\hat{p} = \frac{52 + 105}{52 + 105 + 2 + 24} = \frac{157}{183} = 0.85792. \tag{31.2}$$

This is the overall proportion of students eating most meals off-campus, since we assumed no difference between students living with and not with their parents. Effectively, this proportion has been computed by summing the columns in Table 31.1 and using this combined data to compute the proportion of students eating most meals off-campus.

> ⚠ As with any hypothesis test, the null hypothesis is assumed to be true. For a test comparing two proportions, that implies the proportion in each group is the same, and so the standard errors are computed using the common (overall) proportion.

The *sample* proportions for the two groups (L and N) will vary from sample to sample and so have a sampling distribution. The standard error of the sample proportion for each sample is computed using this common proportion $\hat{p}$, using the same idea as in Equation (22.3):

$$\text{s.e.}(p_L) = \sqrt{\frac{\hat{p} \times (1-\hat{p})}{n_L}} = \sqrt{\frac{0.85792 \times (1-0.85792)}{54}} = 0.047511, \text{and}$$

$$\text{s.e.}(p_N) = \sqrt{\frac{\hat{p} \times (1-\hat{p})}{n_N}} = \sqrt{\frac{0.85792 \times (1-0.85792)}{129}} = 0.030739.$$

⚠ When computing the standard errors as part of a *hypothesis test*, the common or overall proportion is used to compute the standard errors.

The difference between the two proportions will vary from sample to sample too, and hence have a sampling distribution; under certain conditions (Sect. 31.7), this sampling distribution will have a normal distribution. The standard error of this sampling distribution for the *difference* between the proportions is

$$\text{s.e.}(\hat{p}_L - \hat{p}_N) = \sqrt{\text{s.e.}(\hat{p}_L)^2 + \text{s.e.}(\hat{p}_N)^2} = \sqrt{0.047511^2 + 0.030739^2} = 0.056588,$$

which is similar to Equation (31.1), except that a common proportion was used to compute s.e.($\hat{p}_L$) and s.e.($\hat{p}_L$).

Definition 31.2 (Sampling distribution for the difference between two sample proportions for a hypothesis test)**.** When conducting a hypothesis test, the *sampling distribution of the difference between two sample proportions* $\hat{p}_1$ and $\hat{p}_2$ is (when the appropriate conditions are met; Sect. 31.7) described by:

- an approximate normal distribution,
- centred around a sampling mean whose value is $p_1 - p_2$, the difference between the *population* proportions (from H_0),
- with a standard deviation, called the standard error of the difference between the proportions, of s.e.($\hat{p}_1 - \hat{p}_2$).

The standard error for the difference between the proportions is

$$\text{s.e.}(\hat{p}_1 - \hat{p}_2) = \sqrt{\text{s.e.}(\hat{p}_1)^2 + \text{s.e.}(\hat{p}_2)^2},$$

where

$$\text{s.e.}(p_1) = \sqrt{\frac{\hat{p} \times (1-\hat{p})}{n_1}} \quad \text{and} \quad \text{s.e.}(p_2) = \sqrt{\frac{\hat{p} \times (1-\hat{p})}{n_2}},$$

where $\hat{p}$ is the common (overall) sample proportion.

Since the sampling distribution has an approximate normal distribution, the test statistic is

$$z = \frac{(\hat{p}_L - \hat{p}_N) - (p_L - p_N)}{\text{s.e.}(\hat{p}_L - \hat{p}_N)} = \frac{0.14901 - 0}{0.056588} = 2.633.$$

Since the sampling distribution has an approximate normal distribution, the approximate P-value can be computed from normal distribution tables (Sect. 20.6), approximated using the 68–95–99.7 rule, or from software output (Fig. 31.1). The two-tailed P-value reported by software (Fig. 31.1, under the column **p**) is indeed small: 0.008 to three decimal places.

⚠ The test statistic for tests involving proportions is a z-score and *not* a t-score.

A small P-value means strong evidence exists to supporting H_1: the evidence suggests a difference between the *population* proportions. We write:

> The *sample* provides strong evidence ($z = 2.63$; two-tailed $P = 0.008$) that the proportion of students in the *population* of having most meals off-campus is different for students living with their parents (proportion: 0.963, $n = 54$) and students *not* living with their parents (proportion: 0.814, $n = 129$; difference: 0.149; approximate 95% CI from 0.0633 to 0.235, higher for students living with their parents).

The conclusion includes three components (Sect. 28.8): the *answer to the RQ*; the *evidence* used to reach that conclusion ('$z = 2.63$; two-tailed $P = 0.008$'); and some *sample summary statistics* (including the 95% CI for the difference between proportions). The conclusion makes clear which proportion is higher.

31.5 Confidence intervals for comparing odds (for an odds ratio)

A CI can be formed for the OR, as well as for the difference between two proportions. Every sample of students is likely to be different, and hence the odds of students eating off campus will vary from sample to sample (in both groups). Hence, the OR varies also from sample to sample. That is, *sampling variation* exists, so the OR has a *sampling distribution.*

However, the sampling distribution of the sample OR does *not have a normal distribution.*[1] For this reason, the CI for the OR will be taken directly from software output, and the sampling distribution is not discussed.

Software (Fig. 31.2, right panel) gives the sample OR as 5.94, and the (exact) 95% CI as 1.35 to 26.1. The value of the OR is the same as the value computed manually.

We write:

> The odds of students eating most meals off-campus is 5.94, higher for students living with their parents (odds: 26.0; $n = 54$) than for students *not* living with their parents (odds: 4.38; $n = 129$), with the 95% CI from 1.35 to 26.1.

There is a 95% chance that this CI straddles the population OR. Notice that the *meaning* of the OR is explained in the conclusions: the odds of eating most meals *off*-campus, and comparing students living with parents to *not* living with parents.

The CI for an OR is not symmetrical, like the others we have seen;[2] that is, the sample OR of 5.94 is not in the centre of the CI.

> ⓘ Interpreting and explaining ORs can be challenging, so care is needed!

[1]For those interested (this is *optional*): the *logarithm* of the OR has an approximate normal distribution under certain conditions.

[2]For those interested (this is *optional*): this is because the OR has no upper limit, but the lower limit of an OR is zero. The *logarithm* of the limits of the CI form a symmetric interval.

31.6 Hypothesis tests for comparing odds: χ^2-test

31.6.1 Hypotheses

For the 2 × 2 table of counts in Table 31.1, odds can be compared rather than proportions:

> Are the *population odds* of students eating most meals off-campus the same for students *living with* their parents and for students *not living with* their parents?

If the odds are the same in the two groups, this is equivalent to an OR of one. Hence, the RQ could also be written as

> Is the *population OR* of eating most meals off-campus, comparing students who live *with their* parents to students *not living with* their parents, equal to one?

Either way, the *parameter* is the population OR, and the null hypothesis is the 'no difference, no change, no relationship' position:

- H_0: The *population* OR is one, or (equivalently)
  The *population* odds are the same in each group.

This hypothesis proposes that the *sample* odds are not the same in the two groups only due to sampling variation. This is the initial *assumption*. The alternative hypothesis is

- H_1: The *population* OR is not one, or (equivalently)
  The *population* odds are *different* in each group.

⚠ For comparing odds, the alternative hypothesis *is always two-tailed.*

In our example then:

- H_0: The *population* odds of eating most meals off-campus is the *same* for students living with their parents and for students not living with their parents.
- H_1: The *population* odds of eating most meals off-campus is *different* for students living with their parents and for students not living with their parents.

As usual, the decision-making process starts by *assuming* the null hypothesis is true: that the *population* OR is one (i.e., the population odds in each group are equal).

⚠ For two-way tables, RQs can be framed in terms of ORs, comparing odds, comparing proportions, or (for larger two-way table) using associations (or relationships).

For consistency: if the RQ is about odds, the hypotheses and conclusion should be about the odds; if the RQ is about proportions, the hypotheses and conclusion should be about the proportions; and so on.

31.6.2 Finding expected counts

Assuming the null hypothesis is true (which is the initial assumption made) means that the odds are the same in both groups (and the proportions are the same in both groups too). That is, the proportions of students eating most meals off-campus is the same for students *living with* and *not living with* their parents. Let's consider the implication.

From Table 31.1, 157 students out of 183 ate most meals off-campus, so that $157 \div 183 = 0.8579$ of students in the entire sample ate most of their meals off-campus (which is the common proportion found in Equation (31.2)).

If the proportion of students who eat most of their meals off-campus is the *same* for those who live with their parents and those who don't, then we'd *expect* 0.8579 of students in *both* groups to be eating most meals off-campus. In other words, the two *conditional* probabilities would be the same. In that case:

- we would *expect* a proportion of 0.8579 of the 54 students who *live with their parents* (i.e., $0.8579 \times 54 = 46.33$ students) to eat most meals off-campus.
- we would *expect* a proportion of 0.8579 of the 129 students who *don't live with their parents* (i.e., $0.8579 \times 129 = 110.67$ students) to eat most meals off-campus.

In other words, the proportions (and hence the odds) of eating most meals off-campus is the same in each group. Those are the *expected* counts if the proportions (or odds) were exactly the same in each group (Table 31.4), as assumed in H_0.

How close are the *observed* counts (Table 31.1) to the *expected* counts (Table 31.4)? For instance, 46.33 of the 54 students who *live with their parents* are *expected* to eat most meals off-campus; yet we observed 52; 110.67 of the 129 students who *don't live with their parents* are *expected* to eat most meals off-campus; yet we observed 105.

The observed and expected counts are similar, but not the exactly same. The difference between the observed and expected counts *may* be explained by sampling variation (that is, the null hypothesis explanation).

The hypothesis test effectively compares the observed counts to the expected counts (assuming no relationship between the variables) over the whole 2×2 table.

⚠ You *do not* have to compute the expected counts explicitly (software does it in the background, or explicitly if requested). However, seeing how the decision-making process works in this context is helpful.

In previous hypothesis tests, the *sampling distribution* had an approximate normal distribution. However, the sampling distribution of the OR is more complicated[3] so will not be presented. We will use software output only to conduct the test.

TABLE 31.4: Where university students live and eat: expected counts.

	Most off-campus	Most on-campus	Total
Living with parents	46.328	7.672	54
Not living with parents	110.672	18.328	129
Total	157.000	26.000	183

31.6.3 Computing the value of the test statistic

The decision-making process compares what is *expected* if the null hypothesis about the parameter is true (Table 31.4) to what is *observed* in the sample (Table 31.1). Previously, when the sampling distribution was a normal distribution, the test statistic was a t-score or

[3]For those interested: the *logarithm* of the sample ORs have an approximate normal distribution, and hence a *standard error*.

a z-score. However, the sampling distribution for an OR does *not* have a normal distribution, and so a different test statistic is needed.

In this context, the test-statistic is 'chi-squared', written χ^2. The χ^2-score measures the overall size of the differences between the expected counts and observed counts, over the entire 2×2 table.

> The Greek letter χ is pronounced 'kie', as in **kite** (*not* 'chi' as in **China** or in **chin**). The test statistic χ^2 is pronounced as 'chi-squared'.

From the software (Fig. 31.1), $\chi^2 = 6.934$. But what does this value *mean*? Is it 'large' or 'small'? The χ^2-value, for 2×2 tables of counts, has an equivalent z-score, so that a P-value can be estimated using the 68–95–99.7 rule. The χ^2-value is equivalent to

$$z = \sqrt{\chi^2} \qquad \text{for a } 2 \times 2 \text{ table of counts only.}$$

Here then, the χ^2-value is equivalent to a z-score of $\sqrt{6.934} = 2.633$. This is the *same* z-score produced when comparing two proportions (Sec. 31.4; Fig. 31.1), and hence the P-value will be the same also. Using the 68–95–99.7 rule, a small P-value is expected. The two-tailed P-value reported by software (Fig. 31.1, under the column p) is indeed small: 0.008 to three decimals.

> Recall that χ^2-tests always have *two-tailed* alternative hypotheses, so two-tailed P-values are always reported.

31.6.4 Writing conclusions

A very small P-value (0.008 to three decimals) means strong evidence exists to supporting H_1: the evidence suggests a difference in the *population* odds in the two groups. We write:

> The *sample* provides strong evidence ($\chi^2 = 6.934$, $n = 54$; two-tailed $P = 0.008$) that the odds in the *population* of having most meals off-campus is different for students living with their parents (odds: 26) and students *not* living with their parents (odds: 4.375, $n = 129$; OR: 5.94; 95% CI from 1.35 to 26.1).

The conclusion includes three components (Sect. 28.8): the *answer to the RQ*; the *evidence* used to reach that conclusion ('$\chi^2 = 6.934$; two-tailed $P = 0.008$'); and some *sample summary statistics* (including the 95% CI for the OR).

The conclusion makes clear what the odds and the OR *mean*. The odds are described as the 'odds of having most meals off-campus', and the OR as then comparing these odds between 'students living with their parents and students *not* living with their parents'.

31.7 Statistical validity conditions

As usual, these results hold under certain conditions. The CIs and tests above are statistically valid if:

- all *expected* counts are at least five.

Some books may give other (but similar) conditions.

The statistical validity condition refers to the *expected* (not the *observed*) counts. In some software, the *expected* counts must be explicitly requested to see if this condition is satisfied (Fig. 31.3). The units of analysis are also assumed to be *independent* (e.g., from a simple random sample).

If the statistical validity conditions are not met, other similar options include using a Fisher's exact test [Conover, 2003] or using resampling methods [Efron and Hastie, 2021].

Contingency Tables

Live		Meals: Most off-campus	Most on-campus	Total
Living with parents	Observed	52	2	54
	Expected	46.3279	7.6721	54.0000
Not living with parents	Observed	105	24	129
	Expected	110.6721	18.3279	129.0000
Total	Observed	157	26	183
	Expected	157.0000	26.0000	183.0000

FIGURE 31.3: The expected counts, as computed by software.

Example 31.1 (Statistical validity). For the student-eating data, the smallest *observed* count is 2 (living with parents; most meals off-campus), but the smallest *expected* count (see Table 31.4 or Fig. 31.3) is 7.67, which is greater than five. This means the two analyses (comparing proportions; comparing odds) are both statistically valid. The size of the *expected* counts is important for the statistical validity condition.

Usually, you do not compute these expected counts. However, a quick check for the statistical validity is to compute the *smallest* expected counts, using

$$\frac{(\text{Smallest row total}) \times (\text{Smallest column total})}{\text{Overall total}}. \qquad (31.3)$$

If this value is greater than five, the CIs and tests are statistically valid.

31.8 Hypothesis tests of independence more generally: χ^2-tests

Often a table of counts is larger than 2×2. In these situations, the RQ may not be able to be worded in terms of comparing proportions or odds. Instead, the hypotheses can be worded in terms of *independence*, *relationships* or *associations* (but *not* correlations) between the variables:

Is there a relationship (or association) between one qualitative variable and another qualitative variable?

The RQ is answered using a χ^2-test, by extending the ideas in Sect. 31.6; *z-tests and t-tests are not appropriate.*

Example 31.2 (Two-way tables larger than 2×2). [*Dataset:* `RipsID`] Diez-Fernández et al. [2023] studied Spanish people's knowledge of ocean rips (Table 31.5, left table). The table is a 4×2 two-way table. The rows are the age groups, as the age groups are being compared. The RQ is

> Is there a relationship (or association) between age group and people's ability to correctly identify a rip?

TABLE 31.5: Identifying rips. Left: the data by age group. Right: a summary table. The ORs are relative to the 51 to 65 age group.

	Identifying rips	
	Correctly	Incorrectly
18 to 24	41	5
25 to 34	47	12
35 to 50	106	19
51 to 65	52	7

	Correctly identifying rips			
	Odds	OR	Percentage	n
18 to 24	8.200	1.104	89.1	46
25 to 34	3.917	0.527	79.7	59
35 to 50	5.579	0.751	84.8	125
51 to 65	7.429		88.1	59

The odds and percentage of people in each age group that can correctly identify rips can be computed (Table 31.5, right table), but this is not always possible (e.g., for a 3×4 table). ORs compare *pairs* of odds, and the ORs in Table 31.5 (right table) are all relative to those in the 51 to 65 age group (hence, no OR is given for the 51 to 65 age group, which is the *reference level*). For example, the odds of someone aged 18 to 24 correctly identifying a rip is 1.104 times the odds of someone aged 51 to 65 correctly identifying a rip.

For tables larger than 2×2 more generally, the hypothesis are usually worded in terms of associations or relationships (but *not* correlations) between the variables:

- H_0: In the *population*, there *is no association* between correctly identifying a rip and age group.
- H_1: In the *population*, there *is an association* between correctly identifying a rip and age group.

The test statistic is a χ^2-value, which compares the observed and expected counts; the expected counts are found in the same way as in Sect. 31.6.2.

For two-way tables larger than 2×2, the parameter describing the association between the variables is the χ^2-value. When no relationship exists in the sample, the observed and expected counts are the same, and $\chi^2 = 0$. The larger the difference between the observed and expected counts, the larger the value of χ^2. Sampling variation means that the observed counts will vary from sample to sample, so that χ^2 may not be exactly zero, even if there is no association between the variables.

Software computes $\chi^2 = 2.406$, and the two-tailed P-value as $P = 0.492$ (Fig. 31.4, left panel). This P-value means there is not persuasive evidence to support the alternative hypothesis:

> The *sample* provides no evidence ($\chi^2 = 2.406$, $n = 289$; two-tailed $P = 0.492$) of an association between age group and the ability to correctly identify a rip among Spanish people.

⚠ For hypothesis tests involving tables of counts larger than 2×2, the alternative hypothesis *is always two-tailed.*

χ^2 Tests

	Value	df	p
χ^2	2.4062	3	0.492
N	289		

Contingency Tables

		Identification		
AgeGroup		Correct	Incorrect	Total
18 to 24	Expected	39.1557	6.8443	46.0000
25 to 34	Expected	50.2215	8.7785	59.0000
35 to 50	Expected	106.4014	18.5986	125.0000
51 to 65	Expected	50.2215	8.7785	59.0000
Total	Expected	246	43	289

FIGURE 31.4: Software output for the hypothesis test about knowledge of ocean rips.

The statistical validity conditions are the same as in Sect. 31.7: all *expected* counts are at least five. Using Equation (31.3),

$$\frac{\text{(Smallest row total)} \times \text{(Smallest column total)}}{\text{Overall total}} = \frac{46 \times 43}{289} = 6.84$$

(as in Fig. 31.4, right panel), which is larger than five. The test is statistically valid.

31.9 Example: turtle nests

The hatching success of loggerhead turtles on Mediterranean beaches is often compromised by fungi and bacteria. Candan et al. [2021] studied the odds of a nest being infected, comparing relocated nests (relocated due to the risk of tidal inundation), and non-relocated nests (Table 31.6, left table). The researchers were interested in knowing:

> For Mediterranean loggerhead turtles, are the odds of infections the same for natural and relocated nests?

TABLE 31.6: The turtles data (left), and the numerical summary (right).

	Not infected	Infected
Natural	29	10
Relocated	14	8

	Odds infected	Proportion infected	Sample size
Natural	2.90	0.744	39
Relocated	1.75	0.636	22
	OR: 1.66	*Diff.*: 0.107	

Since the RQ is written in terms of odds, the hypotheses should be written using odds also:

- H_0: The odds of a nest being infected is *the same* for natural and relocated nests.
- H_1: The odds of a nest being infected is *not the same* for natural and relocated nests.

Here, N refers to **N**atural nests, and R to **R**elocated nests. The parameter is the odds ratio of a nest being infected, comparing natural to relocated nests. (The equivalent hypotheses written in terms of proportions would be H_0: $p_N - p_R = 0$ and H_1: $p_N - p_R \neq 0$. The hypotheses could also be written in terms of associations.)

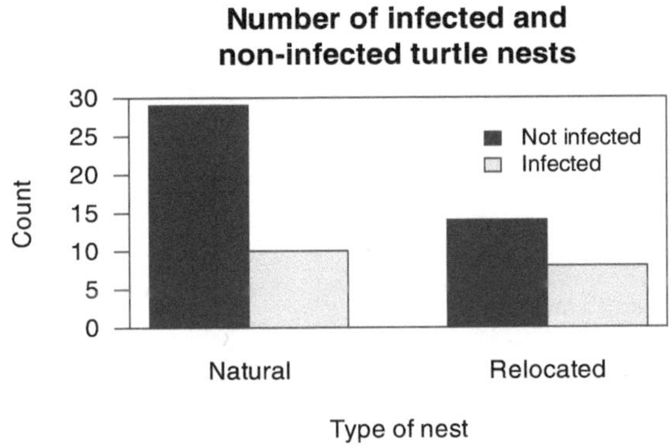

FIGURE 31.5: Bar chart for the turtle-nesting data.

A graphical summary is shown in Fig. 31.5. A numerical summary table (Table 31.6, right table) shows that the odds of natural nest being infected is 1.66 times the odds of a relocated nest being infected. From the software output (Fig. 31.6), the χ^2-value is 0.777. Since the table is a 2×2 table, the equivalent z-score can be found: $z = \sqrt{0.777} = 0.88$. This z-score is very small, so expect a large P-value. (This is the value of the z-score shown in Fig. 31.6 for comparing two proportions.) The P-value is 0.378 on the output (for both tests).

The smallest *expected* count is $(22 \times 18)/61 = 6.49$, which exceeds five, so the test is statistically valid. Since the RQ and hypotheses were written in terms of odds, the conclusion is also written in terms of odds:

> There is no evidence of a difference in the odds of infection (χ^2: 0.777; P-value: 0.378; OR: 1.657; 95% CI: 0.537 to 5.12) between natural nests (odds: 2.90; $n = 39$) and relocated nests (odds: 1.75; $n = 22$).

There is no evidence that relocating the nest (to protect them from tidal inundation) changes the risk of infection.

> ⚠ We *do not* say whether the evidence supports the null hypothesis. We assume the null hypothesis is true, so we state the strength of evidence to change our mind (and hence support the alternative hypothesis). The current sample presents no evidence to contradict the assumption, but future evidence may emerge.

31.10 Example: health of female burros

Johnson et al. [1987] studied 315 introduced female burros (donkeys) in the Mojave Desert (California) to understand management processes. One RQ was:

> For these female burros, is the reproductive status of the burros related to their health?

χ² Tests

	Value	df	p
χ²	0.777	1	0.3779
z test difference in 2 proportions	0.882		0.3779
N	61		

Comparative Measures

		95% Confidence Intervals	
	Value	Lower	Upper
Difference in 2 proportions	0.107 [a]	−0.136	0.351
Odds ratio	1.657	0.537	5.117

[a] Rows compared

FIGURE 31.6: The software output for the turtle-nesting data.

The data (Table 31.7, left table) are given in a 3×3 table of counts. The data are summarised using row proportions in Table 31.7 (right table), and in a graph in Fig. 31.7 (left panel). Software output is shown in Fig. 31.7 (right panel).

TABLE 31.7: Left: the health and reproductive status of female burros. Right: row proportions for the burro data (i.e., rows sum to one). Pregnant and lactating burros were counted with the lactating burros only.

	Health: counts			
	Excellent	**Fair**	**Poor**	**Total**
Barren	16	21	38	75
Pregnant	14	53	62	129
Lactating	4	29	78	111

	Health: row proportions		
	Excellent	**Fair**	**Poor**
Barren	0.213	0.280	0.507
Pregnant	0.109	0.411	0.481
Lactating	0.036	0.261	0.703

The hypotheses must be worded in terms of associations (or *relationships*):

- H_0: *No association* exists between reproductive status and overall health.
- H_1: *An association* exists between reproductive status and overall health.

From the software output (Fig. 31.7, right panel), $\chi^2 = 23.585$. Notice that a comparison of proportions is not possible for tables larger than 2×2. Software reports $P < 0.001$, which suggests very strong evidence in the sample that an association exists between reproductive status and overall health.

The conclusion could be written as

> The sample provides very strong evidence ($\chi^2 = 23.585$; $P < 0.001$; 3×3 table) of an association between reproductive status and overall health of female burros ($n = 315$).

Adding sample summary information to this conclusion is cumbersome. Instead, readers can be pointed to the numerical summary (Table 31.7, right table). Furthermore, CIs are not reported.

While we know there is an association between the variables, we can only speculate on the nature of the association (i.e., for which group(s) the *population* proportions are different).

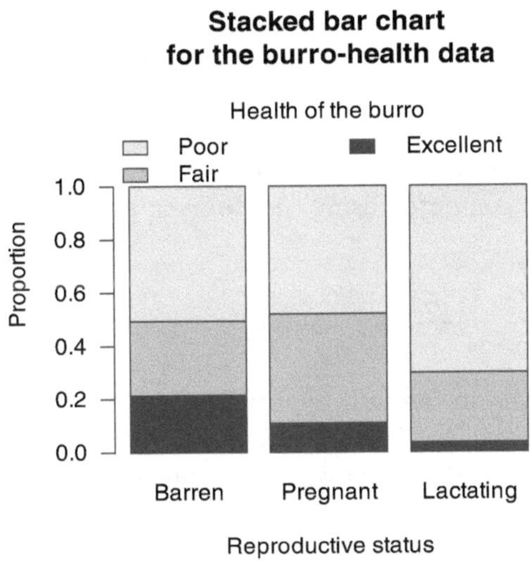

χ^2 Tests

	Value	df	p
χ^2	23.585	4	< .001
z test difference in 2 proportions	NaN [a]		
N	315		

[a] z test only available for 2x2 tables

FIGURE 31.7: Left: a stacked bar chart for the burro-health data. Right: software output for the burro-health data.

Formal methods for doing so requires methods beyond this book, but Fig. 31.7 (left panel) suggests that lactating burros are far more likely to have poor health.

The smallest *expected* value is $75 \times 34/315 = 8.1$, which exceeds 5, so the results are statistically valid.

31.11 Chapter summary

To compare a two-level qualitative variable between two groups, a CI can be formed for the difference between two proportions, or for an OR.

To compute a CI for the difference between two proportions, compute the difference between the two sample proportions, $\hat{p}_1 - \hat{p}_2$, and identify the sample sizes n_1 and n_2. Then the standard error, which quantifies how much the value of $\hat{p}_1 - \hat{p}_2$ varies across all possible samples, is

$$\text{s.e.}(\hat{p}_1 - \hat{p}_2) = \sqrt{\text{s.e.}(\hat{p}_1)^2 + \text{s.e.}(\hat{p}_2)^2},$$

where s.e.($\hat{p}_1$) and s.e.($\hat{p}_2$) are the standard errors of Groups 1 and 2 (Equation (22.3)). The *margin of error* is (multiplier $\times$ standard error), where the multiplier is 2 for an approximate 95% CI (using the 68–95–99.7 rule). Then the CI is:

$$(\hat{p}_1 - \hat{p}_2) \pm (\text{multiplier} \times \text{standard error}).$$

Software is used to compute a CI for the OR, as the sampling distribution does not have a normal distribution.

These steps are used to test a hypothesis about a difference between two population proportions $p_1 - p_2$.

- Write the null hypothesis (H_0) and the alternative hypothesis (H_1); initially *assume* the value of $(p_1 - p_2)$ in the null hypothesis to be true.

- Describe the *sampling distribution*, which describes what to *expect* from the difference between the sample proportions based on this assumption: under certain statistical validity conditions, the difference between the sample proportions vary with:
  - an approximate normal distribution,
  - with sampling mean whose value is the value of $(p_1 - p_2)$ (from H_0), and
  - having a standard deviation of s.e.$(\hat{p}_1 - \hat{p}_2)$ computed using the *common* proportion.
- Compute the value of the *test statistic*:

$$z = \frac{(\hat{p}_1 - \hat{p}_2) - (p_1 - p_2)}{\text{s.e.}(\hat{p}_1 - \hat{p}_2)},$$

  where $p_1 - p_2$ is the hypothesised difference given in the null hypothesis.
- An approximate *P-value* can be estimated using the 68–95–99.7 rule, or an exact P-value found using software. Use the P-value to make a decision, and write a conclusion.
- Check the statistical validity conditions.

These steps are used to test a hypothesis for comparing two odds, or to test for a relationship between two qualitative variables more generally.

- Write the null hypothesis (H_0) and the alternative hypothesis (H_1); initially *assume* no relationship between the two variables.
- Find the value of the *test statistic* (a χ^2-score) on the software output. (For 2×2 tables only, the equivalent z-score is $\sqrt{\chi^2}$.)
- A *P-value* is found using software; use the P-value to make a decision, and write a conclusion.
- Check the statistical validity conditions.

The statistical validity conditions should be checked: all *expected* counts should exceed five.

31.12 Quick review questions

Chen et al. [2023] investigated the relationship between body temperature of patients admitted to hospital following a heart attack (HA), and a having a subsequent HA while in hospital (Table 31.8).

Are the following statements *true* or *false*?

1. From the software output, the P-value is 0.180.
2. The alternative hypothesis two-tailed.
3. There is *no* evidence of a difference in odds of having an in-hospital HA, comparing patients with low and high body temperatures.
4. The CI is not statistically valid, because the CIs for the difference between the proportions has a *negative* value.
5. The CI means that the sample OR is likely to be between 0.330 and 5.568.
6. The χ^2-value is 0.180.
7. Of patients with a low body temperature, $4/27 = 0.148$ had an in-hospital HA.
8. The *odds* that a patient with a low body temperature had an in-hospital HA is $4/23 = 0.174$.

Select the correct answer:

9. The OR in the output is given as 1.357. What does this OR *mean*?

a. The odds of having an in-hospital heart attack is 1.357 times greater for those with a low body temperature.
b. The odds of having an in-hospital heart attack is 1.357 times smaller for those with a low body temperature.
c. The proportion of patients having an in-hospital heart attack is 1.357 times greater for those with a low body temperature.
d. The proportion of patients having an in-hospital heart attack is 1.357 times smaller for those with a low body temperature.

TABLE 31.8: Body temperature of patients, and whether they experienced a heart attack in hospital.

	In-hospital heart attack	
	Yes	**No**
Low body temp.	4	23
High body temp.	5	39

χ^2 Tests

	Value	df	p
χ^2	0.180	1	0.671
z test difference in 2 proportions	0.424		0.671
N	71		

Comparative Measures

		95% Confidence Intervals	
	Value	Lower	Upper
Difference in 2 proportions	0.073 [a]	−0.273	0.420
Odds ratio	1.357	0.330	5.568

[a] Rows compared

FIGURE 31.8: Software output for the heart-attack study.

31.13 Exercises

Answers to odd-numbered exercises are given at the end of the book.

Exercise 31.1. Consider the expected counts in Table 31.4. Confirm that the *odds* of having most meals off-campus is the same for students living with their parents, and for students not living with their parents.

Exercise 31.2. Compute all four expected counts in Table 31.8. Confirm that the corresponding test may not be statistically valid.

Exercise 31.3. Sketch the sampling distribution for the difference between the proportions of students eating most meals off-campus, for those living with parents minus those not living with parents. What is the sampling distribution for the equivalent OR?

Exercise 31.4. Sketch the sampling distribution for the difference between the proportion of non-infected turtle nests, for natural nests minus relocated nests (in Sect. 31.9). What is the sampling distribution for the equivalent OR?

Exercise 31.5. Suppose an analysis of a 2×2 table of counts produces a value of $\chi^2 = 10.66$.

1. What would be the equivalent z-score for comparing the two proportions?
2. What would be the approximate P-value?

Exercise 31.6. Suppose an analysis of a 2×2 table of counts produces a value of $\chi^2 = 4.06$.

1. What would be the equivalent z-score for comparing the two proportions?
2. What would be the approximate P-value?

Exercise 31.7. [*Dataset*: `EVPurchase`] Egbue et al. [2017] studied the adoption of electric vehicles (EVs) by a group of professional Americans (Table 31.9). Software output is shown in Fig. 31.9.

1. Based on the output, how is the difference between the two proportions defined?
2. Write the hypothesis for comparing the *proportions* using this definition of the difference.
3. Use the software output to conduct the test.
4. Use the software output to write down the corresponding CI for the difference in proportions.
5. Based on the output, how is the OR defined?
6. Write the hypothesis for comparing the *odds*, for those with and without post-graduate study.
7. Use the software output to conduct the test.
8. Use the software output to write down the corresponding CI for the OR.
9. Are the CIs and tests statistically valid?

TABLE 31.9: Responses to 'Would you purchase an electric vehicle in the next 10 years?' by education.

	Yes	No
No post-grad	24	8
Post-grad study	51	29

χ^2 Tests

	Value	df	p
χ^2	1.3077	1	0.25282
z test difference in 2 proportions	1.1435		0.25282
N	112		

Comparative Measures

		95% Confidence Intervals	
	Value	Lower	Upper
Difference in 2 proportions	0.1125 [a]	−0.0708	0.2958
Odds ratio	1.7059	0.6792	4.2843

[a] Rows compared

FIGURE 31.9: Software output for the EV study.

Exercise 31.8. Meresa et al. [2023] investigated Ethiopian farmers' adoption of improved soil and water conservation structures on their farms (Table 31.10). Software output is shown in Fig. 31.10.

TABLE 31.10: Adoption of conservation practices by Ethiopian farmers, by farm size.

	Adopter?	
	No	Yes
$<$ 0.5 ha (small)	86	61
$\geq$ 0.5 ha (large)	43	71

χ^2 Tests

	Value	df	p
χ^2	11.0959	1	0.00087
z test difference in 2 proportions	3.3310		0.00087
N	261		

Comparative Measures

		95% Confidence Intervals	
	Value	Lower	Upper
Difference in 2 proportions	0.2078 [a]	0.0884	0.3273
Odds ratio	2.3279	1.4104	3.8422

[a] Rows compared

FIGURE 31.10: Software output for the farming study.

1. Based on the output, how is the difference between the two proportions defined?
2. Write the hypothesis for comparing the proportions, using this definition of the difference.
3. Use the software output to conduct the test.
4. Use the software output to write down the corresponding CI for the difference in proportions.
5. Based on the output, how is the OR defined?

6. Write the hypothesis for comparing the odds, for farmers with small and large farms.
7. Use the software output to conduct the test.
8. Use the software output to write down the corresponding CI for the OR.
9. Are the CIs and tests statistically valid?

Exercise 31.9. [*Dataset*: **CarCrashes**] Wang et al. [2020] recorded information about car crashes in a rural, mountainous county in western China (Table 31.11).

1. Sketch a suitable graph to display the data.
2. Compute the *proportion* of crashes involving a pedestrian in 2011 ($\hat{p}_{2011}$), and in 2015 ($\hat{p}_{2015}$).
3. Compute the *difference between the proportion* of crashes involving a pedestrian from 2011 to 2015, consistent with the definition used in the output (Fig. 31.11).
4. Compute the value of s.e.($\hat{p}_{2011} - \hat{p}_{2015}$), needed for constructing the CI.
5. Construct the *approximate* 95% CI for the difference between the proportions.
6. Write down a 95% CI for the difference between the proportions.
7. Interpret what this CI means.
8. Compute the *odds* of crashes involving a pedestrian in 2011, and also in 2015.
9. Compute the *OR* of crashes involving a pedestrian, comparing 2011 to 2015.
10. Write down the CI for the OR.
11. Construct an appropriate numerical summary table for the data.
12. Compute the value of s.e.($\hat{p}_{2011} - \hat{p}_{2015}$), needed for conducting a hypothesis test.
13. Conduct a hypothesis test to determine if there is a difference between p_{2011} and p_{2015}.
14. Conduct a hypothesis test to determine if there is a difference between the odds of a crash involving a pedestrian for 2011 and 2015.
15. Are the CIs and tests statistically valid?

TABLE 31.11: Type of car crashes in different years.

	Involving pedestrians	Involving vehicles
2011	15	35
2015	37	85

χ^2 Tests

	Value	df	p
χ^2	0.002	1	0.9661
z test difference in 2 proportions	−0.043		0.9661
N	172		

Comparative Measures

		95% Confidence Intervals	
	Value	Lower	Upper
Difference in 2 proportions	−0.003 [a]	−0.154	0.148
Odds ratio	0.985	0.480	2.018

[a] Rows compared

FIGURE 31.11: Software output for the car-crash study.

Exercise 31.10. [*Dataset*: **ScarHeight**] Wallace et al. [2017] compared the heights of scars from burns received by people in Western Australia (Table 31.12). Software was used to analyse the data (Fig. 31.12).

1. Sketch an appropriate graph to summarise the data.
2. Compute the *proportion* of men having a smooth scar, and the *proportion* of women.
3. Compute the *difference between the proportions* of men and women having a smooth scar.
4. Compute the standard error for the difference between the proportions, needed for constructing a CI.
5. Compute the *approximate* 95% CI for the difference between the proportions.
6. Write down the 95% CI for the difference between the proportions, using the software output.
7. Interpret what this CI means.
8. Compute the *odds* of having a smooth scar for men, and for women.
9. Compute the *OR* of having a smooth scar, comparing men to women.

10. Write down the CI for the OR of having a smooth scar, comparing men to women.
11. Compile a numerical summary table.
12. Compute the value of standard error of the difference between the proportions, needed for conducting a hypothesis test.
13. Conduct a hypothesis test to determine if there is a difference between the proportions for men and women.
14. Conduct a hypothesis test to determine if there is a difference between the odds for men and women.
15. Are the CIs and tests statistically valid?

TABLE 31.12: Heights of scars for men and women.

	Smooth (0 mm)	Over 0 mm, up to 1 mm
Men	216	115
Women	99	62

χ^2 Tests

	Value	df	p
χ^2	0.6670	1	0.41410
z test difference in 2 proportions	0.8167		0.41410
N	492		

Comparative Measures

		95% Confidence Intervals	
	Value	Lower	Upper
Difference in 2 proportions	0.0377 [a]	−0.0533	0.1287
Odds ratio	1.1763	0.7966	1.7370

[a] Rows compared

FIGURE 31.12: Software output for the scar-height data.

Exercise 31.11. [*Dataset*: **PetBirds**] Kohlmeier et al. [1992] examined people with lung cancer, and a matched set of controls who did not have lung cancer, and recorded the number in each group that kept pet birds. The data are shown in Table 31.13, and the software output in Fig. 31.13.

Consider this RQ:

> Are the odds of having a pet bird the same for people *with* lung cancer (cases) and for people *without* lung cancer (controls)?

1. Compute the difference between the proportions of people with pet birds, for those with and without lung cancer.
2. Compute the standard error needed to compute the CI for the difference in proportions.
3. Compute the standard error needed to conduct the hypothesis test to compare the proportions.
4. Explain *why* the two standard errors have slightly different values.
5. Compute an approximate 95% CI for the difference between the two proportions.
6. Write down the 95% CI for the difference between the proportions using the output (Fig. 31.13).
7. Interpret the CIs.
8. Conduct a hypothesis test to compare the two proportions.
9. Confirm that the OR in the output is correct.
10. Write down a 95% CI for the OR, and interpret what it means.
11. Perform a hypothesis test to determine if the odds of having a pet bird is the same for people with and without lung cancer.
12. Are the CIs and tests statistically valid?
13. Explain why no cause-and-effect can be reached.

Exercise 31.12. [*Daatset*: **EmeraldAug**] The *Southern Oscillation Index* (SOI) is a standardised measure of the air pressure difference between Tahiti and Darwin, and is related to rainfall in some parts of the world [Stone et al., 1996], and especially Queensland [Stone and Auliciems, 1992].

TABLE 31.13: The pet bird data.

	Adults with lung cancer	Adults without lung cancer	Total
Did not keep pet birds	141	328	469
Kept pet birds	98	101	199
Total	239	429	668

χ^2 Tests

	Value	df	p
χ^2	22.3742	1	<.00001
z test difference in 2 proportions	−4.7301		<.00001
N	668		

Comparative Measures

	Value	95% Confidence Intervals Lower	Upper
Difference in 2 proportions	−0.1918 [a]	−0.2727	−0.1109
Odds ratio	0.4430	0.3151	0.6230

[a] Rows compared

FIGURE 31.13: Software output for the pet-birds data.

The rainfall at Emerald (Queensland) was recorded for Augusts between 1889 and 2002 inclusive [Dunn and Smyth, 2018], where the monthly average SOI was positive, and when the SOI was non-positive (zero or negative), as shown in Table 31.14.

1. Compute the difference between the proportions of Augusts with rain, for months with a positive SOI compared to months with a non-positive SOI.
2. Compute the standard error needed to compute the CI for the difference in proportions.
3. Compute the standard error needed to conduct the hypothesis to compare the proportions.
4. Explain *why* the two standard errors have slightly different values.
5. Compute an approximate 95% CI for the difference between the two proportions.
6. Write down the 95% CI for the difference between the proportions using the output (Fig. 31.14).
7. Interpret the CIs.
8. Conduct a hypothesis test to compare the two proportions.
9. Confirm that the OR in the output is correct.
10. Write down a 95% CI for the OR, and interpret what it means.
11. Perform a hypothesis test to determine if the odds of recoding rain is the same for Augusts with non-positive and positive SOI.
12. Are the CIs and tests statistically valid?

TABLE 31.14: The SOI, and whether rainfall was recorded in Augusts between 1889 and 2002 inclusive.

	Rainfall recorded	No rainfall recorded
Positive SOI	53	7
Non-positive SOI	40	14

Exercise 31.13. [*Dataset*: `HatSunglasses`] Dexter et al. [2019] recorded the number of people at the foot of the Goodwill Bridge, Brisbane, who wore hats between 11:30am to 12:30pm. Of the 366 females observed, 22 wore hats; of the 386 males observed, 79 wore hats.

1. Construct the two-way table for the data.
2. Compute the proportions of females and males wearing a hat, and hence the difference between the proportions.
3. Compute the odds of a female and the odds of a male wearing a hat, and hence the OR.
4. Compute an approximate 95% CI for the difference between the proportions.
5. Write down the 95% CI for the difference between the proportion (Fig. 31.15).
6. Interpret the CIs.
7. Write down, then interpret, a 95% CI for the OR.

χ² Tests

	Value	df	p
χ²	3.8454	1	0.04988
z test difference in 2 proportions	1.9610		0.04988
N	114		

Comparative Measures

		95% Confidence Intervals	
	Value	Lower	Upper
Difference in 2 proportions	0.1426 [a]	0.0003	0.2849
Odds ratio	2.6500	0.9789	7.1735

[a] Rows compared

FIGURE 31.14: Software output for the Emerald-rain data.

8. Perform a hypothesis test to determine if the odds of wearing a hat is the same for females and males.
9. Write down the conclusion.
10. Are the CIs and tests statistically valid?

Comparative Measures

		95% Confidence Intervals	
	Value	Lower	Upper
Difference in 2 proportions	0.145 [a]	0.098	0.192
Odds ratio	4.024	2.448	6.613

[a] Rows compared

χ² Tests

	Value	df	p
χ²	33.763	1	<.0001
z test difference in 2 proportions	5.811		<.0001
N	752		

FIGURE 31.15: Software output for the hats data.

Exercise 31.14. [*Dataset*: **Turbines**] A study of turbine failures [Myers et al., 2002, Nelson, 1982] ran 73 turbines for around 1 800 h, and found that seven developed fissures (small cracks). They also ran a different set of 42 turbines for about 3 000 h, and found that nine developed fissures.

1. Construct the two-way table for the data.
2. Compute the difference between the proportions of fissures at 1 800 h and 3 000 h, and hence the difference between the proportions.
3. Compute the odds of a fissure after 1 800 h and after 3 000 h, and hence the OR.
4. Compute an approximate 95% CI for the difference between the proportions.
5. Write down the 95% CI for the difference between the proportions (Fig. 31.16).
6. Interpret the CIs.
7. Write down, then interpret, a 95% CI for the OR.
8. Test for a relationship.
9. Are the CIs and tests statistically valid?

Exercise 31.15. Witmer and Pipas [2020] compared various types of repellents (including bear faeces) to prevent bears damaging trees in an Idaho forest. Part of the data are summarised in (Table 31.15, left table).

1. Compute the odds of new damage for both repellents, and hence the OR.
2. Compute the proportion of trees with new damage for both repellents, and hence the difference between the proportions.
3. Write the hypothesis for conducting a hypothesis test involving proportions.
4. Write the hypothesis for conducting a hypothesis test involving odds.

χ^2 Tests

	Value	df	p
χ^2	3.120	1	0.0773
z test difference in 2 proportions	−1.766		0.0773
N	115		

Comparative Measures

		95% Confidence Intervals	
	Value	Lower	Upper
Difference in 2 proportions	−0.118 [a]	−0.260	0.023
Odds ratio	0.389	0.133	1.137

[a] Rows compared

FIGURE 31.16: Software output for the turbine data.

5. Software gives χ^2 as 4.4850. What is the equivalent z-score (e.g., for the test of proportions)? Would you expect a large or small P-value?
6. The P-value, from software, is $P = 0.0342$. Write a conclusion, either using odds or proportions.
7. Is the analysis statistically valid?

Exercise 31.16. [*Dataset*: **B12Diet**] Gammon et al. [2012] examined B12 deficiencies in 'predominantly overweight/obese women of South Asian origin living in Auckland'. Some women were on a vegetarian diet and some were not (Table 31.15, right table). One RQ was:

> Among this group of women, are the odds of being vitamin B12 deficient different for women on a vegetarian diet compared to women on a non-vegetarian diet?

1. Compute the odds of B12 deficiency for both diets, and hence the OR.
2. Compute the proportion of women with B12 deficiency for both diets, and hence the difference between the proportions.
3. Write the hypothesis for conducting a hypothesis test involving proportions.
4. Write the hypothesis for conducting a hypothesis test involving odds.
5. Software gives χ^2 as 4.7067. What is the equivalent z-score (e.g., for the test of proportions)? Would you expect a large or small P-value?
6. The P-value, from software, is $P = 0.0305$. Write a conclusion, either using odds or proportions.
7. Is the analysis statistically valid?

TABLE 31.15: Left: the number of trees with new damage by bears, according to different repellents. Right: the number of vegetarian and non-vegetarian women who are (and are not) B12 deficient.

	New damage	No new damage
Bear faeces	6	69
Control (water)	15	60

	B12 deficient	Not B12 deficient
Vegetarians	8	26
Non-vegetarians	8	82

Exercise 31.17. [*Dataset*: **DogWalks**] Naughton et al. [2024] studied the difference between the activities of dogs kept in the city and on farms (Table 31.16). One RQ was:

> For Northern Ireland dogs, is there an association between length of walks, and location?

1. Write down the hypotheses to answer this RQ.
2. Perform a hypothesis to answer the RQ, using the output in Fig. 31.17.

3. Write down the conclusion, in terms of odds, including a CI.
4. Write down the conclusion, in terms of proportions, including a CI.
5. Is the test statistically valid?

TABLE 31.16: The length of walks for dogs, living in the city and farms. ('Varies' means usually short walks, but occasional longer walks.)

	Walk length (in mins)			
	Under 30	30 **to under** 60	60 **to under** 120	**Varies**
City	138	84	13	264
Farm	84	102	33	243

χ^2 Tests

	Value	df	p
χ^2	23.0522	3	0.00004
N	961		

FIGURE 31.17: Software output for the dog-walking data.

Exercise 31.18. [*Dataset*: `Mumps`] Soud et al. [2009] studied the compliance of students with an isolation request following a large mumps outbreak in Kansas in 2006. One RQ was:

> Is there an association between age group, and compliance with the isolation order?

The data are shown in Table 31.17 and the software output in Fig. 31.18.

1. Write down the hypotheses.
2. Compute the proportion of each age group that complied with the isolation request.
3. Compute the odds of each age group that complied with the isolation request.
4. Compute the relevant ORs (using 'Older than 22' as the reference level), and interpret what these mean.
5. Determine the χ^2-value and perform a hypothesis to answer the RQ.
6. Is the test statistically valid?

TABLE 31.17: The compliance of students by age group.

	Complied	**Did not comply**
18 to 19	40	10
20 to 21	37	14
Older than 22	22	9

χ^2 Tests

	Value	df	p
χ^2	1.0989	2	0.57727
N	132		

FIGURE 31.18: Software output for the compliance data.

Exercise 31.19. [*Dataset*: `ShoppingBags`] Choon et al. [2017] studied 400 residents of Klang Valley, Malaysia, to examine residents' approach to waste management. One RQ was:

> For residents of Klang Valley, is age group associated with whether people bring their own bags when shopping?

The data (Table 31.18) are given in a 3×2 table of counts. The software output is shown in Fig. 31.19.

TABLE 31.18: Whether shoppers bring their own bags, and the shoppers age group.

	Brings bags?	
	Yes	**No**
30 and under	126	138
31 to 40	50	32
Over 40	41	13

χ^2 Tests

	Value	df	p
χ^2	16.24	2	< .001
N	400		

FIGURE 31.19: Software output for the shopping-bags data.

1. Compute the odds of someone bringing a shopping bag, for each age group.
2. Compute the OR of bringing a shopping bag (using the 'Over 40' age group as the reference level).
3. Compute the percentage of people bringing a shopping bag, for each age group.
4. Construct the hypotheses for testing for an association between the variables.
5. Use the software output to answer the research question.
6. Write a conclusion.
7. Is the test statistically valid.

Exercise 31.20. [*Dataset*: `CrabShells3`] Hermit crabs place sea anemones on their shells for protection. Brooks [1989] studied the placement of the anemones:

> Is there a relationship between the vertical and horizontal locations of anemones placed by hermit crabs on their shells?

The data are shown in Table 31.19, and output in Fig. 31.20.

1. Perform a hypothesis test to answer the RQ using the 3×3 table (Fig. 31.20, top output).
2. Confirm that the statistical validity conditions are not met when using the 3×3 table.
3. Construct a 2×2 table, recording the location of the anemones as either 'Central' or 'Side' without distinguishing *which* side. Hence, repeat the test using the 2×2 table (Fig. 31.20, bottom output). (These data are in the file `CrabShell2`.)

TABLE 31.19: The location of anemones placed on shells by hermit crabs.

	Column		
	Side 1	**Central**	**Side 2**
Row: Side 1	2	9	9
Row: Central	22	30	37
Row: Side 2	1	0	2

χ^2 Tests

	Value	df	p
χ^2	3.876	4	0.4230
N	112		

χ^2 Tests

	Value	df	p
χ^2	0.237	1	0.6265
N	112		

FIGURE 31.20: Software output for the 3×3 table of crab-shell data (top output), and for the 2×2 table of crab-shell data (bottom output).

Answers to *Quick review* questions: 1. False. **2.** True. **3.** True. **4.** False. **5.** False. **6.** True. **7.** True. **8.** True. **9.** a.

32

Finding sample sizes for CIs

You have learnt to ask an RQ, design a study, classify and summarise the data, construct confidence intervals, and conduct hypothesis tests. **In this chapter**, you will learn to:

- estimate the sample size for producing a CI of given width for a proportion, mean, mean difference, difference between two means, and difference between two proportions.
- explain issues relevant to estimating sample sizes.

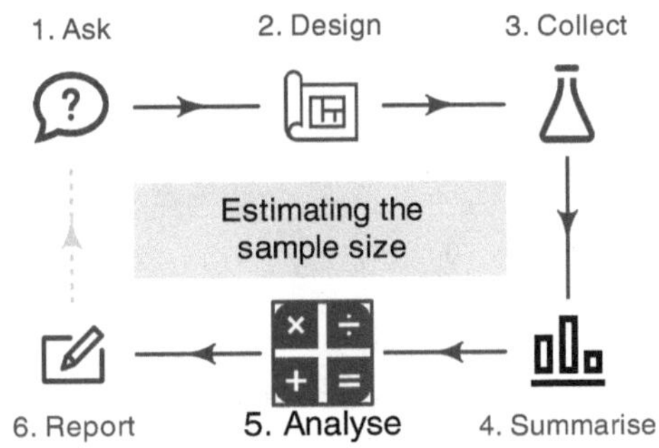

32.1 Introduction

A confidence interval (CI) is an interval which gives a range of values of the parameter that could plausibly have produced the observed value of the statistic. All else being equal, a *larger* sample size gives a *more precise* estimate of the parameter (Sect. 6.3); that is, a *narrower* CI. After all, that's why larger samples are preferred over smaller samples: they provide more *precise* estimates.

Example 32.1 (Impact of sample size on CIs). Suppose we wish to estimate an unknown proportion, and find that $\hat{p} = 0.52$ from a sample of size $n = 25$. The approximate 95% CI is 0.52 ± 0.200 (so the *margin of error* is 0.200).

If the estimate of $\hat{p} = 0.52$ was found from a sample of size $n = 100$ (rather than $n = 25$), a more precise estimate should be expected. The approximate 95% CI is 0.52 ± 0.100; the margin of error is 0.100, so the estimate is indeed more precise.

If the estimate of $\hat{p} = 0.52$ was found from a sample of size $n = 400$, the approximate 95% CI is 0.52 ± 0.050; the margin of error is 0.050 (which is more precise again).

At each step, the sample size was four times as large, but the margin of error was halved.

Figure 32.1 shows the approximate width of the CI for estimating a proportion, for various sample sizes (all else being equal). Observe that:

- greater precision (*smaller* CI width) is obtained using *larger* sample sizes.
- for *small* sample sizes (say, smaller than 15), precision greatly increases with small increases in the sample size.
- for *large* sample sizes (say, greater than 30), precision improves only slightly when the sample size is increased.

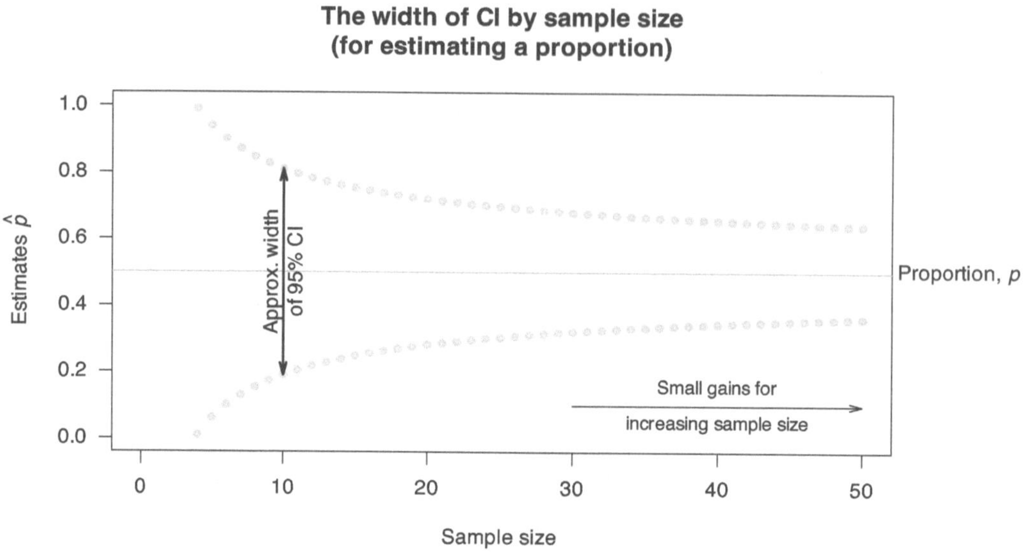

FIGURE 32.1: The approximate width of a 95% CI for a proportion, when various size samples are used.

That is, improving precision gets more difficult as sample sizes get larger: large gains in precision are made by moderately increasing small sample sizes, but only small gains in precision are made by large increases in already-large sample sizes.

⚠ Remember that the sample size is the number of *units of analysis*.

32.2 General ideas

If larger samples give more precise estimates, should the largest sample possible always be used? Not necessarily; using large samples also has disadvantages:

- as seen above, very large sample sizes only slightly improve precision.
- studies with larger samples sizes take longer to complete.
- studies with larger samples sizes are more expensive.
- ethics committees aim to keep sample sizes as small as possible, so that:
  - the environment is impacted as little as possible.
  - the fewest possible animals are harmed.
  - the fewest possible people are harmed or inconvenienced.
  - resources, time and money are not wasted.

Example 32.2 (The cost of research)**.** Farrar et al. [2021] studied the residual effect of organic biochar compound fertilisers (BCFs) *two years* after application. This study required planting turmeric in pots using soil previously treated with BCFs.

After the turmeric was grown, the concentration of potassium, phosphorus and nitrogen—as well as many trace minerals—was determined from the soil in *every* pot. In addition, *every* turmeric plant was analysed for the number of shoots, the leaf mass fraction, and foliar nutrient information.

Every pot that is used has a substantial cost, both in terms of time and money. Using more pots increases precision, but also increases costs and the time to complete the study.

Determining the sample size to use in a study is a trade-off between the advantages of increasing precision, and the challenges of cost, time, and remaining ethical (Chap. 5). In addition, *how* the sample is obtained is important: random samples give more *accurate* samples (Sect. 6.3) than non-random samples. That is, the sample size is not the only issue to consider; *how* the sample is obtained (i.e., random; non-random) is also important.

For these reasons, researchers usually identify a margin of error that is meaningful (i.e., of practical importance) in the context of their study, to help identify an appropriate sample size.

Example 32.3 (Practical importance in sample size calculations). In a weight-loss study, estimating the weight loss with a precision of 1 g is far more precise than is necessary: a weight loss of 1 g has no practical importance, but would require a massive sample size to detect.

In contrast, the sample size needed to detect a mean weight loss with a precision of 50 kg would be far smaller. However, a weight loss so great is of no practical importance either, as most people who are looking to lose weight are hoping to lose far less than 50 kg.

The researchers may decide that a weight loss of 5 kg is sufficient to be of practical importance, and determine the sample size based on this value.

In this chapter, we learn how to compute the (approximate) minimum sample size needed to obtain a given precision (i.e., for a given *margin of error*) for a 95% CI. The estimation of sample sizes for constructing a CI is studied for these situations:

- estimating a proportion in Sect. 32.3.
- estimating a mean in Sect. 32.4.
- estimating a mean difference in Sect. 32.5.
- estimating a difference between two means in Sect. 32.6.
- estimating a difference between two proportions in Sect. 32.7.

The formulas given in this chapter only apply for *forming* 95 % *CIs*, and are very *conservative*: they will probably give *minimum* samples sizes that are a little *too large*, but that is better than being too small.

To ensure that the required targets are met, the results from the sample size calculation should always be rounded up. In addition, sample sizes slightly larger than calculated are often used, to allow for *drop-outs*: animals or plants that die; people who can no longer be contacted; and so on.

⚠ Always *round up* the result of the sample size calculation.

32.3 Sample size for estimating one proportion

In Sect. 22.8, a CI was formed for the *population* proportion of female college students in the United States that drink coffee daily [Kelpin et al., 2018]. From a sample of $n = 360$, the CI was 0.1694 ± 0.0395 (i.e., the *margin of error* is 0.0395), or from 0.130 to 0.209.

To obtain a more precise estimate (i.e., a narrower CI), a larger sample is needed, but how much larger? For instance, suppose we would like a CI with margin of error of 0.02 (rather than 0.0395). What size sample is needed?

Definition 32.1 (Sample size: proportion). Conservatively, the size of the *simple random sample* needed *for a 95% CI for a proportion* with a specified margin of error is *at least*

$$\frac{1}{(\text{Margin of error})^2}.$$

For the coffee-drinking situation above, at least $1 \div (0.02^2) = 2\ 500$ female college students in the US is needed. This is a substantial increase from the original sample size of 360.

Example 32.4 (Sample size calculations for one proportion). To estimate the population proportion of South Africans that smoke, within 0.07 with 95% confidence, at least

$$\frac{1}{(\text{Margin of error})^2} = \frac{1}{0.07^2}$$

people are needed; *at least* $n = 204.08$ people. In practice, *at least* 205 people are needed to achieve this desired level of precision (that is, *always round up* in sample size calculations).

32.4 Sample size for estimating one mean

Estimating a mean depends on the variation in the observations. If the data have a small amount of variation, estimating the mean requires a smaller sample size as most observations are similar.

Definition 32.2 (Sample size: mean). Conservatively, the size of the *simple random sample* needed *for a 95% CI for the mean* with a specified margin of error is *at least*

$$\left(\frac{2 \times s}{\text{Margin of error}}\right)^2,$$

where s is an estimate of the standard deviation in the population.

The formula requires a value for the sample standard deviation, s. But if we don't have a sample yet, how can we have a value for the standard deviation of the *sample* to use? An approximate value for s is used, which can come from:

- the value of s from the results of a pilot study (Sect. 9.2).
- the results of a similar study, where the value s there can be used (see Example 32.5).

Example 32.5 (Sample size estimation for one mean). Sect. 23.6 discusses a study about the mean cadmium concentrations in peanuts in the United States, where $s = 0.0460$ ppm [Blair and Lamb, 2017].

To estimate the mean cadmium concentration in *Canadian* peanuts, within 0.005 ppm with 95% confidence, this value for s can be used. Then:

$$\left(\frac{2 \times 0.0460}{0.005}\right)^2 = 338.56;$$

we would need at least 339 Canadian peanuts.

32.5 Sample size for estimating a mean difference

The ideas in the previous section also work for computing sample sizes for estimating *mean differences*, since the differences can be treated like a single sample.

Definition 32.3 (Sample size: mean difference). Conservatively, the size of the *simple random sample* needed *for a 95% CI for the mean difference* with a specified margin of error is *at least*

$$\left(\frac{2 \times s_d}{\text{Margin of error}}\right)^2,$$

where s_d is an estimate of the standard deviation of the population differences.

Again, an approximate value for s_d can come from a pilot study (Sect. 9.2), or from the results of a similar study.

Example 32.6 (Sample size estimation for mean differences). In Sect. 29.4, a CI is computed for the difference between the distances walked in 6 mins (the six-minute walk test, 6MWT), using a 20 m and 30 m walkway [Saiphoklang et al., 2022], for 50 Thai patients. The *approximate* 95% CI is from 15.80 m to 28.26 m, further for a 30 m walkway (i.e., the margin of error is 6.234 m).

Suppose we wanted to estimate the mean difference in the 6MWT distances for Malaysian patients; we could use the value of s from this study (i.e., $s = 22.03920$). Also, suppose we wanted a precision of 4 m (that is, the margin of error is 4). For this *more precise* estimate, we would need a *larger sample*. So compute:

$$\left(\frac{2 \times 22.03920}{4}\right)^2 = 121.43;$$

we would need at least 122 patients, after rounding up.

32.6 Sample size for estimating a difference between two means

A formula for computing sample sizes for estimating the *difference between two means* is simple if we make two assumptions:

- the sample size in both groups being compared is the same.
- the standard deviation in both groups being compared is the same.

Formulas are available for computing sample sizes without these restrictions, but are more complicated than that given here. Again, an approximate value for s can come from a pilot study (Sect. 9.2), or from the results of a similar study.

Definition 32.4 (Sample size: difference between two means)**.** Conservatively, the size of the *simple random sample* needed *for a 95% CI for the difference between two means* with a specified margin of error is *at least*

$$2 \times \left(\frac{2 \times s}{\text{Margin of error}} \right)^2$$

for *each* sample, where s is an estimate of the common standard deviation in the population for both groups.

Example 32.7 (Sample size estimation for difference between means)**.** In Sect. 30.7, a CI is computed for difference between the mean speeds of cars before and after signage was added [Ma et al., 2019]. Suppose we wanted to estimate the difference between the mean reaction times within $5\,\text{km.h}^{-1}$.

In Sect. 30.7, the two groups (before and after signage added) produced standard deviations of 13.194 and 13.134 (which are very similar). We decide to use $s = 13.15$ in the sample-size calculation as the common value of s:

$$2 \times \left(\frac{2 \times 13.15}{5} \right)^2 = 55.335.$$

We would need to measure the speed of at least 56 cars before signage was added, and another 56 cars after the addition of signage (rounding up the result).

32.7 Sample size for estimating a difference between proportions

A formula for computing sample sizes for estimating the *difference between two proportions* is simple if we assume the sample size in both groups being compared is the same. Formulas are available for computing sample sizes without this restriction, but are more complicated than that given here.

Definition 32.5 (Sample size: difference between two proportions)**.** Conservatively, the size of the *simple random sample* needed *for a 95% CI for the difference between two proportions* with a specified margin of error is *at least*

$$\frac{2}{(\text{Margin of error})^2}$$

for *each* sample.

Example 32.8 (Sample size estimation for difference between proportions). In Sect. 31.9, a CI is computed for difference between the proportion of infected turtles nests, comparing natural and relocated nests [Candan et al., 2021]. Suppose we wanted to estimate the difference between the proportion of infected nests within 0.15.

We compute:

$$\frac{2}{0.15^2} = 88.89.$$

We would need to record data from at least 89 natural nests and 89 relocated nests.

32.8 More details about these sample size calculations

The above calculations form just one part of the information needed to make the final decision about the necessary sample size. For example, the *cost* (time and money) of taking the samples has not been considered.

The calculations in this chapter assume a *simple random sample* will be used, which is often unreasonable. Other, more complex, formulas are available for computing sample sizes for other random-sampling schemes (such as stratified samples). However, the above calculations give an approximate *minimum* sample size required. In addition, the calculations in this chapter are only for producing 95% CI.

In practice, researchers often start with a slightly larger sample than calculated to allow for drop-outs (e.g., plants die, or people withdraw from the study).

32.9 Example: emergency residential aged care

Dwyer et al. [2021] studied residential aged care residents in Australia needing emergency care and recorded, among other information, the average age of such residents ($\bar{x} = 85$; $s = 7.3$) and the proportion of calls related to falls ($\hat{p} = 0.156$).

Suppose a similar study was to be conducted in New Zealand. The aim was to estimate the mean age of residents within 2 years of age, and the proportion of incidents related to falls within 0.10.

Using the value of s from Australia, the sample size required to meet the age requirement is at least

$$n = \left(\frac{2 \times s}{\text{Margin of error}}\right)^2 = \left(\frac{2 \times 7.3}{2}\right)^2 = 53.29,$$

or at least 54 residents (rounding up). The sample size required to meet the falls requirement is at least

$$n = \frac{1}{(\text{Margin of error}^2)} = \frac{1}{0.1^2} = 100.$$

Since the same subjects are needed for both estimates, at least 100 residents are needed.

32.10 Chapter summary

Estimating a sample size is a compromise between the precision of the estimate, and the need to remain ethical and reduce costs. All else being equal, making a sample size four times as large results in a CI half as wide. This means that large gains in precision are made by increasing small sample sizes, but only small gains are made by increasing already-large sample sizes.

32.11 Quick review questions

Are the following statements *true* or *false*?

1. A *larger* sample size produces a *more accurate* estimate of the parameter, all else being equal.
2. A *larger* sample size produces a *more random* sample.
3. We should *always* take the *largest* possible sample size.

32.12 Exercises

Answers to odd-numbered exercises are given at the end of the book.

Exercise 32.1. To obtain a *narrower* CI, is a larger or smaller sample size necessary (all else being equal)?

Exercise 32.2. Does a *narrow* CI imply a *precise* estimate, or an *accurate* estimate of the parameter?

Exercise 32.3. Suppose we need to estimate a population *mean* (with 95% confidence), using $s = 1$ kg.

1. What size sample is needed to estimate the population mean within 0.4 kg?
2. What size sample is needed to estimate the population mean within 0.2 kg (that is, the CI will be *half* as wide as in the first calculation)?
3. What size sample is needed to estimate the population mean within 0.1 kg (that is, the CI will be *a quarter* as wide as in the first calculation)?
4. To get a CI *half* as wide, how many *times* more units of analysis are needed?
5. To get a CI *a quarter* as wide, how many *times* more units of analysis are needed?
6. Would a *smaller* or *larger* sample be needed to estimate the population mean within 0.4 kg, with 99% confidence? Explain.

Exercise 32.4. Suppose we need to estimate a difference between two population *means* (with 95% confidence), using $s = 8$ cm.

1. What size samples are needed to estimate the difference between the population means within 4 cm?
2. What size samples are needed to estimate the difference between the population means within 2 cm (that is, the CI will be *half* as wide as in the first calculation)?
3. What size samples are needed to estimate the difference between the population means within 1 cm (that is, the CI will be *a quarter* as wide as in the first calculation)?
4. To get a CI *half* as wide, how many *times* more units of analysis are needed?
5. To get a CI *a quarter* as wide, how many *times* more units of analysis are needed?

6. Would a *smaller* or *larger* sample be needed to estimate the population mean within 4 cm, with 99% confidence? Explain.

Exercise 32.5. Mann and Blotnicky [2017] studied of the eating habits of university students in Canada (Sect. 22.5). They estimated the proportion of Canadian students that ate a sufficient number of servings of grains each day.

Suppose we wished to repeat the study but for *New Zealand* university students; that is, we seek an estimate of the population proportion of New Zealand students that eat a sufficient number of servings of grains each day (with 95% confidence).

1. What size sample is needed to estimate the proportion within 0.01?
2. What size sample is needed to estimate the proportion within 0.02?
3. What size sample is needed to estimate the proportion within 0.10?
4. Do you think this study would be costly, in terms of time and money?

Exercise 32.6. We wish to estimate the population proportion of Kenyans that smoke.

1. Suppose we wish our 95% CI to have a margin of error of 0.05. How many Kenyans would need to be surveyed?
2. Suppose we wish our 95% CI to have a margin of error of 0.025; that is, we wish to *halve* the width of the interval above. How many Kenyans would need to be surveyed?
3. How many *times* as many Kenyans are needed to *halve* the width of the CI?

Exercise 32.7. Tager et al. [1979] measured the lung capacity of 11-year-old girls in East Boston, using the *forced expiratory volume* (FEV) of the children (Exercise 23.3). Suppose we wished to repeat the study, and find a 95% CI for the mean FEV for 11-year-old *Australian* girls.

Since Australian and American children might be somewhat similar, we could use, as an approximation, the standard deviation from that study: $s = 0.43$ L.

1. What size sample is needed to estimate the mean within 0.02 L?
2. What size sample is needed to estimate the mean within 0.05 L?
3. What size sample is needed to estimate the mean within 0.10 L?
4. Suppose we wished to find 99% (not 95%) CI for the mean FEV for 11-year-old *Australian* girls, within 0.10 L. Would this sample size be *larger* or *smaller* than the sample size found for a 95% CI (also within 0.10 L)?
5. Do you think this study would be costly, in terms of time and money?

Exercise 32.8. Williams and Boyle [2007] asked paramedics ($n = 199$) to estimate the amount of blood loss on four different surfaces. When the actual amount of blood spill on concrete was 1 000 mL, the mean guess was 846.4 mL (with a standard deviation of 651.1 mL). For a different study:

1. how many paramedics are needed to estimate the mean with a precision of 50 mL?
2. how many paramedics are needed to estimate the mean with a precision of 25 mL?
3. how many times greater does the sample size need to be to *halve* the width of the margin of error?

Exercise 32.9. Skypilot is an alpine wildflower native to the Colorado Rocky Mountains (USA). In recent years, a willow shrub has been encroaching on skypilot territory and, because willow often flowers early, Kettenbach et al. [2017] studied whether the willow may 'negatively affect pollination regimes of resident alpine wildflower species' (p. 6 965). Data for both species was collected at 25 different sites, so the data are *paired* by site. The 'first-flowering day' is the number of days since the start of the year (e.g., January 12 is 'day 12') when flowers were first observed.

Suppose a similar paired study was to be conducted on skypilot growing in Sierra Nevada, California. Using the software output in Fig. 13.3:

1. determine the sample size needed to estimate the mean difference in first-flowering day within two days.
2. determine the sample size needed to estimate the mean difference in first-flowering day within three days.

Exercise 32.10. MacGregor et al. [1979] studied treating hypertension with Captopril. Patients had their systolic blood pressure measured (in mm Hg) immediately *before* and two hours *after* being given the drug. A pilot study showed that the difference between the two measurements had a standard deviation of about 9 mm Hg.

1. Determine the sample size needed to estimate the mean reduction in *systolic* blood pressure within 2 mm Hg.
2. Determine the sample size needed to estimate the mean reduction in *diastolic* blood pressure within 1.5 mm Hg.

Exercise 32.11. Agbayani et al. [2020] studied gray whales (*Eschrichtius robustus*) and measured (among other variables) the length of whales at birth. Summary information is shown in Table 30.7. Suppose another research study wanted to study sperm whales, which have an approximately similar size.

1. Determine the sample size needed to estimate the difference between the mean lengths for female and male sperm whales at birth, within 0.15 m.
2. Determine the sample size needed to estimate the difference between the mean lengths for female and male sperm whales at birth, within 0.10 m.
3. Determine the sample size needed to estimate the difference between the mean lengths for female and male *goldfish* at birth, within 1 mm.

Exercise 32.12. Suppose researchers are trialling a new drug to reduce the recovery time (compared to standard treatments) after contracting pneumonia. They conduct a pilot study, and find the standard deviation of the duration of the symptoms, in both groups, is about $s = 1.25$ days.

1. What size sample is needed to estimate the difference between the mean recovery times between the two treatments within 1 day.
2. What size sample is needed to estimate the difference between the mean recovery times between the two treatments within 0.5 days.

Exercise 32.13. Table 31.8 summarises the data from a study of the incidents of in-hospital heart attacks for people admitted following an earlier heart attack. To estimate the difference between the proportion of patients having an in-hospital heart attack (between patients with a low body temperature and patients with a high body temperature) within 0.03, what size samples are needed?

Exercise 32.14. Exercise 31.13 describes a study comparing the proportion of females and males who wore sunglasses in Brisbane, Australia [Dexter et al., 2019]. Suppose we wished to make a similar comparison for people in Auckland, estimating the difference in the proportions within 0.07. How many females and males would be needed?

Answers to *Quick review* questions: 1. False. **2.** False. **3.** False.

33

Correlation and regression: CIs and tests

So far, you have learnt about the research process, including analysing data using confidence intervals and conducting hypothesis tests. **In this chapter**, you will learn to:

- produce confidence intervals for correlation coefficients.
- conduct hypothesis tests for correlation coefficients.
- produce and interpret linear regression equations.
- conduct hypothesis tests for the slope of a regression line.
- produce confidence intervals for the slope of a regression line.
- determine whether the conditions for using these methods apply in a given situation.

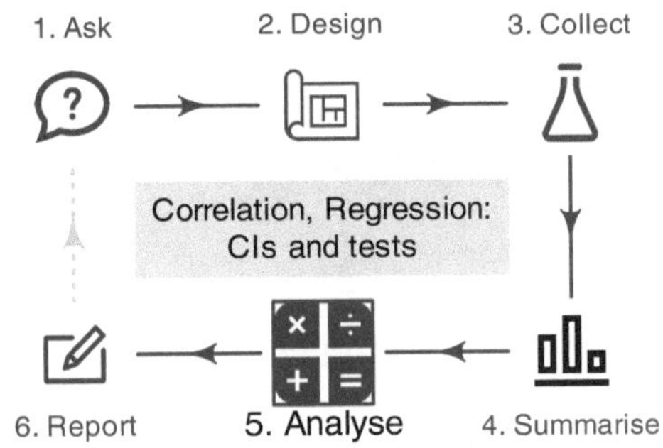

33.1 Introduction: sorghum yield and borers

So far, RQs about single variables (descriptive RQs) and RQs for comparisons (relational and repeated-measures RQs) have been studied. In this chapter, the relationship between two quantitative variables is studied (correlational RQs) *when that relationship is approximately linear.* The strength of the relationship (correlation) and the nature of that relationship (regression) are discussed.

For this chapter, consider this (one-tailed) RQ:

> In sorghum crops (AG1090 hybrid) in Brazil, is a larger sugarcane borer infestation associated with smaller yields?

Souza et al. [2024b] recorded the borer infestation in sorghum crops, for $n = 24$ crops over three years [Souza et al., 2024a], shown in Table 33.1. The data comprises two quantitative variables (Fig. 33.1, left panel).

Knowing the amount of infestation provides some information about the yield: a moderate relationship between the variables seems evident. The relationship also seems somewhat *linear.* The *Pearson correlation coefficient* (Fig. 33.1, right panel) is $r = -0.934$, so $R^2 = (-0.934)^2 = 87.2\%$. This means that the unexplained variation in yield reduces by 78.8% by knowing the amount of infestation.

TABLE 33.1: Sorghum AG1090 yield and sugarcane borer infestation; the first five and last five of $n = 24$ observations.

Infestation (%)	Yield (kg.ha^{-1})	Infestation (%)	Yield (kg.ha^{-1})
10.57	2844.12	⋮	⋮
13.29	3055.69		
18.64	3025.14	35.79	1109.62
11.43	3101.36	27.42	1573.81
5.53	4539.90	27.09	1587.57
⋮	⋮	25.01	2403.90
		23.04	3819.29

⚠ Recall that the *sample* correlation coefficient is denoted by r, and the *population* correlation coefficient is denoted by ρ.

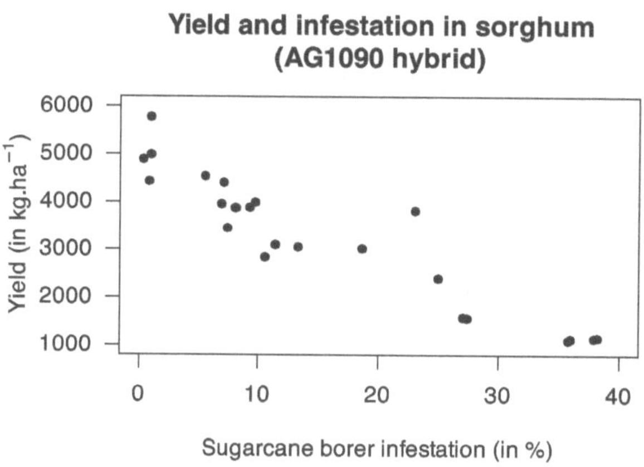

Correlation Matrix

		Yield	Infestation
Yield	Pearson's r	—	
	df	—	
	p-value	—	
	95% CI Upper	—	
	95% CI Lower	—	
Infestation	Pearson's r	-0.9339	—
	df	22	—
	p-value	< .001	—
	95% CI Upper	-0.8511	—
	95% CI Lower	-0.9713	—

FIGURE 33.1: Sorghum yield against borer infestation. Left: scatterplot. Right: correlation output.

33.2 Correlation: CIs and tests for ρ

33.2.1 Correlation: CIs for ρ

The sorghum data in Table 33.1 is only one of the countless possible samples of sorghum crops that could have been studied. The value of r (an estimate of ρ, the *parameter*) will vary from sample to sample; that is, the value of r has a sampling distribution, and sampling variation exists. The sampling distribution of r, however, does *not* have a normal distribution, so CIs for ρ will be taken directly from software output (Fig. 33.1, right panel). For the sorghum data, the 95% CI for ρ is from -0.971 to -0.851. This CI is not symmetrical: the value of r is not halfway between these limits. We write:

> For soghum crops, the correlation coefficient between yield and infestation percentage is -0.934, with a 95% CI from -0.971 to -0.851 ($n = 24$).

In other words, a population with a correlation coefficient ρ between -0.971 and -0.851 could reasonably have produced a sample correlation coefficient of $r = -0.934$ from a sample of size $n = 24$.

Example 33.1 (Correlation)**.** The relationship between the number of cyclones y in the Australian region each year from 1969 to 2005, and a unitless climatological index called the *Ocean Niño Index* (ONI, x), averaged over October, November and December, is shown in Fig. 33.2 (left panel) [Dunn and Smyth, 2018].

The relationship has a *negative* direction, so the value of r is *negative*. From the software output, $r = -0.683$ with a 95% CI from -0.824 to -0.460.

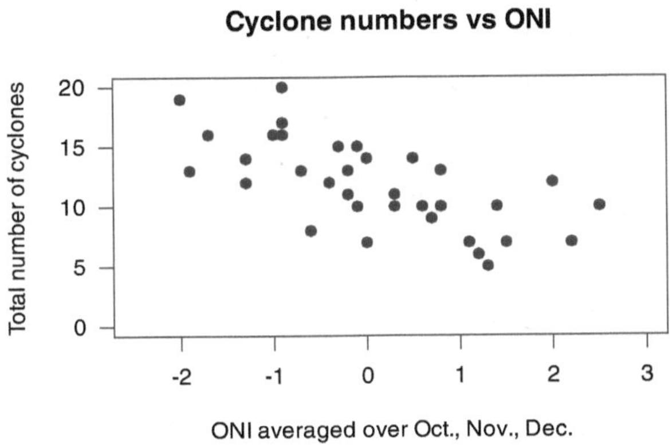

Correlation Matrix

		Total	OND
Total	Pearson's r	—	
	df	—	
	p-value	—	
	95% CI Upper	—	
	95% CI Lower	—	
OND	Pearson's r	−0.6825	—
	df	35	—
	p-value	<.0001	—
	95% CI Upper	−0.4602	—
	95% CI Lower	−0.8242	—

FIGURE 33.2: The number of cyclones in the Australian region each year from 1969 to 2005, and the ONI averaged over October, November, December. Left: scatterplot. Right: software output.

33.2.2 Correlation: hypothesis test for ρ

A hypothesis test can also be conducted regarding ρ, the Pearson correlation coefficient in the *population*. The null hypothesis is, as always, the 'no difference, no change, no relationship' position which is, in this context:

- H_0: $\rho = 0$.

Clearly, the *sample* correlation coefficient r for the data is not zero, and the RQ is effectively asking if sampling variation is the reason for this discrepancy between r and the parameter ρ.

Since the RQ (in Sect. 33.1) is one-tailed (negative direction), the alternative hypothesis is:

- H_1: $\rho < 0$ (*one-tailed* test, based on the RQ).

As usual, initially *assume* that $\rho = 0$ (from H_0), then describe what values of r could be *expected* using the *sampling distribution*, under that assumption, across all possible samples. Then the *observed* value of r is compared to the values expected through sampling variation to determine if the value of r supports or contradicts the assumption.

For a correlation coefficient, the sampling distribution of r does not have a normal distribution.[1] However, the output (Fig. 33.1, right panel) contains the relevant two-tailed P-value for the test: less than 0.001. Hence, the one-tailed P-value for the test is less than 0.0005. *Very strong evidence* exists to support H_1 (that the correlation in the population is negative).

[1]For those interested: the value of r only varies between -1 and 1, so cannot have a normal distribution. A transformation of r *does* exist that has an approximate normal distribution and *standard error*.

We write:

> The sample presents very strong evidence (one-tailed $P < 0.0005$) that the sorghum yield has a negative association with borer infestation percentage ($r = -0.934$ with 95% CI from -0.971 to -0.851; $n = 24$) in the population.

Notice the three features of writing conclusions again: an answer to the RQ, evidence to support the conclusion, and some sample summary information.

> ⓘ If the evidence suggests that the correlation coefficient is *not zero* (in the population), this does *not* necessarily mean a *strong* correlation exists. The correlation may be weak in the population (as estimated by the value of r), but evidence exists that the correlation is *not zero* in the *population.*
>
> That is, the test is about statistical significance, not practical importance.

Example 33.2 (Correlation). The relationship between the number of cyclones y in the Australian region each year from 1969 to 2005, and the ONI x is shown in Fig. 33.2 (left panel). To test for a relationship, use

$$H_0\colon \rho = 0 \qquad \text{against} \qquad H_0\colon \rho \neq 0;$$

software reports that $P < 0.0001$ (Fig. 33.2, right panel). There is very strong evidence of a relationship between the number of cyclones in the Australian region and the ONI (averaged over October, November and December).

33.3 Regression

33.3.1 Introducting regression

Correlation measures the *strength* and *direction* of the *linear* relationship between two quantitative variables x (an explanatory variable) and y (a response variable). Sometimes, however, *describing* the nature of the relationship is useful. This is called *regression.*

The regression relationship is described mathematically using an *equation*, and allows us to:

1. *Predict* the mean value of y from a given value of x (Sect. 33.3.4).
2. *Understand* the relationship between x and y (Sect. 33.3.5).

An example of a *linear* regression equation, describing the linear relationship between the observed values of an explanatory variable x and the observed values of a response variable y, is

$$\hat{y} = -4 + (2 \times x), \qquad \text{usually written} \qquad \hat{y} = -4 + 2x. \tag{33.1}$$

The notation $\hat{y}$ refers to the mean of all the y-values that could be observed for some given value of x. That is, for some value of x, many different values of y could be observed, and $\hat{y}$ is the value that regression equation predicts as the *mean* of all those possible values. This equation describes the connection between the values of x and the corresponding average values of y.

> y refers to the values of the response variable *observed* from individuals. $\hat{y}$ refers to *predicted mean* value of y for given values of x.

> $\hat{y}$ is pronounced as 'why-hat'; the 'caret' above the y is called a 'hat'.

More generally, the equation of a straight line is

$$\hat{y} = b_0 + (b_1 \times x), \qquad \text{usually written} \qquad \hat{y} = b_0 + b_1 x,$$

where the values of b_0 and b_1 are unknown, and estimated from sample data. Notice that b_1 is the number multiplied by x. In Equation (33.1), $b_0 = -4$ and $b_1 = 2$.

Example 33.3 (Regression equations)**.** A report on the growth of Australian children [Pfizer Australia, 2008] found an approximate linear relationship between the age (in years) x and height (in cm) y of girls aged between four and seven. The regression equation was approximately

$$\hat{y} = 73 + 7x.$$

The regression equation is the same if written as

$$\hat{y} = 7x + 73.$$

In both cases, $b_0 = 73$ and $b_1 = 7$. This regression equation describes the connection between ages x and heights y for girls aged between four and seven.

33.3.2 Reviewing linear equations

To introduce, or revise, the idea of a linear equation, consider the (artificial) data in Fig. 33.3 (left panel), with an explanatory variable x and a response variable y. In the graph, a sensible line is drawn on the graph that seems to capture the relationship between x and y. (You may have drawn a slightly different, but similar, line.) The line describes the predicted mean values of y (i.e., values of $\hat{y}$) for various values of x. The relationship is *positive* and *linear*.

A *regression equation* specifies the line. In the regression equation $\hat{y} = b_0 + b_1 x$, the numbers b_0 and b_1 are called *regression coefficients*, where

- b_0 is the *intercept* (or the *y-intercept*), whose value corresponds to the *predicted* mean value of y when $x = 0$.
- b_1 is the *slope*, whose value measures how much the value of $\hat{y}$ changes, on average, when the value of x *increases* by 1.

We will use software to find the values of b_0 and b_1 (as the formulas are tedious to use). However, a rough approximation of the values of b_0 and b_1 can be obtained using the rough straight line drawn on the scatterplot (Fig. 33.3).

A rough approximation of the value of the *intercept* b_0 is the value of $\hat{y}$ when $x = 0$, from the drawn line. When $x = 0$, the regression line suggests the value of $\hat{y}$ is about 2 in Fig. 33.3 (left panel); that is, the value of b_0 is approximately 2.

A rough approximation of the *slope* b_1 is found using

$$\frac{\text{Change in } \hat{y}}{\text{Corresponding } \textit{increase} \text{ in } x} = \frac{\text{rise}}{\text{run}} \qquad (33.2)$$

from the drawn line. This approximation of the slope is the *change* in the value of $\hat{y}$ (the 'rise') divided by the corresponding *increase* in the value of x (the 'run').

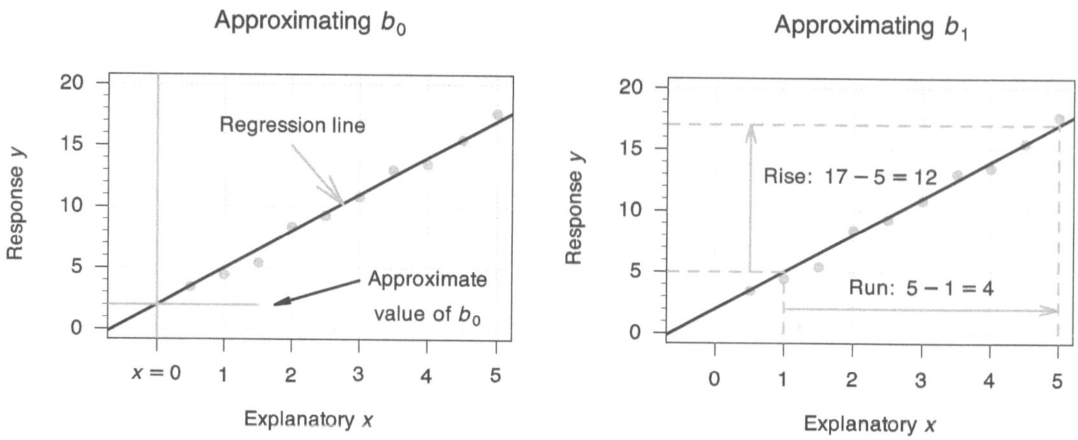

FIGURE 33.3: Estimating the regression equation from a line using an example scatterplot. Left: approximating b_0. Right: approximating b_1 using rise-over-run.

Consider what happens in Fig. 33.3 (right panel) when the value of x increases from 1 to 5 (a *run* of $5 - 1 = 4$). The corresponding value of y changes from about 5 to about 17, a *rise* of $17 - 5 = 12$. So, using Equation (33.2),

$$\frac{\text{rise}}{\text{run}} = \frac{17-5}{5-1} = \frac{12}{4} = 3.$$

The value of b_1 is about 3. When using rise-over-run, better guesses for b_1 are found when using one value of x near the left-side of the scatterplot, and another value of x near the right-side of the scatterplot, but any two values can be used (try using $x = 0$ and $x = 3$).

Combing the rough guesses for b_0 and b_1, the regression line is approximately $\hat{y} = 2 + (3 \times x)$, usually written

$$\hat{y} = 2 + 3x.$$

⚠ The regression equation has $\hat{y}$ (not y) on the left-hand side. That is, the equation *predicts* the *mean* values of y, which may not be equal to any of the observed values (which are denoted by y).

A 'good' regression equation would produce predicted values $\hat{y}$ close to the observed values y; that is, the line passes close to each point on the scatterplot.

The *intercept* b_0 has the same measurement units as the response variable. The measurement units for the *slope* b_1 is the 'measurement units of the response variable', per 'measurement units of the explanatory variable'.

Example 33.4 (Measurement units of regression parameters)**.** In Example 33.3, the regression line for the relationship between the age of Australian girls x (in years) and their height (in cm) y was $\hat{y} = 73 + 7x$ (for girls aged between four and seven years).

In the equation, the intercept is $b_0 = 73\,\text{cm}$ and the slope is $b_1 = 73\,\text{cm/y}$ (the growth *rate*).

Example 33.5 (A rough approximation of the regression equation)**.** For the sorghum data, a rough estimate of the regression line can be drawn on a scatterplot to estimate b_0 and b_1 (Fig. 33.4). The estimate of b_0 (the value of $\hat{y}$ when $x = 0$) is roughly 4 800 kg.ha^{-1}.

The estimate of b_1 can be found using the rise-over-run idea. When $x = 0$, the value of $\hat{y}$ (according to the drawn line) is about 4 800. At the other extreme of the plot, where $x = 40$, the value of $\hat{y}$ is about 1 000. (Any two points on the line can be used, but using two points at each end gives better guesses of the slope.) So, as x increases from 0 to about 40, the value of $\hat{y}$ *reduces* from about 4 800 to about 1 000, a change of about −3 800. That is, for a 'run' of $40 - 0 = 40$, the 'rise' is $4\,800 - 1\,000 = -3\,800$ (i.e., a drop of 3 800), and so a rough estimate of the slope is $-3\,800/40 = -95$. (The relationship is *negative*, so the slope is *negative*.)

The rough guess of the regression line is therefore

$$\hat{y} = 4\,800 - 95x,$$

where x is the infestation percentage, and y is yield (in kg.ha^{-1}). The rough guess of the intercept b_0 is 4 800 kg.ha^{-1}, while the rough guess of the slope b_1 is −95 kg.ha^{-1}/%.

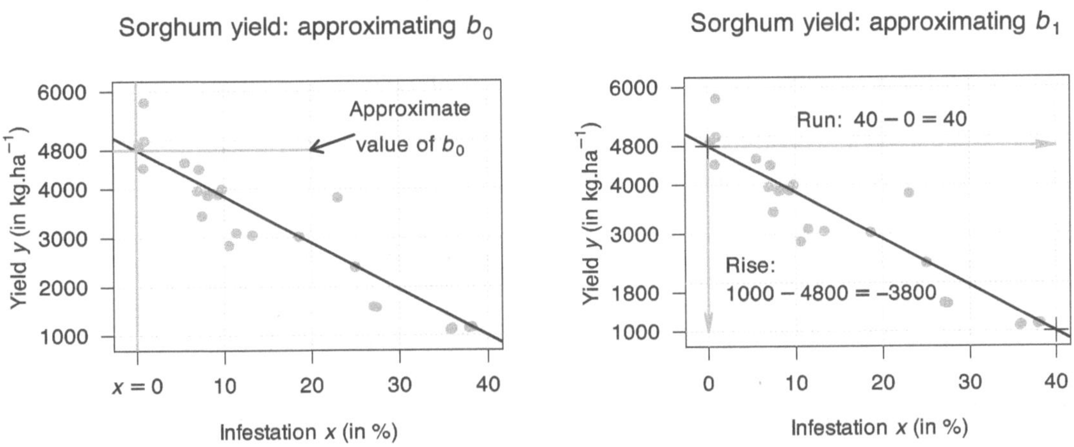

FIGURE 33.4: Obtaining rough guesses for the regression equation for the sorghum data. Left: approximating b_0. Right: approximating b_1 using rise-over-run. The plus signs on the right plot indicate the points used to estimate the slope.

Example 33.6 (Estimating regression parameters)**.** The relationship between the number of cyclones y in the Australian region each year from 1969 to 2005, and the ONI x is shown in Fig. 33.2 (left panel).

To make a guess of the regression coefficients, a sensible line can be drawn through the data (Fig. 33.5). When the value of x is zero, the predicted value of y is about 12, so b_0 is about 12 cyclones. Recall: the intercept is the predicted value of y when $x = 0$, which is *not* at the left of the graph in Fig. 33.5.

To approximate the value of b_1, use the rise-over-run idea. When $x = -2$, the predicted mean value of y is about 17; when $x = 2$, the predicted mean value of y is about 8. The value of x increases by $2 - (-2) = 4$, while the value of $\hat{y}$ changes by $7.5 - 17 = -9.5$ (a *decrease* of about 9.5). Hence, b_1 is approximately $-9.5/4 = -2.375$ cyclones per unit change in ONI. (You may get a slightly different value from a slightly different line.)

The relationship has a *negative* direction, so the slope must be *negative*. The regression line is approximately $\hat{y} = 12 - 2.375x$.

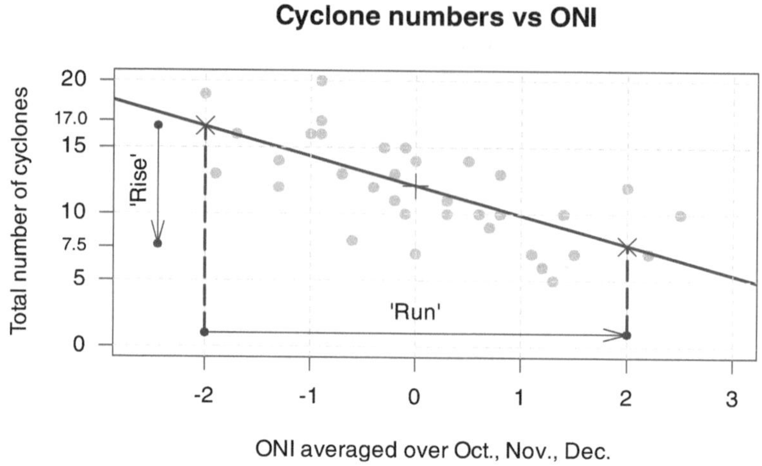

FIGURE 33.5: The number of cyclones in the Australian region each year from 1969 to 2005, and the ONI averaged over October, November, December. An estimate of the reggression line is shown. The plus sign + is located on the line where $x = 0$. The crosses × are located to find rise-over-run.

The above method gives a crude approximation to the values of the intercept b_0 and the slope b_1. In practice, *many* reasonable lines could be drawn through a scatterplot of data, each giving slightly different rough guesses for b_0 and b_1. However, one of those lines is the 'best' line in some sense,[2] and is sometimes called the 'line of best fit'.

33.3.3 Regression: finding equations using software

Software is almost always used to find the estimates for the intercept b_0 and slope b_1, as the formulas are complicated and tedious to use. For the sorghum data (Fig. 33.1), the relevant software output is shown in Fig. 33.6.

In the output, the values of b_0 and b_1 are in the column labelled `Estimate`; the value of the sample y-intercept is $b_0 = 4\,814.1\,\text{kg.ha}^{-1}$, and the value of the sample *slope* is $b_1 = -101.4\,\text{kg.ha}^{-1}/\%$. These are the values of the two *regression coefficients*. The regression equation is, after rounding:

$$\hat{y} = 4\,814.1 + (-101.4 \times x), \qquad \text{usually written as} \qquad \hat{y} = 4\,814.1 - 101.4x. \tag{33.3}$$

These are close to the values obtained using the rough method in Sect. 33.3.2 (which gave $b_0 = 4\,800$ and $b_1 = -95$ approximately).

ⓘ The *sign* of the slope b_1 and the *sign* of correlation coefficient r are always the same. For example, if the slope is negative, the correlation coefficient will be negative. However, the *value* of the slope cannot be deduced solely from the value of the correlation coefficient, nor can the *value* of the correlation coefficient be deduced solely from the value of the slope.

[2]For those interested: the 'line of best fit' is the line for which the sum of the *squared* vertical distances between the observations y and the predicted values $\hat{y}$ (i.e, the regression line) is as small as possible.

Model Coefficients - Yield

Predictor	Estimate	SE	t	p
Intercept	4814.1	163.691	29.41	< .001
Infestation	-101.4	8.279	-12.25	< .001

FIGURE 33.6: The regression output for the sorghum data.

Example 33.7 (Regression coefficients). The regression equation for the cyclone data (Fig. 33.5) is found from the software output (Fig. 33.7) as

$$\hat{y} = 12.1 - 2.23x,$$

where x is the ONI (averaged over October, November, December) and y is the number of cyclones; that is, $b_0 = 12.1$ cyclones and $b_1 = -2.23$ cyclones per unit change in ONI. These values are close to the approximations made in Example 33.6 ($b_0 = 12$ and $b_1 = -2.375$ respectively).

Model Coefficients - Total

			95% Confidence Interval			
Predictor	Estimate	SE	Lower	Upper	t	p
Intercept	12.1389	0.4521	11.2210	13.0567	26.8483	< .0001
OND	-2.2339	0.4044	-3.0549	-1.4130	-5.5241	< .0001

FIGURE 33.7: The software output for the cyclone data.

33.3.4 Regression: making predictions

Regression equations can be used to make *predictions* of the mean value of y for a given value of x. For example, the regression equation for the sorghum data in Equation (33.3) can be used to make *predictions* of the mean yield for a given infestation percentage. For example, the equation can be used to predict the *average* yield of crops with an infestation percentage of 30%. Since x represents the infestation percentage, use $x = 30$ in the regression equation:

$$\begin{aligned} \hat{y} &= 4\,814.1 - (101.4 \times 30) \\ &= 4\,814.1 - 3\,042 = 1\,772.1. \end{aligned}$$

Crops with an infestation percentage of 30% are predicted to have a *mean* yield of 1 772.1 kg.ha^{-1} (though individual crops with an infestation percentage of 30% may have smaller or greater yields). The model predicts that the *mean* yield for crops with an infestation percentage of 30% will be about 1 772.1 kg.ha^{-1}.

⚠ The value of $\hat{y}$ is computed using the estimates b_0 and b_1, which are computed from sample data. Hence, the value of $\hat{y}$ also depends on which one of the countless possible samples is used. This means that $\hat{y}$ also has a sampling distribution and a standard error.

Suppose we were interested in crops with an infestation percentage of 50%; the mean yield is

$$\hat{y} = 4\,814.1 - (101.4.39 \times 50) = -255.9,$$

or about -256 kg.ha^{-1}, which is clearly silly (negative yields are impossible). In the data, the heaviest infestation is about 40%, so no data exists beyond a 40% infestation percentage. As a result, the regression line does not even apply for infestations exceeding 40%. (This means that the relationship must be non-linear after 40%.)

Making predictions outside the range of the available data is called *extrapolation*, and *extrapolation* beyond the data may lead to nonsense predictions.

Definition 33.1 (Extrapolation). *Extrapolation* refers to making predictions outside the range of the available data. Extrapolation beyond the data may lead to nonsense.

Example 33.8 (Extrapolation). The regression equation (Example 33.3) used to predict the mean height of girls $\hat{y}$ from their age x (for girls aged between four and seven) was given as

$$\hat{y} = 73 + 7x.$$

For girls five years-of-age (i.e., $x = 5$), the predicted mean height is

$$\hat{y} = 73 + (7 \times 5) = 73 + 35 = 108.$$

The heights of girls will vary around a *mean* height of 108 cm; some individual girls aged five will be taller than 108 cm, and some will be shorter than 108 cm.

Using the equation to estimate the height of girls aged 21 would predict a mean height of 220 cm. However, this is extrapolation and the prediction is nonsense. Young children grow at a fast rate, but growth rate slows as children age.

33.3.5 Regression: understanding relationships

The regression equation can be used to *understand* the relationship between the two variables. Consider again the sorghum regression equation:

$$\hat{y} = 4\,814.1 - 101.4x. \tag{33.4}$$

What does this equation reveal about the relationship between x and y?

b_0 is the *predicted* value of $\hat{y}$ when $x = 0$ (Sect. 33.3.3). Using $x = 0$ in Equation (33.4) predicts a mean yield of

$$\hat{y} = 4\,814.1 - (101.4 \times 0) = 4\,814.1$$

for crops with an infestation of zero; this is the value of b_1. Sometimes, using $x = 0$ is *extrapolating*, as no data exists near $x = 0$, so sometimes this interpretation of b_0 produces nonsense.

ⓘ The value of the intercept b_0 is sometimes (but not always) meaningless. The value of the slope b_1 is usually of greater interest, as it explains the *relationship* between the two variables.

The slope b_1 quantifies how the value of $\hat{y}$ changes (on average) when the value of x *increases* by one (Sect. 33.3.3). For the sorghum data, b_1 is the change in predicted mean yield for each percentage point[3] increase in borer infestation.

Specifically, each extra percentage point of borer infestation is associated with a mean change in yield of $-101.4\,\text{kg.ha}^{-1}$ (from Equation (33.4)); that is, a *decrease* in yield by a mean of $101.4\,\text{kg.ha}^{-1}$ for each extra percentage point of infestation.

To demonstrate, consider the case where $x = 10$, when the regression equation predicts $\hat{y} = 3\,800.1\,\text{kg.ha}^{-1}$. For infestations one percentage point greater than this (i.e., $x = 11$), the value of the prediction $\hat{y}$ will increase by an average of $-101.4\,\text{kg.ha}^{-1}$ (or, equivalently, *decrease* by an average of $101.4\,\text{kg.ha}^{-1}$). That is, we would predict $\hat{y} = 3\,800.1 - 101.4 = 3\,698.7\,\text{kg.ha}^{-1}$. This is the same prediction made by using $x = 11$ in Equation (33.4).

> If the value of b_1 is *positive*, then the predicted mean values of y *increase* as the values of x *increase*. If the value of b_1 is *negative*, then the predicted mean values of y *decrease* as the values of x *increase*.

This interpretation of b_1 explains the relationship: the predicted mean yield is, on average, about $101.4\,\text{kg.ha}^{-1}$ less for each extra percentage point increase of infestation.

> In general, we say that a change in the value of x is *associated* with a change in the value of $\hat{y}$. Unless the study is experimental (Sect. 4.4), we cannot say that the change in the value of x *causes* the change in the value of $\hat{y}$.

> If the value of the slope is zero, there is *no linear relationship* between x and $\hat{y}$. A slope of zero means that a change in the value of x is associated with a change of zero in the value of $\hat{y}$. In this case, the correlation coefficient is also zero.

33.4 Regression: CIs and t-test for regression parameters

33.4.1 Introduction

A regression equation exists in the *population* that connects the values of x and $\hat{y}$. This regression line is estimated from one of the countless possible samples, and is an estimate of the regression line in the population.

In the *population*, the intercept is denoted by β_0 and the slope by β_1. The values of the parameters β_0 and β_1 are unknown, and are estimated by the statistics b_0 and b_1 respectively.

> The symbol β is the Greek letter 'beta', pronounced 'beater' (as in 'egg beater'). So β_0 is pronounced as 'beater-zero', and β_1 as 'beater-one'.

Every sample is likely to produce slightly different values for both b_0 and b_1 (sampling variation), so both b_0 and b_1 have a sampling distribution and a standard error. The formulas for computing the values of b_0 and b_1 (and their standard errors) are intimidating, so we

[3] A 'percentage point' increase means a change from, say, 10% to 11%, or 35% to 36%.

will use software to perform the calculations. The sampling distributions for b_0 and b_1 have approximate normal distributions under certain conditions (Sect. 33.5).

Usually the slope is of greater interest than the intercept, because the slope explains the *relationship* between the two variables (Sect. 33.3.5). For this reason, the sampling distribution for the slope only is given below, but the sampling distribution for the intercept is analogous.

Definition 33.2 (Sampling distribution of a sample slope)**.** The sampling distribution of the sample regression slope is (when certain conditions are met; Sect. 33.5) described by

- an approximate normal distribution,
- with a mean of β_1, and
- a standard deviation, called the *standard error of the slope* and denoted s.e.(b_1).

A formula exists for finding s.e.(b_1), but is tedious to use, and we will not give it.

33.4.2 CIs for the regression parameters

The sampling distribution describes all possible values of the sample slope from all possible samples, through *sampling variation*. For the sorghum data then, the values of the sample slope across all possible samples is described (Fig. 33.8) as, using Def. 33.2:

- an approximate normal distribution,
- with a sampling mean whose value is β_1, and
- a standard deviation of s.e.$(b_1) = 8.279$ (from software; Fig. 33.6).

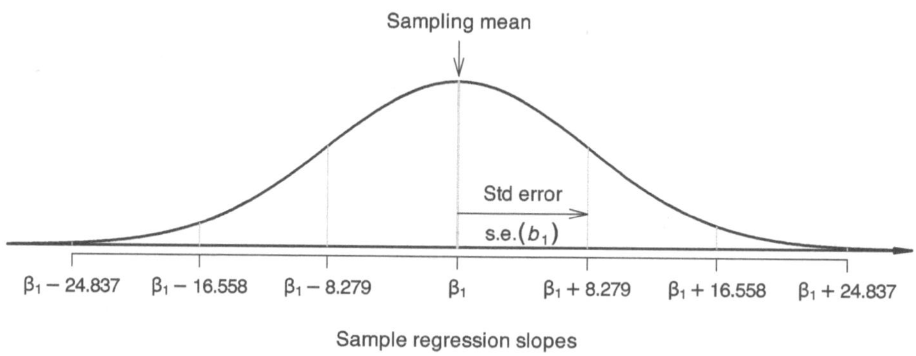

FIGURE 33.8: The distribution of sample slope for the sorghum data, around the population slope β_1.

Since the sampling distribution is an approximate normal distribution, CIs have the form

$$\text{statistic} \pm (\text{multiplier} \times \text{standard error}),$$

where the multiplier is 2 for an *approximate* 95% CI (from the 68–95–99.7 rule). In this context, a CI for the slope is

$$b_1 \pm \big(\text{multiplier} \times \text{s.e.}(b_1)\big).$$

Thus, an approximate 95% CI for the slope is

$$-101.4 \pm (2 \times 8.279) \qquad \text{or} \qquad -101.4 \pm 16.558,$$

which is from -118.0 to $-84.8\,\text{kg.ha}^{-1}$ (after rounding).

Software can be used to produce *exact* CIs too; the exact 95% CI is from -118.6 to $-84.3\,\text{kg.ha}^{-1}$ (Fig. 33.9). The *approximate* and *exact* 95% CIs are very similar. We write:

For each increase of one percentage point in borer infestation, the mean yield *increases* by $-101.4\,\text{kg.ha}^{-1}$ (95% CI: -118.6 to -84.3; $n = 24$).

Alternatively (and equivalently, but easier to understand):

For each increase of one percentage point in borer infestation, the mean yield *decreases* by $101.4\,\text{kg.ha}^{-1}$ (95% CI: 84.3 to 118.6; $n = 24$).

Model Coefficients - Yield

			95% Confidence Interval			
Predictor	Estimate	SE	Lower	Upper	t	p
Intercept	4814.1	163.691	4474.6	5153.55	29.41	<.001
Infestation	-101.4	8.279	-118.6	-84.25	-12.25	<.001

FIGURE 33.9: Output for the sorghum data, including the CIs for the regression parameters.

Example 33.9 (Cyclones). Using the software output (Fig. 33.7) for the cyclone data, $\text{s.e.}(b_1) = 0.404$, so the approximate 95% CI for the regression slope β_1 is

$$-2.23 \pm (2 \times 0.404) \qquad \text{or} \qquad -2.23 \pm 0.808.$$

The interval from -3.04 to -1.42 is likely to straddle the population slope. This approximate CI is very similar to the exact CI shown in Fig. 33.7.

33.4.3 Regression: t-tests for regression parameters

Since the regression line describing the relationship between x and $\hat{y}$ is computed from one of countless possible samples, any relationship observed in the sample may be due to sampling variation; possibly, no relationship actually exists in the population (i.e., $\beta_1 = 0$). In other words, a hypothesis test can be conducted for the slope to determine if sampling variation can explain the discrepancy between β_1 and b_1. (Similar hypothesis tests can be conducted for testing if the intercept is zero, but are usually of less interest.)

The null hypothesis for tests about the slope is the usual 'no relationship' hypothesis. In this context, 'no relationship' means that the slope is zero (Sect. 33.3.5), so the null hypothesis (about the *population*) is H_0: $\beta_1 = 0$. A slope of $\beta_1 = 0$ is equivalent to *no relationship* between the variables. (We would also find $\rho = 0$.)

For the sorghum data, the RQ implies these hypotheses about the slope:

$$H_0\colon \beta_1 = 0 \quad \text{and} \quad H_1\colon \beta_1 < 0.$$

The parameter is β_1, the population slope for the regression equation predicting yield from infestation percentage. The alternative hypothesis is one-tailed, based on the RQ.

Assuming the null hypothesis is true (i.e., that $\beta_1 = 0$), the possible values of the sample slope b_1 can be described (Def. 33.2).

For the sorghum data, the variation in the sample slope across all possible samples when $\beta_1 = 0$ is described (Fig. 33.10) using:

- an approximate normal distribution,
- with a sampling mean whose value is $\beta_1 = 0$ (from H_0), and
- a standard deviation of $\text{s.e.}(b_1) = 8.279$ (from software; Fig. 33.9).

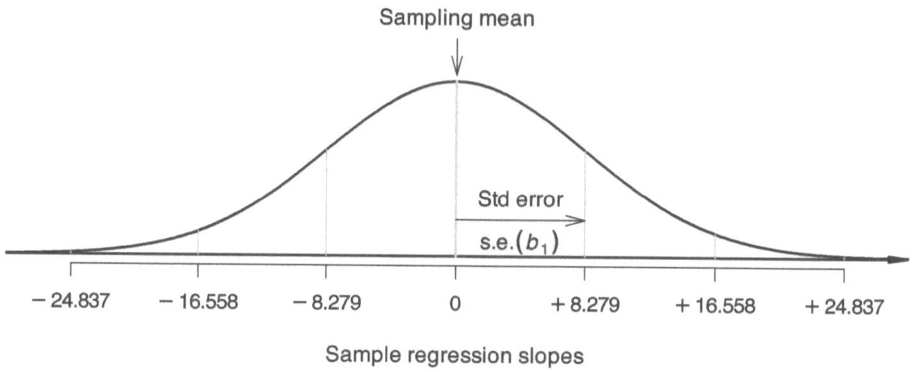

FIGURE 33.10: The distribution of sample slopes for the sorghum data, if the population slope is $\beta_1 = 0$.

The *observed* sample slope for the sorghum data is $b_1 = -101.4$. Locating this value on Fig. 33.10 shows that it is *very* unlikely that any of the many possible samples would produce such a slope, just through sampling variation, if the population slope really was $\beta_1 = 0$. The *test statistic* is found using the usual approach when the sampling distribution has an approximate normal distribution, using a t-score:

$$\begin{aligned} t &= \frac{\text{observed value} - \text{mean of the distribution of the statistic}}{\text{std deviation of the distribution of the statistic}} \\ &= \frac{b_1 - \beta_1}{\text{s.e.}(b_1)} = \frac{-101.4 - 0}{8.279} = -12.25, \end{aligned}$$

where the values of b_1 and $\text{s.e.}(b_1)$ are taken from the software output (Fig. 33.9). This t-score is the same value reported by the software.

To determine if the statistic is *consistent* with the null hypothesis, the P-value can be approximated using the 68–95–99.7 rule, approximated using tables, or taken from software output (Fig. 33.9). Since $t = -12.25$, the P-value will be very small; software shows the *two*-tailed P-value is $P < 0.001$ (so the one-tailed P-value is $P < 0.0005$).

We write:

> The sample presents very strong evidence ($t = -11.23$; one-tailed $P < 0.0005$) that, in the population, the yield of sorghum decreases as infestation percentage increases (slope: -101.4; 95% CI from -84.3 to -118.6; $n = 24$).

Notice the three features of writing conclusions: an answer to the RQ; evidence to support the conclusion (a t-score and P-value); and sample summary information (including a CI).

⚠ The P-value for a test of H_0: $\rho = 0$ will be the same as the P-value from a test of H_0: $\beta_1 = 0$. The test are effectively equivalent, both testing if the relationship observed in the sample can be explained by sampling variation.

Example 33.10 (Hypothesis testing). For the cyclone data (Example 33.7), the RQ is:

> In the Australian region, is there a relationship between ONI and the number of cyclones?

This RQ implies these hypotheses:

$$H_0\colon \beta_1 = 0 \quad \text{and} \quad H_1\colon \beta_1 \neq 0.$$

From the output (Fig. 33.7), $t = -5.52$ and the P-value is small: $P < 0.0001$. We write:

> The sample presents very strong evidence ($t = -5.52$; two-tailed $P < 0.0001$) that, in the population, the number of cyclones is related to the ONI (slope: -2.23; 95% CI from -3.04 to -1.42; $n = 37$).

33.5 Statistical validity conditions

As usual, these results hold under certain conditions. The conditions for which the CIs and tests are statistically valid are:

1. The relationship is approximately linear (necessary for the (Pearson) correlation coefficient and regression line to be appropriate).
2. The variation in the response variable is approximately constant for all values of the explanatory variable.
3. The sample size is at least 25.

The sample size of 25 is a rough figure; some books give other values. The units of analysis are also assumed to be *independent* (e.g., from a simple random sample).

If the relationship non-linear but is increasing-only or decreasing-only, alternatives to the Pearson correlation coefficient include the Spearman or Kendall correlation coefficients [Conover, 2003]. Depending on which statistical validity conditions are not met, other regression-like options may be available. For example, generalised linear models [Dunn and Smyth, 2018] may be appropriate for some non-linear relationships and/or relationships with non-constant variation in y.

Example 33.11 (Statistical validity). For the sorghum data, the scatterplot (Fig. 33.1, left panel) shows the relationship is approximately linear, so using a (Pearson) correlation coefficient and a regression line is appropriate. For the hypothesis test, the variation in yield doesn't seem to be obviously getting consistently larger or smaller for heavier infestations, and the sample size is only just smaller than 25 (with $n = 24$). The CIs and tests are very likely to be statistically valid.

Example 33.12 (Cyclones). The scatterplot for the cyclone data (Fig. 33.5) shows the relationship is approximately linear, that the variation in the number of cyclones seems reasonably constant for different values of the ONI, and the sample size is larger than 25 ($n = 37$). The CIs (Examples 33.1 and 33.9) and the tests (Example 33.2 and 33.10) are statistically valid.

33.6 Example: removal efficiency

In wastewater treatment facilities, air from biofiltration is passed through a membrane and dissolved in water, and is transformed into harmless by-products. The removal efficiency y (in %) may depend on the inlet temperature x (in °C). Chitwood and Devinny [2001] asked:

> In treating biofiltration wastewater, is the removal efficiency linearly associated with the inlet temperature?

The scatterplot of the $n = 32$ observations was shown (and described) in Sect. 16.6, and repeated here (Fig. 33.11); the relationship is positive and approximately linear.

The output (Fig. 33.12) shows that the sample correlation coefficient is $r = 0.891$ (with a 95% CI from 0.79 to 0.95), and so $R^2 = (0.891)^2 = 79.4\%$. This means that the unexplained variation in removal efficiency reduces by about 79.4% by knowing the inlet temperature.

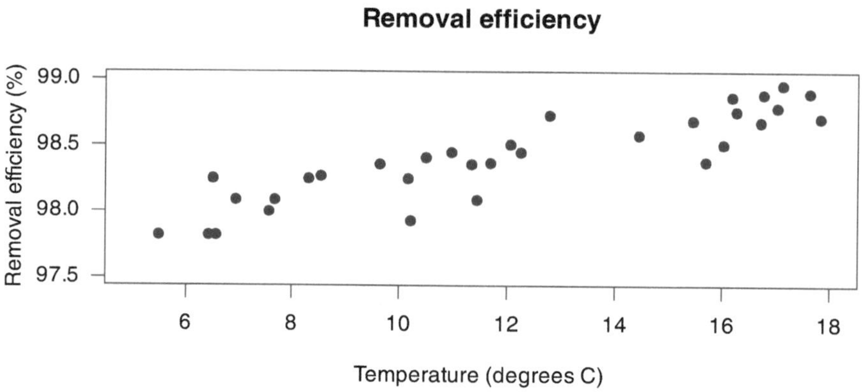

FIGURE 33.11: The scatterplot showing the relationship between removal efficiency and inlet temperature.

Correlation Matrix

		Removal	Temp
Removal	Pearson's r	—	
	df	—	
	p-value	—	
	95% CI Upper	—	
	95% CI Lower	—	
Temp	Pearson's r	0.8909	—
	df	30	—
	p-value	< .0001	—
	95% CI Upper	0.9458	—
	95% CI Lower	0.7865	—

Model Coefficients - Removal

Predictor	Estimate	SE	t	p
Intercept	97.4986	0.0889	1096.1693	< .0001
Temp	0.0757	0.0070	10.7425	< .0001

FIGURE 33.12: The software output exploring the relationship between removal efficiency and inlet temperature.

As always, the RQ is about the parameter, the correlation between the removal efficiency and inlet temperature in the population ρ. To test if a linear relationship exists in the population, write:

$$H_0\colon \rho = 0 \quad \text{and} \quad H_1\colon \rho \neq 0.$$

The alternative hypothesis is two-tailed (as implied by the RQ). The software output (Fig. 33.11, right panel) shows that $P < 0.001$.

The scatterplot of the data (Fig. 33.11) shows the relationship is approximately linear, so a regression line is appropriate. From the software output (Fig. 33.12), $b_0 = 97.5$ and $b_1 = 0.0757$; hence

$$\hat{y} = 97.5 + 0.0757x$$

for x and y defined above. The slope quantifies the relationship, so we can test

$$H_0\colon \beta_1 = 0 \qquad \text{and} \qquad H_1\colon \beta_1 \neq 0.$$

From the output, $t = 10.7$ which is huge; the P-value is small as expected: $P < 0.001$. The output does not include the CI for the slope, but since s.e.$(b_1) = 0.0070$, the *approximate* 95% CI is

$$0.0757 \pm (2 \times 0.0070), \quad \text{or} \quad 0.0757 \pm 0.0140.$$

We write:

> Very strong evidence exists ($t = 10.7$; $P < 0.001$) that inlet temperature is linearly related to removal efficiency (slope: 0.0757; approximate 95% CI: 0.0616 to 0.0898).

The CI and test are statistically valid: the relationship is approximately linear, the variation in y is approximately constant for all values of x, and $n = 32$.

33.7 Chapter summary

The CI for the correlation coefficient is found from software output. These steps are used to test a hypothesis about a correlation between two variables in the population, ρ.

- Write the null hypothesis (H_0: $\rho = 0$) and the alternative hypothesis (H_1); initially *assume* the value of ρ in the null hypothesis to be true (usually zero).
- Find the P-value for the test from software.
- Use the P-value to make a decision, and write a conclusion.
- Check the statistical validity conditions.

Regression mathematically describes the relationship between two *quantitative* variables: the response variable y, and the explanatory variable x. The linear relationship between x and y (the *regression equation*), in the sample, is

$$\hat{y} = b_0 + b_1x,$$

where b_0 is a number (the *intercept*), b_1 is a number (the *slope*), and the 'hat' above the y indicates that the equation gives a *predicted mean* value of y for a given x-value. Software provides the values of b_0 and b_1.

The *intercept* is the predicted mean value of y when the value of x is zero. The *slope* is how much the predicted mean value of y changes, on average, when the value of x *increases* by 1.

The regression equation can be used to make *predictions* or to *understand* the relationship between the two variables. Predictions made with values of x outside the values of x used to create the regression equation (called *extrapolation*) may not be reliable.

To compute a CI for the population slope of a regression equation β_1, software provides the standard error of b_1; then, the CI is

$$b_1 \pm \big(\text{multiplier} \times \text{s.e.}(b_1)\big).$$

The *margin of error* is (multiplier × standard error), where the multiplier is 2 for an approximate 95% CI (using the 68–95–99.7 rule).

These steps are used to test a hypothesis about a population slope β_1:

- Write the null hypothesis (H_0: $\beta_1 = 0$) and the alternative hypothesis (H_1); initially *assume* the value of β_1 in the null hypothesis to be true.
- Describe the *sampling distribution*, which describes what to *expect* from the sample slope under this assumption: under certain statistical validity conditions, the sample slope varies with:
  - an approximate normal distribution,
  - with sampling mean whose value is $\beta_1 = 0$ (from H_0), and
  - having a standard deviation of s.e.(b_1).
- Compute the value of the *test statistic*:

  $$t = \frac{b_1 - \beta_1}{\text{s.e.}(b_1)},$$

  where b_1 is sample slope.
- The t-value is like a z-score, and so an approximate *P-value* can be approximated using the 68–95–99.7 rule, or found using software. Use the P-value to make a decision, and write a conclusion.
- Check the statistical validity conditions.

33.8 Quick review questions

Telford and Cunningham [1991] examined the relationship between the height and weight of $n = 37$ rowers at the *Australian Institute of Sport* (AIS; Fig. 33.13). The regression equation is $\hat{y} = -138 + 1.2x$, and $P < 0.0001$ for the two-tailed P-value for a test of the correlation.

Are the following statements *true* or *false*?

1. The x-variable is the height of the rower.
2. Since the P-value is small, the correlation is quite strong.
3. The relationship is a *positive* relationship.
4. Based on the scatterplot, 'weight of the rower' is considered the y-variable.
5. Using the rise-over-run idea, a very rough estimate of the value of the slope is 1.2.
6. The measurements units for the slope are kg.
7. The measurements units for the intercept are kg.
8. The standard error of the slope is 0.112, so the value of the *test statistic* to test if the population slope is zero is $t = 10.7$.
9. The P-value for this test will be *very small.*
10. Predicting the mean weight of a 220 cm-tall rower would be *extrapolation.*

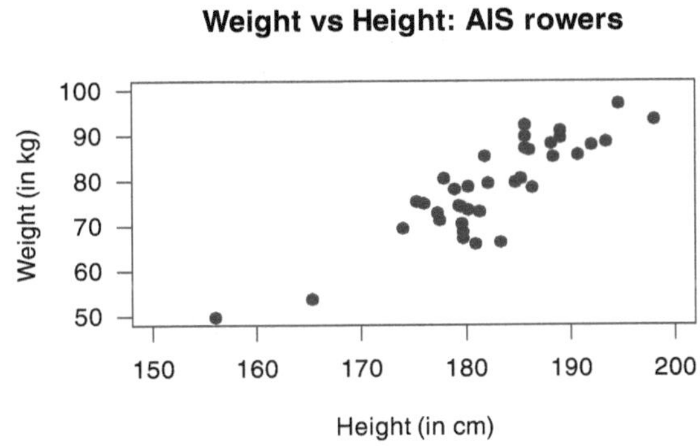

FIGURE 33.13: Scatterplot of weight against height for rowers at the AIS.

Select the correct answer:

11. What does the 'hat' above the y mean?

    a. That the weights are not measured accurately.
    b. That the weights are population values.
    c. That the regression model gives *poor* estimates.
    d. That the regression model gives *good* estimates.
    e. That the regression model estimates the weight for a given height.
    f. That the regression model estimates the mean weight for a given height.

12. What mean weight is predicted for a rower who is 180 cm tall?
    a. −24 624 kg; **b.** 78 kg; **c.** 138 kg.

33.9 Exercises

Answers to odd-numbered exercises are given at the end of the book.

Exercise 33.1. For each of the plots in Fig. 33.14, where appropriate:

1. estimate the value of r (this is hard!).
2. estimate the intercept of the regression line.
3. estimate the slope of the regression line, using the rise-over-run idea.
4. write down the estimated regression equation.

Exercise 33.2. [*Dataset*: `Throttle`] Amin and Mahmood-ul-Hasan [2019] measured the throttle angle (x) and the manifold air pressure (y), as a fraction of the maximum value, in gas engines.

1. The value of r is given in the article as 0.972986604. Comment on this, and what it means.
2. Comment on the use of a regression model, based on the scatterplot (Fig. 33.15; reconstructed from Amin and Mahmood-ul-Hasan [2019]).
3. The authors fitted the following regression model: $y = 0.009 + 0.458x$. Identify errors that the researchers have made when giving this regression equation.
4. Critique the researchers' approach.

Exercise 33.3. In a correlation analysis, the researchers find that $P = 0.0002$. Which (if any) of these statements are *consistent* with this P-value?

1. $r = 0.89$.
2. $r = -0.891$.
3. $r = 0.04$.
4. $r = -0.06$.

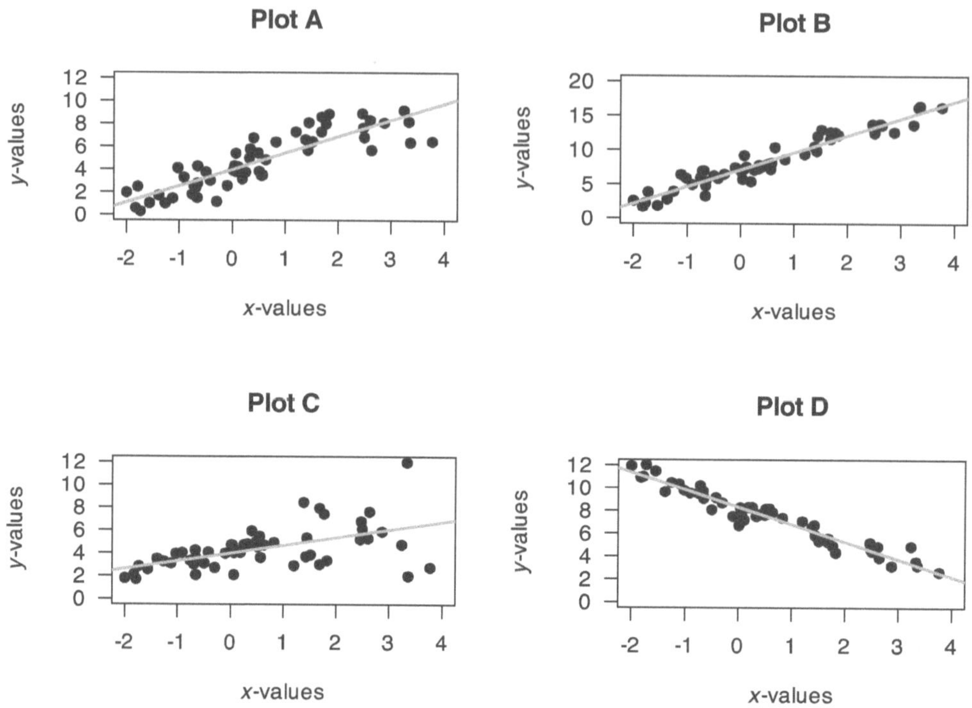

FIGURE 33.14: Four scatterplots.

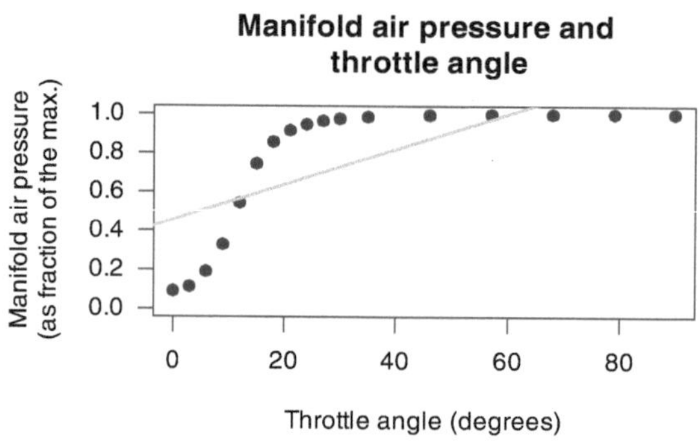

FIGURE 33.15: Manifold air pressure and throttle angle for an internal-combustion gas engine.

Exercise 33.4. In a correlation analysis, the researchers find that $r = 0.36$. Which (if any) of these statements are *consistent* with this value of the correlation coefficient?

1. The P-value is very small.
2. The P-value is very large.
3. The P-value is 0.36.
4. The P-value is 0.36^2, or 13%.

Exercise 33.5. For each regression equation below, identify the values of b_0 and b_1.

1. $\hat{y} = 3.5 - 0.14x$.
2. $\hat{y} = -0.0047x + 2.1$.
3. $\hat{y} = -25.2 - 0.95x$.
4. $\hat{y} = -0.22x + 0.15$.

Exercise 33.6. For each regression equation below, identify the values of b_0 and b_1.

1. $\hat{y} = -1.03 + 7.2x$.
2. $\hat{y} = -1.88x - 0.46$.
3. $\hat{y} = 201x + 16$.
4. $\hat{y} = 3.04x - 0.032$.

Exercise 33.7. Draw the regression line $\hat{y} = 5 + 2x$ for values of x between 0 and 10.

1. Add some points to the scatterplot such that the correlation is approximately $r = 0.9$.
2. Add some more points to the scatterplot such that the correlation is approximately $r = 0.3$.

Exercise 33.8. Draw the regression line $\hat{y} = 20 - 3x$ for values of x between 0 and 5.

1. Add some points to the scatterplot such that the correlation is approximately $r = -0.95$.
2. Add some more points to the scatterplot such that the correlation is approximately $r = -0.2$.

Exercise 33.9. LeBlanc et al. [2005] studied $n = 30$ paramedicine students, using *correlations* to study the relationship between the amount of stress experienced while performing drug-dose calculations (measured using the State–Trait Anxiety Inventory, STAI), and length of work experience.

1. Write the hypotheses for testing if a relationship exists between the STAI score and the length of work experience.
2. The article gives the correlation coefficient as $r = 0.346$ and $P = 0.18$. What do you conclude?
3. What must be *assumed* for the test to be statistically valid?

Exercise 33.10. Einsiedel et al. [2024] used *correlations* to study the relationship between amount of pesticide residue reported on a variety of fresh fruits and vegetables, and various weather measurements. One pesticide studied was perchlorate.

1. Write the hypotheses for testing if a relationship exists between the perchlorate residue and *maximum* temperature at the growing location.
2. The article gives the correlation coefficient as $r = -0.059$ and $P = 0.035$. What do you conclude?
3. Write the hypotheses for testing if a relationship exists between the perchlorate residue and *minimum* temperature at the growing location.
4. The article gives the correlation coefficient as $r = -0.025$ and $P = 0.365$. What do you conclude?
5. What must be *assumed* for the tests to be statistically valid?

Exercise 33.11. [*Dataset*: `SDrink`] A study examined the time taken to deliver soft drinks to vending machines [Montgomery and Peck, 1992] using a sample of size $n = 25$ (Fig. 33.16, left panel). To test if a linear relationship exists, are the statistical validity conditions met?

Exercise 33.12. [*Dataset*: `Mandible`] Royston and Altman [1994] examined the mandible length and gestational age for $n = 167$ foetuses from the 12th week of gestation onward (Fig. 33.16, right panel). To test if a linear relationship exists, are the statistical validity conditions met?

Exercise 33.13. Heerfordt et al. [2018] studied the relationship between the time (in minutes) spent on sunscreen application x, and the amount (in g) of sunscreen applied y, using $n = 31$ people. The fitted regression equation was $\hat{y} = 0.27 + 2.21x$.

1. Interpret the meaning of b_0 and b_1. Do they seem sensible?
2. What are the units of measurement for the slope and intercept?
3. According to the article, a hypothesis test for testing H_0: $\beta_0 = 0$ produced a P-value *much* larger than 0.05. What does this mean?
4. For people who spend 8 mins applying sunscreen, how much sunscreen would they use, on average?
5. The article reports that $R^2 = 0.64$. Interpret this value.
6. What is the value of the correlation coefficient?

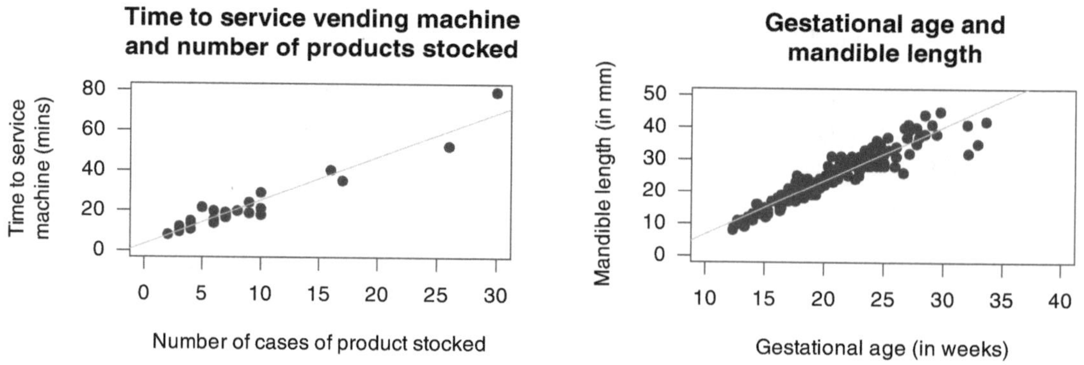

FIGURE 33.16: Two scatterplots. Left: the time taken to deliver soft drinks to vending machines. Right: gestational age and mandible length. In both plots, the solid line displays the linear relationship.

Exercise 33.14. Bhargava et al. [1985] stated (p. 1 617):

> In developing countries [...] logistic problems prevent the weighing of every newborn child. A study was performed to see whether other simpler measurements could be substituted for weight to identify neonates of low birth weight and those at risk.

One relationship they studied was between infant chest circumference (in cm) x and birth weight (in grams) y. The regression equation was given as:

$$\hat{y} = -3440.2403 + 199.2987x.$$

The correlation coefficient was $r = 0.8696$ with $P < 0.001$.

1. Critique the way in which the regression equation and correlation coefficient are reported.
2. Based on the *correlation* information, could chest circumference be used as a useful predictor of birth weight? Explain.
3. Interpret the intercept and the slope of the regression equation.
4. What are the units of measurement for the intercept and slope?
5. Predict the mean birth weight of an infant with a chest circumference of 30 cm.

Exercise 33.15. [*Dataset*: `Dogs`] Quan et al. [2017] studied Phu Quoc Ridgeback dogs (*Canis familiaris*), and recorded many measurements of the dogs, including body length and body height. The scatterplot displaying this relationship and the software output are shown in Fig. 33.17. In this example, it does not matter which variable is used as x or y.

1. Describe the relationship.
2. *Taller* dogs might be expected to be *longer*. To test this, write the hypotheses in terms of correlations.
3. Perform the test, using the output. Write a conclusion.
4. Is the test statistically valid?

Exercise 33.16. [*Dataset*: `Soils`] The *California Bearing Ratio* (CBR) value is used to describe soil sub-grade for flexible pavements (such as in the design of air field runways). Talukdar [2014] examined the relationship between CBR and other properties of soil, including the plasticity index (PI, a measure of the plasticity of the soil). The scatterplot and software output from 16 different soil samples from Assam, India, are shown in Fig. 33.18.

1. Describe the plot in words
2. Find and interpret the value of R^2.
3. Write down the CI for the correlation coefficient.
4. Conduct a hypothesis test for ρ.
5. Would the test be statistically valid?

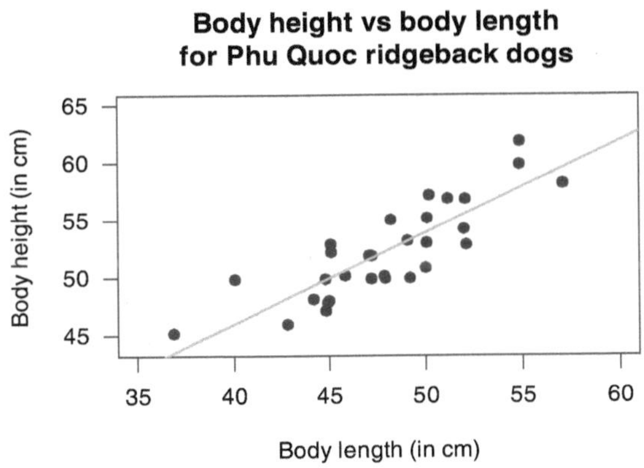

Correlation Matrix

		BH	BL
BH	Pearson's r	—	
	df	—	
	p-value	—	
	95% CI Upper	—	
	95% CI Lower	—	
BL	Pearson's r	0.8374	—
	df	28	—
	p-value	< .0001	—
	95% CI Upper	0.9201	—
	95% CI Lower	0.6832	—

FIGURE 33.17: Phu Quoc ridgeback dogs. Left: a scatterplot of the body height vs length. Right: software output.

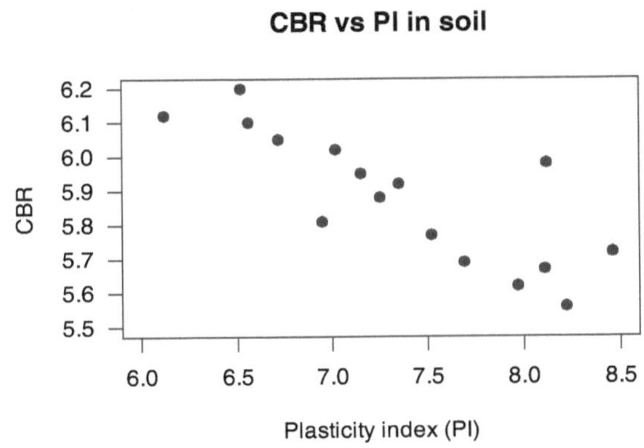

Correlation Matrix

		PI	CBR
PI	Pearson's r	—	
	df	—	
	p-value	—	
	95% CI Upper	—	
	95% CI Lower	—	
CBR	Pearson's r	−0.8190	—
	df	14	—
	p-value	0.0001	—
	95% CI Upper	−0.5442	—
	95% CI Lower	−0.9351	—

FIGURE 33.18: The relationship between CBR and PI in 16 soil samples. Left: scatterplot. Right: software output.

Exercise 33.17. [*Dataset*: `OSA`] de Carvalho et al. [2020] studied obstructive sleep apnoea (OSA) in 60 adults with Down Syndrome. The response variable y is OSA severity. The explanatory variable x is the average number of episodes of sleep disruption (according to specific criteria) per hour of sleep, the Respiratory Event Index (REI). One RQ is:

> Among Down Syndrome adults, is there a linear relationship between REI and neck size?

The data are plotted in Fig. 33.19 (left panel).

1. Using the software output (Fig. 33.19), determine the value of r.
2. Interpret the value of R^2.
3. Write down the values of the intercept and the slope, and hence the regression equation.
4. Explain what the slope in the regression equation means.
5. Find an approximate 95% CI for the slope.
6. Perform a hypothesis to test if a relationship exists between the variables.
7. Are the test and CI statistically valid?

Exercise 33.18. [*Dataset*: `EDpatients`] Brunette et al. [1991] studied the relationship between the number of emergency department (ED) patients and the number of days following the distribution of monthly welfare monies, from 1986 to 1988 in Minneapolis, MN (Fig. 33.20).

1. Write down the estimated regression equation.
2. Interpret the slope in the regression equation.

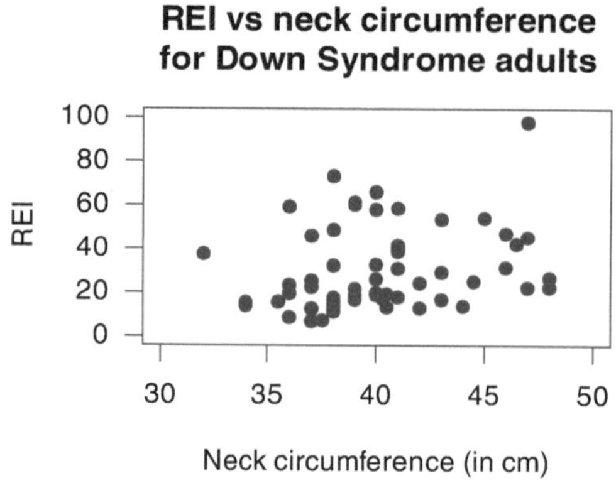

Model Fit Measures

Model	R	R²
1	0.26440	0.069907

Model Coefficients - REI

Predictor	Estimate	SE	t	p
Intercept	-24.4728	26.37586	-0.92785	0.35733
Neck	1.3663	0.65441	2.08790	0.04121

FIGURE 33.19: Neck circumference vs REI for Down Syndrome adults. Left: scatterplot. Right: software output.

3. Find an approximate 95% CI for the slope.
4. Conduct a hypothesis test for the slope, and explain what the result means.
5. What is the value of the correlation coefficient?

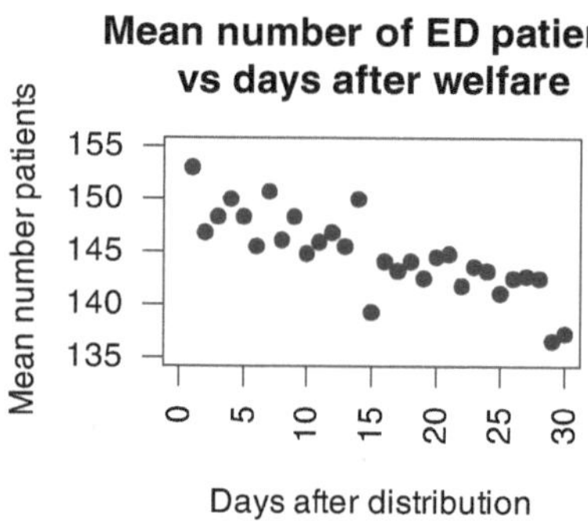

Model Coefficients - Mean ED patients

Predictor	Estimate	SE	t	p
Intercept	150.18575	0.828561	181.2610	< .00001
Days since welfare	-0.34790	0.046672	-7.4541	< .00001

FIGURE 33.20: The number of emergency department patients, and the number of days since distribution of welfare. Left: scatterplot. Right: software output.

Exercise 33.19. [*Dataset*: **Bitumen**] Panda et al. [2018] made $n = 42$ observations of hot mix asphalt, and measured the volume of air voids and the bitumen content by weight (Fig. 33.21).

1. Describe the plot in words.
2. For the data, $R^2 = 99.29\%$. Determine, and interpret, the value of r.
3. Write down the regression equation using the software output.
4. Interpret what the regression equation means.
5. Perform a test to determine if there is a relationship between the variables.
6. What is the P-value for testing H_0: $\rho = 0$?
7. Predict the mean percentage of air voids by volume when the percentage bitumen is 5.0%. Do you expect this to be a good prediction? Why or why not?
8. Predict the mean percentage of air voids by volume when the percentage bitumen is 6.0%. Do you expect this to be a good prediction?
9. Would the test be statistically valid?

Exercise 33.20. [*Dataset*: **Possums**] Williams et al. [2022] studied Leadbeater's possums in the Victorian Central Highlands. They recorded, among other information, the body weight of the possums (in g) and their location, including the elevation (in m; `DEM`), as shown in Fig. 33.22.

1. The value of R^2 is 23.0%. Determine, and interpret, the value of r.
2. Write down the regression equation.
3. Determine if there is a relationship between the possum weight and the elevation.

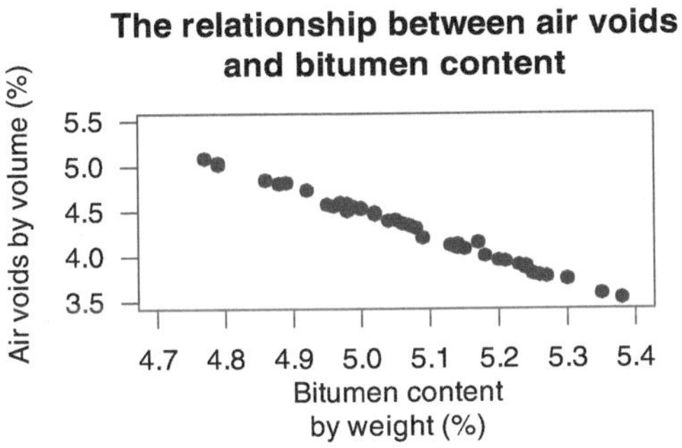

Model Coefficients - AirVoids

Predictor	Estimate	SE	t	p
Intercept	17.47	0.1757	99.4	< .001
Bitumen	-2.59	0.0346	-74.9	< .001

FIGURE 33.21: Air voids in bitumen. Left: scatterplot. Right: software output

4. What is the P-value for a test of H_0: $\rho = 0$?
5. Interpret the meaning of the slope.
6. Predict the mean weight of male possums at an elevation of 1 000 m. Do you expect this to be a good prediction? Why or why not?
7. Predict the mean weight of male possums at an elevation of 200 m. Do you expect this to be a good prediction? Why or why not?

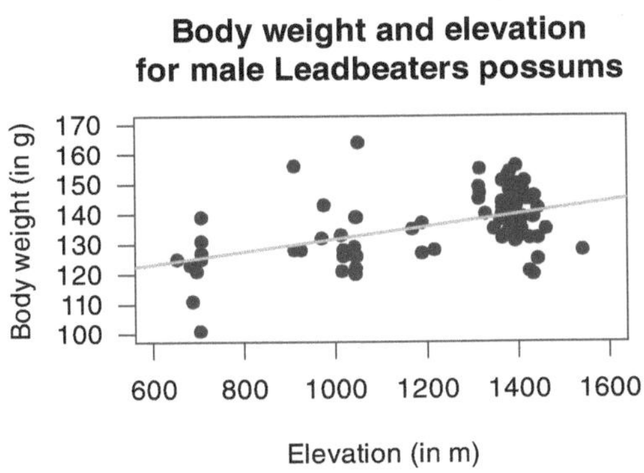

Model Coefficients - Wgt

Predictor	Estimate	SE	t	p
Intercept	110.5867	5.34802	20.68	<.001
DEM	0.0210	0.00427	4.92	<.001

FIGURE 33.22: The relationship between weight of possums and the elevation of their location. Left: scatterplot. Right: software output.

Exercise 33.21. [*Dataset*: **Gorillas**] Wright et al. [2021] examined 25 gorillas and recorded their chest-beating rates and size (the breadth of the gorillas' backs). The relationship is shown in Fig. 33.23. Use the software output (Fig. 33.24) to study the relationship.

1. Determine the value of r and R^2.
2. Perform a hypothesis test for the slope, and write a conclusion.
3. Find the regression equation.

Exercise 33.22. [*Dataset*: **Corollas**] On 25 June 2014, I searched *Gum Tree* (an Australian online marketplace), for **Toyota Corolla** in the 'Cars, Vans & Utes' category. I recorded the age and the price of each (second-hand) car from the first two pages of results that were returned.

I restricted the data to cars less than 14 years old at the time, removed one 13-year-old Corolla advertised for sale for $390 000, then produced the scatterplot in Fig. 33.25 (left panel).

1. Describe the relationship displayed in the graph, in words.
2. What else could influence the price of a second-hand Corolla besides the age?
3. Consider a seven-year-old Corolla selling for $15 000. Would this be cheap or expensive? Explain.
4. As stated, I removed one observation: a 13-year-old Corolla for sale at $390 000. What do you think the price was meant to be listed as, by looking at the scatterplot? Explain.

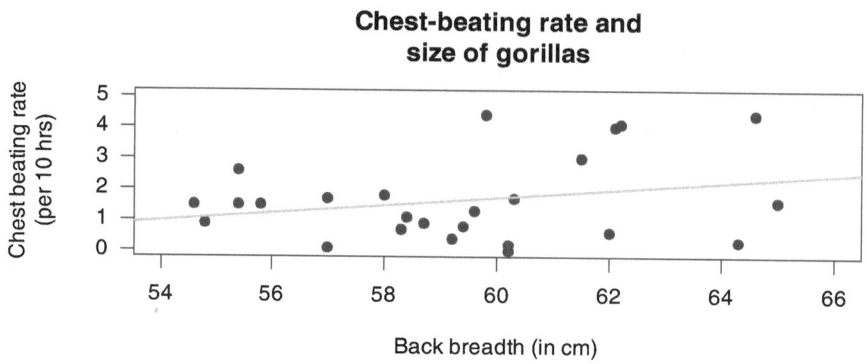

FIGURE 33.23: The scatterplot for the chest-beating data.

Correlation Matrix

		ChestBeatRate	BackBreadth
ChestBeatRate	Pearson's r	—	
	p-value	—	
BackBreadth	Pearson's r	0.271	—
	p-value	0.190	—

Model Coefficients - ChestBeatRate

Predictor	Estimate	SE	t	p
Intercept	−5.647	5.4008	−1.05	0.307
BackBreadth	0.123	0.0909	1.35	0.190

FIGURE 33.24: Software regression output for the gorilla data.

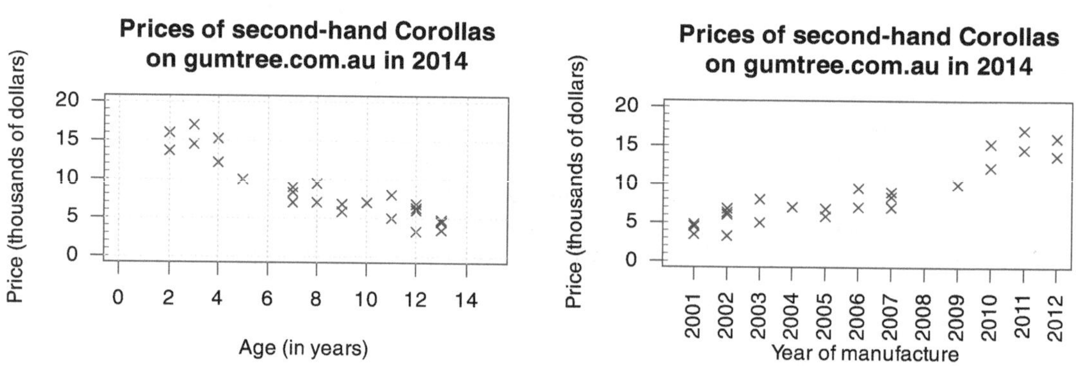

FIGURE 33.25: The price of second-hand Toyota Corollas ($n = 38$) as advertised on *Gum Tree* on 25 June 2014, plotted against age (left) and year of manufacture (right).

5. With a ruler or another straight edge (such as a book), draw an estimate of the regression line on the scatterplot. Then, *estimate* the value of b_0 (the intercept) from the line you drew. What does this mean? Do you think this value is meaningful?
6. *Estimate* the value of b_1 (the slope) from the line you drew. What does this mean? Do you think this value is meaningful?
7. From the line you drew above, write down an *estimate* of the regression equation.
8. What are the units of the intercept and the slope?
9. Use the software output (Fig. 33.26) relating the price (in thousands of dollars) to age to write down the regression equation.
10. Using the software output, write down the value of r. Using this value of r, compute the value of R^2. What does this mean?

Correlation Matrix

		PriceThous	Age
PriceThous	Pearson's r	—	
	df	—	
	p-value	—	
Age	Pearson's r	−0.922	—
	df	33	—
	p-value	<.0001	—

Model Coefficients - PriceThous

Predictor	Estimate	SE	t	p
Intercept	16.406	0.666	24.616	<.0001
Age	−0.958	0.070	−13.714	<.0001

FIGURE 33.26: The jamovi output, analysing the Corolla data

11. Use the regression equation from the software output to estimate the sale price of a Corolla that is 20-years-old, and explain your answer.
12. Using the software output, perform a suitable hypothesis test to determine if there is evidence that lower prices are associated with older Corollas.
13. Compute an approximate 95% CI for the population slope (use the software output).
14. I could have drawn a scatterplot with Price on the vertical axis and Year of manufacture on the horizontal axis (Fig. 33.25, right panel). For this graph:
    a. What is the value of the correlation coefficient?
    b. How would the value of R^2 change (if at all)?
    c. How would the value of the slope change (if at all)?
    d. How would the value of the intercept change (if at all)?

Exercise 33.23. [*Dataset*: **Elephants**] Weighing elephants is not easy due to their size. Height (to the shoulder) is easier to measure, and may be a useful proxy for the mass [Lalande et al., 2022a]. Two scatterplots of some relevant data [Lalande et al., 2022b] are shown in Fig. 33.27.

1. Which graph do you think is for males and which for female elephants? Explain.
2. Which plot has a correlation coefficient closest to one? Explain.
3. Use software to find the correlation coefficients for each sex.
4. For which sex is the height likely to be better for estimating mass? Explain.
5. Use software to find the regression equations for predicting mass from height (one for each sex).
6. Test to confirm the relationship between mass and height, for each sex.
7. Use the regression lines to predict the mass of an elephant with a height of 225 m, for each sex.
8. Discuss the statistical validity conditions.

Exercise 33.24. [*Dataset*: **Jeans**] Diehm and Thomas [2018] recorded data on the size of front pockets in men's and women's jeans. This exercise considers the correlation between the maximum widths and maximum heights of front pockets (Fig. 33.28).

1. The correlation for all jeans is $r = 0.38$, with $P = 0.00051$. What does this mean?
2. For men's jeans only, the correlation is $r = -0.09$, with $P = 0.59$. What does this mean?
3. For women's jeans only, the correlation is $r = 0.14$, with $P = 0.38$. What does this mean?
4. Compute the means for both variables for the combined data, for men's jeans only, and for women's jeans only.

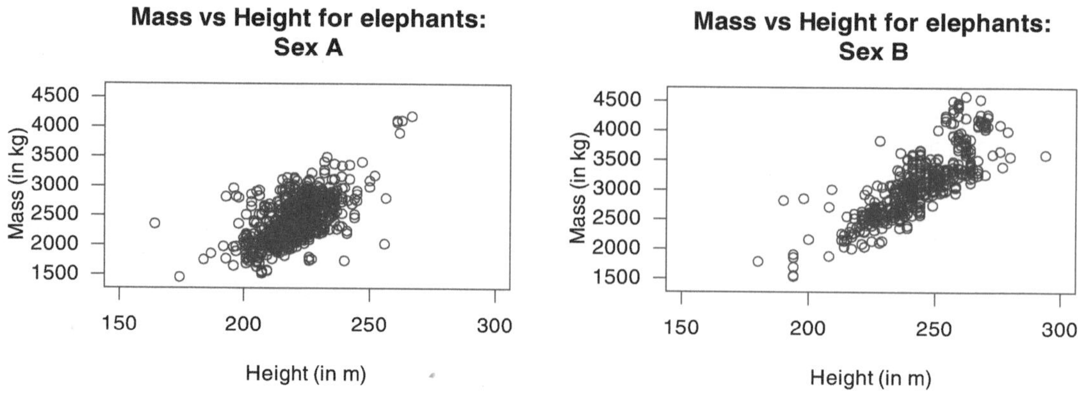

FIGURE 33.27: Mass and height of elephants.

5. From the last four questions, how would you describe the relationship between the maximum widths and maximum heights of the front pockets of jeans?

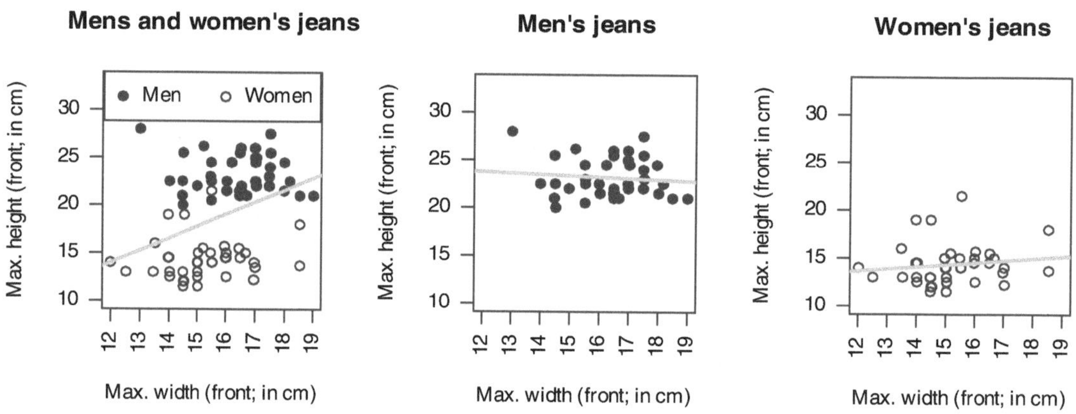

FIGURE 33.28: The relationships between minimum and maximum heights of front pockets for all jeans (left), men's jeans only (centre) and women's jeans only (right).

Exercise 33.25. [*Dataset*: `DogsLife`] The `DogsLife` dataset gives the average breed weight and average breed lifetime for 73 breeds of dogs. Determine if a relationship exists between breed weight and breed lifetime.

Exercise 33.26. [*Dataset*: `Typing`] The `Typing` dataset contains four variables: typing speed (`mTS`), typing accuracy (`mAcc`), age (`Age`), and sex (`Sex`) for 1 301 students [Pinet et al., 2022]. Is there evidence of a linear relationship between a person's mean typing speed and mean accuracy? Explain.

Answers to *Quick review* questions: 1. True. **2.** Not necessarily true. **3.** True. **4.** True. **5.** True (*very* roughly). **6.** False: kg/cm. **7.** True. **8.** True. **9.** True. **10.** True. **11.** f. **12.** b. 78 kg.

34

Selecting an analysis

So far, you have learnt about the research process, including analysing data using confidence intervals and conducting hypothesis tests. **In this chapter**, you will learn to:

- select the correct analysis.

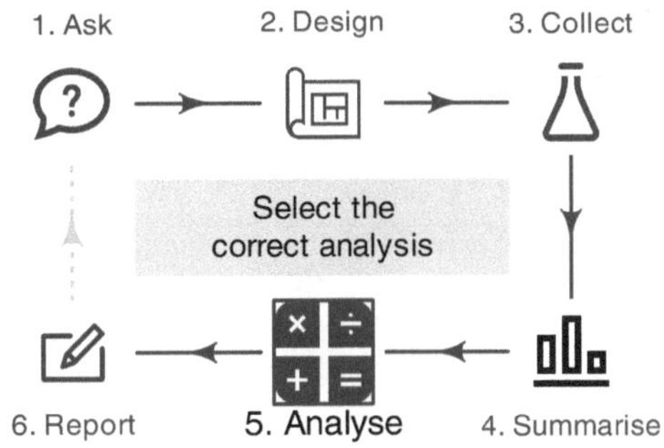

34.1 About selecting an appropriate analysis

Selecting the correct CI or hypothesis test can be challenging, and this book only describes a few possible scenarios. For the situations studied in this book, identifying the *type* of RQ (e.g., descriptive or correlational), and the *number* and *type* of variables (qualitative or quantitative) is important (Table 34.1). *Appendix C* may also prove useful.

Example 34.1 (Selecting an analysis). Bjornsson et al. [2021] studied whether the 'presence of a prehospital physician improves survival from cardiac arrest' (p. 227). They studied 471 cardiac arrests: 200 treated by prehospital physicians (2004 to 2007), and 271 treated by emergency medical technicians (2008 to 2014).

For each cardiac admission (the unit of analysis), two variables are recorded. *Whether a prehospital physician is present* (the explanatory variable) is qualitative with two levels (Yes; No). *Whether a patient survived* (the response variable) is qualitative with two levels (Yes; No). They compared the survival proportions for the two scenarios; this is a *relational RQ*.

To study the *proportion* of survivors for each scenario, a z-test for the difference between proportions (and corresponding CI) would be used. Alternatively, a χ^2-test for comparing the odds of survival (and a CI for the OR) could also be used.

Example 34.2 (Selecting an analysis). Lyons et al. [2023] studied the relationship between the ball release speed (BRS) and the height of female cricket players, and BRS and arm length. Since they are exploring the relationship between two *pairs of quantitative variables*, both RQs are correlational RQs.

If the relationships are approximately linear (determined by examining the two scatterplots), a test for correlations (and corresponding CI) would be used, for one each relationship to be studied. Alternatively, a linear regression model could be fitted (one for each relationship),

TABLE 34.1: Analysis scenarios studied.

Summarise		Analyse	
Graphical display	**Numerical summary**	**Confidence interval**	**Hypothesis test**
Descriptive RQ: proportion in one sample (i.e., one qualitative variable)			
Bar chart Pie chart Dot chart	Counts Percentages Odds	For one proportion	One-sample z-test for p
(Chap. 16)	(Chap. 16)	(Chap. 22)	(Chap. 26)
Descriptive RQ: mean of one sample (i.e., one quantitative variable)			
Histogram Stemplot Dot chart	Means, medians Std dev., IQR Outliers	For one mean	One-sample t-test for μ
(Chap. 11)	(Chap. 11)	(Chap. 23)	(Chap. 27)
Repeated-measures RQ: paired quantitative data			
Histogram of differences Case-profile	Mean, median of diffs. Std dev., IQR of diffs. Outliers, etc.	For mean difference	t-test for mean differences μ_d
(Chap. 13)	(Chap. 13)	(Chap. 29)	(Chap. 29)
Relational RQs: comparing quantitative variables			
Boxplot Dot chart Error bar chart	Diff. between means Std error of difference Summary of both groups	For difference between two means	t-test for difference between two means $\mu_1 - \mu_2$
(Chap. 14)	(Chap. 14)	(Chap. 30)	(Chap. 30)
Relational RQs: comparing qualitative variables			
Side-by-side bar Stacked bar Dot chart	Odds Odds ratio (OR) Proportions Percentages	For ORs For difference between two proportions	χ^2-test for OR z-test for difference between two proportions $p_1 - p_2$
(Chap. 15)	(Chap. 15)	(Chap. 31)	(Chap. 31)
Correlational RQs			
Scatterplot	Correlation R^2	For correlation For regression parameters	Correlation test t-test for regression parameters
(Chap. 16)	(Chap. 16)	(Chap. 33)	(Chap. 33)

and a test for the *slope* of the fitted regression equation (and corresponding CI) could be conducted.

Example 34.3 (Selecting an analysis). Hitt et al. [2023] studied the impact of soil lead levels in New Orleans (USA) neighbourhoods on northern mockingbirds (p. 2):

> We tested the hypothesis that nestling mockingbird lead levels in blood and feathers differ with respect to neighborhood soil lead levels...

They compared the mean lead concentration in blood, for birds in neighbourhoods with *low* and with *high* lead levels. They also compared the mean lead concentration in feathers, for birds in neighbourhoods with *low* and *high* lead levels. These are both *relational RQs*.

For each bird (the unit of analysis), three variables were recorded. The *lead levels* (the explanatory variable) is qualitative with two levels: high lead-level neighbourhoods, and low lead-level neighbourhoods. The *blood lead concentrations* (one response variable) is quantitative (continuous). The *lead concentrations in feathers* (another response variable) is quantitative (continuous).

To study the difference between the mean lead concentrations in the two groups, a two-sample t-test for the difference between the means (and the corresponding CI) is needed (provided the statistical validity assumptions are met). One test is needed for comparing blood lead concentrations, and another for comparing concentrations in feathers.

34.2 Exercises

Answers to odd-numbered exercises are given at the end of the book.

Exercise 34.1. Identify which of these statistics *do not* have a sampling distribution well-modelled by a normal distribution. Explain your answer.

1. The difference between two sample means $\bar{x}_1 - \bar{x}_2$, with samples of size $n_1 = 55$ and $n_2 = 61$, but slightly right-skewed distributions of the data for each sample.
2. The sample slope in a regression equation b_1, with an approximate linear relationship between the variables, approximately constant variation in the values of y, and $n = 24$.
3. The sample OR, with both samples of size $n = 43$.

Exercise 34.2. Identify which of these statistics *do not* have a sampling distribution well-modelled by a normal distribution. Explain your answer.

1. The sample mean of a set of differences $\bar{d}$, with a sample of $n = 32$ difference, but the distribution of the differences are slightly right-skewed.
2. The sample correlation coefficient r, with an approximate linear relationship between the variables, approximately constant variation in the values of y, and $n = 29$.
3. The sample proportion $\hat{p}$, with $n = 26$ and $\hat{p} = 0.154$.

Exercise 34.3. Suppose researchers compare the mean number of hours of exercise per week for the same British office workers, both in summer and in winter, to study the mean change.

What methods would be a suitable for creating a summary and performing analyses?

Exercise 34.4. Castro-Maqueda et al. [2019] estimated the difference between the mean number of hours of sunlight exposure per day for physical education teachers and non-physical education teachers in Spain.

What methods would be a suitable for creating a summary and performing analyses?

Exercise 34.5. Suppose researchers wanted to study the proportion of koalas that live in regions with tree canopies of different heights (classified as Class 1 (highest canopy height) to Class 4 (lowest canopy height)) in the 'core' areas (areas of intensive use and feeding) and non-core areas (based on Mitchell et al. [2023]).

What methods would be a suitable for creating a summary and performing analyses?

Exercise 34.6. Chen et al. [2018] studied the relationship between the mass and the length of crocodile eggs.

What methods would be a suitable for creating a summary and performing analyses?

Exercise 34.7. Meadley et al. [2021] studied the *relationship* between maximal aerobic capacity (VO_{2peak}) while swimming, and the maximal aerobic capacity while running, in helicopter rescue paramedics.

What methods would be a suitable for creating a summary and performing analyses?

Exercise 34.8. Suppose researchers are wanting to *estimate* the difference between the mean number of hours spent on social media for Indian people aged over 30, to people aged 30 and under.

What methods would be a suitable for creating a summary and performing analyses?

Part VII

Reporting and reading research

35

Reporting and writing research

So far, you have learnt the about the process of research: asking an RQ, designing a study, collecting data, describing and summarising the data, and analysing the data. **In this chapter**, you will learn to:

- report research effectively and clearly.
- appropriately structure research writing.

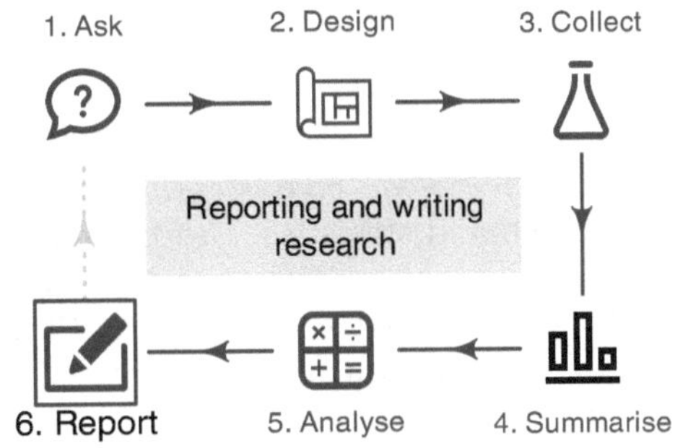

35.1 Introduction

Research needs to be effectively communicated and shared, so the results can be used, evaluated and built on by others. The purpose of writing about research is to effectively and clearly communicate.

Research may be shared using face-to-face or online presentations (Sect. 35.4) or written documents (Sect. 35.5). The style and expectations vary widely between these two formats, between disciplines, and even between journals in the same discipline. Hence, this chapter gives general comments about writing, rather than specific requirements. Formal guidelines for writing about research exist, for both *experimental* [Hopewell et al., 2022] and *observational* studies [Von Elm et al., 2007], though we will not discuss these specifically. Since different disciplines and journals have their own styles, read articles from your discipline or target journal for examples of the required style and formatting.

35.2 General writing advice

The purpose of writing about research is to effectively and clearly communicate the research. With this in mind, some general advice is given below.

- *Write carefully and precisely*. Use simple, clear but technically-correct language. Use the Glossary if necessary. Carefully choose every word you use to ensure it conveys the correct and intended meaning.

  Oppenheimer [2006] concluded, from experiments, that students often believe that using fancy words makes them appear smarter. However, he recommended students 'write clearly and simply if you can, and you'll be more likely to be thought of as intelligent' (p. 153).

- *Use correct spelling, grammar, punctuation and formatting.* Use (but do not rely upon) a spell checker and grammar checker; use a dictionary. Specifically:
  - do *not* confuse similar words (there/their/they're; your/you're; affect/effect; chose/choose; etc.).
  - capitalise correctly.
  - use apostrophes correctly. For example, *it's* is only ever an abbreviation for *it is*.
- *Be inclusive.* Unless specifically referring to men or women, use inclusive language (e.g., 'fire-fighter', not 'fireman'; 'nurse' rather than 'male nurse').
- *Take care using comparative terms.* For example, writing 'this treatment is *better*' must be clarified. Better than *what*? And 'better' in what sense: cost? ease of use? patient outcomes?
- *Use terminology consistently.* Different words may be used for the same concept in research and statistics. Use the term that is common in your discipline; most of all, be consistent.
- *Be clear, concise and complete.* Place material in an Appendix (Sect. 35.5.8) if it will interrupt the flow of the narrative. Often, this material can be made available online if too lengthy in printed form.
- *Ensure pronouns clearly identify the nouns they refer to.* For example, consider this sentence: 'When the weeds and crops were sprayed, its growth rate reduced by 80%'. The word *its* may refers to the growth rate of the weeds, the crops, or both.
- *Ensure verbs and nouns agree.* Both the nouns and verbs in a sentence should be singular or plural. For example, 'the rats *was* weighed' should be 'the rats *were* weighed'. Usually, 'data' is considered plural ('datum' is the singular; 'dataset' is also singular), so write 'the data *were* right skewed' rather than 'the data *was* right skewed' (but the latter use is becoming more common). In any case, be consistent.
- *Avoid leaps of logic, and reaching conclusions unsupported by the evidence.* Ensure your conclusions are consistent with the evidence in the study.

  For example, a student project found that the proportion of provisional drivers (those yet to get a full licence) was *higher* in the free university car park, compared to paid car parks. They concluded that provisional drivers seek to 'save money by parking in free car parks'. This *may* be true, but is not supported by the evidence. The evidence simply shows a difference in proportions, but does not explain *why*.
- *Present the facts in an unbiased manner, and avoid promoting personal opinions.* For example, do not describe results as 'exciting'. Because academic writing generally shuns personal opinions, writing in third person ('the fertiliser was applied') is usually (but not always) preferred over writing in first person ('I applied the fertiliser').

Writing well is difficult; editing can be painful; revising is time-consuming. Revise your document carefully as many times as necessary; having someone else read and comment on your writing can be useful.

⚠ Many authors have stated variations of this phrase:

> Don't write so that you *can* be understood; write so that you *can't* be misunderstood.

Be unambiguous: say what you mean, and mean what you say.

Example 35.1 (Write what you mean). A student project at my university asked:

> Are dark-coloured car owners more likely to park undercover?

They actually meant:

> Are drivers of dark-coloured cars more likely to park undercover?

Don't just be understood; avoid being *mis*understood!

35.3 Ethics when writing

As always, ethical practice is important (Sect. 5), including when writing about research. Some relevant issues are given below.

- *Producing reproducible research.* When possible, research should be *reproducible* (Sect. 5.3). This includes describing the protocol, and making available any data (when possible; sometime this is not ethical or permitted) and any instructions or code used to analyse the data.
- *Authorship.* Ensure everyone who has made an intellectual contribution is listed as an author. Brand et al. [2015] suggests authorship be considered for those involved with:
  - conceptualisation.
  - methodology.
  - software.
  - data analysis.
  - investigation.
  - resourcing.
  - data curation.
  - creating images or taking photographs.
  - writing, including writing drafts, reviewing and editing.
  - visualization.
  - supervision.
  - project administration.
  - funding acquisition.
- *Acknowledgements.* An optional *Acknowledgements* section is used to acknowledge research funding bodies, and people who supported the research. Avoid writing 'The authors would like to thank...'; instead, thank them: 'We thank...'. Reviewers of the article, when appropriate (who are almost always volunteers), are usually thanked also.
- *Use of artificial intelligence (AI).* Any use of AI in the study should be disclosed. This includes using AI *during* the research (e.g., generating figures or research design) or when writing *about* the research. The description should indicate where AI was used, which AI systems (such as ChatGPT) were used, and how they were used. AI also may make mistakes, so any material generated using AI should be verified by the authors.
- *Plagiarism.* Writing about research almost always refers to, and builds on, others' work: to formulate the research question, to establish ideas and to explain the background of the research. However, *plagiarism* (using other people's words and ideas without acknowledgement) *must* be avoided. All sources used when writing research should be acknowledged.

  Plagiarism is a serious offence: it is theft of intellectual property. *Do not plagiarise*; use quotes if necessary and cite the work of others as needed. Plagiarism applies to words, text, images, photographs, ideas, etc.

Example 35.2 (Plagiarism). Shamim [2014] published an article to discourage plagiarism. Later, the article was retracted because parts of the article were plagiarised.

35.4 Preparing presentations

Presentations are often used to share progress reports of research, or give an overview of completed research. They are used at conferences, workshops, and progress meetings, and may be given to peers, stakeholders, funding bodies, small groups of other researchers, or work teams. Presentations should be adapted to suit the time allocated and the audience: a conference presentation to your research peers should be different from a presentation to a progress meeting.

Presentations are mostly a *verbal* (speaking) and *visual* (preparing slides) medium.

As a *verbal* medium, speak slowly, clearly, loudly, and with expression. Use eye contact, and practice beforehand. Ensure you keep to your allocated time. Ensure technical or unusual words are pronounced correctly; aids to correct pronunciation of many unfamiliar terms have been given in this book.

As a *visual* medium, presentations usually omit technical details and give the audience an overview of the major points and processes; sharing tedious technical details is unlikely to produce an engaging presentation. Presentations usually focus on the *why* and the *what* of the research. Presentations may encourage audience members to learn more by reading your written documents (Sect. 35.5).

Presentations also tend to use graphs, images, short sentences, and minimal text. Presentation software encourages the use of fancy fonts, transitions and animations, but these are usually more distracting than informative; avoid. Ensure your fonts and colours are readable from a distance (especially in tables and graphs).

Using bullet points on slides, while common, is not necessary; short sentences are fine. Slides should *not* contain information that you simply *read* to the audience; a good presenter adds important details around the structure provided by information on the slides. The slides *guide*, but do not have to *tell*, the story of your research.

35.5 Writing articles

Written documents are more likely to be formally written and prepared than presentations. Unlike presentations, written documents tend to provide details of *how* the research was conducted. Written documents may be journal articles, progress reports, reports to stakeholders, or funding applications; these are all referred to as 'articles' in what follows, for brevity.

⚠ Journal articles, and most other written documents too, should contain sufficient details so that other professionals can repeat the study (Chap. 9.2); i.e., the research should, as far as possible, be reproducible (Sect. 5.3).

Articles usually have a more formal structure than presentations. Sometimes the acronym AIMRaD is used to remember these sections:

- ***A****bstract*.
- ***I****ntroduction*.
- ***M****ethods*.
- ***R****esults*.
- ***D****iscussion* (or *Summary*, or *Conclusions*).

These components capture the six-step research process used in this book (Fig. 35.1).

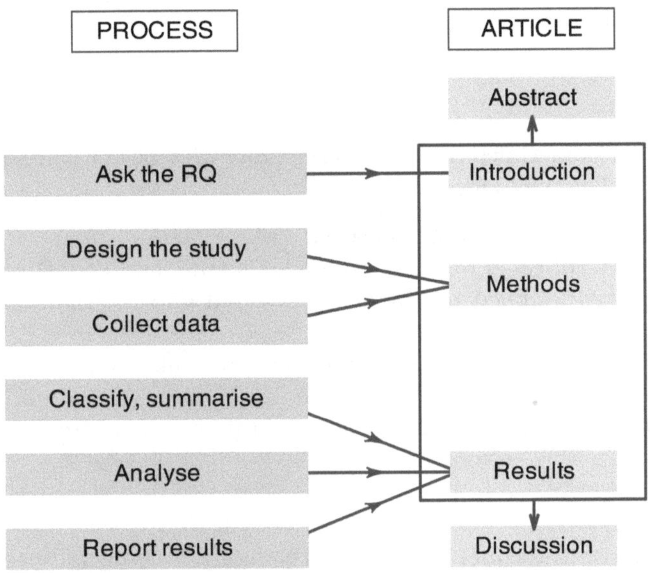

FIGURE 35.1: The connection between the article and the steps studied. The *Abstract* briefly covers all aspects of the study, and the *Discussion* explains what has been learnt through the process, and discusses the results.

35.5.1 Article titles

Titles are important: poor titles can discourage a reader from reading an article. A title should clearly describe the main purpose of the article. Sometimes this is achieved by posing questions in the title ('Do warning lights and sirens reduce ambulance response times?'; Brown et al. [2000]) or providing answers in the title ('No harm from five year ingestion of oats in coeliac disease'; Janatuinen et al. [2002]).

As far as possible, avoid overly-specific technical language and uncommon abbreviations in the title.

35.5.2 Abstract

The *Abstract* (or *Summary*, or *Overview*) is a short section at the start of an article summarising the *whole* article, including the results; it is *not* an introduction. The *Abstract* is the most important part of any article: it is the only part that many people will read. Some journals require a *structured abstract*, with specific sub-headings (for example, *Introduction*, *Methods*, *Results*, and *Conclusion* (or *Discussion*)).

35.5.3 Introduction

The purpose of the *Introduction* is to:

- show how the research fills a gap in existing knowledge, by discussing existing literature (sometimes the *Literature review* forms a separate section).
- gain the interest of readers, and encourage them to read more of the article.
- establish the context and background.
- define the language, acronyms and definitions used in the study.
- introduce the theoretical groundwork of the subject.
- state the purpose of the article: why it was written, and what the authors hope to learn.
- summarise the structure of what follows.

35.5.4 Methods

The *Methods* section (sometimes called *Materials and Methods* or similar) explains how the data were obtained. This includes:

- how the *sample* was identified and located.
- how the data were *collected* from the individuals (the data collection *protocol*).
- how the study was designed to maximise external and internal validity, and manage confounding.
- how the data were *analysed*, including the software (and version number) used, and the statistical methods used.
- what specialised equipment was used (pencils, rulers, paper, etc. are not listed).

35.5.5 Results

The *Results* summarise the conclusions from the analysis, especially regarding the initial RQ. The *Results* section:

- shows all the relevant findings from the research.
- presents a summary of the data: the number of observations, the number of missing values, and a verbal description of important variables.
- presents tabular, numerical and/or graphical summaries of the data and relationships of importance.
- gives a brief verbal interpretation of these summaries.
- gives the results from any hypothesis tests and CIs.
- identifies trends, consistencies, anomalies, etc.
- does *not* provide interpretations or explanations of the results (that is the purpose of the *Discussion*).

Unless the dataset is small, the data itself is usually not given (though may appear in an Appendix or online).

⚠ Cutting-and-pasting software output into reports is rarely acceptable, except for carefully-prepared graphs (Chap. 17; Sect. 35.6.1).

35.5.6 Discussion

No new information should be presented in this section. This section:

- summarises the results.
- gives a short evaluation of the results.
- answers the stated RQ (i.e., makes a conclusion).
- discusses limitations (Chap. 8), strengths, weaknesses, problems, challenges.
- tries to anticipate and respond to potential questions about the research.

Readers should reach the conclusions based on the *evidence* presented. Sometimes, articles have separate *Conclusion* and *Discussion* sections; sometimes they are combined.

35.5.7 Referencing

The *References* (or *Bibliography*) section gives the full citations of any work referenced, in the required format (such as APA, Harvard, etc.). Most journals have strict guidelines for how references should be listed and formatted (which must be followed).

35.5.8 Appendices

Sometimes an *Appendix* is included, which contains important material that would otherwise break the flow of the article's narrative. The *Appendix* may include large tables, data, images, discussions of technical details, mathematical development, and so on. Sometimes, this material is placed online.

35.6 Specific advice

35.6.1 Constructing tables, graphs and images

Good figures and tables take time and care to prepare (Chap. 17). Their purpose should always be to *display information in the simplest, clearest possible way*, and to highlight the important information. In general, tables, graphs and images:

- *should* be discussed (not just presented) and referred to in the text.
- *should* be clear and uncluttered.
- *should* include units of measurement (such as kg or inches) where appropriate.
- *should* be able to be understood without reference to the article, as far as possible.
- *should* use easy-to-read fonts and colours: for example, ensure the font size is sufficiently large when placed in the article at the final size.
- *should* avoid using different colours, line types or fonts unless these have a purpose (i.e., to differentiate between groups in the study); if they are used, their purpose should be explained (e.g., using a figure legend or caption).
- *should not* include *chart junk* (unnecessary additions), such as artificial third dimensions for graphs (Sect. 17.2) and unnecessary lines in tables.

Figures and images typically have captions *below*, while tables typically have captions *above*. The source of images (e.g., the photographer) should be acknowledged (as ethical practice), when appropriate. Tables should use very few horizontal lines, and no vertical lines.

35.6.2 Presenting numbers

Any numbers presented should be rounded appropriately. Software may report more decimal places (or more significant figures) than necessary. When appropriate, ensure units of measurement are given.

Be consistent and careful with decimal numbers. Some journals require numbers to be written with a leading zero (e.g., $P = 0.024$), and some do not (e.g., $P = .024$). Counts are usually written in words when fewer than ten (or sometimes twelve), and otherwise presented using digits. However, usually numbers are written in full when starting a sentence ('Thirty-seven people volunteered').

> ⚠ Numbers taken from software output may need to be sensibly rounded before being included in a report (including in tables and graphs), and units of measurement added.

35.6.3 Lexically ambiguous words

Readers should not be able to misinterpret your meaning, so write *carefully* and *precisely*. One potential source of confusion is words with a different meaning in research compared to every-day use or in other disciplines (called *lexical ambiguity*; Dunn et al. [2016]). Some specific words where care is needed are given below.

- *Average*: struct Use the specific word 'mean' or 'median' when that is what you intend.
- *Confidence*: In research, 'confidence' is usually used in the phrase 'confidence interval' (Sect. 24.4). Take care when using 'confidence' in other contexts to avoid confusion.
- *Control*: In research, a 'control' is usually used in the context of a control group (Def. 2.17), but may have other meanings in other disciplines.
- *Correlation*: In research, correlation describes the (often linear) relationship between two *quantitative* variables (Sect. 16.4.1). In general usage, 'correlation' may mean any 'association' between any two variables.
- *Estimate*: In research, 'estimating' means to *calculate* an estimate for an unknown population parameter using sample information. In general usage, 'estimate' often means to make an educated guess.
- *Experiment*: In research, an experiment is a specific type of research study (Sect. 4.4). The word 'study' can be used to talk about research more generally.
- *Independent*: This word has many different uses in statistics and research, in science, and in general usage. The word 'independent' in this book refers to events that do not impact each other in a probabilistic sense (Sect. 18.4).
- *Intervention*: In research, an 'intervention' (Sect. 2.7) is specifically when the researchers can manipulate the comparison.
- *Normal*: In research, 'normal' often refers to the 'normal distribution' (Chap. 20.3). If this is *not* the meaning you intend to convey, consider using the word 'usual' or similar.
- *Odds*: In research, 'odds' has a specific meaning (Sect. 12.5) and is not the same as probability. In general usage, 'probability' and 'odds' are often used interchangeably.

- *Population*: In research, the 'population' refers to a larger group of interest (Def. 2.1). In general usage, 'population' often refers to groups of people specifically.
- *Random*: In research, 'random' has a specific meaning: based on impersonal chance. In general usage, it often means 'haphazard' or 'without structure'.
- *Regression*: In research, 'regression' refers to the mathematical (often linear) relationship between two quantitative variables (Chap. 33).
- *Sample*: In research, we usually have '*one* sample of 30 hyenas'; in some disciplines, this could be described as 'taking 30 samples of hyenas'.
- *Significant*: In research, 'significance' is usually understood to refer to 'statistical significance' (Sect. 28.6). If this is *not* the meaning you intend to convey, consider using the word 'substantial' or similar.
- *Variable*: In research, a 'variable' is a characteristic that can vary from individual to individual (Def. 2.9).

Some *symbols* may also have different meanings in other disciplines. Ensure the meaning of symbols and notation is clearly defined.

35.7 Chapter summary

Communicating research is a vital step in the research process. Writing clearly is important.

Presentations are a verbal and visual medium, and usually focus on the major points and conclusions rather than the *how*.

Written documents are usually formal, and include details of *what* was done. They should be written carefully and precisely, using the appropriate technically-correct words. Use short sentences for easier reading and omit unnecessary words.

Remember: 'Don't write so that you *can* be understood; write so that you *can't* be misunderstood'.

35.8 Quick review questions

Are these statements true or false?

1. Using long, obscure words makes writing sound more scientific.
2. Presentations generally focus on the details of how the study was done.
3. The *Introduction* should explain why the study was done.
4. Numbers should be given to as many decimal places as possible, for the greatest accuracy.
5. The design of the study should be explained in detail in the *Methods* section.

35.9 Exercises

Answers to odd-numbered exercises are given at the end of the book.

Exercise 35.1.

1. Select the correct word to use to complete this sentence: *to*, *too* or *two*?
   'Liquid fertiliser was applied [__________] pots each morning at 9am.'
2. Select the correct word to use to complete this sentence: *its* or *it's*?
   'Each kangaroo was observed for signs that [__________] tracking device caused discomfort.'
3. What are the problems with this sentence?
   'We took 50 samples of students; the average age of the 50 students was 26.2.'
4. What is the problem with this text?
   'The subjects are not blinded. Because the subjects would clearly know they were in a study.'

Exercise 35.2.

1. Select the correct word to use to complete this sentence: *there*, *their* or *they're*?
   'The subjects were told to eat [__________] snacks at about 8am.'
2. What is the problem with this text?
   'The sample of pedestrians were all taken on a Thursday.'
3. Select the correct word to use to complete this sentence: *affect* or *effect*?
   'The [__________] of the diet was to increase the blood pressure.'
4. What is the problem with this sentence?
   'The new formulation produces better concrete'.

Exercise 35.3.

1. Explain how this sentence can be misinterpreted, and write an improved version:
   'There was one rat in the cage that was male.'
2. Explain how this sentence can be misinterpreted, and write an improved version:
   'The research assistant recorded the pH of the lake water in the beaker after removing weeds.'

Exercise 35.4.

1. Explain how this sentence can be misinterpreted, and write an improved version:
   'Fertiliser was applied to one of the fields that was in liquid form.'
2. Explain how this sentence can be misinterpreted, and write an improved version:
   'The new diet lost more weight, on average, than the traditional diet.'

Exercise 35.5.

1. Explain how this statement can be improved:
   'A significant change in the weight gain of the pigs is expected to be found'.
2. Explain how this statement can be improved:
   'The data is summarised in Table 2.'

Exercise 35.6.

1. Explain how this statement can be improved:
   'There is a correlation between sex of the person and chance of contracting the disease'.
2. Explain how this statement can be improved:
   'The group were asked to sign a consent form.'

Exercise 35.7. Oyerinde et al. [2019] state (p. 1):

> The regression correlation coefficients of 0.999996066 and 0.999653453 were obtained for the temperatures and speeds respectively [as associated with the time the engine had been running].

What is the problem with this statement?

Exercise 35.8. David et al. [2007] published an article entitled 'Are patients with self-inflicted injuries more likely to die?' What is the problem with this title?

Exercise 35.9. In a student project, students compared the mean reading speed for people when reading text displayed in one of two different fonts. Their RQ was:

> Which font allows [...] students to read a pangram the fastest, between a default and what is considered to be a 'easy to read' font.

(A pangram is a sentence that uses every letter of the alphabet at least once.) In their *Abstract*, the conclusion was given as:

> The Georgia font [...] is therefore the faster of the two.

1. Explain why this is a poorly-worded RQ. Rewrite the RQ.
2. Explain what is wrong with the conclusion. Rewrite the statement.

Exercise 35.10. In a student project, the heights that students could jump vertically were compared, starting from a squat or standing position. Every student in the study performed both jumps. Critique the *numerical summary* produced by the research team (Table 35.1).

TABLE 35.1: A numerical summary of the data, showing how much higher the standing jump height is compared to the squat jump.

n	Mean	Standard deviation	Standard error	Confidence interval 95%	t value	P value
50	7.48	4.674	0.661	6.152 to 8.808	11.316	0.000

Exercise 35.11. The aim of a student project was 'to determine if the proportion of males and females that use disposable (coffee) cups on campus is the same'. The two variables observed on each person in the study were (a) whether the person used a disposable cup, and (b) the sex of the person. In reporting the results in their *Abstract*, the students state:

> Based on the sample results, the 95% confidence interval for the population mean number of disposable cups used by males and females is between 0.690 and 1.625. Meaning that the population mean is likely to fall between those two intervals.

Critique this statement.

Exercise 35.12. The aim of a student project was 'to determine if the average hang time is different between two types of paper plane designs'. The two variables in the study were: design type (Basic Dart; Hunting Flight), and the hang time of the flight of the plane (in seconds). In reporting the results in their *Abstract*, the students state:

> Very strong evidence proving a difference ($P = .000$) between the Basic Dart mean hang time (881.84 ± 140.73 ms) and the Hunting Flight mean hang time (1504.19 ± 699.86 ms). 95% CI for the means of The Basic Dart (829.29 – 934.39) and the Hunting Flight (1242.86 – 1765.52).

Critique this statement.

Answers to *Quick review* questions: 1. False. **2.** False. **3.** True. **4.** False. **5.** True.

36

Reading and critiquing research

So far, you have learnt the about the process of research: asking an RQ, designing a study, collecting data, describing and summarising the data, and analysing the data. You have learnt how to write about your research. **In this chapter**, you will learn to:

- read and critique research.

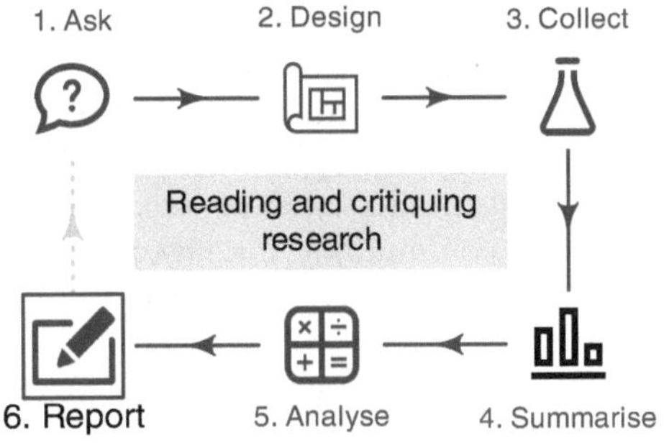

36.1 Introduction

All academic disciplines change and adapt. Staying current in your discipline requires reading, critiquing and understanding the research of others, as communicated in *journal articles* or *presentations* (Chap. 35). (*Critiquing* means to evaluate: identifying what is good, and what can be improved.)

At some time during your studies or employment, you will need to read research articles:

- to understand current practices in your discipline.
- to know *why* your discipline uses the current procedures and practices.
- to learn about new procedures and practices that may be adopted.
- to critique the evidence for current or new practices.
- to identify open or unresolved questions in your discipline.
- to help answer these open or unresolved questions.

Familiarity with the language and concepts of research is important for understanding these articles, even if you will not be conducting your own research.

Reading research articles can be challenging. Rather than reading articles thoroughly from start to finish, first read the *Abstract* to obtain an overview of the whole article, without becoming lost in the details. Then, read the *Discussion*, which highlights the important findings. Next, skim the rest of the article (perhaps focusing on graphs and tables of results). Finally, if necessary, read the article for details.

⚠ Terminology and notation varies widely in research (Sect. 35.6.3). When reading research, check the terminology and notation being used if you are unsure!

The six steps of the research process (Sect. 1.4) can guide the research critique.

1. *Asking the RQ.*

- What RQ is the research answering?
- Why is this RQ important?
- To what population will the results apply?
- What are the units of analysis and units of observation?
- How are important terms defined?

2. *Designing the study.*

- Is the study observational or experimental?
- Is the study well-designed? What is not explained or clear?
- What design features and used, and why?
- How many individuals are in the study?
- How was the sample obtained? What are the implications for external validity?
- Is the study designed to maximise internal validity?
- How is confounding managed?
- What are the design limitations?
- Are there ethical concerns?
- What is the source of funding? Are there potential conflicts of interest?

3. *Collecting the data.*

- How were the data collected?
- Are the necessary details provided so the study can be approximately replicated?

4. *Classifying and summarising the data.*

- Is the data summary appropriate, complete and clear?
- What does the data summary reveal about the data?
- What do the tables and graphs reveal about the data and relationships?

5. *Analysing the data.*

- What types of confidence intervals (CIs) and/or hypothesis tests were used?
- Is the analysis appropriate, accurate, valid and clear?
- What do the results mean?
- What software was used?
- Are the results statistically valid?

6. *Reporting the results.*

- What are the main conclusions, and how do they answer the RQ?
- Are the conclusion consistent with the results?
- Are the results accurate, appropriate and well-reported?
- Are the results of practical importance?
- Are the study limitations acknowledged, and their implications discussed?
- What other questions have emerged?

36.2 Example: walking while texting

Sajewicz and Dziuba-Słonina [2023] studied the impact of texting (on a smartphone) on how students walk (including walking speed). In this section, the article will be briefly discussed.

36.2.1 The abstract

Part of the unstructured abstract for the article reads:

> The aim of this experiment was to investigate whether using a cell phone while walking affects walking velocity [...] in young people. Forty-two subjects (20 males, 22 females; mean age: 20.74 ± 1.34 years; mean height: 173.21 ± 8.07 cm; mean weight: 69.05 ± 14.07 kg) participated in the study. The subjects were asked to walk on an FDM–1.5 dynamometer platform four times at a constant comfortable velocity and a fast velocity of their choice. They were asked to continuously type one sentence on a cell phone while walking at the same velocity. The results showed that texting while walking led to a significant reduction in velocity compared to walking without the phone.

As this is the *abstract*, many details are absent (but explained in the article itself). Nonetheless, a lot can be learnt about the study from the abstract.

- *Asking* the RQ:
  - this is a repeated-measures RQ: data are collected from the same students 'four times' (which are explained more fully in the article).
  - the *population* is 'young people'.
  - the numbers that follow the $\pm$ are not explained: are they CI limits, standard deviations, IQRs, ranges, standard errors?
  - the *units of analysis* are the students in the study, each with four measurements.
  - the main outcome is the (average; presumably mean) 'walking velocity'.
- *Designing* the study:
  - the sampling method is not stated, but likely to be voluntary-response.
  - the sample size is $n = 42$ (20 males; 22 females).
- *Analyse* the data:
  - a quantitative variable (walking velocity) is being compared *within* individuals, so paired t-tests are a likely method of analysis (Chap. 29).
- *Report the results*:
  - details of the analysis are not given (e.g., P-values or CIs).
  - nonetheless, the conclusion is that 'texting led to a significant reduction in velocity compared to walking without the phone'.

36.2.2 Introduction

The *Introduction* section introduces the context for the study, and establishes what is known about the topic. The *aim* of the study is (p. 1):

> ... to analyze how the use of a cell phone while walking at different velocities affects gait parameters, i.e., velocity, cadence, stride width, and stride length.

(Cadence refers to the tempo or rhythm of the walking, measured in steps per minute.)

36.2.3 Methods

The *Methods* (or *Materials and Methods*) section provides details of the study.

- The *sample* comprises students from the University School of Physical Education (Poland) studying a course in gait analysis. This sample may not represent any general population, though the conclusions may possibly apply to non-students and not-Poles.
- Exclusion criteria (e.g., using lower limb prostheses) and inclusion criteria (e.g., daily use of a cell phone while walking) were given.

- Extraneous variables collected included age, height, weight and sex.
- The study was ethical, with permission sought from the students, and approval given by the Senate Committee on Research Ethics at the university.
- Control variables included the temperature and humidity: 'the air temperature was constant at 22 degrees Celsius, and the air humidity was 47%' (p. 3).
- Details of the specialised equipment used was given: 'The experiment was conducted using an FDM–1.5 Zebris dynamography platform'.
- Further details of the protocol used were given.
- Each subject participated in four tasks. In each, the subject (without footwear) made as many passes on the platform as possible in one minute:
  1. The subject walked at a constant *comfortable* velocity (i.e., the velocity, chosen by the subject, that the person walks most naturally).
  2. The subject walked at a constant *fast* velocity (i.e., as fast as the subject could comfortably walk, chosen by the subject).
  3. Task 1 was repeated, with subjects continuously typing a sentence on a cell phone.
  4. Task 2 was repeated, with subjects continuously typing a sentence on a cell phone.
- Five response variables were used: left-side stride length, right-side stride length, cadence, stride width, and walking velocity. These appear to have been measured *objectively.*
- The 'sentences' to be typed were defined as 'tongue twisters... not used in everyday conversation', but were not provided.
- The analysis was given as 'paired Student's t-test' in most cases, or the Wilcoxon test if the statistical validity conditions were not satisfied.
- The software (and version) used was stated: TIBCO Statistica® 13.3.0 (StatSoft Poland).

36.2.4 Results

The *Results* section provides the results of the analyses.

- The data are not immediately available (probably due to ethics concerns) but 'are available upon request' (p. 7). Details of the analysis were not available.
- Case-profile plots were produced for the five response variables, showing the means for each task rather than for all 42 individuals (much like Fig. 13.2).
- Correlations were computed between the five response variables. The correlations between stride width and the other variables were negative; all other correlations were positive. Since the relationships were non-linear, Spearman correlations were used rather than the Pearson correlations studied in Chap. 33. All the corresponding P-values were less than 0.05 apart from the correlation between stride width and cadence.
- The following comparisons were made for each response variable:
  1. Task 1 and Task 3 were compared: this explored the impact of texting on a smartphone when walking at a comfortable velocity.
  2. Task 2 and Task 4 were compared: this explored the impact of texting on a smartphone when walking at a fast velocity.
  3. Task 3 and Task 4 were compared: this explored the difference between texting and not texting when walking at a fast velocity.
- The mean and standard deviation of the three response variables were provided for each task. However, the numerical summary for the *differences* were not provided; a P-value only was provided.
- Fifteen hypothesis tests were conducted: the three between-tasks comparison above, for the five response variables. This increases the chance of making a Type I error.

- 'Statistically significant' was defined as a P-value smaller than 0.017, rather than the commonly-used $P < 0.05$. The reason was to reduce the chance of making a Type I error (Sect. 28.7), but the details are beyond the scope of this book.
- The results are summarised as (p. 5):

> ... right and left single step length and gait velocity, were found to be statistically significant in each comparison. The value of the change in step width was statistically significant only when comparing trials 1 and 3, and cadence showed statistical significance when comparing trials 2 and 4, as well as 3 and 4.

- The researchers also made qualitative observations; they observed 'moments when the subject took their eyes off the phone in order to assess the direction of the path' (p. 5).

36.2.5 Discussion

The *Discussion* section makes the following observations.

- The researchers concluded that 'the use of a cell phone while walking significantly affects gait parameters, causing a decrease in walking velocity and a reduction in stride length' (p. 6), thus providing answers to the RQs.
- The researchers state (p. 6) that 'This proves that texting on a cell phone has a major impact on gait'. However, the research does *not* prove anything based on one of countless possible samples.
- The study had poor ecological validity, as the study 'was conducted in a measurement workshop room, the air temperature was constant at 22 degrees Celsius, and the air humidity was 47%'.
- The researchers listed limitations of the study, including:
  - that the sentence typed by the subjects was unchanged in both Trials 3 and 4; hence, Trial 4 may have been easier than the previous trial as the sentence was familiar (the carryover effect).
  - that the step width differed between comfortable walking speed and walking at the *same* speed with the phone. The researchers attribute this to the carryover effect, as 'walking with the phone at comfortable velocity and at fast velocity directly followed one after the other', and so 'subjects were able to respond to the new conditions by adapting to them during the first transition'.
- The researchers made recommendations: 'that the selection of the order of trials and the sentence to be typed on the cell phone be randomized' (p. 7).

Overall, the study was conducted well and reported well.

36.3 Chapter summary

The six steps of research can be used as a scaffold for critiquing research articles. Starting by reading the *Abstract* (or *Summary*, or *Overview*) for an overview, then the *Discussion*, and then skim the rest of the article (perhaps focusing on graphs and tables of results). If necessary, read the article for details.

36.4 Quick review questions

Are these statements true or false?

1. Reading an article thoroughly, from start to finish, is the best approach.
2. The six steps of research are a useful scaffold for critiquing an article.
3. Critiquing an article means to focus on finding all the problems.

36.5 Exercises

Answers to odd-numbered exercises are given at the end of the book.

Exercise 36.1. Duncan et al. [2018] examined the accuracy of step counts, as recorded on iPhones. The article states that participants

> ... were recruited through word of mouth and posters displayed around the [researcher's] university. Participants were eligible if they were ambulatory, $\geq$ 18 years of age, and owned an iPhone 6 [...] or newer model.

1. How would you describe the sampling method? What is the implication?
2. How would you describe the information given about the subjects needing to be ambulatory and 18 years of age or over?

Although 33 participants were selected, the authors note some parts of the study used a smaller sample size because one subject lost their phone, while others chose to withdraw from the study.

3. Why did the authors discuss these changes in sample size for some parts of the study?

The article notes that previous studies have been able to:

> ... demonstrate the accuracy of the iPhone pedometer function in laboratory test conditions. However, no studies have attempted to evaluate evidence [...] in the field.

4. Describe the issue that the authors raise with previous studies, using the language in this book.
5. Among many other comparisons, the researchers compared the *mean difference* between the number of step counts recorded by manually counting steps (mean: 92.6) and the iPhone-recorded number of steps (mean: 85.4). What statistical test would be appropriate?
6. What hypotheses are being tested?
7. While walking at 2.5 km.h^{-1}, the above statistical test resulted in $t = 2.95$. What is the approximate P-value? Interpret the results.
8. The sample size for the analysis mentioned above was $n = 32$. Is the test statistically valid?

Exercise 36.2. Mohammadpoorasl et al. [2019] studied the relationship between hearing loss, and headphone and earphone use in Iranian students, using a non-directional study. The article states:

> ... 890 students were randomly selected from five schools at QUMS [...] using a proportional cluster sampling method...

Only 866 of the 890 students agreed to participated in the study; of these, 745 used *ear*phones. The participants completed a hearing test and a Hearing Loss Questionnaire (HLQ; values between 17 and 34; higher scores indicating more severe hearing loss).

1. What is the population?
2. Is this an observational or experimental study?
3. Critique the sampling method. What is the implication for interpreting the results of the study?

One question in the HLQ is:

> Does a hearing problem cause you difficulty when listening to TV or radio?

4. What is a potential problem with this question?
5. Compute the 95% CI for the proportion of students who had used earphones.

Some results are presented in Table 36.1.

6. What statistical test was appropriate for comparing the mean scores for males and females?
7. What are the hypotheses being tested?
8. What is the standard error for the *difference* between the means?
9. Perform the hypothesis tests; what do the results mean?
10. Compute the approximate 95% CI for the difference between the means.
11. Are the test and the CI statistically valid?

Table 36.1 also compares the HLQ scores for the frequency of *earphone* use specifically.

12. What are the hypotheses being tested?
13. Why is the sample size for this comparison only 791 and not 845?
14. Interpret the *P*-value for this test; what do the results mean?

Table 36.1 also compares the HLQ scores for those who use and do not use *earphones*.

15. Form an approximate 95% CI for the mean hearing loss score for students who use earphones.
16. Compute the standard error of the *difference* between the mean hearing loss score for students who use and do earphones.
17. Perform a hypothesis test to compare the *difference* between the mean hearing loss score for students who use and do not use earphones, and confirm that the *P*-value is indeed very small.

TABLE 36.1: The Hearing Loss Questionnaire scores for various demographic variables.

	Levels	Sample size	HLQ Mean	HLQ Std dev.	*P*-value
Sex					
	Female	543	19.37	2.91	0.009
	Male	302	19.99	3.51	
Frequency of earphone use					
	0, 1 times/day	194	19.20	2.87	0.001
	2 to 3 times/day	319	19.60	2.66	
	More than 3 times/day	278	20.20	3.54	
Earphone use					
	Yes	745	19.80	3.08	< 0.001
	No	100	19.00	1.71	

Exercise 36.3. Mesrkanlou et al. [2023] studied the effect of an earthquake on pregnant mothers in Varzaghan, Iran (p. 2), using:

> ... 1000 cases of pregnant women living in urban and rural areas of Varzaghan city that consisted of 550 pre-earthquake and 450 post-earthquake cases.

The researchers compared the mothers in the two groups (pre- and post-earthquake) on various measurements. For example, the mean age of mothers in the *pre*-group was 25.82 y, and the mean age of the mothers in the *post*-group was 26.71 y; the difference has a *P*-value of 0.084.

1. What does this result *mean*?
2. *Why* did the researchers make this comparison between the mothers' ages in two groups?
3. What type of hypothesis test was used to make this conclusion?

The researchers also compared the mean birth weights of the babies born to the mothers in the two groups. In the *pre*-group, the mean birth weight was 3.25 kg ($s = 0.52$) and in the *post*-group the mean birth weight was 3.18 kg ($s = 0.54$).

4. Compute the standard error for comparing the *difference* between the two means.
5. Perform a hypothesis test to compare the mean birthweights. Interpret the results.
6. The two-tailed P-value for this test as given as 0.001. Is this consistent with your calculations?

The researchers also compared the percentage of babies with a Low Birth Weight (LBW; less than 2.5 kg). For the *pre*-group, the percentage was 6.01%; for the *post*-group, the percentage was 8.92%.

7. What *type* of definition is given for LBW?
8. Construct the 2×2 table for displaying these data.
9. What type of test was probably used for this comparison?
10. For the test, $\chi^2 = 3.052$. Deduce the equivalent z-score and the approximate P-value.
11. What limitations can you identify for this study?

Exercise 36.4. Tracy et al. [1990] studied the selenium (Se) concentration in irrigation and stock water sources in California. For drinking water, the maximum recommended concentration was $10\,\mu\text{g.L}^{-1}$; for irrigation water, the maximum recommended concentration was $20\,\mu\text{g.L}^{-1}$.

Part of the study examined the area within 5 km of wells. When Pliocene rocks were within this radius, the relationship between the Se concentration y in the water and the electrical conductivity of the water x (in deciSiemens per meter, dS.m^{-1}) was $\hat{y} = -3.1 + 7.0x$, where $R^2 = 27\%$.

1. Interpret the meaning of R^2.
2. What is the value of the correlation coefficient?
3. The P-value for testing the slope is $P < 0.001$. Interpret what this means in this context.
4. What are the measurement units of the slope?

For the $n = 151$ wells in the study, Table 36.2 shows the Se concentration of the water and the geology within 5 km of the well.

5. What hypotheses are being tested by the table?
6. The article states that $\chi^2 = 31.5$. What is the equivalent z-score for the test?
7. What is the approximate P-value for the test? Interpret what this means.

TABLE 36.2: Number of wells with dissolved selenium (Se) concentration above $2\,\mu\text{g.L}^{-1}$, and the geology within 5 km.

	Plioscene rocks present	**Plioscene rocks not present**
Se concentration $\le 2\ \mu\text{g.L}^{-1}$	78	15
Se concentration $> 2\,\mu\text{g.L}^{-1}$	23	35

Exercise 36.5. Russell [2023] compared the larvae of two types of mosquitoes: *Ae. albopictus* (invasive) and *Cx. pipiens* (native). One study compared the survival rates of the larvae at two temperatures, and in the presence or absence of predator (a small crustacean, called copepods).

At 15°C and 25°C, the survival rates were 86.8% and 86.1%, respectively, for the control group (i.e., no copepods). The papers quoted a P-value of $P = 0.8076$.

1. What type of test was probably used?
2. Interpret what the P-value means in this context.

The researchers also compared the size of the surviving larvae in the control groups for both temperatures. In comparing *Cx. pipiens* to *Ae. albopictus* larvae, the paper gives this information:

mean±SD: *Cx. pipiens* = 1.64±0.18 mm, *Ae. albopictus* = 1.36±0.13 mm; *p*-value =< .0001.

The two sample sizes are $n = 410$ and $n = 498$ respectively.

3. How would these results be interpreted?
4. What type of test would probably have been used?
5. Compute the standard error for the difference between the two types of mosquitoes.
6. Compute the t-score and approximate P-value for the test. What does the mean?
7. Is the P-value in the article consistent with your calculations?
8. Is the test statistically valid?

The length of the surviving larvae from both species were compared for the two temperatures also.. For surviving *Cx. pipiens*, the paper reports:

> mean ± SD: 15°C = 1.66 ± 0.01 mm, 25°C = 1.60 ± 0.02 mm; *p*-value = .0065.

For surviving *Ae. albopictus*, the paper reports:

> mean ± SD: 15°C = 1.35 ± 0.01 mm, 25°C = 1.36 ± 0.01 mm; *p*-value = .4343.

9. How would these results be interpreted?
10. What type of test would probably have been used?

Megacyclops viridis (a copepod) preys on the larvae. The association between predation efficiency (y; a percentage) and predator–prey size-ratio (x; no units) was (using $n = 45$) $\hat{y} = -19.56 + 31.64x$. The standard errors of the regression coefficients were 17.92 (intercept) and 13.88 (slope).

11. Find an approximate 95% CI for each regression parameter.
12. Estimate the P-value for testing if the population slope is zero. Interpret what this means.
13. Is this test statistically valid?
14. Interpret the meaning of the slope.
15. The value of R^2 was given as 0.087 (i.e., 8.7%). Interpret this value.
16. Find the value of the correlation coefficient, r.

Exercise 36.6. Li et al. [2017] studied the maximum mouth opening (MMO; in mm) for 452 Chinese adults aged from 20 to 35.

1. Would the individuals in the study have been blinded? Explain. What are the implications?

The correlation between height and MMO was given as $r = 0.54$ with $P < 0.001$.

2. What does this *mean*?
3. Compute and interpret the value of R^2.

The regression equation relating the height x (in cm) and MMO y was given as $\hat{y} = 0.36x - 10.15$.

4. Interpret the estimates of the regression parameters.
5. Use the regression equation to predict the MMO for a person 179 cm tall.

The mean MMO of males was 54.18 mm ($s = 5.21$), and for females was 49.62 mm ($s = 3.69$).

6. What *type* of hypothesis tests was used to compare the mean MMO for males and females?
7. The t-score for comparing MMO for males and females is $t = 10.63$. What is the P-value?
8. Is this result statistically valid?
9. What is the meaning of this comparison?
10. Is gender likely to be a confounding variable in this regression analysis? Explain carefully.

The authors state one of the limitations as:

> First, participants were recruited from a pool of people who were undergoing regular medical examinations in our hospital [...]

11. What does this mean? What are the implications? Are there other limitations?

Exercise 36.7. Drinkwater et al. [1995] compared tomatoes growing on conventional (CNV; $n = 14$) and organic (ORG; $n = 17$) farms. Between 1989 and 1990, the researchers sampled tomato fields during April and September (p. 1 100). An area between 0.04 and 0.1 ha was set aside within each field for collecting data. Each area was divided into 20 sections, then a 1 m row was selected at random within each section to be sampled.

1. Explain what type of sampling is being used.

One important measure of soil health is the number of actinomycetes. When comparing ORG and CNV, the researchers found that the (p. 1 103):

> ... total numbers of actinomycetes [...] were significantly larger in the ORG soils [...] (Student's t test, $t = 5.4$, $P = 0.006$)...

2. Explain what the results mean.
3. Are the results statistically valid?

The researchers also found that (p. 1 103):

> ... starch hydrolyzing actinomycetes were more numerous in CNV [...] (Student's t test, $t = 4.0$, $P = 0.005$).

4. Explain what these results mean.

They also found that (p. 1 103):

> Total actinomycete abundance [was] negatively correlated with corky root [a disease] severity ($r = -0.76$, $P = 0.08\ldots$).

5. Explain what these results mean.
6. Compute and interpret the value of R^2.

Exercise 36.8. Teo et al. [2022] studied pregnant Malaysian women with sleeping disruptions in the last month of pregnancy. The 56 patients were (p. 1):

> ... randomized to the use of eye-mask and earplugs or "sham" headbands during night sleep (both introduced as sleep aids).

Thus, two groups were used: one using eye-masks and earplugs (treatment group, T; $n = 29$) and one using sham headbands (control or placebo group, P; $n = 27$).

1. What was the purpose of using 'sham' headbands if it was an ineffective intervention?
2. What type of study is this: experimental or observational? Explain.

Sleep duration was measured in Week 1 (no intervention) and again in Week 2 (with the allocated intervention) for each subject, using a 'wrist actigraphy monitor'.

3. Why is using a 'wrist actigraphy monitor' better than self-reported sleep duration?

The women in the two groups were compared. For example, the mean age of the women was 30.6 y ($s = 3.6$) (T) and 30.1 y ($s = 3.3$) (P); the P-value for the comparison was given as $P = 0.56$.

4. *Why* was this comparison made?
5. Compute the standard error for the difference between the two mean sleep durations.
6. Compute the t-score for the test.
7. Is the quoted P-value consistent with your calculations? What do these results mean?
8. Is the result statistically valid?

Another comparison was the room 'condition' where the women slept: in the treatment group, 13 had a room with a fan (16 had air conditioning), while in the control group 10 women had a fan (and 17 air conditioning). The P-value for the comparison was given as $P = 0.60$.

9. *Why* was this comparison made?
10. Construct the 2×2 table summarising the data.
11. The χ^2-score for the test is 0.35064. Compute the equivalent z-score. Interpret the results.
12. Is the quoted P-value consistent with your calculations?
13. Is the result statistically valid?

In the *treatment* group, the mean sleep duration in Week 1 was 279.0 mins ($s = 18.9$) and in Week 2 was 303.6 mins ($s = 18.8$). The *increase* was 24.7 mins ($s = 14.9$).

14. Test if sleep duration *increased* in the treatment group. Interpret the results mean.

In the *control* group, the mean sleep duration in Week 1 was 286.3 mins ($s = 20.9$) and in Week 2 was 301.9 mins ($s = 21.8$; $n = 26$). The *increase* was 18.1 mins ($s = 17.3$).

15. Test if sleep duration *increased* in the control group. Interpret the results.
16. Why would sleep duration *increase*, if the control group used an ineffective intervention?

The *increase* in sleep duration can be compared for the two groups.

17. Compute the standard error for difference between the mean increases s.e.$(\bar{x}_T - \bar{x}_T)$.
18. Compare the increase in sleep duration for the two groups. Interpret the results.
19. Is the test statistically valid?

Answers to *Quick review* questions: 1. False. **2.** True. **3.** False.

A

Datasets

Most datasets used in this book are available in the **R** package **SRMData**, available free from CRAN.[1] Most datasets used in this book can also be downloaded from the online version of this book (`https://bookdown.org/pkaldunn/SRM-Textbook/`).

In the list below (alphabetical within chapters), all datasets are from the **SRMData** package except when noted (in parentheses). Other packages listed are also available from CRAN.

Chapter 7

- `Placebos`

Chapter 10

- `Orthoses` (Exercise)

Chapter 11

- `BabyBoom`
- `Cyclones`
- `faithful` (in **R**)
- `Gorillas`
- `MaryRiver`
- `Perm`
- `WaterAccess`

- `CherryRipe` (Exercise)
- `FriesWt` (Exercise)
- `Jeans` (Exercise)
- `Lime` (Exercise)
- `NHANES` (Exercise; in **NHANES** package)
- `Orthoses` (Exercise)

Chapter 12

- `WaterAccess`

- `BabyBoom` (Exercise)
- `LungCap` (Exercise)

Chapter 13

- `IgE`
- `Running`
- `Tape`

- `Captopril` (Exercise)
- `Flowering` (Exercise)
- `Insulation` (Exercise)
- `Jumping` (Exercise)
- `PainRelief` (Exercise)
- `Running` (Exercise)
- `Stress` (Exercise)
- `WCTennis` (Exercise)

Chapter 14

- `Gorillas`
- `Jellyfish`
- `WaterAccess`

- `AISsub` (Exercise)
- `Deceleration` (Exercise)
- `Dental` (Exercise)
- `ForwardFall` (Exercise)
- `NHANES` (Exercise; in **NHANES** package)
- `Snakes` (Exercise)
- `Speed` (Exercise)
- `Typing` (Exercise)

Chapter 15

- `KStones`
- `WaterAccess`

- `EmeraldAug` (Exercise)
- `PremierL` (Exercise)

Chapter 16

- `LungCap`
- `RedDeer`
- `Removal`
- `Sanddollars`
- `YieldDen`

- `BoneQuality` (Exercise)
- `Cyclones` (Exercise)
- `Gorillas` (Exercise)
- `Lime` (Exercise)
- `Mandible` (Exercise)
- `Peas` (Exercise)
- `SDrink` (Exercise)
- `SoilCN` (Exercise)
- `StudentWt` (Exercise)
- `Windmill` (Exercise)

Chapter 17

- `BodyTemp`
- `DanishLC`
- `NHANES` (in **NHANES** package)
- `WaterAccess`

- `AISsub` (Exercise)
- `HCrabs` (Exercise)
- `NHANES` (Exercise; in **NHANES** package)
- `NMiner` (Exercise)
- `Typing` (Exercise)

Chapter 18

- `HatSunglasses`
- `QSchools`

[1] https://CRAN.R-project.org

Chapter 20

- `Diabetes`
- `Possums`

Chapter 22

- `HatSunglasses` (Exercise)

Chapter 23

- `Fluoro`

- `LungCap` (Exercise)
- `NHANES` (Exercise; in **NHANES** package)
- `PizzaSize` (Exercise)

Chapter 26

- `PremierL` (Exercise)

Chapter 27

- `BodyTemp`

- `CherryRipe` (Exercise)
- `LHconc` (Exercise)
- `PizzaSize` (Exercise)

Chapter 28

- `Battery` (Exercise)

Chapter 29

- `Flowering`
- `SixMWT`

- `Anorexia` (Exercise)
- `Captopril` (Exercise)
- `Ferritin` (Exercise)
- `Fruit` (Exercise)
- `Jumping` (Exercise)
- `SoilCN` (Exercise)
- `Stress` (Exercise)
- `StudentWt` (Exercise)
- `WCTennis` (Exercise)

Chapter 30

- `Lime`
- `Snakes`
- `Speed`

- `Anorexia` (Exercise)
- `BMI` (Exercise)
- `BodyTemp` (Exercise)
- `Deceleration` (Exercise)
- `Dental` (Exercise)
- `ForwardFall` (Exercise)
- `Lime` (Exercise)
- `NHANES` (Exercise; in **NHANES** package)
- `ReactionTime` (Exercise)

Chapter 31

- `Burros`
- `PetBirds`
- `RipsID`
- `StudentsEat`

- `B12Diet` (Exercise)
- `CarCrashes` (Exercise)
- `CrabShells2` (Exercise)
- `CrabShells3` (Exercise)
- `DogWalks` (Exercise)
- `EmeraldAug` (Exercise)
- `EVpurchase` (Exercise)
- `EVPurchase` (Exercise)
- `HatSunglasses` (Exercise)
- `Mumps` (Exercise)
- `PetBirds` (Exercise)
- `RipsID` (Exercise)
- `ScarHeight` (Exercise)
- `ShoppingBags` (Exercise)
- `Turbines` (Exercise)

Chapter 32

- `Diabetes`

Chapter 33

- `AISsub`
- `Borers`
- `Cyclones`
- `Removal`

- `Bitumen` (Exercise)
- `Corollas` (Exercise)
- `Dogs` (Exercise)
- `DogsLife` (Exercise)
- `EDpatients` (Exercise)
- `Elephants` (Exercise)
- `Gorillas` (Exercise)
- `Jeans` (Exercise)
- `Mandible` (Exercise)
- `OSA` (Exercise)
- `Possums` (Exercise)
- `SDrink` (Exercise)
- `Soils` (Exercise)
- `Throttle` (Exercise)
- `Typing` (Exercise)

B

z-score tables

This appendix contains z-score tables.

These tables provide the area *to the left* of a given z-score associated with a normal distribution: Appendices B.1 (for negative values of z) and B.2 (for positive values of z).

The online version of this book (https://bookdown.org/pkaldunn/SRM-Textbook) has online tables, which are easier to use.

Using these tables

Details for using these tables are provided in Sect. 20.6. Here we give a short summary.

To find the probability that z is less than -2.43, follow these steps:

1. Use Appendix B.1 to find -2.4 in the *left* margin of the table (see image below).
2. Then, find the second decimal place (in this case, 3) in the *top* margin of the table.
3. Where these intersect is the area (or probability) *less than* the z-score of -2.43.

The probability of finding a z-score less than $z = -2.43$ is 0.0075, or about 0.75%.

	0.00	0.01	0.02	0.03	0.04
−3.5	0.0002	0.0002	0.0002	0.0002	0.0002
−3.4	0.0003	0.0003	0.0003	0.0003	0.0003
−3.3	0.0005	0.0005	0.0005	0.0004	0.0004
−3.2	0.0007	0.0007	0.0006	0.0006	0.0006
−3.1	0.0010	0.0009	0.0009	0.0009	0.0008
−3.0	0.0013	0.0013	0.0013	0.0012	0.0012
−2.9	0.0019	0.0018	0.0018	0.0017	0.0016
−2.8	0.0026	0.0025	0.0024	0.0023	0.0023
−2.7	0.0035	0.0034	0.0033	0.0032	0.0031
−2.6	0.0047	0.0045	0.0044	0.0043	0.0041
−2.5	0.0062	0.0060	0.0059	[illegible]	0.0055
−2.4	0.0082	0.0080	0.0078	0.0075	0.0073
−2.3	0.0107	0.0104	0.0102	0.0099	0.0096

B.1 Normal distribution: negative z-values probabilities

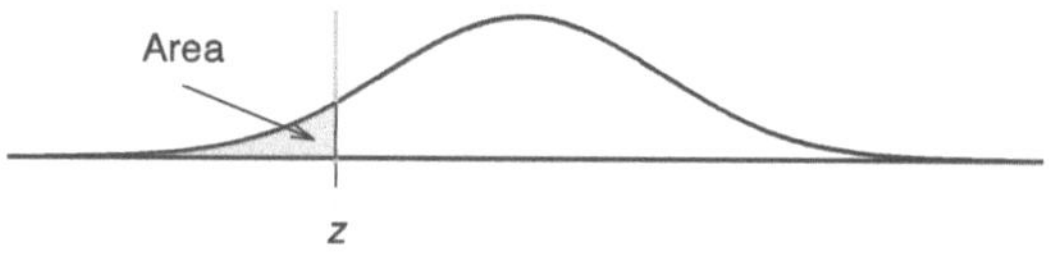

The table gives the probability (area) that a z-score is *less* than a given value. For example: the area *less than* $z = -1.38$ is 0.0838, or 8.38%.

	0.00	0.01	0.02	0.03	0.04	0.05	0.06	0.07	0.08	0.09
−3.5	0.0002	0.0002	0.0002	0.0002	0.0002	0.0002	0.0002	0.0002	0.0002	0.0002
−3.4	0.0003	0.0003	0.0003	0.0003	0.0003	0.0003	0.0003	0.0003	0.0003	0.0002
−3.3	0.0005	0.0005	0.0005	0.0004	0.0004	0.0004	0.0004	0.0004	0.0004	0.0003
−3.2	0.0007	0.0007	0.0006	0.0006	0.0006	0.0006	0.0006	0.0005	0.0005	0.0005
−3.1	0.0010	0.0009	0.0009	0.0009	0.0008	0.0008	0.0008	0.0008	0.0007	0.0007
−3.0	0.0013	0.0013	0.0013	0.0012	0.0012	0.0011	0.0011	0.0011	0.0010	0.0010
−2.9	0.0019	0.0018	0.0018	0.0017	0.0016	0.0016	0.0015	0.0015	0.0014	0.0014
−2.8	0.0026	0.0025	0.0024	0.0023	0.0023	0.0022	0.0021	0.0021	0.0020	0.0019
−2.7	0.0035	0.0034	0.0033	0.0032	0.0031	0.0030	0.0029	0.0028	0.0027	0.0026
−2.6	0.0047	0.0045	0.0044	0.0043	0.0041	0.0040	0.0039	0.0038	0.0037	0.0036
−2.5	0.0062	0.0060	0.0059	0.0057	0.0055	0.0054	0.0052	0.0051	0.0049	0.0048
−2.4	0.0082	0.0080	0.0078	0.0075	0.0073	0.0071	0.0069	0.0068	0.0066	0.0064
−2.3	0.0107	0.0104	0.0102	0.0099	0.0096	0.0094	0.0091	0.0089	0.0087	0.0084
−2.2	0.0139	0.0136	0.0132	0.0129	0.0125	0.0122	0.0119	0.0116	0.0113	0.0110
−2.1	0.0179	0.0174	0.0170	0.0166	0.0162	0.0158	0.0154	0.0150	0.0146	0.0143
−2.0	0.0228	0.0222	0.0217	0.0212	0.0207	0.0202	0.0197	0.0192	0.0188	0.0183
−1.9	0.0287	0.0281	0.0274	0.0268	0.0262	0.0256	0.0250	0.0244	0.0239	0.0233
−1.8	0.0359	0.0351	0.0344	0.0336	0.0329	0.0322	0.0314	0.0307	0.0301	0.0294
−1.7	0.0446	0.0436	0.0427	0.0418	0.0409	0.0401	0.0392	0.0384	0.0375	0.0367
−1.6	0.0548	0.0537	0.0526	0.0516	0.0505	0.0495	0.0485	0.0475	0.0465	0.0455
−1.5	0.0668	0.0655	0.0643	0.0630	0.0618	0.0606	0.0594	0.0582	0.0571	0.0559
−1.4	0.0808	0.0793	0.0778	0.0764	0.0749	0.0735	0.0721	0.0708	0.0694	0.0681
−1.3	0.0968	0.0951	0.0934	0.0918	0.0901	0.0885	0.0869	0.0853	0.0838	0.0823
−1.2	0.1151	0.1131	0.1112	0.1093	0.1075	0.1056	0.1038	0.1020	0.1003	0.0985
−1.1	0.1357	0.1335	0.1314	0.1292	0.1271	0.1251	0.1230	0.1210	0.1190	0.1170
−1.0	0.1587	0.1562	0.1539	0.1515	0.1492	0.1469	0.1446	0.1423	0.1401	0.1379
−0.9	0.1841	0.1814	0.1788	0.1762	0.1736	0.1711	0.1685	0.1660	0.1635	0.1611
−0.8	0.2119	0.2090	0.2061	0.2033	0.2005	0.1977	0.1949	0.1922	0.1894	0.1867
−0.7	0.2420	0.2389	0.2358	0.2327	0.2296	0.2266	0.2236	0.2206	0.2177	0.2148
−0.6	0.2743	0.2709	0.2676	0.2643	0.2611	0.2578	0.2546	0.2514	0.2483	0.2451
−0.5	0.3085	0.3050	0.3015	0.2981	0.2946	0.2912	0.2877	0.2843	0.2810	0.2776
−0.4	0.3446	0.3409	0.3372	0.3336	0.3300	0.3264	0.3228	0.3192	0.3156	0.3121
−0.3	0.3821	0.3783	0.3745	0.3707	0.3669	0.3632	0.3594	0.3557	0.3520	0.3483
−0.2	0.4207	0.4168	0.4129	0.4090	0.4052	0.4013	0.3974	0.3936	0.3897	0.3859
−0.1	0.4602	0.4562	0.4522	0.4483	0.4443	0.4404	0.4364	0.4325	0.4286	0.4247
−0.0	0.5000	0.4960	0.4920	0.4880	0.4840	0.4801	0.4761	0.4721	0.4681	0.4641

For $z = -4$, the probability is 0.00003. For $z = -5$, the probability is 0.0000003.

B.2 Normal distribution: positive z-values probabilities

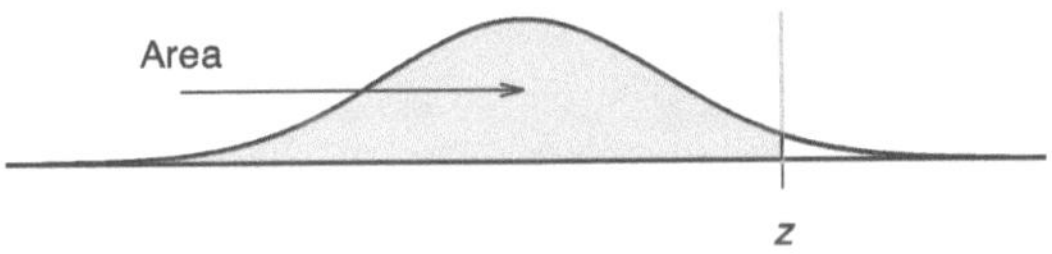

The table gives the probability (area) that a z-score is *less* than a given value. For example: the area *less than* $z = 1.87$ is 0.9693, or 96.93%.

	0.00	0.01	0.02	0.03	0.04	0.05	0.06	0.07	0.08	0.09
0.0	0.5000	0.5040	0.5080	0.5120	0.5160	0.5199	0.5239	0.5279	0.5319	0.5359
0.1	0.5398	0.5438	0.5478	0.5517	0.5557	0.5596	0.5636	0.5675	0.5714	0.5753
0.2	0.5793	0.5832	0.5871	0.5910	0.5948	0.5987	0.6026	0.6064	0.6103	0.6141
0.3	0.6179	0.6217	0.6255	0.6293	0.6331	0.6368	0.6406	0.6443	0.6480	0.6517
0.4	0.6554	0.6591	0.6628	0.6664	0.6700	0.6736	0.6772	0.6808	0.6844	0.6879
0.5	0.6915	0.6950	0.6985	0.7019	0.7054	0.7088	0.7123	0.7157	0.7190	0.7224
0.6	0.7257	0.7291	0.7324	0.7357	0.7389	0.7422	0.7454	0.7486	0.7517	0.7549
0.7	0.7580	0.7611	0.7642	0.7673	0.7704	0.7734	0.7764	0.7794	0.7823	0.7852
0.8	0.7881	0.7910	0.7939	0.7967	0.7995	0.8023	0.8051	0.8078	0.8106	0.8133
0.9	0.8159	0.8186	0.8212	0.8238	0.8264	0.8289	0.8315	0.8340	0.8365	0.8389
1.0	0.8413	0.8438	0.8461	0.8485	0.8508	0.8531	0.8554	0.8577	0.8599	0.8621
1.1	0.8643	0.8665	0.8686	0.8708	0.8729	0.8749	0.8770	0.8790	0.8810	0.8830
1.2	0.8849	0.8869	0.8888	0.8907	0.8925	0.8944	0.8962	0.8980	0.8997	0.9015
1.3	0.9032	0.9049	0.9066	0.9082	0.9099	0.9115	0.9131	0.9147	0.9162	0.9177
1.4	0.9192	0.9207	0.9222	0.9236	0.9251	0.9265	0.9279	0.9292	0.9306	0.9319
1.5	0.9332	0.9345	0.9357	0.9370	0.9382	0.9394	0.9406	0.9418	0.9429	0.9441
1.6	0.9452	0.9463	0.9474	0.9484	0.9495	0.9505	0.9515	0.9525	0.9535	0.9545
1.7	0.9554	0.9564	0.9573	0.9582	0.9591	0.9599	0.9608	0.9616	0.9625	0.9633
1.8	0.9641	0.9649	0.9656	0.9664	0.9671	0.9678	0.9686	0.9693	0.9699	0.9706
1.9	0.9713	0.9719	0.9726	0.9732	0.9738	0.9744	0.9750	0.9756	0.9761	0.9767
2.0	0.9772	0.9778	0.9783	0.9788	0.9793	0.9798	0.9803	0.9808	0.9812	0.9817
2.1	0.9821	0.9826	0.9830	0.9834	0.9838	0.9842	0.9846	0.9850	0.9854	0.9857
2.2	0.9861	0.9864	0.9868	0.9871	0.9875	0.9878	0.9881	0.9884	0.9887	0.9890
2.3	0.9893	0.9896	0.9898	0.9901	0.9904	0.9906	0.9909	0.9911	0.9913	0.9916
2.4	0.9918	0.9920	0.9922	0.9925	0.9927	0.9929	0.9931	0.9932	0.9934	0.9936
2.5	0.9938	0.9940	0.9941	0.9943	0.9945	0.9946	0.9948	0.9949	0.9951	0.9952
2.6	0.9953	0.9955	0.9956	0.9957	0.9959	0.9960	0.9961	0.9962	0.9963	0.9964
2.7	0.9965	0.9966	0.9967	0.9968	0.9969	0.9970	0.9971	0.9972	0.9973	0.9974
2.8	0.9974	0.9975	0.9976	0.9977	0.9977	0.9978	0.9979	0.9979	0.9980	0.9981
2.9	0.9981	0.9982	0.9982	0.9983	0.9984	0.9984	0.9985	0.9985	0.9986	0.9986
3.0	0.9987	0.9987	0.9987	0.9988	0.9988	0.9989	0.9989	0.9989	0.9990	0.9990
3.1	0.9990	0.9991	0.9991	0.9991	0.9992	0.9992	0.9992	0.9992	0.9993	0.9993
3.2	0.9993	0.9993	0.9994	0.9994	0.9994	0.9994	0.9994	0.9995	0.9995	0.9995
3.3	0.9995	0.9995	0.9995	0.9996	0.9996	0.9996	0.9996	0.9996	0.9996	0.9997
3.4	0.9997	0.9997	0.9997	0.9997	0.9997	0.9997	0.9997	0.9997	0.9997	0.9998
3.5	0.9998	0.9998	0.9998	0.9998	0.9998	0.9998	0.9998	0.9998	0.9998	0.9998

For $z = 4$, the probability is 0.99997. For $z = 5$, the probability is 0.9999997.

C

Symbols, formulas, statistics and parameters

C.1 Symbols and standard errors

- The following table lists the statistics used to estimate unknown population parameters.
- When the sampling distribution is approximately normally distributed, under appropriate statistical validity conditions, this is indicated by ✔.
- The value of the mean of the sampling distribution (the *sampling mean*) is:
  - unknown, for *confidence intervals*.
  - assumed to be the value given in the null hypothesis, for *hypothesis tests*.

TABLE C.1: Sample statistics used to estimate population parameters. Some statistics have appproximately normally-distributed sampling distributions under appropriate (statistical validity) conditions, as indicated using a ✔.

		Sampling distribution			
	Statistic	**Parameter, and sampling mean**	**Normal distn?**	**Standard error**	**Ref.**
Proportion	$\hat{p}$	p	✔	CI: $\sqrt{\frac{\hat{p} \times (1-\hat{p})}{n}}$	Ch. 22
				HT: $\sqrt{\frac{p \times (1-p)}{n}}$	Ch. 26
Mean	$\bar{x}$	μ	✔	$\frac{s}{\sqrt{n}}$	Chs. 23, 27
Mean difference	$\bar{d}$	μ_d	✔	$\frac{s_d}{\sqrt{n}}$	Ch. 29
Difference between means	$\bar{x}_1 - \bar{x}_2$	$\mu_1 - \mu_2$	✔	$\sqrt{\text{s.e.}(\bar{x}_1)^2 + \text{s.e.}(\bar{x}_2)^2}$	Ch. 30
Difference between proportions	$\hat{p}_1 - \hat{p}_2$	$p_1 - p_2$	✔	CI: $\sqrt{\text{s.e.}(\hat{p}_1)^2 + \text{s.e.}(\hat{p}_2)^2}$	Ch. 31
				HT: $\sqrt{\text{s.e.}(\hat{p}_1)^2 + \text{s.e.}(\hat{p}_2)^2}$ using *common* proportion $\hat{p}$	
Odds ratio (OR)	Sample OR	Pop. OR	✗	(Not given)	Ch. 31
Correlation	r	ρ	✗	(Not given)	Ch. 33
Regression: slope	b_1	β_1	✔	s.e.(b_1) (value from software)	Ch. 33
Regression: intercept	b_0	β_0	✔	s.e.(b_0) (value from software)	Ch. 33

C.2 Confidence intervals

For statistics whose sampling distribution has an approximate normal distribution, *confidence intervals (CIs)* have the form

$$\text{statistic} \pm \big(\text{multiplier} \times \text{s.e.(statistic)}\big).$$

Notes:

- The multiplier is *approximately* 2 to create an *approximate* 95% CI (based on the 68–95–99.7 rule).
- The quantity 'multiplier × s.e.(statistic)' is called the *margin of error.*
- Software uses *exact* multipliers to form *exact* confidence intervals.
- When the sampling distribution for the statistic does *not* have an approximate normal distribution (e.g., for ORs and correlation coefficients), *this formula does not apply* and the CIs are taken directly from software output when available.

C.3 Hypothesis testing

For statistics whose sampling distribution has an approximate normal distribution, the *test statistic* has the form:

$$\text{test statistic} = \frac{\text{statistic} - \text{parameter}}{\text{s.e.(statistic)}},$$

where s.e.(statistic) is the standard error of the statistic. The test-statistic is a t-score for most hypothesis tests in this book when the sampling distribution is described by a normal distribution, but is a z-score for a hypothesis test involving one or two *proportions.*

Notes:

- If the test-statistic is a z-score, the P-value can be found using tables (Appendices B.1 and B.2), or *approximated* using the 68–95–99.7 rule.
- If the test-statistic is a t-score, the P-value can be *approximated* using tables (Appendices B.1 and B.2), or *approximated* using the 68–95–99.7 rule (since t-scores are similar to z-scores; Sect. 28.4).
- When the sampling distribution for the statistic does not have an approximate normal distribution (e.g., for ORs and correlation coefficients), *this formula does not apply* and P-values are taken from software when available.
- A hypothesis test about ORs uses a χ^2 test statistic. For 2×2 tables only, the χ^2-value is equivalent to a z-score with a value of $\sqrt{\chi^2}$.

C.4 Sample size estimation

The following formulas compute the *approximate* minimum (i.e., conservative) sample size needed to produce a 95% CI with a specified margin of error (i.e., the 'give-or-take' amount).

- To estimate the sample size needed for *estimating a proportion* (Sect. 32.3), use:

$$n = \frac{1}{(\text{Margin of error})^2}.$$

- To estimate the sample size needed for *estimating a mean* (Sect. 32.4) use:

$$n = \left(\frac{2 \times s}{\text{Margin of error}} \right)^2$$

  for some estimate s of the standard deviation of the data.

- To estimate the sample size needed for *estimating a mean difference* (Sect. 32.5) use:

$$n = \left(\frac{2 \times s_d}{\text{Margin of error}} \right)^2$$

  for some estimate s_d of the standard deviation of the differences.

- To estimate the sample size needed for *estimating the difference between two means* (Sect. 32.6) use:

$$n = 2 \times \left(\frac{2 \times s}{\text{Margin of error}} \right)^2$$

  for *each* group being compared, where s is an estimate of the common standard deviation in the population for both groups. This formula assumes:

  - the sample size for each group will be the same; and
  - the standard deviation in each group is the same.

- To estimate the sample size needed for *estimating the difference between two proportions* (Sect. 32.7) use:

$$n = \frac{2}{(\text{Margin of error})^2}$$

  for *each* group being compared. This formula assumes the sample size in each group will be the same.

Notes:

- In *sample size* calculations, *round up* the sample size found from the above formulas.

C.5 Other formulas

- To *calculate z-scores* (Sect. 20.4), use

$$z = \frac{\text{value of variable} - \text{mean of the distribution of the variable}}{\text{standard deviation of the distribution of the variable}}.$$

  t-scores are like z-scores. When the 'variable' is a sample estimate (such as $\bar{x}$), the 'standard deviation of the distribution' is a standard error (such as s.e.$(\bar{x})$).
- The *unstandardising formula* (Sect. 20.8) is $x = \mu + (z \times \sigma)$.
- The *interquartile range* (IQR) is $Q_3 - Q_1$, where Q_1 and Q_3 are the first and third quartiles respectively (or, equivalently, the 25th and 75th percentiles).
- The smallest expected value (for assessing statistical validity when forming CIs and conducting hypothesis tests with proportions or ORs) is

$$\frac{(\text{Smallest row total}) \times (\text{Smallest column total})}{\text{Overall total}}.$$

- The *regression equation* in the *sample* is $\hat{y} = b_0 + b_1 x$, where b_0 is the sample intercept and b_1 is the sample slope.

C.6 Other symbols and abbreviations used

Symbol or abbreviation	Meaning	Reference
RQ	Research question	Chap. 2
s	Sample standard deviation	Sect. 11.7.2
σ	Population standard deviation	Sect. 11.7.2
s_d	Sample standard deviation of differences	Sect. 11.7.2
σ_d	Population standard deviation of differences	Sect. 11.7.2
R^2	R-squared	Sect. 16.4.2
H_0	Null hypothesis	Sect. 28.2
H_1	Alternative hypothesis	Sect. 28.2
CI	Confidence interval	Chap. 24
s.e.	Standard error	Def. 19.4
n	Sample size	Def. 2.21
χ^2	The chi-squared test statistic	Sect. 31.6.3
$\pm$	Plus-or-minus (give-or-take)	Sect. 22.3

Glossary

68–95–99.7 rule For *any* bell-shaped distribution, *approximately* 68% of values lie within one standard deviation of the mean, 95% of values lie within two standard deviations of the mean, and 99.7% of values lie within three standard deviations of the mean. Also called the *empirical rule*. See also *Normal distribution*.

Accuracy *Accuracy* refers to how close a *sample* estimate is likely to be to the *population* value, on average. See also *Precision*.

Alternative hypothesis The *alternative hypothesis* H_1 proposes that the discrepancy between the proposed value of the parameter and the observed value of the statistic cannot be explained by *sampling variation*. It proposes that the value of the parameter is not the value claimed in the null hypothesis. See also *Hypothesis test*, *Null hypothesis*.

Bell-shaped distributions See *Normal distribution*.

Between-individual comparisons See *Comparison (between individuals)*, *Comparison*.

***Between*-individuals variables** *Between*-individuals variables vary from one individual to another individual. See also *Variables*, *Within-individual variables*.

Bias *Bias* refers to any systematic misrepresentation of the target population or a parameter caused by the sampling or the study design.

Blinding *Blinding* occurs when those involved in the study do not know information about the study. A study can blind the *researcher* to knowing what comparison group the individuals are in, the *participants* to knowing what comparison group they are in, and/or the *analysts* to knowing what comparison group the individuals are in during analysis.

Blocking *Blocking* occurs when units of analysis are analysed as separate groups of similar units (called *blocks*).

Carryover effect The *carryover effect* occurs when the influence of one treatment or condition on the response variable influences the response variable for subsequent treatments or conditions (in a repeated-measures study).

Cases *Cases* are the individual units in the population; the *units of analysis*. Also called *individuals*, or (when the individuals are people) *subjects*.

Categorical data See *Qualitative data*.

Cherry-picking *Cherry-picking* is a non-random sampling method where individuals are specifically chosen to reach the conclusion that the researchers want.

Chi-square (χ^2) score The *chi-square (χ^2) score* is the value of the test-statistic used to study the relationship between two qualitative variables. The χ^2-statistic measures the overall size of the differences between the expected counts and observed counts, over the entire two-way table.

Classical approach to probability In the *classical approach to probability*, the probability of an event occurring is the number of elements of the sample space included in the event, divided by the total number of elements in the sample space, *when all outcomes are equally likely* (i.e., no reason exists to expect one event to occur more often than the others). See also *Relative-frequency approach to probability*, *Sample space*, *Subjective approach to probability*.

Cluster sampling *Cluster sampling* is a random sampling method where the population is split into a large number of small groups called *clusters*, then a *simple random sample* of clusters is selected and *every* member of the chosen small groups is part of the sample. See also *Simple random sampling*.

Comparison In an RQ, a *comparison* may be *within* individuals, or *between* groups of individuals. See also *Comparison (between individuals)*, *Comparison (within individuals)*.

Comparison (between individuals) The *between-individuals comparison* in the RQ identifies the small number of groups of different individuals for which the outcome is compared. See also *Comparison*, *Comparison (within individuals)*.

Comparison (within individuals) The *within-individuals comparison* in the RQ identifies the small number of different, distinct situations for which the outcome is compared for each individual. See also *Comparison*, *Comparison (between individuals)*.

Compound event A *compound event* is any combination of *simple events*. See also *Event*, *Simple event*.

Conceptual definition A *conceptual definition* articulates precisely *what* words or phrases mean in a study. See also *Operational definition*.

Conditions The *conditions* are the values of the comparison that those in the *observational* study have or experience, but are not manipulated or imposed by the researchers. See also *Observational studies*, *Treatments*.

Confidence interval A CI is an interval which contains the unknown value of the *parameter* a given percentage of the time (over repeated sampling). Informally: a *confidence interval* (CI) is an interval likely to contain the unknown value of the *parameter*. We studied CIs in specific situations (see Sect. C.1); there are hundreds more.

Confounding *Confounding* is when a third variable influences the observed relationship between the response and explanatory variable.

Confounding variable A *confounding variable* (or a *confounder*) is an extraneous variable associated with the response *and* explanatory variables. See also *Confounding variable*, *Extraneous variable*.

Continuous data *Continuous* quantitative data has (at least in theory) an infinite number of possible values between any two given values. See also *Discrete data*, *Quantitative data*.

Control A *control* is a unit of analysis without the treatment or condition of interest, but as similar as possible in every other way to other units of analysis.

Control variable *Control (or controlled) variables* are extraneous variables whose values are fixed for the study.

Convenience sampling *Convenience sampling* is a non-random sampling method where individuals are selected because they are convenient for the researcher.

Correlation *Correlation* refers to the association between two variables, measured by a correlation coefficient.

Correlation coefficient The (Pearson) *correlation coefficient* (r for a sample; ρ for a population) measures the *strength* and *direction* of the *linear* relationship between two quantitative variables. Its value is always between -1 and 1. (Other types of correlation coefficients also exist.)

Correlational research question *Correlational RQs* explore the relationship between two quantitative variables.

Data *Data* refers to information (observations or measurements), such as numbers, labels, recordings, videos, text, etc. (such as height of seedlings, or the type of medication given).

Dataset A *dataset* refers to an organised and structured collection of data.

Descriptive research question *Descriptive RQs* have a population and an outcome.

Descriptive study *Descriptive studies* answer descriptive research questions.

Discrete data *Discrete* quantitative data have a countable number of possible values between any two given values of the variable. See also *Continuous data*, *Quantitative data*.

Distribution The *distribution* of a variable describes what values are present in the data, and how often those values appear. See also *Normal distribution*.

Ecological validity A study is *ecologically valid* if the study methods, materials and context closely approximate the real situation of interest.

Event An *event* is any combination of the elements in the sample space. See also *Compound event*, *Sample space*, *Simple event*.

Exclusion criteria *Exclusion criteria* are characteristics that disqualify potential individuals from being included in the study. See also *Inclusion criteria*.

Empirical rule See the 68–95–99.7 *rule*.

Experimental studies (or Experiments) *Experimental studies* (or *experiments*) study relationships *with* an intervention. See also *Intervention*, *Observational studies*.

Experimenter effect See *Observer effect.*

Explanatory variable An *explanatory variable* may (partially) explain or be associated with changes in another variable of interest (the response variable). In an experimental study, it is the variable that can be manipulated by the researchers. See also *Response variable.*

External validity *External validity* refers to the ability to generalise the results of the study to the rest of the population, beyond just those in the studied sample. For a study to be truly externally valid, the sample must be a random sample from the population. See also *Internal validity.*

Extraneous variable An *extraneous* variable is any variable associated with the response variable, but is not the explanatory variable. See also *Confounding variable*, *Lurking variable.*

Extrapolation *Extrapolation* refers to making a prediction outside the range of the available data. Extrapolation beyond the data may lead to nonsense.

Hawthorne effect The *Hawthorne effect* is the tendency of individuals to change their behaviour if they know (or think) they are being observed.

Hypothesis A *hypothesis* is a possible answer to a (research) question. See also *Alternative hypothesis*, *Hypothesis test*, *Null hypothesis.*

Hypothesis test A *hypothesis test* is a way to formally answer questions about a population, based on information obtained from a sample. In this book, we studied specific hypothesis tests (see Sect. C.1); hundreds more exist.

Inclusion criteria *Inclusion criteria* are characteristics that individuals must meet explicitly to be included in the study. See also *Exclusion criteria.*

Independence Two events are *independent* if the probability of one event doesn't change depending on whether or not other event has happened.

Individuals *Individuals* are the units in the population from which the observations of interest could be made; the *units of analysis*. Also called *Cases*, or *Subjects* when the individuals are people. See also *Units of analysis.*

Internal validity *Internal validity* refers to the extent to which a cause-and-effect relationship can be established in a study. A study with *high* internal validity shows that the changes in the response variable can be (at least partially) attributed to changes in the explanatory variables; other explanations have been ruled out. See also *External validity.*

Intervention An *intervention* is present when researchers can manipulate (or impose) the values of the explanatory variable on the individuals to determine the impact on the response variable.

IQR The *IQR* is a measure of variation. The *IQR* is the range in which the middle 50% of the data lie; the difference between the third and the first quartiles. See also *Quartiles.*

IQR rule for identifying outliers The *IQR rule* is a way to identify outliers. The *IQR rule* can identify outliers as either *extreme* (observations 3 × IQR more unusual than Q_1 or Q_3) or *mild* (observations 1.5 × IQR more unusual than Q_1 or Q_3, that are not extreme outliers).

Jittering *Jittering* is when a small amount of randomness is added in either the horizontal or vertical direction (or sometimes both) to separate points that would otherwise be overplotted. See also *Overplotting*, *Stacking.*

Judgement sampling *Judgement sampling* is a non-random sampling method where individuals are selected, based on the researchers' judgement, depending on whether the researcher thinks they are likely to be agreeable or helpful.

Levels of a qualitative variable The *levels* (or the *values*) of a qualitative variable refer to the names of the distinct categories of the variable.

Lurking variable A *lurking variable* is an extraneous variable associated with the response *and* explanatory variables (that is, a *confounding* variable), but whose values *are not* recorded in the study data. See also *Confounding variable*, *Extraneous variable.*

Mean The *mean* ($\bar{x}$ for a sample; μ for a population) is one way to measure the 'average' value of quantitative data. The *arithmetic mean* is the 'balance point' of the data. The positive and negative distances from the mean add to zero. See also *Median.*

Median The *median* is one way to measure the 'average' value of some data. A *median* is a value such that half the values are larger than the median, and half the values are smaller than the median. See also *Mean.*

Mode A *mode* is the level (or levels) of a qualitative variable with the most observations.

Multi-stage sampling *Multi-stage sampling* is a random sampling method where large groups are selected using a *simple random sample*, then smaller groups within those large groups are selected using a *simple random sample*. The simple randomly sampling can continue for as many levels as necessary. See also *Simple random sampling.*

Nominal variable A *nominal* qualitative variable is a qualitative variable where the levels *do not* have a natural order. See also *Ordinal variable*, *Qualitative data.*

Non-response bias *Non-response bias* occurs when chosen participants do not respond. See also *Bias.*

Normal distribution A *normal distribution* is symmetrical distribution, with most values near the centre of the distribution (the mean). The normal distribution is described by its *mean* and *standard deviation.* A picture of a normal distribution is shown below. Normal distributions are also called *bell-shaped* distributions. See also 68–95–99.5 rule.

Null hypothesis The *null hypothesis* H_0 proposes that *sampling variation* explains the discrepancy between the proposed value of the parameter, and the observed value of the statistic. See also *Alternative hypothesis*, *Hypothesis test.*

Objective data *Objective data* refers to facts and measurable evidence.

Observational studies *Observational studies* study relationships *without* an intervention. See also *Experimental studies.*

Observer effect The *observer effect* occurs when the researchers (unconsciously) change their behaviour to conform to expectations because they know what values of the explanatory variable apply to the individuals. This may then cause the *individuals* to change their behaviour or reporting also.

Odds The *odds* are the number (or proportion, or percentage) of results of interest, divided by the remaining number (or proportion, or percentage) of results. See also *Probability.*

Odds ratio (OR) The *odds ratio (OR)* is how many *times* greater the odds of an event are in one group, compared to the odds of the *same* event in a *different* group. See also *Odds.*

Operational definition An *operational definition* articulates exactly *how* something will be identified, measured, observed or assessed. See also *Conceptual definition.*

Ordinal variable An *ordinal* qualitative variable is a qualitative variable where the levels *do* have a natural order. See also *Nominal variable*, *Qualitative data.*

Overplotting *Overplotting* occurs when observations in a scatterplot or dot plot have the same, or nearly the same, values, and so are plotted at the same, or nearly the same, places on the graph. See also *Jittering*, *Stacking.*

Outcome The *outcome* in an RQ is the result, output, consequence or effect of interest in a study, numerically summarised for a group of individuals.

Outliers An *outlier* is an observation that is 'unusual' (either larger or smaller) compared to the bulk of the data. Rules for identifying outliers are arbitrary. See also *IQR rule for identifying outliers*, *Standard deviation rule for identifying outliers.*

P-value A *P-value* is the probability of observing the sample results (or something even more extreme) over repeated sampling, under the assumption that the null hypothesis about the population is true. P-values are used in decision-making. See also *Hypothesis testing.*

Parameter A *parameter* is a number, usually unknown, describing some feature of a population, and estimated by a statistic. See also *Statistic.*

Paired data *Paired data* occurs when the outcome in repeated-measures studies is compared for two different, distinct situations for each unit of analysis.

Percentage A *percentage* is a proportion, multiplied by 100. In this context, percentages are numbers between 0% and 100%. See also *Proportion.*

Percentiles The pth *percentile* of the data is a value separating the smallest $p\%$ of the data from the rest. See also *Quartiles.*

Pilot study A *pilot study* is a small test run of the study used to check that the protocol is appropriate and practical, and to identify (and hence fix) possible problems with the research design or protocol.

Placebo A *placebo* is a treatment with no intended effect or active ingredient, but appears to be the real treatment.

Placebo effect The *placebo effect* occurs when individuals report perceived or actual effects despite not receiving an active treatment or condition, in experimental studies. See also *Placebo.*

Plagiarism *Plagiarism* is using other people's ideas and research to develop new conclusions, or confirm existing conclusions. All sources used when writing research should be acknowledged to avoid plagiarism.

Population A *population* is a group of individuals (or cases, or subjects if the individuals are people) from which the total set of observations of interest could be made, and to which the results will generalise. See also *Individuals, Sample, Units of analysis.*

Precision *Precision* refers to how similar the sample estimates from different samples are likely to be to each other (that is, how much variation is likely in the sample estimates). See also *Accuracy.*

Probability A *probability* is a number between zero and one inclusive (or between 0% and 100% inclusive) that quantifies the likelihood that a certain *event* will occur. A probability of zero (or 0%) means the event is 'impossible' (will *never* occur), and a probability of one (or 100%) means that the event is *certain* to happen (will always occur). Most events have a probability between the extremes of 0% and 100%. See also *Odds.*

Proportion A *proportion* is a fraction out of a total. Proportions ($\hat{p}$ for a sample; p for a population) are numbers between 0 and 1. See also *Percentage.*

Protocol A *protocol* is a predefined procedure detailing the design and implementation of studies, and for data collection.

Qualitative data *Qualitative data* are not *mathematically* numerical data: they comprise mutually exclusive (and usually exhaustive) categories or labels (even if those labels are numbers). Also called *Ordinal variable, Qualitative data, Quantitative data.*

Quantitative data *Quantitative data* are *mathematically* numerical: the numbers have numerical meaning, and represent quantities or amounts. Quantitative data generally arise from counting or measuring. Also called *Continuous data, Discrete data, Qualitative data.*

Quantitative research *Quantitative research* summarises and analyses data (quantitative or qualitative data) using numerical methods, such as producing averages and percentages.

Quartiles *Quartiles* describe the variation and shape of data. The first quartile Q_1 is a value that separates the smallest 25% of observations from the largest 75%; it is like the median of the *smaller* half of the data, halfway between the minimum value and the median.

The second quartile Q_2 is a value that separates the smallest 50% of observations from the largest 50% (and is also the *median*).

The third quartile Q_3 is a value that separates the smallest 75% of observations from the largest 25%; it is like the median of the *larger* half of the data, halfway between the median and the maximum value. See also *Median, Percentiles.*

Quasi-experiment In a *quasi-experiment*, the researchers (1) allocate treatments to groups of individuals (i.e., allocate the values of the explanatory variable to the individuals, as it is an experiment), but (2) do *not* determine who or what is in those groups. See also *True experiment.*

Questionnaire A *questionnaire* is a set of questions for respondents to answer.

Random *Random* means 'determined completely by impersonal chance'. See also *Simple random sampling.*

Random procedure A *random procedure* is a sequence of well-defined steps that (a) can be repeated, in theory, indefinitely under essentially identical conditions; (b) has well-defined results; and (c) has results that are unpredictable for any individual repetition.

Random sample In a *random sample*, each individual in the population can be selected; and each individual is chosen on the basis of *impersonal* chance. See also: *Simple random sampling, Representative sampling.*

Range The *range* is a measure of variation. The *range* is the maximum value *minus* the minimum value.

Relational research question *Relational RQs* have a population, outcome, and a *between*-individuals comparison.

Relative frequency approach to probability In the *relative frequency approach to probability*, the probability of an event is approximately the number of times the outcomes of interest has appeared in the past, divided by the number of 'attempts' in the past. This produces an *approximate* probability. See also *Classical approach to probability, Subjective approach to probability.*

Repeated-measures research question *Repeated-measures RQs* have a population, outcome and a *within*-individuals comparison.

Representative sample In a *representative sample*, those *in* the sample are not likely to be different from those *not in* the sample, at least for the variables of interest. A representative sample is *not* a random sample. See also: *Random sample.*

Research design *Research design* refers to the decisions made by the researchers to maximise *external validity* and *internal validity.*

Response bias *Response bias* occurs when participants provide *incorrect or false information.*

Response variable A *response variable* records the result, output, consequence or effect of interest from changes in another variable (the explanatory variable). See also *Explanatory variable.*

Sample A *sample* is a subset of individuals from the population. The data are collected from the sample. Usually, countless possible samples could be obtained from a population. See also *Population, Sample size.*

Sample size The sample size n is the number of units of analysis. See also *Population, Sample, Random sample.*

Sample space The *sample space* is a list of all possible and mutually exclusive (distinct) results after administering a random procedure once. See also *Event.*

Sampling distribution A *sampling distribution* is the distribution of a statistic, showing how its value varies across all possible samples. See also *Sampling mean, Standard error.*

Sampling frame The *sampling frame* is a list of all the individuals in the population.

Sampling mean The *sampling mean* is the mean of the sampling distribution of a statistic: the mean of the values of the statistic from all possible samples. See also *Sampling distribution, Sampling mean, Sampling variation, Standard error.*

Sampling variation *Sampling variation* refers to how the sample estimates (statistics) vary from sample to sample, because every possible sample is different. See also *Sampling distribution, Sampling mean, Standard error.*

Selection bias *Selection bias* is the tendency of a sample to over- or under-estimate a population quantity. See also *Bias.*

Scale data See *Quantitative data.*

Simple event A *simple event* is a single element of the sample space. See also *Compound event, Event, Sample space.*

Simple random sampling *Simple random sampling* is a random sampling method where *every* possible sample of a given size has *same* chance of being selected.

Stacking *Stacking* is when points are plotted above other points with similar values, to separate points that would otherwise be overplotted. See also *Jittering, Stacking.*

Standard deviation The *standard deviation* (s for a sample; σ for a population) is a measure of variation. The *standard deviation* is, approximately, the mean distance of observations from the mean.

Standard deviation rule for identifying outliers The *standard deviation rule* is a way to identify outliers For approximately symmetric distributions, any observation more than three standard deviations from the mean can be considered an outlier.

Standard error A *standard error* is the standard deviation of all possible values of the sample estimate (from samples of a certain size): the standard deviation of the values of the statistic from all possible samples. Any quantity estimated from a sample has a standard error. See also *Sampling distribution*, *Sampling mean*, *Sampling variation*.

Stratified sampling *Stratified sampling* is a random sampling method where the population is split into a small number of large (usually similar) groups called *strata*, then cases are selected using a *simple random sample* from *each* stratum. See also *Simple random sampling*.

Statistic A *statistic* is a number describing some feature of a sample (to estimate the unknown value of the population parameter). See also *Parameter*.

Statistical validity A result is *statistically valid* if the conditions for the underlying mathematical calculations to be approximately correct are met, such as the sampling distribution having an approximate normal distribution. Every confidence interval and hypothesis test has statistical validity conditions.

Subjective approach to probability In the *subjective approach to probability*, various factors are incorporated subjectively to determine the probability of an event occurring. See also *Relative-frequency approach to probability*, *Subjective approach to probability*.

Subjective data *Subjective data* refers to opinions, feelings, and interpretations (by the subjects or the researchers).

Subjects The individual units in the population when the units are people; the *units of analysis*. Also called *individuals* or *cases*; however, those two terms do not refer exclusively to people. See also *Units of analysis*.

Systematic sampling *Systematic sampling* is a random sampling method where the first case is *randomly* selected; then, every nth individual is selected thereafter.

t-score A *t-score* measures how many standard deviations a value is from the mean. A t-score is similar to a z-score. See also *z-score*.

Treatments The *treatments* are the values of the explanatory variable that the researchers can manipulate and impose upon the individuals in the *experimental* study. See also *Conditions*, *Experiments*.

True experiment In a *true experiment*, the researchers (1) allocate treatments to groups of individuals (i.e., values of the explanatory variable to the individuals), *and* (2) determine who or what is in those groups. While the steps may not happen *explicitly*, they happen *conceptually*. See also *Quasi-experiment*.

Unit of observation The *unit of observation* is the entity that is observed, from or about which measurements are taken and data collected. See also *Unit of analysis*.

Unit of analysis The *unit of analysis* are the smallest collection of units of observations (and perhaps the units of observations themselves) about which conclusions are made; the smallest distinct, *independent* elements of the population for which information is analysed. In an *experimental study*, the unit of analysis is the smallest collection of units of observations that can be randomly allocated to separate treatments. See also *Individuals*, *Unit of observation*.

Unstandardising formula When the z-score is known, the *unstandardising formula* determines the corresponding value of the observation x: $x = \mu + z(z \times \sigma)$. See also *$z$-score*.

Values of a qualitative variable See *Levels*.

Variables A *variable* is a single aspect or characteristic associated with the individuals, whose values can vary from individual to individual.

Voluntary response (self-selecting) sampling *Voluntary response (or self-selecting) sampling* is a non-random sampling method where individuals participate if they wish to.

Within-individuals comparison See *Comparison (within individuals)*, *Comparison*.

***Within*-individuals variables** *Within*-individuals variables vary from one recording or measurement to another *within* the same individuals. See also *Between-individual variables*, *Variables*.

z-score A *z-score* measures how many standard deviations a value is from the mean. In symbols:

$$z = \frac{\text{value} - \text{mean of the distribution}}{\text{standard deviation of the distribution}} = \frac{x - \mu}{\sigma}$$

where x is the value, μ is the mean of the distribution, and σ is the standard deviation of the distribution. See also *t-score*.

Answers to odd-numbered exercises

Chap. 1: Research an introduction

Ex. 1.1. 1. From many people: type of tourniquet; time to apply. **2. Quant**itative.
Ex. 1.3. 1. For many Egyptians: whether side effects experienced after medication. **2. Quant**itative.
Ex. 1.5. Qualitative

Chap. 2: Research questions

Ex. 2.1. 1. *Percentage* of vehicles that crash. **2.** *Average* jump height. **3.** *Average* number of tomatoes per plant.
Ex. 2.3. 1. Whether diet is vegan. **2.** Whether coffee is caffeinated **3.** Num. iron tablets/day.
Ex. 2.5. 1. *Between*-individuals. Outcome: percentage wearing hats. **2.** *Between*-intervals. Outcome: average yield (in kg/plant, tomatoes/plant, etc.).
Ex. 2.7. 1. Correlational. **2.** Neither variable is explanatory, response.
Ex. 2.9. 1. P: Danish Uni students; O: *average* resting diastolic blood pressure; C: between students who regularly drive, ride their bicycles to uni. **2.** No intervention. **3.** Relational. **4.** Decision-making. **5.** *Conceptual*: 'regularly'; 'university student' (on-campus? undergraduate?). *Operational*: how 'resting diastolic blood pressure' measured. **6.** Resting diastolic blood pressure; whether they regularly drive, ride to uni. **7.** Both: Danish uni students.
Ex. 2.11. 1. Probably relational. **2.** Two-tailed. **3.** Probably not, but possible. **4.** How *individual* people using phones ('Talking'; 'texting'). **5.** Walking speed. **6.** *Average* walking speed. **7.** Observation: person. Analysis: individuals not in a group, as people in group may not walk independently of others (i.e., they keep pace with each other).
Ex. 2.13. 1. Animal. **2.** *Pen*: food allocated to pen. Animals in same pen *not* independent: compete for same space, food, resources, have similar environments. **3.** Between diets.
Ex. 2.15. Ten adults is sample. Unclear how many (or which) fonts compared. Perhaps: 'Among Australian adults, does average time taken to read passage of text differ when Arial font used compared to Times Roman font?'
Ex. 2.17. 1. Analysis: person; Observation: individual nose hairs. Each unit of analysis has 50 units of observation. **2.** $n = 2$.
Ex. 2.19. 1. P: American adults; individuals: American adults. **2.** O: average number recorded steps. **3.** Response: number steps recorded for individuals. Explanatory: location of accelerometer. **4.** *Within* individuals.
Ex. 2.21. 1. Relational; decision-making. **2.** Correlational; estimation. Intervention unlikely.
Ex. 2.23. $n = 27$; unit of analysis: emu.
Ex. 2.25. Unit of observation: tyre. Unit of analysis: car (brand allocated to car; each car gets one brand). Tyres on any car exposed to same day-to-day use, drivers, distances, etc. Each unit of analysis produces four units of observations. *Sample size*: 10 cars (40 observations).
Ex. 2.27. 1. Board. **2.** 5. **3.** 10. **4.** 10.. **5.** Within-board variation smaller (except first). **6.** 1.

Chap. 3: Overview of research design

Ex. 3.1. 1. Arsenic conc. **2.** Distance of lake from mine. **3.** No: recorded. **4.** Yes: may be related to response, explanatory variables. **5.** Probably confounding.
Ex. 3.3. *Response*: perhaps 'whether woman develops cancer of digestive system'. *Explanatory*: 'whether participants drank green tea at least 3 times per week'. *Lurking*: 'health consciousness of participants' (appears unrecorded).
Ex. 3.5. 1. Sex; number siblings. **2.** No. **3.** Yes.
Ex. 3.7. Age of person.
Ex. 3.9. 1. Soil quality; climate; size. Control var.: perhaps farm size (e.g., farms over certain size).

Chap. 4: Types of research studies

Ex. 4.1. **1.** Between-individuals. **2.** Relational. **3.** Most likely. **4.** Estimation. **5.** Intervention: experiment. Likely true experiment.

Ex. 4.3. True experiment.

Ex. 4.5. Quasi-experiment.

Ex. 4.7. **1.** Answers vary. **2.** Researchers *intervene*: *give*, *not give* subjects pet. **3.** Researchers *do not intervene*: subjects already *do*, *do not* own pet.

Chap. 5: Ethics in research

Ex. 5.1. Answers vary.

Ex. 5.3. Answers vary.

Chap. 6: External validity: sampling

Ex. 6.1. c. Externally-valid study more likely.

Ex. 6.3. **1.** Under: practicality; intentional. **2.** Under; deception; probably intentional. **3.** Over; deception; intentional (cherry-picking).

Ex. 6.5. **1.** Every 7th day is same day of week. **2.** Maybe select days at random over three-months.

Ex. 6.7. **1.** Multi-stage. **2.** Stratified (floor), then convenience. **3.** Convenience. **4.** Part stratified (floors), then convenience. Use first.

Ex. 6.9. Random sampling (schools), then, self-selecting.

Ex. 6.11. Stratified by zone; then convenience.

Ex. 6.13. Stage 3 selection *not* random.

Ex. 6.15. **1.** Households in Santiago. **2.** (c): all households in Santiago. **3.** Voluntary response. **4.** Multi-stage.

Chap. 7: Internal validity

Ex. 7.1. All false.

Ex. 7.3. All but random samples (*external*).

Ex. 7.5. Also possible in observational studies.

Ex. 7.7. In case hive size a confounder.

Ex. 7.9. Statements 1, 3, 4 and 8 true. ‘Sex’, ‘Initial weight’ *possible* confounders.

Ex. 7.11. **1.** Observational. **2.** *Response*: amount sunscreen used; *explanatory*: time applying sunscreen. **3.** Potential extraneous, confounding variables. **4.** To determine if sex a *confounder*. **5.** Participants, researchers blinded.

Ex. 7.13. Random allocation; exclusion criteria; blinding; comparing two groups; ethical.

Ex. 7.15. **1.** A: no blinding; B: double-blind. **2.** Hawthorne effect impacting internal validity. **3.** B; no Hawthorne effect.

Ex. 7.17. Randomly allocate type of water to subjects (or the *order* that subjects taste *each* drink.) Subjects blind to water type. Person providing water and receiving ratings blind to water type. Random or representative sampling is good (but hard). Use third-party if possible.

Ex. 7.19. Carryover effect; observer effect.

Ex. 7.21. **1.** Floor area. **2.** Hours labour. **3.** Extraneous. **4.** Analysis **5.** F. **6.** T. **7.** F. **8.** F.

Chap. 8: Research design limitations

Ex. 8.1. External.

Ex. 8.3. Population is students; external validity if applies to students at UniX not wider.

Ex. 8.5. Sample not random; researchers (rightly) state results may not *generalise* to all hospitals. Data collected at night; not *ecologically valid*?

Ex. 8.7. Observational study: people with severe cough may take more cough drops.

Ex. 8.9. Lacks *ecological validity*.

Chap. 9: Collecting data

Ex. 9.1. No place for 18-year-olds.

Ex. 9.3. Best: second. First: *leading* (*concerned* cat owners...) Third: *leading* (Do you *agree*...)

Ex. 9.5. First fine; ‘seldom’ (for instance) may have different meanings to different people; possible recall bias. Second: overlapping options (both 1 h and 2 h in two categories).

Chap. 10: Classifying data and variables

Ex. 10.1. Quant. continuous. Qual. nominal. Quant. continuous. Qual. nominal.

Ex. 10.3. F, T, F.

Ex. 10.5. Nominal; qualitative.

Ex. 10.7. Sex of person

Ex. 10.9. 1. Quant. continuous. **2.** Qual. nominal. **3.** Qual. ordinal. **4.** Quant. discrete.

Ex. 10.11. 1. Qual. nominal. **2.** Quant. discrete. **3.** Qual. ordinal (perhaps quant. discrete). **4.** Qual. nominal. **5.** Quant. continuous.

Ex. 10.13. *Gender*: qual. nominal. *Age*: quant. continuous. *Height*: quant. continuous. *Weight*: quant. continuous. GMFCS: qual. ordinal.

Ex. 10.15. *Kangaroo response*: qual. ordinal (perhaps nominal?). *Drone height*: quant.; four values used; probably treated as qual. ordinal. *Mob size*: quant. discrete. *Sex*: qual. nominal.

Chap. 11: Summarising quantitative data

Ex. 11.1. Shape: skewed *left*. Average: perhaps 70–100? Variation: most between 30, 80. Outliers: none; 'bump' at lower ages.

Ex. 11.3. 1. Probably median (but mean probably OK). **2.** Slightly right skewed; average near 1.5 mmol/L; most between 3, 4 mmol/L; some large outliers.

Ex. 11.5. 1. 3.7. **2.** 3.5. **3.** 1.888562. **4.** $5 - 2 = 3$.

Ex. 11.7. Plot not shown. **1.** −2.42. **2.** 0.8. **3.** 29.6 (−19.8 to 9.8). **4.** 9.831172; about 9.83. **5.** IQR: $4.95 - (-11.4) = 16.35$ (*not* including median in each half). (No units of measurement.)

Ex. 11.9. 1. In cm: 127.4; 129.0; 14.4; 24 from software. Manually (*without* median in each half): $Q_1 = 113$, $Q_3 = 138$, IQR is 25. **2.** Don't know. **3.–5.** Not shown. **6.** Hard to describe with standard language; approx. symmetric?.

Ex. 11.11. In seconds: Average about 10?; variation 0 to 30 perhaps; skewed right. Value between 35 and 40 perhaps outlier.

Ex. 11.13. 1. Men's: about 50%; women's: about 100%. **2.** Men's: about 0%; women's: about 50%.

Ex. 11.15. D; C; A; D.

Chap. 12: Summarising qualitative data

Ex. 12.1. 1. Graph not shown; no commonly-observed social group include M. **2.** Mode: many F plus offspring; median inappropriate.

Ex. 12.3. None *bad*. I'd prefer bar chart; any OK.

Ex. 12.5. 1. Gender: nominal; others ordinal. **2.–4.** Gender: modes are F, M; no median. Place: mode is city > 100 000 residents; median is city 20 000 to 100 000 residents Response: mode is 'Agree'; median is 'Neutral. **5.** 5.12: respondents about 5 times more likely to come from city than rural. **6.** 0.613: respondents about 0.61 times as likely to agree, strongly disagree than choose other option. **7.** 1: respondents as likely to be M as F.

Ex. 12.7. 1. Walking; Bus **2.** Bus. **3.** No. **4.** 0.102, 0.123; 0.246, 0.412, 0.116. **5.** 3.44; i.e., students 3.44 times as likely to use motorised transport than active. **6.** 0.141; i.e, for every 100 students that *do not walk*, $100 \times 0.141 = 14.1$ *do walk*. **7.** Not shown.

Ex. 12.9. 1. Not shown. **2.** Gender: M; smoking: Non-smoking. **3.** Gender: percentage, odds M: 51.4%, 1.06; smoking: percentage, odds non-smoking: 9.9%, 0.11.

Chap. 13: Comparing quantitative data within individuals

Ex. 13.1. 1. House. **2.** Each house has before, after. Graph, table not shown.

Ex. 13.3. Graph, table not shown.

Ex. 13.5. Not shown.

Ex. 13.7. 1. How much further people jump in shoes. Graph and table not shown.

Chap. 14: Comparing quantitative data between individuals

Ex. 14.1. 1. In general, DB smaller cost over-runs. **2.** Tricky: DB: 2 cm; DBB: 3 cm. **3.** Tricky: DB: 2 cm; DBB: 3 cm.

Ex. 14.3. I: B (mean; standard deviation). II: A (median; IQR). III: C (median; IQR).

Ex. 14.5. 1. 0.61; 0.40; 0.42 panels/min. **4.** Worker 2 faster, more consistent (using IQR); Worker 1 slower. Plots not shown.

Ex. 14.7. 1. Error bar chart. **2.** Not shown.

Ex. 14.9. 1. Not shown. **2.** Not shown.

Ex. 14.11. 1. `mAcc`: highly *left* skewed; `Age`: highly *right* skewed; `mTS`: slightly right skewed. Perhaps medians, IQRs for summarising (mean, std dev. probably OK for `mTS`). **2.** Not shown. **3.** Little diff between M, F in *sample*.

Ex. 14.13. 1. Very similar mean SVL. **2.** Crayfish regions: smaller mean SVL. **3.** Crayfish regions: larger mean SVL. **4.** Confounding.

Chap. 15: Comparing qualitative data between individuals

Ex. 15.1. Zero

Ex. 15.3. 1. *Vomited*: 0.50 beer, wine; 0.50 wine only. *Didn't vomit*: 0.738 beer, wine, 0.262 wine only. Prop. drank various drinks, among those who did, didn't vomit. **2.** *Beer, wine*: 8.8% vomited, 91.2% didn't. *Wine only*: 21.4% vomited, 78.6% didn't. Percentage that vomited, for each drinking type. **3.** $(6+6)/(6+6+62+22) = 0.125$. **4.** 0.2727. **5.** 0.09677. **6.** 2.82. **7.** 0.354. **8.** −0.176. **9.** Higher percentage vomited after beer+wine, compared to beer only.

Ex. 15.5. 1. About 18.4%. **2.** About 25.9%. **3.** About 11.7%. **4.** About 0.226. **5.** 0.35. **6.** About 0.132. **7.** About 2.7. **8.** Odds no August rainfall in Emerald 2.7 times higher in months with non-positive SOI.

Ex. 15.7. Not shown.

Ex. 15.9. 1. *Prop.* F skipped: $\hat{p}_F = 0.359$. **2.** *Prop.* M skipped: $\hat{p}_M = 0.284$. **3.** Odds(Skips breakfast, F): 0.5598; **4.** Odds(Skips breakfast, M): 0.3966. **5.** OR: 1.41. **6.** Odds F skipping 1.41 *times* odds M skipping. **7.** Not shown.

Ex. 15.11. 1. Not shown. **2.** 74.6. **3.** 60.9%. **4.** 2.487. **5.** 1.558. **6.** 1.596. **7.** 0.626. **8.** Not shown.

Ex. 15.13. OR(W; home) $= 4/6 = 0.667$; OR(W; away) $= 7/4 = 1.75$. OR $= 0.6667/1.75 = 0.381$.

Chap. 16: Correlations between quantitative variables

Ex. 16.1. Answers vary.

Ex. 16.3. You cannot be precise. Software: $r = 0.71$. Realistically: 'reasonably high, positive r'.

Ex. 16.5. 1. A tree. **2.** *Form*: starts straight-ish, then hard to describe. *Direction*: biomass increases as age increases (on average). *Variation*: small-ish for small ages; large-ish for older trees (after 60). **3.** No.

Ex. 16.7. 1. Approximately linear; positive relationship; variation larger for more cases. **2.** A delivery. **3.** Non-constant variation: no.

Ex. 16.9. No relationship.

Ex. 16.11. Approx. linear; positive; strong.

Ex. 16.13. $R^2 = (-0.682)^2 = 0.465$: about 46.5% of the unknown variation in number cyclones explained by knowing value of ONI.

Chap. 17: More details about tables and graphs data

Ex. 17.1. Scatterplot; histogram of diffs; side-by-side bar.

Ex. 17.3. Individual variables: *bar chart* for origin; *histogram* for others. Between biomass, origin: *boxplot*. Between biomass, other variables: *scatterplot*. (On scatterplot, could encode origins with different colours, symbols.)

Ex. 17.5. Plotting symbols unexplained. Axis labels unhelpful. Vertical axis could stop at 20.

Ex. 17.7. 1. Response: *change* in MADRS (quant. cont.). **2.** Explanatory: treatment group (qual. nominal, 3 levels). **3.** Response: histogram. Explanatory: bar chart. Relationship: boxplot.

Ex. 17.9. Plots not shown. *Speed*: average: around 60 wpm; variation: about 30 to 120 wpm. Slightly right skewed; no obvious outliers. *Accuracy*: average: around 85%; variation: about 65% to 95%. Left skewed; no obvious outliers. *Age*: average: 25; variation: about 15 to 35. *Very* right skewed, perhaps unseen large outliers. *Sex*: about twice as many F as M. *Speed* and *Sex*: not big difference between M, F. *Accuracy* and *Age*: hard to see relationship; no older people very slow.

Average speed, accuracy vary by age, not sex. How data collected (self-reported? Or measured how?). How students obtained: a random, somewhat representative or self-selecting sample?

Chap. 18: Probability

Ex. 18.1. 1. Subjective. **2.** Rel. frequency.

Ex. 18.3. 1. Just **Kings** and **Aces**. **2.** $8/52 = 2/13$. **3.** Picture cards. **4.** $16/52 = 4/13$. **5.** **Ace**, **King**, **Queen**, **Jack** of ♠. **6.** $4/52 = 1/13$. **7.** Any ♡, ♢ or ♣. **8.** $39/52 = 3/4$. **9.** $4/16 = 1/4$. **10.** 4/13.

Ex. 18.5. 1. Yes. **2.** Yes. **3.** 1/2. **4.** 1/2. **5.** HH, HT, TH, TT (Coin A listed first).

Ex. 18.7. 1. 4/6. **2.** 5. **3.** Yes: die outcome won't change coin outcome. **4.** 1/2. **5.** 1/6. **6.** 1/3.

Ex. 18.9. 1. In order drawn: BB, BR, RB, RR. **2.** Equally-likely outcomes, so 1/2. **3.** 1/2. **4.** Yes.

Ex. 18.11. 1. 0.087. **2.** 0.708. **3.** F: prob FN: 0.107; M: prob FN: 0.108; close to independent. **4.** F: prob FN: 0.040; M: prob FN: 0.035; close to independent. **5.** Gov: prob FN: 0.107; NGov: prob FN: 0.040; not independent. **6.** Gov: prob FN: 0.108; NGov: prob FN: 0.035; not independent. **7.** Regardless of sex, First Nations children more likely to be at government school.

Ex. 18.13. 1. *Not independent*: less likely to walk in rain. **2.** *Not independent*: smoker far more likely to suffer from lung cancer than non-smoker. **3.** *Not independent*: if it rains, I won't water garden.

Ex. 18.15. Reasoning assumes three *equally likely* outcomes (HH, TT, HT); untrue. Consider tossing 20-c coin (lower-case, normal) and 1-coin (capitals, **bold**). *Four* outcomes: h**H**, h**T**, t**H** t**T**.

Chap. 19: Sampling variation

Ex. 19.1. **1.** Std dev. **2.** Std error (of mean).
Ex. 19.3. **1.** No. **2.** Yes. **3.** Yes.
Ex. 19.5. **1.** Reasonable, if fair. **2.** Almost impossible, if fair. **3.** Unlikely (not impossible), if fair. **4.** Highly unlikely, if fair.
Ex. 19.7. *Std error of the mean* describes how sample mean varies from sample to sample.

Chap. 20: Models and normal distributions

Ex. 20.1. Only **1.** and **2.** false.
Ex. 20.3. **1.** 0.9671. **2.** 0.0183. **3.** Close to zero. **4.** Close to one.
Ex. 20.5. About 2.5% of girls under 100 cm tall.
Ex. 20.7. **1:** C; **2:** A; **3:** B; **4:** D.
Ex. 20.9. 68.26%; very close to 68%.
Ex. 20.11. **1.** $z = -0.30$; about 38.2%. **2.** $z = 0.07$; about 47.2%. **3.** $z = -0.67$ and $z = 0.44$; about 41.9%. **4.** About $z = 1.04$; diameter about 11.6 inches. **5.** About $z = -0.67$; diameter about 7.0 inches.
Ex. 20.13. **1.** $z = -0.61$; 72.9%. **2.** $z = -1.83$; 3.4%. **3.** $z = -4.878$ and $z = -1.83$; 3.4%. **4.** $z = 1.64$ (or 1.65); 5% *longer* than 42.7 weeks. **5.** z-score: -1.28; 10% *shorter* than 37.9 weeks.
Ex. 20.15. $z = 2.05$. IQ: 130.75. IQ > 130.
Ex. 20.17. Use *number minutes from (say) 5:30pm.* Std dev.: 120 mins, plus $0.28 \times 60 = 16.8$ mins $=$ 136.8 mins. **1.** 9pm; 210mins$ from 5:30pm; $z = 1.54$; 6.2%. **2.** $z = -0.22$; 41.3%. **3.** $z_1 = -0.22$ and $z_2 = 0.22$; $0.5871 - 0.4129$; 17.4%. **4.** z-score: 0.52; $x = 71.136$ mins after 5pm; about one hour and 11 mins after 5:30pm, or 6:41pm. **5.** z-score: -1.04; $x = -141.272$, or 141.272 mins *before* 5:30pm; about two hours and 21 mins before 5:30pm, or 3:09pm.

Chap. 22: Confidence intervals: one proportion

Ex. 22.1. $\hat{p} = 0.81944$, $n = 864$. s.e.$(\hat{p}) = 0.01309$; approx. 95% CI: $0.819 \pm (2 \times 0.0131)$ or 0.793 to 0.845. Stat. valid. (Many decimal places used for working; final answers rounded.)
Ex. 22.3. $z = 1.96$.
Ex. 22.5. $\hat{p} = 0.051948$; s.e.$(\hat{p}) = 0.00178$; approx. 95% CI: 0.0519 ± 0.0358. Stat. valid.
Ex. 22.7. $\hat{p} = 0.317059$; $n = 6882$. s.e.$(\hat{p}) = 0.0056092$. CI: 0.317 ± 0.011. Stat. valid.
Ex. 22.9. $\hat{p} = 0.241$. s.e.$(\hat{p}) = 0.010984$. Approx. 95% CI: 0.219 to 0.263.
Ex. 22.11. $\hat{p} = 0.3182$. s.e.$(\hat{p}) = 0.0702175$. Approx. 95% CI: 0.178 to 0.459.
Ex. 22.13. $\hat{p} = 0.13431$. s.e.$(\hat{p}) = 0.012434$. Approx. 95% CI: 0.109 to 0.159.

Chap. 23: Confidence intervals: one mean

Ex. 23.1. **1.** *Parameter*: pop. mean weight of American black bear, μ. **2.** s.e.$(\bar{x}) = 3.756947$. **3.** Normal; mean μ; std dev: 3.757, **4.** 77.4 to 92.4 kg. **5.** Approx. 95% confident *population mean* weight of male American black bears between 77.4 and 92.4 kg. **6.** Stat. valid: $n \geq 25$.
Ex. 23.3. s.e. $= 0.06410$. *Approx.* 95% CI: 2.72 L to 2.98 L.
Ex. 23.5. Approx. 95% CI: 29.9 to 36.1 s.
Ex. 23.7. *None* acceptable. **1.** CIs not about observations, but *statistics*. **2.** CIs not about observations, but *statistics*. **3.** *Samples* don't vary between values; *statistics* do. (CIs about populations anyway.) **4.** *Populations* don't vary between values. **5.** Parameters do not vary. **6.** *Know* $\bar{x} = 1.3649$ mmol/L. **7.** *Know* $\bar{x} = 1.3649$ mmol/L.
Ex. 23.9. s.e.$(\bar{x}) = 5.36768$; approx. 95% CI: 50.56 to 72.04 s. Stat. valid.
Ex. 23.11. **1.** *One* observation $x = 44$; claimed *population mean* is $\mu = 45$. **2.** OK to have decimal value as *mean*. **3.** $\bar{x} = 44.9$; $\mu = 45$: different things; why should they be same? **4.** CI allows for sampling variation. **5.** 44.850 to 44.950. **6.** Possibly lying; not certain. **7.** Sample mean; $\bar{x} = 44.9$. **8.** Population mean; value unknown, but *claimed* to be $\mu = 45$.

Chap. 24: More details about CIs

Ex. 24.1. **1.** CIs are intervals for unknown *parameters*, not known *statistics*. **2.** CIs for proportion (or percentage), not *number* of trees. (The CI is 68% anyway, not 95%.)
Ex. 24.3. **1.** CIs *not* about individuals. **2.** CIs *not* about *sample* means.
Ex. 24.5. Intervals for different things. First: 95% CI for *mean* weight. Second: *not* CI; for weights of *individuals* possums.

Chap. 25: Making decisions

Ex. 25.1. **1.** Yes; very likely (can't be sure). **2.** Assuming fair die, *not* expect ⚅ ten times in row.
Ex. 25.3. Seems unlikely.
Ex. 25.5. **1.** $p = 0.36$. **2.** $\hat{p} = 0.433$; probably not. **3.** $\hat{p} = 0.1$; probably.

Chap. 26: Hypothesis tests: one proportion

Ex. 26.1. In tests, p assumed known. In CI, have no value for p to use.
Ex. 26.3. 0.38 is *sample* proportion; RQ asks about *pop.* proportion of 1/6.
Ex. 26.5. Tests *not* about sample value (we *know* value of $\hat{p}$), but about unknown pop. value (i.e., p).
Ex. 26.7. **1.** One-in-five: 0.2. **2.** H_0: $p = 0.2$; H_1: $p > 0.2$. **3.** One-tailed. **4.** Normal; mean 0.2, std deviation s.e.$(\hat{p}) = 0.0444$. **5.** $\hat{p} = 0.6173$; $z = 9.39$: P *very small*: Very strong evidence people do better-than-guessing at identifying placebo.
Ex. 26.9. H_0: $p = 0.5$; H_1: $p \neq 0.5$. $\hat{p} = 0.39726$; s.e.$(\hat{p}) = 0.05727$; $z = -1.794$. P not that small. No evidence of difference.
Ex. 26.11. H_0: $p = 0.0602$; H_1: $p < 0.602$ (*one*-tailed). $\hat{p} = 0.5008489$; $n = 589$: s.e.$(\hat{p}) = 0.0201689$, so $z = -5.015$. P very small. Strong evidence prop. F using machines lower than prop. F in uni pop.
Ex. 26.13. H_0: $p = 0.5$; H_1: $p > 0.5$ (one-tailed). $\hat{p} = 0.802817$; $n = 71$: s.e.$(\hat{p}) = 0.0593391$, so $z = 5.10$. P very small. Strong evidence majority like breadfruit pasta (for pop. represented by sample anyway).
Ex. 26.15. H_0: $p = 1/16 = 0.625$; H_1: $p \neq 0.0625$. $\hat{p} = 0.139535$ and s.e.$(\hat{p}) = 0.018457$; $z = 26.3$: *massive*; P very small. Very strong evidence pop. proportion not 1/16; borers *not* resistant.
Ex. 26.17. $\hat{p} = 0.56$; $z = 1.91$. Slight evidence of bias.

Chap. 27: Hypothesis tests: one mean

Ex. 27.1. **1.** μ, pop. mean speed (km.h^{-1}). **2.** H_0: $\mu = 90$; H_1: $\mu > 90$ (one-tailed). **3.** s.e.$(\bar{x}) = 0.6937$. **5.** $t = 9.46$. **6.** t-score *huge*; (one-tailed) P very small: very strong evidence mean speed of vehicles on road greater than 90 km.h^{-1}. **7.** Stat. valid.
Ex. 27.3. H_0: $\mu = 50$; H_1: $\mu > 50$ (one-tailed). s.e.$(\bar{x}) = 4.701076$. $t = 7.23$: P very small. *Very* strong evidence ($P < 0.001$) mean mental demand *greater* than 50.
Ex. 27.5. H_0: $\mu = 14$; H_1: $\mu \neq 14$ (two-tailed). s.e.$(\bar{x}) = 0.092493$. t-score: 10.35: *huge*; P very small. *Very* strong evidence ($P < 0.001$) mean weight of Fun Size *Cherry Ripe* bar not 14 g. SD: the variation in weight of individual bars. SE: the variation in sample means for $n = 67$.
Ex. 27.7. H_0: $\mu = 10$ (or $\mu \geq 10$) and H_1: $\mu < 10$. F: s.e.$(\bar{x}) = 0.05924742$; $t = -25.32$. M: s.e.$(\bar{x}) = 0.0700152$; $t = -19.42$. Both P extremely small. For both boys and girls, very strong evidence mean sleep time on weekend less than 10 h.
Ex. 27.9. **1.** μ: population mean pizza diameter. **2.** $\bar{x} = 11.486$; $s = 0.247$. **3.** 0.02205479. **4.** H_0: $\mu = 12$; H_1: $\mu \neq 12$. **5.** Two-tailed; RQ asks if diameter is 12 inches, or not. **6.** Normal distribution, mean 12 and std dev. of s.e.$(\bar{x}) = 0.02205$. **7.** $t = -23.3$. **8.** P *really* small. **9.** Very strong evidence mean diameter not 12 inches. **10.** n much larger than 25; stat. valid.

Chap. 28: More details about hypothesis tests

Ex. 28.1. Use 68–95–99.7 rule and diagram: **1.** Very small; certainly less than 0.003 (99.7% between −3 and 3). **2.** Very small; bit bigger than 0.003 (99.7% between −3 and 3). **3.** Bit smaller than 0.05 (95% between −2 and 2). **4.** *Very* small; *much* smaller than 0.003.
Ex. 28.3. *Half* values in Ex. 28.1. **1.** Very small; certainly less than 0.0015. **2.** Very small; bit bigger than 0.0015. **3.** Bit smaller than 0.025. **4.** *Very* small; *much* smaller than 0.0015.
Ex. 28.5. P *just larger* than 0.05; 'slight evidence' to support H_1. P *just smaller* than 0.05; 'moderate evidence' to support H_1. The difference between 0.0501 and 0.0499 trivial.
Ex. 28.7. **1.** Hypotheses about *parameters.* **2.** RQ two-tailed. **3.** 36.8052 is sample mean; hypothesis written *before* data collected. **4.** Hypotheses about parameters; 36.8052 is sample mean. These test if *sample* mean is 36.8052; we *know* it is. **5.** Hypothesis written *before* data collected. **6.** Hypotheses about parameters.
Ex. 28.9. **1.** Conclusion about pop. **mean** energy intake. **2.** Conclusions *never* about statistics. **3.** Conclusion about pop. **mean** energy intake. **4.** P *is* 0.018, not *less than* 0.018.
Ex. 28.11. Statements 1 and 3 consistent.

Chap. 29: CIs and tests: mean differences (paired data)

Ex. 29.1. 1. Paired. **2.** Paired.

Ex. 29.3. How much longer task takes on the PC.

Ex. 29.5. CI calculated as: −3.24 to 0.52 days. Meaning, interpretation same as in Sect. 29.7.

Ex. 29.7. 1. *Analysis*: farm. *Observation*: individual fruits. **2.** Pairs have same farm management, soil, etc. **3.** Not shown. **4.** Not shown. **5.** Mean increase in fruit weight from normal (2015) to dry (i.e., normal minus dry). **6.** H_0: $\mu_d = 0$ and H_1: $\mu_d \neq 0$. **7.** Not shown. **8.** $t = 0.205$. **9.** P large; from software, $P = 0.839$. **10.** 19.53 g to 23.99 g heavier in normal year (2015). **11.** Probably stat. valid; n just less than 25. **12.** No evidence ($t = 0.205$; two-tailed $P = 0.839$) of mean increase in weight of squash from dry to normal years (mean change: 2.23 g (95% CI −19.53 to 23.99 g), heavier in normal year).

Ex. 29.9. 1. Not shown. **2.** How much tastier broccoli is with dip. **3.** $t = 1.699$; *approx.* one-tailed P between 16% and 2.5%; not sure if P larger than 0.05, but likely (t-score quite a distance from $z = 1$). Evidence *probably* doesn't support H_1. **4.** Approx. 95% CI: −0.92 to 11.32. **5.** Stat. valid.

Ex. 29.11. $\bar{d} = -424.25$; s.e.$(\bar{d}) = 467.9404$; $n = 20$: $t = -0.907$. $P = 0.376$: evidence doesn't support H_1. 95% CI: −1403.7 to 555.2. Test perhaps not stat. valid $n < 25$); but histogram of data suggests population *might* have normal distribution); P so large probably makes little difference.

Ex. 29.13. 1. *Diffs* are *during* minus *before*: *positive* diffs means *during* value higher. **2.** s.e.$(\bar{d}) = 3.5150$. **3.** $t = 0.762$; P is large; no evidence of change; −4.35 to 9.71 mins.

Ex. 29.15. 0.89 to 4.64 pounds. Possibly not practically important.

Ex. 29.17. Exact CI: 1.204 to 0.069 cm greater barefoot.

Chap. 30: CIs and tests: comparing two means

Ex. 30.1. How much greater the mean lymphocytes cell diameter is compared to tumour cells.

Ex. 30.3. Normal; mean $\mu_B - \mu_A$; std dev.: 2.965 km.h^{-1}.

Ex. 30.5. 1. Parameter: $\mu_F - \mu_M$. Estimate: $\bar{x}_F - \bar{x}_M = -0.06$ m. **2.** Not shown. **3.** 0.092487 **4.** Not shown. **5.** −1.94 to 0.24 m. **6.** $H_0 : \mu_F - \mu_M = 0$; $H_1 : \mu_F - \mu_M \neq 0$. **7.** $t = 0.65$; P very large. **8.** No evidence ($t = 0.65$; two-tailed $P > 0.10$) in sample that mean length of adult gray whales is diff in pop. for F (mean: 4.66 m; std dev.: 0..38 m) and M (mean: 4.60 m; std dev.: 0.30 m; 95% CI for the diff: −1.94 m to 0.24 m). **9.** Yes.

Ex. 30.7. 1. $\mu_P - \mu_E$, reduction in mean duration for those using echinacea. **2.** 0.2728678 days; echinacea: 0.2446822 days **3.** 0.3665054. **4.** Not shown. **5.** −0.203 to 1.263 days. **6.** 5.85 to 6.83 days. **7.** H_0: $\mu_P - \mu_E = 0$; H_1: $\mu_P - \mu_E > 0$ (one-tailed). **8.** 0.3665054. **9.** $t = 1.45$; one-tailed P between 0.025 and 0.32; using z-tables, P approx. 0.074. **10.** Slight evidence of diff. **11.** Not given. **12.** Yes. **13.** Probably not practically important (diff 0.53 days).

Ex. 30.9. 1. Amount of DMFT greater in non-industrialised countries, as upper table has negative mean. **2.** H_0: $\mu_I - \mu_{NI} = 0$; H_1: $\mu_I - \mu_{NI} \neq 0$. **3.** 11.9 to 22.5, greater for industrialised. **4.** Very strong evidence in sample ($P < 0.001$) mean annual sugar consumption per person diff for industrialised (mean: 41.8 kg/person/y) and non-industrialised (mean: 24.6 kg/person/y) countries (95% CI for the diff 11.9 to 22.5). **5.** Yes.

Ex. 30.11. 1. $\mu_Y - \mu_O$: mean amount younger women can lean further forward than older. **2.** Boxplot. **3.** Not shown. **4.** Approx.: 10.2 to 18.8°C. Exact CI (Row 2): 9.1 to 19.9°C. Different: sample sizes small. **5.** One. **6.** H_0: $\mu_Y - \mu_O = 0$; H_1: $\mu_Y - \mu_O > 0$. **7.** $t = 6.69$ (*second row*); $P < 0.001/2$ as one-tailed; i.e., $P < 0.0005$. **8.** Very strong evidence in sample ($t = 6.69$; one-tailed $P < 0.0005$) that pop. mean one-step fall recovery angle for healthy women *greater* for young women (mean: 30.7°C; std dev.: 2.58°C; $n = 10$) compared to older women (mean: 16.20°C; std dev.: 4.44°C; $n = 5$; 95% CI for the diff: 9.1°C to 19.9°C). **9.** Probably not stat. valid.

Ex. 30.13. H_0: $\mu_M - \mu_F = 0$; H_1: $\mu_M - \mu_F \neq 0$. From output, $t = -2.29$; (two-tailed) $P = 0.024$. Moderate evidence ($P = 0.024$) mean internal body temperature diff for F (mean: 36.9°C) and M (mean: 36.7°C). Diff between the means (0.2 of degree) of little *practical* importance.

Ex. 30.15. 1. 2.76 kg. **2.** CB: 0.227 to 5.79 kg; Control: −3.68 to 2.78 kg. **3.** −0.68 to 7.59 kg, greater for CB. **4.** $t = 1.67$; Two-tailed $P = 0.095$. Slight evidence of diff.

Ex. 30.17. $t = 2.631$; one-tailed $P = 0.0055$; evidence of diff.

Chap. 31: CIs and tests: comparing two odds or proportions

Ex. 31.1. Both odds: 6.04.

Ex. 31.3. Normal; mean $p_P - p_N$ and std dev. 0.0428. For OR: not normal distribution.

Ex. 31.5. 1. $z = 3.26$. **2.** Very small.

Ex. 31.7. 1. No-PG proportion (Row 1), minus PG (Row 2). **2.** H_0: $p_{\text{With}} - p_{\text{W'out}} = 0$; H_1: $p_{\text{With}} - p_{\text{W'out}} \neq 0$. **3.** $z = 1.14$; $P = 0.253$; no evidence of diff. **4.** −0.071 to 0.296, larger intent to purchase for those without PG study. **5.** Odds yes (Col 1), comparing no-PG (Row 1) to PG (Row 2). **6.** One option: H_0: OR = 1; H_1: OR ≠ 1. **7.**

$\chi^2 = 1.31$; $P = 0.253$; no evidence of diff. **8.** OR between 0.68, 4.28. **9.** Smallest expected: 10.6; yes.

Ex. 31.9. **1.** Not shown. **2.** 0.3000; 0.3033. **3.** −0.0033. **4.** s.e.$(\hat{p}_1) = 0.0648074$; s.e.$(\hat{p}_2) = 0.0416170$; s.e. for diff.: 0.077019. **5.** -0.0033 ± 0.154: −0.157 to 0.151. **6.** −0.154 to 0.148. **7.** Not given. **8.** 0.429; 0.436. **9.** $0.429/0.436 = 0.985$. **10.** 0.480 to 2.02. **11.** Not shown **12.** $p = 0.4333$; s.e. for diff: 0.0832097. **13.** $z = (-0.0033 - 0)/0.0832097 = -0.040$; very small P; no evidence of diff. **14.** $\chi^2 = 0.002$ (output); $P = 0.966$; no evidence of diff.

Ex. 31.11. **1.** −0.1918. **2.** 0.04643. **3.** 0.04202. **4.** Second uses common p: test assumes p same in both groups. **5.** $-0.1918 \pm (2 \times 0.04643)$: −0.285 to −0.099. **6.** −0.273 to −0.111. **7.** Not given. **8.** $z = (-0.1918 - 0)/0.04202 = -4.56$; small P; evidence of diff. **9.** 0.443. **10.** 0.315 to 0.623. **11.** $P < 0.0001$; small P; evidence of diff. **12.** Yes. **13.** Observational.

Ex. 31.13. **1.** Not shown. **2.** F: 0.060, M: 0.205; diff: 0.145. **3.** 0.0640, 0.256; 0.250. **4.** s.e. for diff: 0.0243; 0.096 to 0.193. **5.** 0.098 to 0.192. **6.** Not given. **7.** 2.45 to 6.61. **8.** $P < 0.0001$. **9.** Evidence of a diff. **10.** Yes.

Ex. 31.15. **1.** OR: 0.3478261. **2.** Diff: −0.12. **3.** Proportions equal; not equal. **4.** Odds equal; not equal. **5.** $z = 2.12$; P 'small'. **6.** Some evidence of diff. **7.** Yes.

Ex. 31.17. **1.** H_0: No association; H_1: Association. **2.** 23.0522; $P = 0.00004$. **3.** Very strong evidence of association. **4.** Yes.

Chap. 32: Finding sample sizes for CIs

Ex. 32.1. Larger.

Ex. 32.3. **1.** At least 25. **2.** At least 100 (4 times as many). **3.** At least 400 (16 times as many). **4.** Halve width: 4 times as many. **5.** Quarter width: 16 times as many. **6.** More needed for greater precision.

Ex. 32.5. **1.** At least 10 000. **2.** At least 2 500. **3.** At least 1 000. **4.** Expensive (time *and* money): 10 000 and 2 500 probably unrealistic.

Ex. 32.7. Use $s = 0.43$. **1.** At least 1 849. **2.** At least 296. **3.** At least 74. **4.** Expensive (time *and* money); 74 more realistic.

Ex. 32.9. Use, say, $s = 13$. **1.** At least 81 pairs. **2.** At least 76 pairs.

Ex. 32.11. Use, say, $s = 0.35$. **1.** At least 44 in each group. **2.** At least 98 in each group. **3.** Info not relevant to goldfish.

Ex. 32.13. 2 223.

Chap. 33: Correlation and regression

Ex. 33.1. Answers *very* approximate. **1.** r moderate, positive; $\hat{y} = 4 + 1.5x$ **2.** r reasonably strong, positive; $\hat{y} = 6 + 2.3x$. **3.** r not apt: variation in y increases as x increases. **4.** r reasonably strong, negative; $\hat{y} = 8 - 1.5x$.

Ex. 33.3. Any could be; can't tell.

Ex. 33.5. **1.** $b_0 = 3.5$; $b_1 = -0.14$. **2.** $b_0 = 2.1$; $b_1 = -0.0047$. **3.** $b_0 = -25.2$; $b_1 = -0.95$. **4.** $b_0 = 0.15$; $b_1 = -0.22$.

Ex. 33.7. Not shown.

Ex. 33.9. **1.** H_0: $\rho = 0$; H_1: $\rho \neq 0$. **2.** No evidence of relationship. **3.** Approx. linear; variation in STAI same for diff levels of experience. $n \geq 25$.

Ex. 33.11. No: variation increasing.

Ex. 33.13. **1.** b_0: *no* time spent on application, mean 0.27 g applied; nonsense. b_1: each extra minute adds average of 2.21 g sunscreen. **2.** Slope: g/min; intercept: g. **3.** β_0 could be zero; makes sense. **4.** $\hat{y} = 18$ g. **5.** 64% reduction in unexplained variation using time. **6.** $r = 0.8$; strong positive correlation.

Ex. 33.15. **1.** Probably linear; increasing; approx. constant variance in y as x increases. **2.** H_0: $\rho = 0$; H_0: $\rho > 0$. **3.** $r = 0.837$; $P < 0.00005$. Very strong evidence of positive relationship. **4.** Yes.

Ex. 33.17. **1.** $r = 0.264$. **2.** $R^2 = 6.99\%$; using neck circumference reduces unknown variation by about 7%. **3.** $\hat{y} = -24.47 + 1.36x$: y is REI; x is neck circum. (in cm). **4.** Each 1 cm increase in neck circum. increases REI by average of 1.36. **5.** Approx. CI: 0.0575 to 2.675. **6.** $t = 2.09$, $P = 0.041$: slight evidence of relationship. **7.** Stat. valid.

Ex. 33.19. **1.** Very strong, negative linear relationship. **2.** $r = -\sqrt{0.9929} = -0.9964$; must be negative. **3.** $\hat{y} = 17.47 - 2.59x$: x is percentage bitumen by wt; y is percentage air voids by volume. **4.** *Slope*: increase in bitumen wt by one percentage point *decreases* average percentage air voids by volume by 2.59 percentage points. *Intercept*: extrapolation: 0% bitumen content by wt, percentage air voids by volume 17.47%. **5.** $t = -74.9$: massive; extremely strong evidence ($P < 0.001$) of relationship. **6.** $P < 0.001$, as for slope. **7.** $\hat{y} = 4.5027$, or 4.5%; good prediction, as relationship strong. **8.** $\hat{y} = 1.909$, or 1.9%; perhaps poor: extrapolation. **9.** Yes.

Ex. 33.21. **1.** $r = 0.271$; $R^2 = 7.3\%$. **2.** $t = 1.35$; $P = 0.190$; no evidence of relationship. **3.** $\hat{y} = -5.647 + 0.123x$.

Ex. 33.23. **1.** Left probably F. **2.** Sex B. **3.** A: $r = 0.600$; B: $r = 0.815$. **4.** B (M). **5.** A: $\hat{y} = -2289 + 21.31x$; B: $\hat{y} = -3621 + 27.63x$. **6.** Both: $P < 0.0001$ (probably use one-tailed test). **7.** A: 2 503 kg; B: 2 596 kg. **8.** Sample sizes, linearity OK; for Sex B, perhaps increasing variation.

Ex. 33.25. $r = -0.712$ and $P < 0.001$. Regression: $\hat{y} = 13.39 - 0.093x$.

Chap. 34: Selecting an analysis

Ex. 34.1. Only **3. 2.** Almost certainly; $n = 24$ very close to $n = 25$.

Ex. 34.3. Summary of mean *diffs*; histogram of diffs. Paired samples t-test; CI for mean diff.

Ex. 34.5. Comparing two proportions (or ORs); stacked, side-by-side bar chart. CI for diff in proportions (or CI for OR).

Ex. 34.7. Correlation or regression, if linear.

Chap. 35: Writing and reporting research

Ex. 35.1. 1. *to.* **2.** *its.* **3.** One sample of 50 individuals; use 'mean' or 'median', not 'average'; units of age is 'years'. **4.** Should all be one sentence.

Ex. 35.3. 1. Ambiguous; sound like cage is M; passive. 'The cage contained one male rat.' **2.** Seaweed removed from beaker, or from lake water? 'The research assistant recorded the pH of the lake water (after removing weeds) in the beaker.'

Ex. 35.5. 1. 'Substantial' if a *large* change is expected (quote statistics (e.g., P-value) if *statistically* significant intended). **2.** 'The data *are...*'

Ex. 35.7. Number decimal places ridiculous.

Ex. 35.9. RQ: P, O, C and I unclear; fonts should be identified. Perhaps better: for students, is mean reading speed for text in Georgia font same as for Calibri font? **Abstract** statement poor (*fonts* are not fast or slow). Perhaps: sample provided evidence mean reading speeds different ($P =$?), when comparing text in Georgia font (mean: ?) and Calibri font (mean: ?; 95% CI for diff: ? to ?).

Ex. 35.11. Variables *qualitative*: means inappropriate; use OR; values almost certainly refer to CI for OR. Without more information, we can't be sure what the OR means.

Chap. 36: Reading and critiquing research

Ex. 36.1. 1. Convenience; self-selected. Those in study *may* be different than those not in. **2.** Inclusion criteria. **3.** Ethical (drop-outs happen); accurate description of study. **4.** Not ecologically valid. **5.** Paired t-test. **6.** Null: no mean diff between counts on phone, manually counted; alternative: a diff. **7.** P small; evidence mean diff in step-count between methods cannot be explained by chance: likely a diff. **8.** Valid.

Ex. 36.3. 1. Only some evidence of diff in mean age. **2.** Comparing the two groups; age possible confounder. **3.** Two-sample t-test. **4.** 0.03376. **5.** $t = 2.07$; small P; evidence of diff. **6.** Probably, given std errors rounded. **7.** Conceptual. **8.** Not shown. **9.** χ^2. **10.** $z = 1.75$; P between 5% and 32%: not helpful. **11.** Observational: not cause-and-effect; no confounders noted; very restricted population.

Ex. 36.5. 1. χ^2-test to compare proportions. **2.** No evidence of diff in survival rates at temps. **3.** Evidence surviving *Cx.* had larger mean size compared to surviving *Ae.*. **4.** Two-sample t-test. **5.** 0.010628. **6.** $t = 26.3$; very small P; very strong evidence of diff in mean lengths. **7.** Yes. **8.** Yes. **9.** *Cx.*: evidence mean sizes at temps diff; *Ae.*: no evidence mean sizes at temps diff. **10.** Two-sample t-tests. **11.** Intercept: -55.40 to 16.28; slope: 3.88 to 59.40. **12.** $t = 2.28$; expect small P; evidence of linear association. **13.** Need scatterplot to be sure, but $n \geq 25$. **14.** When predator–size ratio increases by one, predation efficiency increases by 31.64 percentage points. **15.** Unknown variation reduces by 8.7% using predator–size ratio. **16.** $r = 0.294$.

Ex. 36.7. 1. Stratified? **2.** Strong evidence the mean number of actinomycetes diff. **3.** Possibly not; sample sizes small. **4.** Very strong evidence mean number higher in CNV. **5.** Larger actinomycetes numbers linearly associated with lower corky root severity. **6.** $R^2 = 57.8\%$; unknown variation decreases by 57.8% using actinomycete abundance.

Bibliography

Dario M. Aedo-Ortiz, Eldon D. Olsen, and Loren D. Kellogg. Simulating a harvester-forwarder softwood thinning: a software evaluation. *Forest Products Journal*, 47(5):36–41, 1997.

Carolina M. Affonso and Mladen Kezunovic. Probabilistic assessment of electric vehicle charging demand impact on residential distribution transformer aging. In *2018 IEEE International Conference on Probabilistic Methods Applied to Power Systems (PMAPS)*, pages 1–6. IEEE, 2018.

Selina Agbayani, Sarah M. E. Fortune, and Andrew W. Trites. Growth and development of North Pacific gray whales (*Eschrichtius robustus*). *Journal of Mammalogy*, 101(3):742–754, 2020.

Alan Agresti and Christine A. Franklin. *Statistics: The Art and Science of Learning from Data*. Pearson Education Limited, third edition, 2007.

Michael Alary and J. R. Joly. Risk factors for contamination of domestic hot water systems by legionellae. *Applied and Environmental Microbiology*, 57(8):2360–2367, 1991.

Ilona Alberca. Kinetic and temporal parameters calculated from raw data collected via wireless instrumented wheel for measuring 3D pushrim kinetics of a racing wheelchair. *DANS Data Station Life Sciences*, 1, 2022. doi: 10.17026/dans-xjf-bs8v.

Ilona Alberca, Félix Chénier, Marjolaine Astier, Eric Watelain, Jean-Marc Vallier, Didier Pradon, and Arnaud Faupin. Sprint performance and force application of tennis players during manual wheelchair propulsion with and without holding a tennis racket. *PLOS ONE*, 17(2):e0263392, 2022. doi: 10.1371/journal.pone.0263392.

Nesrin Alharthy, Raghad Almotairy, Rahaf Aldulhum, Albatool Alghamdi, Reem Aquil, Ghada Alkharaan, Sara Alsuwais, and Abdullah Alshibani. Knowledge and experience of paramedics concerning patients with hearing and visual disability. *BMC Emergency Medicine*, 23(1):91, 2023.

Alicia M. Allen, Nermine M. Abdelwahab, Samantha Carlson, Tyler A. Bosch, Lynn E. Eberly, and Kola Okuyemie. Effect of brief exercise on urges to smoke in men and women smokers. *Addictive Behaviors*, 77:34–37, 2018.

Stephanie Alley, Pauline Wellens, Stephanie Schoeppe, Hein de Vries, Amanda L. Rebar, Camille E. Short, Mitch J. Duncan, and Corneel Vandelanotte. Impact of increasing social media use on sitting time and body mass index. *Health Promotion Journal of Australia*, 28:91–95, 2017.

Ghada AlTarawneh and Simon Thorne. A pilot study exploring spreadsheet risk in scientific research. In *Proceedings of the EuSpRIG 2016 Conference 'Spreadsheet Risk Management'*, 2016. URL www.eusprig.org.

Douglas G. Altman. *Practical Statistics for Medical Research*. Chapman & Hall, 1991.

Arslan Ahmed Amin and Khalid Mahmood-ul-Hasan. Robust active fault-tolerant control for internal combustion gas engine for air–fuel ratio control with statistical regression-based observer model. *Measurement and Control*, 52(9–10):1179–1194, 2019. doi: 10.1177/0020294018823031.

E. B. Andersen. Multiplicative Poisson models with unequal cell rates. *Scandinavian Journal of Statistics*, 4:153–158, 1977.

Anonymous. Green tea cuts risk of cancer. *The Sunday Mail*, page 19, 04 November 2012.

Joyce Augustino, Fabiola Moshi, Angelina Joho, and Joanes Faustinea Kihulya Mageda. Dataset comparing the effectiveness of perineal cold pack application over oral paracetamol 1000 mg on postpartum perineal pain among women after spontaneous vaginal delivery in Dodoma region. *Data in Brief*, 51:1–9, 2023. doi: 10.1016/j.dib.2023.109766.

Nikolaus Axmann, Torben Fischer, Kevin Keller, Kevin Leiby, Daniel Stein, and Paul Wang. Access and adoption of hybrid seeds: evidence from Uganda. *Journal of African Economies*, 29(3):215–235, 2020.

Nur Farhanatul Syasya Mohd Azwari and Abdul Azeez Kadar Hamsa. Evaluating actual speed against the permissible speed of vehicles during free-flow traffic conditions. *Jurnal Kejuruteraan*, 33(2):183–191, 2021.

Z. Bacho, F. J. E. Lajangang, N. Y. Khin, S. S. Shah, Y. K. Chia, E. Jalil, C. C. S. Kelvin, and D. M. Ag Daud. The effects of comprehensive core body resistance exercise on lower extremity motor function among stroke survivors. *Journal of Physics: Conference Series*, 1358(1):012025, 2019.

Jane Bambauer. Defending the dog. *Oregon Law Review*, 91:1203, 2012.

Nigel Barr, Mark Holmes, Anne Roiko, Peter Dunn, and Bill Lord. Self-reported behaviors and perceptions of Australian paramedics in relation to hand hygiene and gloving practices in paramedic-led health care. *American Journal of Infection Control*, 45(7):771–778, 2017.

Bruce Barrett, Roger Brown, Dave Rakel, Marlon Mundt, Kerry Bone, Shari Barlow, and Tola Ewers. Echinacea for treating the common cold: a randomized trial. *Annals of Internal Medicine*, 153(12): 769–777, 2010.

Tad M. Bartareau. Estimating the live body weight of American black bears in Florida. *Journal of Fish and Wildlife Management*, 8(1):234–239, 2017.

Sahas Barve and André A. Dhondt. Elevational replacement of two Himalayan titmice: interspecific competition or habitat preference? *Journal of Avian Biology*, 48(9):1189–1194, 2017.

David F. Bauer. Constructing confidence sets using rank statistics. *Journal of the American Statistical Association*, 67(339):687–690, 1972.

Lori Beaman, Dean Karlan, Bram Thuysbaert, and Christopher Udry. Profitability of fertilizer: experimental evidence from female rice farmers in Mali. *American Economic Review*, 103(3):381–86, 2013.

Heather Becker, Alexa K. Stuifbergen, and Dolores Sands. Development of a scale to measure barriers to health promotion activities among persons with disabilities. *American Journal of Health Promotion*, 5(6): 449–454, 1991.

Vassiliki Benetou, Afroditi Kanellopoulou, Eleftheria Kanavou, Anastasios Fotiou, Myrto Stavrou, Clive Richardson, Philippos Orfanos, and Anna Kokkevi. Diet-related behaviors and diet quality among school-aged adolescents living in Greece. *Nutrients*, 12(12):3804, 2020.

S. K. Bhargava, S. Ramji, Arun Kumar, M. A. N. Mohan, Jasbir Marwah, and H. P. Sachdev. Mid-arm and chest circumferences at birth as predictors of low birth weight and neonatal mortality in the community. *BMJ*, 291(6509):1617–1619, 1985.

Anthony R. Bird, Michelle S. Vuaran, Roger A. King, Manny Noakes, Jennifer Keogh, Matthew K. Morell, and David L. Topping. Wholegrain foods made from a novel high-amylose barley variety (*Himalaya 292*) improve indices of bowel health in human subjects. *British Journal of Nutrition*, 99:1032–1040, 2008.

Hjalti Mar Bjornsson, Gudrun G. Bjornsdottir, Hronn Olafsdottir, Brynjolfur Arni Mogensen, Brynjolfur Mogensen, and Gestur Thorgeirsson. Effect of replacing ambulance physicians with paramedics on outcome of resuscitation for prehospital cardiac arrest. *European Journal of Emergency Medicine*, 28(3):227–232, 2021.

Benjamin F. Blair and Marshall C. Lamb. Evaluating concentrations of pesticides and heavy metals in the U.S. peanut crop in the presence of detection limits. *Peanut Science*, 44:124–133, 2017.

Berglind Soffia Blondal, O. G. Geirsdottir, T. I. Halldorsson, A. M. Beck, P. V. Jonsson, and A. Ramel. HOMEFOOD randomised trial—six-month nutrition therapy in discharged older adults reduces hospital readmissions and length of stay at hospital up to 18 months of follow-up. *The Journal of Nutrition, Health and Aging*, 27(8):632–640, 2023.

David E. Bock, Paul F. Velleman, and Richard D. De Veaux. *Stats: Modeling the World.* Addison-Wesley, 2010. ISBN 9780321570444.

Andrew D. Boltuch, Michael A. Marcotte, Christopher M. Treat, and Anthony L. Marcotte. The palmaris longus and its association with carpal tunnel syndrome. *Journal of Wrist Surgery*, 9(06):493–497, 2020.

Alyne M. Botelho, Anice M. de Camargo, Moira Dean, and Giovanna M. R. Fiates. Effect of a health reminder on consumers' selection of ultra-processed foods in a supermarket. *Food Quality and Preference*, 71:431–437, 2019.

Anthony A. Braga, John M. MacDonald, and Lisa M. Barao. Do body-worn cameras improve community perceptions of the police? Results from a controlled experimental evaluation. *Journal of Experimental Criminology*, pages 1–32, 2021.

Amy Brand, Liz Allen, Micah Altman, Marjorie Hlava, and Jo Scott. Beyond authorship: attribution, contribution, collaboration, and credit. *Learned Publishing*, 28(2):151–155, 2015.

H. J. Brockmann. Satellite male groups in horseshoe crabs, *Limulus polyphemus. Ethology*, 102:1–21, 1996.

William R. Brooks. Hermit crabs alter sea anemone placement patterns for shell balance and reduced predation. *Journal of Experimental Marine Biology and Ecology*, 132(2):109–121, 1989.

Lawrence Brown, Christa L. Whitney, Richard C. Hunt, Michael Addario, and Try Hogue. Do warning lights and sirens reduce ambulance response times? *Prehospital Emergency Care*, 4(1):70–74, 2000.

Douglas D. Brunette, John Kominsky, and Ernest Ruiz. Correlation of emergency health care use, 911 volume, and jail activity with welfare check distribution. *Annals of Emergency Medicine*, 20(7):739–742, 1991.

Elizabeth Brunton, Jessica Bolin, Javier Leon, and Scott Burnett. Fright or flight? Behavioural responses of kangaroos to drone-based monitoring. *Drones*, 3(2):41, 2019.

Stephanie Budgett, Maxine Pfannkuch, Matt Regan, and Chris J. Wild. Dynamic visualizations and the randomization test. *Technology Innovation in Statistics Education*, 7(2), 2013.

Erwin Bulte, Gonne Beekman, Salvatore Di Falco, Joseph Hella, and Pan Lei. Behavioral responses and the impact of new agricultural technologies: evidence from a double-blind field experiment in Tanzania. *American Journal of Agricultural Economics*, 96(3):813–830, 2014.

Ahmet Yavuz Candan, Yusuf Katılmış, and Çağrı Ergin. First report of *fusarium* species occurrence in loggerhead sea turtle (*caretta caretta*) nests and hatchling success in Iztuzu Beach, Turkey. *Biologia*, 76: 565–573, 2021.

Guillermo De Castro-Maqueda, José V. Gutierrez-Manzanedo, Jorge R. Fernandez-Santos, Mario Linares-Barrios, and Magdalena De Troya Martín. Sun protection habits and sun exposure of physical education teachers in the south of Spain. *Photochemistry and Photobiology*, 95(6):1468–1472, 2019.

Centers for Disease Control and Prevention (CDC). Behavioral Risk Factor Surveillance System Survey Questionnaire (BRFSS), 2021–2023.

Centers for Disease Control and Prevention (CDC). National Center for Health Statistics. National Health and Nutrition Examination Survey Data, 2024.

Centers for Disease Control and Prevention (CDC). National Center for Health Statistics. Third National Health and Nutrition Examination Survey, 1988–1994, NHANES III Laboratory Data File. Public Use Data File Documentation Number 76200, 1996. URL https://wwwn.cdc.gov/nchs/data/nhanes3/1a/readme.txt.

Esther Chan, Simone Taylor, Jennifer Marriott, and Bill Barger. Exploration of attitudes and barriers to bringing patient's own medications to the Emergency Department: a survey of paramedics. *Australasian Journal of Paramedicine*, 6(4), 2008.

Colin A. Chapman. Association patterns of spider monkeys: the influence of ecology and sex on social organisation. *Behavioral Ecology and Sociobiology*, 26:409–414, 1990.

D. Chapman, J. Peiffer, C. R. Abbiss, and P. B. Laursen. A descriptive physical profile of Western Australian male paramedics. *Journal of Emergency Primary Health Care*, 5(1), 2007.

C. R. Charig, D. R. Webb, S. R. Payne, and J. E. A. Wickham. Comparison of treatment of renal calculi by open surgery, percutaneous nephrolithotomy, and extracorporeal shockwave lithotripsy. *British Medical Journal*, 292:879–882, 1986. doi: 10.1136/bmj.292.6524.879.

William Checkley, Robert H. Gilman, Robert E. Black, Andres G. Lescano, Lilia Cabrera, David N. Taylor, and Lawrence H. Moulton. Effects of nutritional status on diarrhea in Peruvian children. *Journal of Pediatrics*, 140(2):210–218, 2002. URL http://www.ncbi.nlm.nih.gov/pubmed/11865273.

Kai Chen, Mengjuan Wu, Ying Zhang, Fang Zhang, Huan Wang, Juanjuan Liang, Peng Yan, En Li, Liang Yao, Jinwang Xu, et al. Two introduced crocodile species had changed reproductive characteristics in China. *Animal Reproduction Science*, 196:150–159, 2018.

Shih-Hao Chen, Hung-Chieh Chang, Po-Wei Chiu, Ming-Yuan Hong, I-Chen Lin, Chih-Chun Yang, Chien-Te Hsu, Chia-Wei Ling, Ying-Hsin Chang, Ya-Yun Cheng, et al. Triage body temperature and its influence on patients with acute myocardial infarction. *BMC Cardiovascular Disorders*, 23(1):388, 2023.

Annie Chih, Elke Rudloff, Cheryl Waldner, and Andrew K. J. Linklater. Incidence of hypochloremic metabolic alkalosis in dogs and cats with and without nasogastric tubes over a period of up to 36 hours in the intensive care unit. *Journal of Veterinary Emergency and Critical Care*, 28(3):244–251, 2018.

Derek E. Chitwood and Joseph S. Devinny. Treatment of mixed hydrogen sulfide and organic vapors in a rock medium biofilter. *Water Environment Research*, 73(4):426–435, 2001.

Shay-Wei Choon, Siow-Hooi Tan, and Lee-Lee Chong. The perception of households about solid waste management issues in Malaysia. *Environment, Development and Sustainability*, 19:1685–1700, 2017.

Rashel L. Clark, Oluremi A. Famodu, Ida Holásková, Aniello M. Infante, Pamela J. Murray, I. Mark Olfert, Joseph W. McFadden, Marianne T. Downes, Paul D. Chantler, Matthew W. Duespohl, et al. Educational intervention improves fruit and vegetable intake in young adults with metabolic syndrome components. *Nutrition Research*, 62:89–100, 2019.

Peter Collett and Gregory O'Shea. Pointing the way to a fictional place: a study of direction giving in Iran and England. *European Journal of Social Psychology*, 6(4):447–458, 1976.

W. J. Conover. *Practical Nonparametric Statistics*. John Wiley & Sons Australia, Limited, third edition, 2003.

Giselle Corbie-Smith. The continuing legacy of the Tuskegee Syphilis Study: considerations for clinical investigation. *The American Journal of the Medical Sciences*, 317(1):5–8, 1999.

François-Xavier Coudert. Correcting the scientific record: retraction practices in chemistry and materials science. *Chemistry of Materials*, 31(10):3593–3598, 2019.

N. A. C. Cressie, L. J. Sheffield, and H. J. Whitford. Use of the one sample t-test in the real world. *Journal of Chronic Diseases*, 37(2):107–114, 1984.

Jamie Cross, Tommy Lam, Joel Arndell, John Quach, Buck Reed, Liz Thyer, and Paul M. Simpson. Impact of hand dominance on effectiveness of chest compressions in a simulated setting: a randomised, crossover trial. *Australasian Journal of Paramedicine*, 16, 2019.

G. K. D. Crozier and Albrecht I. Schulte-Hostedde. Towards improving the ethics of ecological research. *Science and Engineering Ethics*, 21(3):577–594, 2015.

Ewa Czarniecka-Skubina, Marlena Pielak, Piotr Sałek, Renata Korzeniowska-Ginter, and Tomasz Owczarek. Consumer choices and habits related to coffee consumption by Poles. *International Journal of Environmental Research and Public Health*, 18(8):3948, 2021.

Siddharta R. Dala, Edward B. Fowlkes, and Bruce Hoadley. Risk analysis of the space shuttle: pre-Challenger prediction of failure. *Journal of the American Statistical Association*, 84(408):945–957, 1989.

Louise Danielsson, Ilias Papoulias, Eva Lisa Petersson, Jane Carlsson, and Margda Waern. Exercise or basic body awareness therapy as add-on treatment for major depression: a controlled study. *Journal of Affective Disorders*, 168:98–106, 2014.

Jean-Stephane David, Benedicte Gelas-Dore, Kenji Inaba, Albrice Levrat, Bruno Riou, Pierre-Yves Gueugniaud, and Anne-Marie Schott. Are patients with self-inflicted injuries more likely to die? *Journal of Trauma and Acute Care Surgery*, 62(6):1495–1500, 2007.

Marko Davidovic, Lidija Djokic, Aleksandra Cabarkapa, Andrej Djuretic, Vladan Skerovic, and Miomir Kostic. Drivers' preference for the color of LED street lighting. *IEEE Access*, 7:72 850–72 861, 2019.

Rachel Davis, Alexander M. Hirsch, Christian C. Morrill, Ahmad Haffar, Mahir Maruf, Joseph Cheaib, Phillip Pierorazio, and Heather N. Di Carlo. Higher prevalence of benign tumors in men with testicular tumors and history of treated cryptorchidism. *Urologic Oncology: Seminars and Original Investigations*, 42(2):33.e1–33.e6, 2024.

Anderson Albuquerque de Carvalho, Fabio Ferreira Amorim, Levy Aniceto Santana, Karlo Jozefo Quadros de Almeida, Alfredo Nicodemos Cruz Santana, and Francisco de Assis Rocha Neves. STOP-Bang questionnaire should be used in all adults with Down Syndrome to screen for moderate to severe obstructive sleep apnea. *PLOS ONE*, 15(5):e0232596, 2020.

Lori J. Delaney, Marian J. Currie, Hsin-Chia Carol Huang, Violeta Lopez, and Frank Van Haren. "They can rest at home": an observational study of patients' quality of sleep in an Australian hospital. *BMC Health Services Research*, 18(1):524, 2018.

Julien Delarue, A.-C. Brasset, F. Jarrot, and Florent Abiven. Taking control of product testing context thanks to a multi-sensory immersive room. A case study on alcohol-free beer. *Food Quality and Preference*, 75:78–86, 2019.

Rajeev Devaraj, Jonathan Jordan, Christophe Gerber, and Ayodele Olofinjana. Exploring the effects of the substitution of freshly mined sands with recycled crushed glass on the properties of concrete. *Applied Sciences*, 11(8):3318, 2021.

Jay L. Devore and Kenneth N. Berk. *Modern Mathematical Statistics with Applications*. Thomson Higher Education, 2007.

Ben Dexter, Rachel King, Simone L. Harrison, Alfio V. Parisi, and Nathan J. Downs. A pilot observational study of environmental summertime health risk behavior in central Brisbane, Queensland: opportunities to raise sun protection awareness in Australia's sunshine state. *Photochemistry and Photobiology*, 95(2): 650–655, 2019.

C. E. Dexter, R. G. Appleby, J. Scott, J. P. Edgar, and D. N. Jones. Individuals matter: predicting koala road crossing behaviour in south-east Queensland. *Australian Mammalogy*, 40(1):67–75, 2018.

Iman Dianat, Nasibeh Sorkhi, Aida Pourhossein, Arezou Alipour, and Mohammad Asghari-Jafarabadi. Neck, shoulder and low back pain in secondary schoolchildren in relation to schoolbag carriage: should the recommended weight limits be gender-specific? *Applied Ergonomics*, 45:437–442, 2014.

Jan Diehm and Amber Thomas. Women's pockets are inferior. *The Pudding*, 2018. URL https://pudding.cool/2018/08/pockets/.

Pelayo Diez-Fernández, Brais Ruibal-Lista, Fernando Lobato-Alejano, and Sergio López-García. Rip current knowledge: do people really know its danger? Do lifeguards know more than the general public? *Heliyon*, 9(7), 2023.

Mehmet Dokur, Emine Petekkaya, and Mehmet Karadağ. Media-based clinical research on selfie-related injuries and deaths. *Ulus Travma Acil Cerrahi Derg*, 24(2):129–135, 2018.

Omid Doosti-Irani, Mahmood Reza Golzarian, Mohammad Hossein Aghkhani, Hasan Sadrnia, and Mahboobe Doosti-Irani. Development of multiple regression model to estimate the apple's bruise depth using thermal maps. *Postharvest Biology and Technology*, 116:75–79, 2016.

L. E. Drinkwater, D. K. Letourneau, F. A. H. C. Workneh, A. H. C. Van Bruggen, and C. Shennan. Fundamental differences between conventional and organic tomato agroecosystems in California. *Ecological Applications*, 5(4):1098–1112, 1995.

Markus J. Duncan, Kelly Wunderlich, Yingying Zhao, and Guy Faulkner. Walk this way: validity evidence of iPhone health application step count in laboratory and free-living conditions. *Journal of Sports Sciences*, 36(15):1695–1704, 2018.

P. M. Dunn. Ignac Semmelweis (1818–1865) of Budapest and the prevention of puerperal fever. *Archives of Disease in Childhood-Fetal and Neonatal Edition*, 90(4):F345–FF348, 2005.

Peter K. Dunn. A simple dataset for demonstrating common distributions. *Journal of Statistics Education*, 7(3), 1999. Published electronically.

Peter K. Dunn. Bootstrap confidence intervals for predicted rainfall quantiles. *International Journal of Climatology*, 21(1):89–94, 2001.

Peter K. Dunn. Assessing claims made by a pizza chain. *Journal of Statistical Education*, 20(1), 2012. URL www.amstat.org/publications/jse/v20n1/dunn.pdf.

Peter K. Dunn. Comparing the lifetimes of two brands of batteries. *Journal of Statistical Education*, 21(1), 2013.

Peter K. Dunn. Generalized linear models. In Robert J. Tierney, Fazal Rizvi, and Kadriye Erkican, editors, *International Encyclopedia of Education*, pages 583–589. Elsevier, 2023.

Peter K. Dunn and Gordon K. Smyth. *Generalized Linear Models With Examples in R*. Springer, 2018.

Peter K. Dunn, Michael D. Carey, Alice M. Richardson, and Christine McDonald. Learning the language of statistics: challenges and teaching approaches. *Statistics Education Research Journal*, 15(1), 2016.

Rosamond A. Dwyer, Belinda J. Gabbe, Thach Tran, Karen Smith, and Judy A. Lowthian. Residential aged care homes: why do they call '000'? A study of the emergency prehospital care of older people living in residential aged care homes. *Emergency Medicine Australasia*, 33(3):447–456, 2021.

Bradley Efron and Trevor Hastie. *Computer Age Statistical Inference, Student Edition: Algorithms, Evidence, and Data Science*, volume 6. Cambridge University Press, 2021.

Ona Egbue, Suzanna Long, and V. A. Samaranayake. Mass deployment of sustainable transportation: evaluation of factors that influence electric vehicle adoption. *Clean Technologies and Environmental Policy*, 19(7):1927–1939, 2017.

Daniel Einsiedel, Sara-Lena Welk, Nevena Zujko, Yvonne Pfeifer, and Christian Krupitzer. Investigating the correlation of analytical data on pesticide residues in fruits and vegetables with local climatic condition. *Environmental Research*, 252:118743, 2024.

Claire Emerson, Dylan Morrissey, Mark Perry, and Rosy Jalan. Ultrasonographically detected changes in Achilles tendons and self reported symptoms in elite gymnasts compared with controls—an observational study. *Manual Therapy*, 15(1):37–42, 2010.

Paul Enck, Ulrike Bingel, Manfred Schedlowski, and Winfried Rief. The placebo response in medicine: minimize, maximize or personalize? *Nature Reviews Drug Discovery*, 12(3):191, 2013.

Mario Estévez-Báez, Claudia Carricarte-Naranjo, Javier Denis Jas-García, Evelyn Rodríguez-Ríos, Calixto Machado, Julio Montes-Brown, Gerry Leisman, Adam Schiavi, Andrés Machado-García, Claudia Sánchez Luaces, et al. Influence of heart rate, age, and gender on heart rate variability in adolescents and young adults. *Advances in Medicine and Medical Research*, pages 19–33, 2019.

H. J. Eysenck. Were we really wrong? *American Journal of Epidemiology*, 133(5):429–433, 1991.

M. Falk and C. D. Anderson. Influence of age, gender, educational level and self-estimation of skin type on sun exposure habits and readiness to increase sun protection. *Cancer Epidemiology*, 37:127–132, 2013.

Michael B. Farrar, Helen M. Wallace, Cheng-Yuan Xu, Thi Thu Nhan Nguyen, Ehsan Tavakkoli, Stephen Joseph, and Shahla Hosseini Bai. Short-term effects of organo-mineral enriched biochar fertiliser on ginger yield and nutrient cycling. *Journal of Soils and Sediments*, pages 1–15, 2018.

Michael B. Farrar, Helen M. Wallace, Cheng-Yuan Xu, Stephen Joseph, Thi Thu Nhan Nguyen, Peter K. Dunn, and Shahla Hosseini Bai. Biochar compound fertilisers increase plant potassium uptake 2 years after application without additional organic fertiliser. *Environmental Science and Pollution Research*, pages 1–15, 2021.

Flávia Fayet-Moore, Véronique Peters, Andrew McConnell, Peter Petocz, and Alison L. Eldridge. Weekday snacking prevalence, frequency, and energy contribution have increased while foods consumed during snacking have shifted among Australian children and adolescents: 1995, 2007 and 2011–12 National Nutrition Surveys. *Nutrition Journal*, 16(65):1–14, 2017.

Yang-chun Feng, Yan-chun Huang, and Xiu-min Ma. The application of Student's t-test in internal quality control of clinical laboratory. *Frontiers in Laboratory Medicine*, 1(3):125–128, 2017.

Arlene Fink. *The Survey Handbook. The Survey Kit.* SAGE Publications, Incorporated, 1995.

F. Fraboni, V. Marín Puchades, M. De Angelis, L. Pietrantoni, and Gabriele Prati. Red-light running behavior of cyclists in Italy: an observational study. *Accident Analysis & Prevention*, 120:219–232, 2018.

Erika Friedmann and Sue Thomas. Health benefits of pets for families. *Marriage & Family Review*, 8(3-4): 191–203, 1985.

Susan N. Friel, Frances R. Curcio, and George W. Bright. Making sense of graphs: critical factors influencing comprehension and instructional implications. *Journal for Research in Mathematical Education*, pages 124–158, 2001.

Juliana R. Fritts, Clara Fort, Anne Quinn Corr, Qihan Liang, Laurie Alla, Terri Cravener, John E. Hayes, Barbara J. Rolls, Christopher D'Adamo, and Kathleen L. Keller. Herbs and spices increase liking and preference for vegetables among rural high school students. *Food Quality and Preference*, 68:125–134, 2018.

R. H. Frost and N. Murtagh. Encouraging planting in urban front gardens: a focus group study. *Perspectives in Public Health*, 143(2):80–88, 2023.

K. J. Froud, R. M. Beresford, and N. C. Cogger. Impact of kiwifruit bacterial canker on productivity of cv. Hayward kiwifruit using observational data and multivariable analysis. *Plant Pathology*, 67(3):671–681, 2018.

Robert W. Furness and David M. Bryant. Effect of wind on field metabolic rates of breeding northern fulmars. *Ecology*, 77(4):1181–1188, 1996.

Cheryl S. Gammon, Pamela R. von Hurst, Joan Coad, Rozanne Kruger, and Welma Stonehouse. Vegetarianism, vitamin B12, and insulin resistance in a group of predominately overweight/obese South Asian women. *Nutrition*, 28:20–24, 2012.

Umberto C. Gatti, Giovani C. Migliaccio, Susan M. Bogus, and Suzanne Schneider. An exploratory study of the relationship between construction workforce physical strain and task level productivity. *Construction Management and Economics*, pages 1–17, 2013.

Jerzy Gębski, Marzena Jezewska-Zychowicz, Julita Szlachciuk, and Małgorzata Kosicka-Gębska. Impact of nutritional claims on consumer preferences for bread with varied fiber and salt content. *Food Quality and Preference*, 76:91–99, 2019.

Andrew Gelman and Deborah Nolan. You can load a die, but you can't bias a coin. *The American Statistician*, 56(4):308–311, 2002.

Niko Gentile. Improving lighting energy efficiency through user response. *Energy and Buildings*, pages 1–12, 2022. doi: 10.1016/j.enbuild.2022.112022.

Stefan Gerber. Material properties of bamboo flooring. Compliance Report 2-04, Queensland Department of Primary Industries, for Bamboo Flooring Pty Ltd, Indooroopilly, Queensland, January 2004.

Safiyeh Ghasemi and Asiyeh Pirzadeh. Effectiveness of educational physical activity intervention for preventive of musculoskeletal disorders in bus drivers. *International Journal of Preventive Medicine*, 10, 2019.

Lauren Giandomenico, Maya Papineau, and Nicholas Rivers. A systematic review of energy efficiency home retrofit evaluation studies. *Annual Review of Resource Economics*, 14:689–708, 2022.

Benoît Gillet, Mickaël Begon, Marine Diger, Christian Berger-Vachon, and Isabelle Rogowski. Shoulder range of motion and strength in young competitive tennis players with and without history of shoulder problems. *Physical Therapy in Sport*, 31:22–28, 2018.

Adrián F González-Acosta, Carlos H. Rábago-Quiroz, Gorgonio Ruiz-Campos, Juan Antonio García-Borbón, María del Carmen Alejo-Plata, Francisco J. Barrón-Barraza, et al. Length-weight and length-length relationships of 39 demersal fish species of an estuarine-coastal ecosystem from the northwestern of the Baja California Peninsula, Mexico. *Journal of Applied Ichthyology*, 2024(1), 2024. doi: 10.1155/2024/6408697.

Belen Gonzalez-Fonteboa and Fernando Martinez-Abella. Shear strength of recycled concrete beams. *Construction and Building Materials*, 21(4):887–893, 2007.

J. Grabosky and N. Bassuk. Seventeen years' growth of street trees in structural soil compared with a tree lawn in New York City. *Urban Forestry & Urban Greening*, 16:103–109, 2016.

Sander Greenland, Stephen J. Senn, Kenneth J. Rothman, John B. Carlin, Charles Poole, Steven N. Goodman, and Douglas G. Altman. Statistical tests, *p* values, confidence intervals, and power: a guide to misinterpretations. *European Journal of Epidemiology*, 31(4):337–350, 2016.

Eric T. Greenlee, Patricia R. DeLucia, and David C. Newton. Driver vigilance in automated vehicles: hazard detection failures are a matter of time. *Human Factors*, 60(4):465–476, 2018.

Carla Greier, Clemens Drenowatz, Herbert Riechelmann, and Klaus Greier. Objective and subjective physical activity levels in Austrian middle school students. *Advances in Physical Education*, 11(4):448–459, 2021.

Donald R. Griffin, Frederic A. Webster, and Charles R. Michael. The echolocation of flying insects by bats. *Animal Behaviour*, 8(3-4):141–154, 1960.

Lluis Guirao, C. Beatriz Samitier, Maria Costea, Josep Maria Camos, Maria Majo, and Eulogio Pleguezuelos. Improvement in walking abilities in transfemoral amputees with a distal weight bearing implant. *Prosthetics and Orthotics International*, 4(26–32), 2017.

Gokhan Hacisalihoglu, Nicole S. Beisel, and A. Mark Settles. Characterization of pea seed nutritional value within a diverse population of *pisum sativum*. *PLOS ONE*, 16(11):e0259565, 2021. doi: 10.1371/journal.pone.0259565.

A. Hald. *Statistical Theory with Engineering Applications*. John Wiley & Sons, New York, 1952.

Darren Hale, Pramen P. Shrestha, G. Edward Gibson, and Giovanni C. Migliaccio. Empirical comparison of Design/Build and Design/Bid/Build project delivery methods. *Journal of Construction Engineering*, 135: 579–587, 2009.

David Hammond, Jessica L. Reid, and Sara Zukowski. Adverse effects of caffeinated energy drinks among youth and young adults in Canada: a Web-based survey. *CMAJ Open*, 6(1):E19, 2018.

D. J. Hand, F. Daly, A. D. Lunn, K. J. McConway, and E. Ostrowski. *A Handbook of Small Data Sets*. Chapman and Hall, London, 1996.

Wolfgang Karl Härdle et al. *Smoothing techniques: with implementation in S*. Springer Science & Business Media, 1991.

Clare Haselgrove, Leon Straker, Anne Smith, Peter O'Sullivan, Mark Perry, and Nick Sloan. Perceived school bag load, duration of carriage, and method of transport to school are associated with spinal pain in adolescents: an observational study. *Australian Journal of Physiotherapy*, 54(3):193–200, 2008.

Johann Jakob Häußermann, Moritz J. Maier, Thea C. Kirsch, Simone Kaiser, and Martina Schraudner. Social acceptance of green hydrogen in Germany: building trust through responsible innovation. *Energy, Sustainability and Society*, 13(1):22, 2023.

Kim Hébert-Losier, Caleb Boswell-Smith, and Ivana Hanzlíková. Effect of footwear versus barefoot on double-leg jump-landing and jump height measures: a randomized cross-over study. *International Journal of Sports Physical Therapy*, 18(4):845, 2023.

Ida M. Heerfordt, Linnea R. Torsnes, Peter A. Philipsen, and Hans Christian Wulf. Photoprotection by sunscreen depends on time spent on application. *Photodermatology, Photoimmunology & Photomedicine*, 34(2):117–121, 2018.

Harold V. Henderson and Paul F. Velleman. Building multiple regression models interactively. *Biometrics*, pages 391–411, 1981.

Elisa Henning, Thamires Ferreira Schubert, and Andinara Ceccatto Maciel. Modelling of university student transport mode choice in Joinville: a binary logistic model for active modes. *Journal of Sustainable Development of Energy, Water and Environment Systems*, 8(4):678–691, 2020.

Jennifer Hieger. Portrait of a homebuyer household: 2 kids and a PC. Technical report, Orange County Register, 27 July 2001.

J. M. Hirst and O. J. Stedman. The epidemiology of apple scab (Venturia inaequalis (Cke.) Wint.) III. The supply of ascospores. *Annals of Applied Biology*, 50(3):551–567, 1962.

Lauren G. Hitt, Sarah Khalil, Annelise Blanchette, Myra E. Finkelstein, Erik N. K. Iverson, Stephanie C. McClelland, Renata Durães Ribeiro, and Jordan Karubian. Lead exposure is correlated with reduced nesting success of an urban songbird. *Environmental Research*, 227:115711, 2023.

Lucius Kang Hua Ho, Valerie Jia Wei Tong, Nicholas Syn, Niranjan Nagarajan, Elizabeth Huiwen Tham, Stacey K. Tay, Shefaly Shorey, Paul Anantharajah Tambyah, and Evelyn Chung Ning Law. Gut microbiota changes in children with autism spectrum disorder: a systematic review. *Gut Pathogens*, 12(1):1–18, 2020.

David C. Hoaglin, Frederick Mosteller, and John W. Tukey. *Exploring Data Tables, Trends, and Shapes*. John Wiley & Sons, 2011.

P. Holgate. Fitting a straight line to data from a truncated population. *Biometrics*, 21(3):715–720, 1965.

Sally Hopewell, Isabelle Boutron, An-Wen Chan, Gary S. Collins, Jennifer A. de Beyer, Asbjørn Hróbjartsson, Camilla Hansen Nejstgaard, Lasse Østengaard, Kenneth F. Schulz, Ruth Tunn, et al. An update to SPIRIT and CONSORT reporting guidelines to enhance transparency in randomized trials. *Nature Medicine*, 28(9):1740–1743, 2022.

Adam James Houben, Rebecca D'Onofrio, Steven V. Kokelj, and Jules M. Blais. Factors affecting elevated arsenic and methyl mercury concentrations in small shield lakes surrounding gold mines near the Yellowknife, NT, (Canada) region. *PLOS ONE*, 11(4):e0150960, 2016.

Qiuyun Huang, Minyan Yang, Hao-Ann Jane, Shuhua Li, and Nicole Bauer. Trees, grass, or concrete? the effects of different types of environments on stress reduction. *Landscape and Urban Planning*, 193:103654, 2020.

Pekka Huhtanen, Mohammad Ramin, and E. H. Cabezas-Garcia. Effects of ruminal digesta retention time on methane emissions: a modelling approach. *Animal Production Science*, 56(3):501–506, 2016.

E. C. Huskisson. Simple analgesics for arthritis. *British Medical Journal*, 4:196–200, 1974.

Sung Yeon Hwang, Seung Mok Ryoo, Jong Eun Park, You Hwan Jo, Dong-Hyun Jang, Gil Joon Suh, Taegyun Kim, Youn-Jung Kim, Seonwoo Kim, Hyun Cho, et al. Combination therapy of vitamin C and thiamine for septic shock: a multi-centre, double-blinded randomized, controlled study. *Intensive Care Medicine*, 46:2015–2025, 2020.

Felice N. Jacka, Peter J. Kremer, Eva R. Leslie, Michael Berk, George C. Patton, John W. Toumbourou, and Joanne W. Williams. Associations between diet quality and depressed mood in adolescents: results from the Australian Healthy Neighbourhoods Study. *Australian and New Zealand Journal of Psychiatry*, 44(5):435–442, 2010.

E. K. Janatuinen, T. A. Kemppainen, R. J. K. Julkunen, V. M. Kosma, M. Mäki, M. Heikkinen, and M. I. J. Uusitupa. No harm from five year ingestion of oats in coeliac disease. *Gut*, 50(3):332–335, 2002.

Ashley Jardina. Why people love 'assistance to the poor' but hate 'welfare', 2018. URL https://talkpoverty.org/2018/01/29/people-love-assistance-poor-hate-welfare/index.html. Accessed: 2024-01-31.

Lauren C. Jenner, Jeanette M. Rotchell, Robert T. Bennett, Michael Cowen, Vasileios Tentzeris, and Laura R. Sadofsky. Detection of microplastics in human lung tissue using μFTIR spectroscopy. *Science of The Total Environment*, 831:154907, 2022. doi: 10.1016/j.scitotenv.2022.154907.

G. Joglekar, J. H. Scheunemyer, and V. LaRiccia. Lack-of-fit testing when replicates are not available. *The American Statistician*, 43:135–143, 1989.

Danika Johnson, Robert Mead, Korey Kennelty, and David Hahn. Menthol cough drops: cause for concern? *The Journal of the American Board of Family Medicine*, 31(2):183–191, 2018.

Emily Johnson, Seán R. Millar, and Frances Shiely. The association between BMI self-selection, self-reported BMI and objectively measured BMI. *HRB Open Research*, 4(37):37, 2021.

Robert A. Johnson, Steven W. Carothers, and Thomas J. McGill. Demography of feral burros in the Mohave Desert. *The Journal of Wildlife Management*, 51(4):916–920, 1987.

Brian L. Joiner. Lurking variables: some examples. *The American Statistician*, 35(4):227–233, 1981.

Steven A. Julious and Mark A. Mullee. Confounding and Simpson's paradox. *BMJ*, 309(6967):1480–1481, 1994.

Michael Kahn. An exhalent problem for teaching statistics. *Journal of Statistical Education*, 13(2), 2005. Available on-line at http://www.amstat.org/publications/jse/v13n2/datasets.kahn.html.

Dae-Wook Kang, James B. Adams, Devon M. Coleman, Elena L. Pollard, Juan Maldonado, Sharon McDonough-Means, J. Gregory Caporaso, and Rosa Krajmalnik-Brown. Long-term benefit of Microbiota Transfer Therapy on autism symptoms and gut microbiota. *Scientific Reports*, 9(1):1–9, 2019.

Roya Kelishadi, Nafiseh Mozafarian, Mostafa Qorbani, Mohammad Esmaeil Motlagh, Saeid Safiri, Gelayol Ardalan, Mojtaba Keikhah, Fatemeh Rezaei, and Ramin Heshmat. Is snack consumption associated with meal skipping in children and adolescents? The CASPIAN-IV study. *Eat Weight Disorders*, 22:321–328, 2017.

Sydney S. Kelpin, Thomas B. Moore, Lynn C. Hull, Pamela M. Dillon, Bridget L. Perry, Leroy R. Thacker, Linda Hancock, and Dace S. Svikis. Alcohol use and problems in daily and non-daily coffee drinking college females. *Journal of Substance Use*, 23(6):574–578, 2018.

Jessica A. Kettenbach, Nicole Miller-Struttmann, Zoë Moffett, and Candace Galen. How shrub encroachment under climate change could threaten pollination services for alpine wildflowers: a case study using the alpine skypilot, *polemonium viscosum*. *Ecology and Evolution*, 7(17):6963–6971, 2017.

Abdul Khair, Lucky Herawati, Noraida Noraida, and Munawar Raharja. The use of earthworms and household organic waste composting length of time. *Kesmas: National Public Health Journal*, 10(2):62–66, 2015.

Mai E. Khalaf, Aqdar Akbar, Qoot Alkhubaizi, and Muawia Qudeimat. Caries among adult patients with controlled celiac disease: a cross-sectional study. *Special Care in Dentistry*, 40(5):457–463, 2020.

Kyong Young Kim and Kyoung Min Kim. Similarities and differences between bone quality parameters, trabecular bone score and femur geometry. *PLOS ONE*, 17(1):e0260924, 2022.

Sung-Soo Kim, Hyun Woo Han, Unyeong Go, and Hai Won Chung. Sero-epidemiology of measles and mumps in Korea: impact of the catch-up campaign on measles immunity. *Vaccine*, 23(3):290–297, 2004.

Donald T. Kirkendall, Sheldon E. Jordan, and William E. Garrett. Heading and head injuries in soccer. *Sports Medicine*, 31(5):369–386, 2001.

Bob Kizer, Gleb Haynatzki, Harry Lazarte, Reshma Patel, John O'Brien, and Asma Dajani. Digestive diseases in the American Indian population of the U.S. Central Plains: disparities in prevalence, screening, and treatment: 1023. *Official Journal of the American College of Gastroenterology*, 101:S402, 2006.

Mariel Gullian Klanian, Mariana Delgadillo Diaz, Javier Aranda, and Carolina Rosales Juárez. Integrated effect of nutrients from a recirculation aquaponic system and foliar nutrition on the yield of tomatoes *Solanum lycopersicum L.* and *Solanum pimpinellifolium*. *Environmental Science and Pollution Research*, pages 1–13, 2018.

Wen-Ru Ko, Wei-Te Hung, Hui-Chin Chang, and Long-Yau Lin. Inappropriate use of standard error of the mean when reporting variability of study samples: a critical evaluation of four selected journals of obstetrics and gynecology. *Taiwanese Journal of Obstetrics and Gynecology*, 53(1):26–29, 2014.

Jöran Köchling, Berit Geis, Stefan Wirth, and Kai O. Hensel. Grape or grain but never the twain? A randomized controlled multiarm matched-triplet crossover trial of beer and wine. *The American Journal of Clinical Nutrition*, 109(2):345–352, 2019.

John Paul Koenen. *An Analysis of Smoking and Gambling Among Las Vegas Visitors*. University of Nevada, Las Vegas, 1995.

L. Kohlmeier, G. Arminger, S. Bartolomeycik, B. Bellach, J. Rehm, and M. Thamm. Pet birds as an independent risk factor for lung cancer: case-control study. *British Medical Journal*, 305(6860):986–989, 1992.

Christine Laine, Steven N. Godman, Michael E. Griswold, and Harold C. Sox. Reproducible research: moving toward research the public can really trust. *Annals of Internal Medicine*, 146(6):450–453, 2007.

Lucas D. Lalande, Virpi Lummaa, Htoo H. Aung, Win Htut, U. Kyaw Nyein, Vérane Berger, and Michael Briga. Sex-specific body mass ageing trajectories in adult Asian elephants. *Journal of Evolutionary Biology*, 35(5):752–762, 2022a.

Lucas D. Lalande, Virpi Lummaa, Htoo H. Aung, Win Htut, U. Kyaw Nyein, Vérane Berger, and Michael Briga. Sex-specific body mass aging trajectories in adult Asian elephants. *Dryad*, 2022b. doi: 10.5061/dryad.5dv41ns59.

Suzanne M. Lambie, Paul L. Mudge, and Bryan A. Stevenson. Microbial community composition and activity in paired irrigated and non-irrigated pastures in New Zealand. *Soil Research*, 60(4):337–348, 2021.

P. W. Lane. Generalized linear models in soil science. *European Journal of Soil Science*, 53:241–251, 2002.

Leila M. Larson, Shruthi Cyriac, Eric W. Djimeu, Mduduzi N. N. Mbuya, and Lynnette M. Neufeld. Can double fortification of salt with iron and iodine reduce anemia, iron deficiency anemia, iron deficiency, iodine deficiency, and functional outcomes? Evidence of efficacy, effectiveness, and safety. *The Journal of Nutrition*, 151(Supplement 1):15S–28S, 2021.

Vicki R. LeBlanc, Russell D. MacDonald, Brad McArthur, Kevin King, and Tom Lepine. Paramedic performance in calculating drug dosages following stressful scenarios in a human patient simulator. *Prehospital Emergency Care*, 9(4):439–444, 2005.

Linda Yin-king Lee, Evangeline Pui-wah Lam, Chiu-kiu Chan, Sum-yi Chan, Man-ki Chiu, Wing-hei Chong, Kin-wai Chu, Man-sze Hon, Lok-ki Kwan, Kit-lam Tsang, et al. Practice and technique of using face mask amongst adults in the community: a cross-sectional descriptive study. *BMC Public Health*, 20(1):1–11, 2020.

Victor R. Lee, Denise Pope, Sarah Miles, and Rosalía C. Zárate. Cheating in the age of generative AI: a high school survey study of cheating behaviors before and after the release of ChatGPT. *Computers and Education: Artificial Intelligence*, 7:100253, 2024. doi: 10.1016/j.caeai.2024.100253.

Yu-Mi Lee, Se-A Kim, In-Kyu Lee, Jung-Guk Kim, Keun-Gyu Park, Ji-Yun Jeong, Jae-Han Jeon, Ji-Yeon Shin, and Duk-Hee Lee. Effect of a brown rice based vegan diet and conventional diabetic diet on glycemic control of patients with Type 2 diabetes: a 12-week randomized clinical trial. *PLOS ONE*, 11(6), 2016.

Sara Grace Leuchtenberger, Maris Daleo, Peter Gullickson, Andi Delgado, Carly Lo, and Michael T. Nishizaki. The effects of temperature and pH on the reproductive ecology of sand dollars and sea urchins: impacts on sperm swimming and fertilization. *PLOS ONE*, 17(12):e0276134, 2022.

Mark S. Levenson. Amusement ride-related injuries and deaths in the United States: 2005 update. *Bethesda, MD: Consumer Product Safety Commission*, 2005.

David Levitsky. DASL: Data and story library. https://dasl.datadescription.com/datafile/freshman-15/. Accessed: 2023-10-01.

David A. Levitsky, Craig A. Halbmaier, and Gordana Mrdjenovic. The freshman weight gain: a model for the study of the epidemic of obesity. *International Journal of Obesity*, 28(11):1435–1442, 2004.

Xiao-Yan Li, Cheng Jia, and Zi-Chuan Zhang. The normal range of maximum mouth opening and its correlation with height or weight in the young adult Chinese population. *Journal of Dental Sciences*, 12 (1):56–59, 2017.

Rongli Lian, Song Zhou, Yuan Guo, Haiyan Liang, Jing Lin, Dongni Li, Wenping Wu, Yuan Rao, Daxing Shao, Peici Zheng, et al. The effect of ice-cold water spray following the model for symptom management on postoperative thirst in patients admitted to intensive care unit: a randomized controlled study. *Intensive and Critical Care Nursing*, 81:103571, 2024.

Yi-Ching Lin, Meng-Che Tsai, Chung-Ying Lin, and Amir H. Pakpour. Sleep duration among preschoolers in Taiwan: a longitudinal study. *Sleep Epidemiology*, 1:100015, 2021.

R. E. London and H. A. Slagter. Statement of retraction: Effects of transcranial direct current stimulation over left dorsolateral pFC on the attentional blink depend on individual baseline performance. *Journal of Cognitive Neuroscience*, 27(12):1, 2021. doi: 10.1162/jocn_x_01680.

María J. López-Serrano, Juan F. Velasco-Muñoz, José A. Aznar-Sánchez, and Isabel M. Román-Sánchez. Farmers' attitudes towards irrigating crops with reclaimed water in the framework of a circular economy. *Agronomy*, 12(2):435, 2022.

Bill Lord, James Cui, and Anne-Maree Kelly. The impact of patient sex on paramedic pain management in the prehospital setting. *The American Journal of Emergency Medicine*, 27(5):525–529, 2009.

Katrin Lorenz, Thomas Hoffmann, Christian Heumann, and Barbara Noack. Effect of toothpaste containing amine fluoride and stannous chloride on the reduction of dental plaque and gingival inflammation. A randomized controlled 12-week home-use study. *International Journal of Dental Hygiene*, 2019.

James B. Lothian, Vijaylaxmi Grey, and Larry C. Lands. Effect of whey protein to modulate immune response in children with atopic asthma. *International Journal of Food Science and Nutrition*, 57(3/4): 204–211, 2006.

Bertrand Loyeung, John Lee, Carolyn Michaeil, and Christopher Zaslawski. An experimental study in distinguishing an authentic herbal substance from sham herbal substances. *Complementary Therapies in Medicine*, 39:92–96, 2018.

A. Daniel. Lunn and D. R. McNeil. *Computer-Interactive Data Analysis*. John Wiley and Sons, Chichester, 1991.

Charlotte Lyons, Paul Felton, and Carla McCabe. Female cricket pace bowling: kinematic and anthropometric relationships with ball release speed. *South African Journal of Sports Medicine*, 35(1), 2023.

Yongfeng Ma, Wenbo Zhang, Xin Gu, and Jiguang Zhao. Impacts of experimental advisory exit speed sign on traffic speeds for freeway exit ramp. *PLOS ONE*, 14(11):e0225203, 2019.

Paul W. Macdermid, Anna Coppelmans, and Darryl Cochrane. The validity and reliability of a Global Navigation Satellite System in canoe slalom. *Biomechanics*, 2(1):20–29, January 2022. doi: 10.3390/biomechanics2010003.

Marissa MacDonald. *Is enough really enough?: Evaluation of an alcohol awareness campaign at ECU Joondalup*. PhD thesis, Edith Cowan University, 2008. URL https://ro.ecu.edu.au/theses_hons/1032.

Russell D. MacDonald, Vicki LeBlanc, Brad McArthur, and Adam Dubrowski. Performance of resuscitation skills by paramedic personnel in chemical protective suits. *Prehospital Emergency Care*, 10(2):254–259, 2006.

Graham A. MacGregor, N. D. Markandu, J. E. Roulston, and J. C. Jones. Essential hypertension: effect of an oral inhibitor of angiotensin-converting enzyme. *British Medical Journal*, 2:1106–1109, 1979.

I. D. M. Macgregor and A. J. Rugg-Gunn. Survey of toothbrushing duration in 85 uninstructed English schoolchildren. *Community Dentistry and Oral Epidemiology*, 7(5):297–298, 1979.

Ian D. M. Macgregor and Andrew J. Rugg-Gunn. Toothbrushing duration in 60 uninstructed young adults. *Community Dentistry and Oral Epidemiology*, 13(3):121–122, 1985.

Philip A. Mackowiak, Steven S. Wasserman, and Myron M. Levine. A critical appraisal of 98.6°F, the upper limit of the normal body temperature, and other legacies of Carl Reinhold August Wunderlich. *Journal of the American Medical Association*, 268(12):1578–1580, 1992.

Hotaka Maeda. *Introducing portable pedal machines inside a university library to reduce sedentary behavior.* PhD thesis, East Carolina University, 2013.

Stephan Mages, Ole Hensel, Antonia Maria Zierz, Torsten Kraya, and Stephan Zierz. Experimental provocation of 'ice-cream headache' by ice cubes and ice water. *Cephalalgia*, 37(5):464–469, 2017.

Abdulai Abdul Majeed, Samwel Oyier Zephaniah, Gaurav Mehta, and Steven Jones. Field-based saturation headway model for planning level applications. *International Journal of Traffic and Transportation Engineering*, 3(5):207–215, 2014.

Javier Manjarrez, Constantino Macias Garcia, and Hugh Drummond. Morphological convergence in a Mexican garter snake associated with the ingestion of a novel prey. *Ecology and Evolution*, 7(18):7178–7186, 2017.

Bryan F. J. Manly and Jorge A. Navarro Alberto. *Introduction to Ecological Sampling.* CRC Press, 2014.

Linda Mann and Karen Blotnicky. Influences of physical environments on university student eating behaviors. *International Journal of Health Sciences*, 5(2):42–52, 2017.

Francisco Manzano, Ana-María Pérez, Manuel Colmenero, María-Mar Aguilar, Emilio Sánchez-Cantalejo, Ana-María Reche, Juan Talavera, Francisca López, Sonia Frías-Del Barco, and Enrique Ferdández-Mondejar. Comparison of alternating pressure mattresses and overlays for prevention of pressure ulcers in ventilated intensive care patients: a quasi-experimental study. *Journal of Advanced Nursing*, 69(9): 2099–2106, 2013.

Duane T. March, Kimberly Vinette-Herrin, Andrew Peters, Ellen Ariel, David Blyde, Doug Hayward, Les Christidis, and Brendan P. Kelaher. Hematologic and biochemical characteristics of stranded green sea turtles. *Journal of Veterinary Diagnostic Investigation*, 2018.

Martine Maron. Threshold effect of eucalypt density on an aggressive avian competitor. *Biological Conservation*, 136:100–107, 2007.

Margaret Marshman and Peter K. Dunn. Teaching statistics with experiential learning: a visual experience. *Mathematics Teacher Education and Development*, to appear, 2025.

Jørgen Jerstad Martnes and Elling Bere. Physical activity when riding an electric-assisted bicycle with and without cargo. *Frontiers in Sports and Active Living*, 5, 2023.

Alison Maunder, Erica Bessell, Romy Lauche, Jon Adams, Amanda Sainsbury, and Nicholas R. Fuller. Effectiveness of herbal medicines for weight loss: a systematic review and meta-analysis of randomized controlled trials. *Diabetes, Obesity and Metabolism*, 22(6):891–903, 2020.

Michael P. McLaughlin. Alcohol-associated illness and injury and ambulance calls in a midwestern college town: a four-year retrospective analysis. *Prehospital Emergency Care*, 14(4):485–490, 2010. doi: 10.3109/10903127.2010.497897.

Samuel E. McLinn, Michael Moskal, Johanna Goldfarb, Frank Bodor, Gerson Aronovitz, Richard Schwartz, Pamela Self, and Michael J. Ossi. Comparison of cefuroxime axetil and amoxicillin-clavulanate suspensions in treatment of acute otitis media with effusion in children. *Antimicrobial Agents and Chemotherapy*, 38 (2):315–318, 1994.

R. Mead. Plant density and crop yield. *Applied Statistics*, 19(1):64–81, 1970.

Ben Meadley, Ella Horton, David B. Pyne, Luke Perraton, Karen Smith, Kelly-Ann Bowles, and Joanne Caldwell. Comparison of swimming versus running maximal aerobic capacity in helicopter rescue paramedics. *Ergonomics*, 64(10):1243–1254, 2021.

Mengistu Meresa, Menfese Tadesse, and Negussie Zeray. Effect of soil and water conservation structures on smallholder farmers' livelihood: Wenago district, Southern Ethiopia. *Cogent Social Sciences*, 9(2): 2272305, 2023. doi: 10.1080/23311886.2023.2272305.

Hossein Amarpoor Mesrkanlou, Seyed Jamal Ghaemmaghami Hezaveh, Sanaz Tahmasebi, Zeinab Nikniaz, and Leila Nikniaz. The effect of an earthquake experienced during pregnancy on maternal health and birth outcomes. *Disaster Medicine and Public Health Preparedness*, 17:e157, 2023.

N. Mine, S. H. Wai, T. C. Lim, and W. Kang. An observational study on the productivity of formwork in building construction. In ISARC. *Proceedings of the International Symposium on Automation and Robotics in Construction*, volume 32, page 1. IAARC Publications, 2015.

Dave L. Mitchell, Mariela Soto-Berelov, and Simon D. Jones. Remote sensing shows south-east Queensland koalas (*phascolarctos cinereus*) prefer areas of higher tree canopy height within their home ranges. *Wildlife Research*, 50(11):939–953, 2023.

Asghar Mohammadpoorasl, Mohammad Hajizadeh, Soudabeh Marin, Payam Heydari, and Mehran Ghalenoei. Prevalence and pattern of using headphones and its relationship with hearing loss among students. *Health Scope*, 8(1), 2019.

D. C. Montgomery and E. A. Peck. *Introduction to Regression Analysis.* Wiley, New York, 1992.

Hortensia Morón-Monge, Soraya Hamed, and María del Carmen Morón Monge. How do children perceive the biodiversity of their nearby environment: an analysis of drawings. *Sustainability*, 13(6):3036, 2021.

Ronita Mukherjee, Rittik Deb, and Soubadra M. Devy. Diversity matters: effects of density compensation in pollination service during rainfall shift. *Ecology and Evolution*, 9(17):9701–9711, 2019.

Raymond H. Myers, Douglas C. Montgomery, and G. Geoffrey Vining. *Generalized Linear Models With Applications in Engineering and the Sciences.* Wiley, 2002.

Peter Nagele. Misuse of standard error of the mean (SEM) when reporting variability of a sample. A critical evaluation of four anaesthesia journals. *British Journal of Anaesthesia*, 90(4):514–516, 2003.

Violetta Naughton, Teresa Grzelak, and Patrick J. Naughton. Association between household location (urban versus rural) and fundamental care provided to domestic dogs (*Canis familiaris*) in Northern Ireland. In *Nutrition and Metabolism of Dogs and Cats*, pages 217–236. Springer, 2024.

Keith E. Naugle, Jason Hackett, Dania Aqeel, and Kelly M. Naugle. Effect of different Kinesio tape tensions on experimentally-induced thermal and muscle pain in healthy adults. *PLOS ONE*, 16(11):e0259433, 2021.

Wayne Nelson. *Applied Life Data.* Wiley, 1982.

Michael T. Nishizaki, Sara Grace Leuchtenberger, Maris Daleo, Peter Gullickson, Andi Delgado, and Carly Lo. Echinoderm sperm swimming and fertilization. *Dryad*, 2022. doi: 10.5061/dryad.jwstqjqbz.

Carmen L. Nochera and Diane Ragone. Development of a breadfruit flour pasta product. *Foods*, 8(3):110, 2019.

D. Ryan Norris. Carry-over effects and habitat quality in migratory populations. *Oikos*, 109(1):178–186, 2005.

Eugene J. O'Brien, Longwei Zhang, Hua Zhao, and Donya Hajializadeh. Probabilistic bridge weigh-in-motion. *Canadian Journal of Civil Engineering*, 45(8):667–675, 2018.

Michael C. Oca, Leo Meller, Katherine Wilson, Alomi O. Parikh, Allison McCoy, Jessica Chang, Rasika Sudharshan, Shreya Gupta, and Sandy Zhang-Nunes. Bias and inaccuracy in AI chatbot ophthalmologist recommendations. *Cureus*, 15(9), 2023. doi: 10.7759/cureus.45911.

B. A. O'Connor, J. Carman, K. Eckert, G. Tucker, R. Givney, and S. Cameron. Does using potting mix make you sick? Results from a *legionella longbeachae* case-control study in South Australia. *Epidemiology and Infection*, 135:34–39, 2007.

Catherine A. Offord and Heidi C. Zimmer. Home gardens contribute to conservation of the critically endangered Wollemi Pine: evaluation of a botanic garden-led horticultural release programme. *Plants, People, Planet*, 2023.

B. Olesen and M. Bjerregaard Feldthaus. Do we see the Hawthorne effect in adherence of the general public to self-protection guidelines during the COVID-19 pandemic? A Danish observational study. *Journal of Hospital Infection*, 110:209, 2021.

Luis Oliveira, Arun Ulahannan, Matthew Knight, and Stewart Birrell. Wireless charging of electric taxis: understanding the facilitators and barriers to its introduction. *Sustainability*, 12(21):8798, 2020.

J. Adam Oostema, Todd Chassee, and Mathew Reeves. Emergency dispatcher stroke recognition: associations with downstream care. *Prehospital Emergency Care*, 22(4):466–471, 2018.

Daniel M. Oppenheimer. Consequences of erudite vernacular utilized irrespective of necessity: problems with using long words needlessly. *Applied Cognitive Psychology: The Official Journal of the Society for Applied Research in Memory and Cognition*, 20(2):139–156, 2006.

A. Y. Oyerinde, S. O. Bamisaye, and Ubong Akpan Essien. Investigation of the effects of air-conditioning system on the temperature and speed of automobile engine using paired t-test and regression analysis. *Open Access Library Journal*, 6, 2019.

R. Pamphlett. Exposure to environmental toxins and the risk of sporadic motor neuron disease: an expanded Australian case–control study. *European Journal of Neurology*, 19:1343–1348, 2012.

R. P. Panda, Sudhanshu Sekhar Das, and P. K. Sahoo. Relation between bitumen content and percentage air voids in semi dense bituminous concrete. *Journal of The Institution of Engineers (India): Series A*, 99(2):327–332, 2018.

Ray Panko. What we don't know about spreadsheet errors today: the facts, why we don't believe them, and what we need to do. In *Proceedings of the EuSpRIG 2015 Conference 'Spreadsheet Risk Management'*, 2015. URL www.eusprig.org.

Nick R. Parsons, M. Dawn Teare, and Alice J. Sitch. Unit of analysis issues in laboratory-based research. *Elife*, 7:e32486, 2018. doi: 10.7554/eLife.32486.

Mehrdad Pasha, Colin Hare, Mojtaba Ghadiri, Alfeno Gunadi, and Patrick M. Piccione. Effect of particle shape on flow in discrete element method simulation of a rotary batch seed coater. *Powder Technology*, 296:29–36, 2016.

Louis Pasteur. Des générations spontanées (7 April 1864). *Oeuvres de Pasteur*, 2:328–346, 1922.

Warren L. Paul and Peter A. Taylor. A comparison of occupant comfort and satisfaction between a green building and a conventional building. *Building and Environment*, 43:1858–1870, 2008.

Pfizer Australia. *Girls 2–18 years*, April 2008.

Svetlana Pinet, Christelle Zielinski, F.-Xavier Alario, and Marieke Longcamp. Typing expertise in a large student population. *Cognitive Research: Principles and Implications*, 7(1):77, 2022.

V. Poovaragavan, Raghvendra Kumar Vidua, Ashok Kumar, P. Ponmani, Atul S. Keche, Niranjan Sahoo, and Puneet K. Agarwal. Estimation of time since death using biochemical markers in synovial fluid. *The American Journal of Forensic Medicine and Pathology*, 44(3):183–187, 2023.

G. S. Prinz and C. D. Murray. On the pullout strength of human nasal hair (vibrissae). *Materialia*, 2023. Available at SSRN 3552446.

Randall Pruim. *NHANES: data from the US National Health and Nutrition Examination Study*, 2015. URL https://CRAN.R-project.org/package=NHANES. R package version 2.1.0.

V. Dimitra Pyrialakou, Christos Gkartzonikas, J. Drew Gatlin, and Konstantina Gkritza. Perceptions of safety on a shared road: driving, cycling, or walking near an autonomous vehicle. *Journal of Safety Research*, 72:249–258, 2020.

Quoc-Dang Quan, Hoang-Dung Tran, and Anh-Dung Chung. The relation of body score (body height/body length) and haplotype E on Phu Quoc Ridgeback dogs (*Canis familiaris*). *Journal of Entomology and Zoology Studies*, 5:388–394, 2017.

R Core Team. *R: A Language and Environment for Statistical Computing*. R Foundation for Statistical Computing, Vienna, Austria, 2018. URL https://www.R-project.org/.

Patricia Raab and Franz X. Bogner. Conceptions of university students on microplastics in Germany. *PLOS ONE*, 16(9):e0257734, 2021.

M. Rafraf, M. Zemestani, and M. Asghari-Jafarabadi. Effectiveness of chamomile tea on glycemic control and serum lipid profile in patients with Type 2 diabetes. *Journal of Endocrinological Investigation*, 38: 163–170, 2015.

Priya Ranganathan and Rakesh Aggarwal. Study designs: Part 1–an overview and classification. *Perspectives in Clinical Research*, 9(4):184, 2018.

David Reilly, David L. Neumann, and Glenda Andrews. Gender differences in self-estimated intelligence: exploring the male hubris, female humility problem. *Frontiers in Psychology*, 13:812483–812483, 2022.

Claire Ridgewell, Neil Sipe, and Nick Buchanan. School travel modes: factors influencing parental choice in four Brisbane schools. *Urban Policy and Research*, 27(1):43–57, 2009.

Joelle E. Romanchik-Cerpovicz, Megan J. A. Jeffords, and Ann C. Onyenwoke. College student acceptance of chocolate bar cookies containing puree of canned green peas as a fat-ingredient substitute. *Journal of Culinary Science & Technology*, 17(6):1–12, 2018. doi: 10.1080/15428052.2018.1492480.

Cristina Romero-Blanco, Julián Rodríguez-Almagro, María Dolores Onieva-Zafra, María Laura Parra-Fernández, María Del Carmen Prado-Laguna, and Antonio Hernández-Martínez. Physical activity and sedentary lifestyle in university students: changes during confinement due to the COVID-19 pandemic. *International Journal of Environmental Research and Public Health*, 17(18):6567, 2020.

Jessica A. Rowland, Natalie J. Briscoe, and Kathrine A. Handasyde. Comparing the thermal suitability of nest-boxes and tree-hollows for the conservation-management of arboreal marsupials. *Biological Conservation*, 209:341–348, 2017.

Patrick Royston and Douglas G. Altman. Regression using fractional polynomials of continuous covariates: parsimonious parametric modelling. *Journal of the Royal Statistical Society, Series C*, 43(3):429–467, 1994.

Priscila Rubbo, Caroline Lievore Helmann, Celso Bilynkievycz dos Santos, and Luiz Alberto Pilatti. Retractions in the engineering field: a study on the web of science database. *Ethics & Behavior*, 29(2):141–155, 2019.

J. Russell, V. Flood, H. Yeatman, and P. Mitchell. Prevalence and risk factors of food insecurity among a cohort of older Australians. *Journal of Nutrition, Health and Aging*, 18(1):3–8, 2014.

Marie C. Russell. A difference in larval mosquito size allows a biocontrol agent to target the invasive species. *Ecology and Evolution*, 13(7):e10294, 2023.

Tracey C. Russell, Catherine A. Herbert, and James L. Kohen. High possum mortality on urban roads: implications for the population viability of the common brushtail and the common ringtail possum. *Australian Journal of Zoology*, 57:391–397, 2009.

Frank M. Sacks, George A. Bray, Vincent J. Carey, Steven R. Smith, Donna H. Ryan, Stephen D. Anton, Katherine McManus, Catherine Champagne, Louise M. Bishop, Nancy Laranjo, Meryl S. Leboff, Jennifer C. Rood, Lilian de Jonge, Frank L. Greenway, Catherine M. Loria, Eva Obarzanek, and Donald A. Williamson. Comparison of weight-loss diets with different compositions of fat, protein, and carbohydrates. *The New England Journal of Medicine*, 360(9):859–873, 2009.

Narongkorn Saiphoklang, Apiwat Pugongchai, and Kanyada Leelasittikul. Comparison between 20 and 30 meters in walkway length affecting the 6-minute walk test in patients with chronic obstructive pulmonary disease: a randomized crossover study. *PLOS ONE*, 17(1):e0262238, 2022.

Julia Sajewicz and Alicja Dziuba-Słonina. Texting on a smartphone while walking affects gait parameters. *International Journal of Environmental Research and Public Health*, 20(5):4590, 2023. doi: 10.3390/ijerph20054590.

Ingvild West Saxvig, Bjørn Bjorvatn, Mari Hysing, Børge Sivertsen, Michael Gradisar, and Ståle Pallesen. Sleep in older adolescents. Results from a large cross-sectional, population-based study. *Journal of Sleep Research*, 30(4):e13263, 2021.

Dmitry Schepaschenko, Anatoly Shvidenko, Vladimir A. Usoltsev, Petro Lakyda, Yunjian Luo, Roman Vasylyshyn, Ivan Lakyda, Yuriy Myklush, Linda See, Ian McCallum, Steffen Fritz, Florian Kraxner, and Michael Obersteiner. A dataset of forest biomass structure for Eurasia. *Scientific Data*, 4:1–11, 2017.

Dmitry Schepaschenko, Anatoly Shvidenko, Vladimir A. Usoltsev, Petro Lakyda, Yunjian Luo, Roman Vasylyshyn, Ivan Lakyda, Yuriy Myklush, Linda See, Ian McCallum, Steffen Fritz, Florian Kraxner, and Michael Obersteiner. Biomass plot data base. *PANGAEA*, 2017. doi: 10.1594/PANGAEA.871465. In supplement to: Schepaschenko, D. et al. (2017): A dataset of forest biomass structure for Eurasia. Scientific Data, 4, 170070, doi:10.1038/sdata.2017.70.

Annina B. Schmid, James M. Elliott, Mark W. Stridwick, Mary Little, and Michel W. Coppeiters. Effect of splinting and exercise on intraneural edema of the median nerve in Carpal Tunnel Syndrome—an MRI study to reveal therapeutic mechanisms. *Journal of Orthopaedic Research*, 30(8):1343–1350, 2012.

J. B. Schorling, J. Roach, M. Siegel, N. Baturka, D. E. Hunt, T. M. Guterbock, and H. L. Stewart. A trial of church-based smoking cessation interventions for rural African Americans. *Preventative Medicine*, 26 (1):92–101, 1997. Data obtained from http://biostat.mc.vanderbilt.edu/DataSets.

Jong Wook Seo, In Ok Lee, Jung Cheol Kim, and Jae Eun Chung. The role of port site local anesthetic injection in laparoendoscopic single site surgery: a prospective randomized study. *Obstetrics & Gynecology Science*, 63(3):387–394, 2020.

Thorakkal Shamim. Development of a guideline to approach plagiarism in Indian scenario. *Indian Journal of Dermatology*, 59(5):473, 2014.

Allen L. Shoemaker. What's normal? — Temperature, gender, and heart rate. *Journal of Statistics Education*, 4(2), 1996. URL http://jse.amstat.org/v4n2/datasets.shoemaker.html.

Blair D. Siegfried, Murugesan Rangasamy, Haichuan Wang, Terence Spencer, Chirakkal V. Haridas, Brigitte Tenhumberg, Douglas V. Sumerford, and Nicholas P. Storer. Estimating the frequency of Cry1F resistance in field populations of the European corn borer (Lepidoptera: Crambidae). *Pest Management Science*, 70 (5):725–733, 2014.

Michael Siegrist. The use or misuse of three-dimensional graphs to represent lower-dimensional data. *Behaviour & Information Technology*, 15(2):96–100, 1996.

Emmanuel João Nogueira Leal Silva, Nancy Kudsi Carvalho, Marina C. Prado, Mayara Zanon, Plinio Mendes Senna, Erick M. Souza, and Gustavo De-Deus. Push-out bond strength of injectable pozzolan-based root canal sealer. *Journal of Endodontics*, 42(11):1656–1659, 2016.

S. G. Silverman, K. Tuncali, D. F. Adams, R. D. Nawfel, K. H. Zou, and P. F. Judy. CT fluoroscopy-guided adbominal interventions: techniques, results, and radiation exposure. *Radiology*, 212:673–681, 1999.

Janet E. Simons and Daniel T. Holmes. Reproducible research and reports with R. *Journal of Applied Laboratory Medicine*, 4(3):471–473, 2019.

Knut R. Skulberg, Knut Skyberg, Kristian Kruse, Wijnand Eduard, Per Djupesland, Finn Levy, and Helge Kjuus. The effect of cleaning on dust and the health of office workers: an intervention study. *Epidemiology*, 15(1):71–78, 2004.

Craig P. Smith, Elliott Fullerton, Liam Walton, Emelia Funnell, Dimitrios Pantazis, and Heinz Lugo. The validity and reliability of wearable devices for the measurement of vertical oscillation for running. *PLOS ONE*, 17(11):e0277810, 2022.

J. M. Snowden and O. Basso. Causal inference in studies of preterm babies: a simulation study. *BJOG: An International Journal of Obstetrics & Gynaecology*, 125(6):686–692, 2018.

Paul R. Solomon, Felicity Adams, Amanda Silver, Jill Zimmer, and Richard De Veaux. ginkgo for memory enhancement. *Journal of the American Medical Association*, 288(7):835–840, 2002.

F. A. Soud, M. M. Cortese, A. T. Curns, P. J. Edelson, R. H. Bitsko, H. T. Jordan, A. S. Huang, J. M. Villalon-Gomez, and G. H. Dayan. Isolation compliance among university students during a mumps outbreak, Kansas 2006. *Epidemiology & Infection*, 137(1):30–37, 2009.

Camila Souza, Bruno Souza, Marcos Fadini, Josélia França, Cícero Menezes, Priscilla Nascimento, and Simone Mendes. What is the potential of sugarcane borer in reducing sorghum fitness and grain production? *Mendeley Data*, 2024a. doi: 10.17632/b6s9wnxgfm.2.

Camilla da Silva Fernandes Souza, Bruno Henrique Sardinha de Souza, Marcos Antônio Matiello Fadini, Josélia Carvalho Oliveira França, Cícero Bezerra de Menezes, Priscilla Tavares Nascimento, and Simone Martins Mendes. What is the potential of sugarcane borer in reducing sorghum fitness and grain production? *Journal of Applied Entomology*, 148(7):818–826, 2024b.

Julia Stafford, Mike Daube, and Peter Franklin. Second hand smoke in alfresco areas. *Health Promotion Journal of Australia*, 21(2):99–105, 2010.

S. Steele. Babies by the dozen for Christmas: 24-hour baby boom. *The Sunday Mail*, page 7, December 21 1997.

Naftali Stern, Assaf Buch, Rebecca Goldsmith, Lesley Nitsan, Miri Margaliot, Ronit Endevelt, Yonit Marcus, Gabi Shefer, and I. Grotto. The role of caloric intake in the association of high salt intake with high blood pressure. *Scientific Reports*, page 15803, 2021.

S. O. Sterndale, D. W. Miller, J. P. Mansfield, J. C. Kim, and J. R. Pluske. Increasing dietary tryptophan and decreasing other large neutral amino acids increases weight gain and feed intake in weaner pigs infected with escherichia coli. *Animal Production Science*, 57(12):2410–2410, 2017.

Simon C. Stirrat. Age structure, mortality and breeding in a population of agile wallabies (*macropus agilis*). *Australian Journal of Zoology*, 56:431–439, 2008.

Roger C. Stone and A. Auliciems. SOI phase relationships with rainfall in eastern Australia. *International Journal of Climatology*, 12:625–636, 1992.

Roger C. Stone, Graeme L. Hammer, and Torben Marcussen. Prediction of global rainfall probabilities using phases of the southern oscillation index. *Nature*, 384:252–255, 1996.

David L. Strayer and William A. Johnston. Driven to distraction: dual-task studies of simulated driving and conversing on a cellular telephone. *Psychological Science*, 12(6):462–466, 2001.

Louise M. Streeting, Deborah S. Bower, Martin L. Dillon, Phil Spark, Michael Gough, Adam Skidmore, Paul G. McDonald, Hannah Delaney, Adrienne Burns, Sandy Watson, et al. Optimising the hatching success of artificially incubated eggs for use in a conservation program for the western saw-shelled turtle (*myuchelys bellii*). *Australian Journal of Zoology*, 70(2):74–82, 2022.

Yu-Sung Su. It's easy to produce chartjunk using Microsoft® Excel 2007 but hard to make good graphs. *Computational Statistics & Data Analysis*, 52(10):4594–4601, 2008.

Jennifer Sutherland, Phil Edwards, Bhavani Shankar, and Alan D. Dangour. Fewer adults add salt at the table after initiation of a national salt campaign in the UK: a repeated cross-sectional analysis. *British Journal of Nutrition*, 2012.

Eva Swinnen, Jean-Pierre Baeyens, Benjamin Van Mulders, Julian Verspecht, and Marc Degelaen. The influence of the use of ankle-foot orthoses on thorax, spine, and pelvis kinematics during walking in children with cerebral palsy. *Prosthetics and Orthotics International*, 42(2):208–213, 2018.

Ira B. Tager, Scott T. Weiss, Bernard Rosner, and Frank E. Speizer. Effect of parental cigarette smoking on the pulmonary function of children. *American Journal of Epidemiology*, 110(1):15–26, 1979.

Dilip Kumar Talukdar. A study of correlation between california bearing ratio (CBR) value with other properties of soil. *International Journal of Emerging Technology and Advanced Engineering*, 4(1):559–562, 2014.

Mark Patrick Taylor, Danielle Camenzuli, Louise Jane Kristensen, Miriam Forbes, and Sammy Zahran. Environmental lead exposure risks associated with children's outdoor playgrounds. *Environmental Pollution*, 178:447–454, 2013.

Eric Teillet, Christine Urbano, Sylvie Cordelle, and Pascal Schlich. Consumer perception and preference of bottled and tap water. *Journal of Sensory Studies*, 25(3):463–480, 2010.

Richard D. Telford and Ross B. Cunningham. Sex, sport, and body-size dependency of hematology in highly trained athletes. *Medicine and Science in Sports and Exercise*, 23(7):788–794, 1991.

Ik Hui Teo, Jesrine Hong, Peng Chiong Tan, and Boon Kiong Lim. Eye masks and earplugs to improve night sleep duration in nulliparas: a randomized trial. *Cureus*, 14(12), 2022.

C. W. Thane, C. J. Bates, and A. Prentice. Zinc and vitamin A intake and status in a national sample of British young people aged 4–18 y. *European Journal of Clinical Nutrition*, 58:363–375, 2004.

The jamovi Project. *jamovi (Version 2.3.28) [Computer Software]*, 2022. URL https://www.jamovi.org.

The Open University. *MDST242 Statistics in Society, Unit A0: Introduction.* The Open University, 1983.

J. E. Tracy, J. D. Oster, and R. J. Beaver. Selenium in the southern coast range of California: well waters, mapped geological units, and related elements. *Jornal of Environmental Quality*, 19(1):46–50, 1990.

Catrine Tudor-Locke, Tiago V. Barreira, and John M. Schuna Jr. Comparison of step outputs for waist and wrist accelerometer attachment sites. *Medicine and Science in Sports and Exercise*, 47(4):839–842, 2015.

Maximilian Ueberham, Uwe Schlink, Martin Dijst, and Ulrike Weiland. Cyclists' multiple environmental urban exposures—comparing subjective and objective measurements. *Sustainability*, 11(5):1412, 2019.

US Department of Transportation. 2009 National Household Travel Survey User's Guide, 2011.

J-B van Helmont. On the necessity of leavens in transformations. In J. L. Conte, editor, *Les oevres de Jean-Baptiste van Helmont.* Lyon, 1671.

Jan P. Vandenbroucke, Erik von Elm, Douglas G. Altman, Peter C. Gøtzsche, Cynthia D. Mulrow, Stuart J. Pocock, Charles Poole, James J. Schlesselman, and Matthias Egger. Strengthening the reporting of observational studies in epidemiology (STROBE): explanation and elaboration. *International Journal of Surgery*, 12(12):1500–1524, 2014.

Paolo Verdecchia, Giuseppe Schillacl, Claudia Borgioni, Antonella Ciucci, Ivano Zampi, Massimo Battistelli, Roberto Gattobigio, Nicola Sacchi, and Carlo Porcellati. Cigarette smoking, ambulatory blood pressure and cardiac hypertrophy in essential hypertension. *Journal of Hypertension*, 13(10):1209–1215, 1995.

Erik Von Elm, Douglas G. Altman, Matthias Egger, Stuart J. Pocock, Peter C. Gøtzsche, and Jan P. Vandenbroucke. The Strengthening the Reporting of Observational Studies in Epidemiology (STROBE) statement: guidelines for reporting observational studies. *The Lancet*, 370(9596):1453–1457, 2007.

Hilary J. Wallace, Mark W. Fear, Margaret M. Crowe, Lisa J. Martin, and Fiona M. Wood. Identification of factors predicting scar outcome after burn in adults: a prospective case-control study. *Burns*, 43: 1271–1283, 2017.

Jeffery P. Walters, Jessica Kaminsky, and Claudio Huepe. Factors influencing household solar adoption in Santiago, Chile. *Journal of Construction Engineering and Management*, 144(6):05018004, 2018.

Liyang Wang, Ruimin Li, Changjun Wang, and Zhiyong Liu. Driver injury severity analysis of crashes in a western China's rural mountainous county: taking crash compatibility difference into consideration. *Journal of Traffic and Transportation Engineering (English edition)*, 2020.

S. Watanabe, M. Ohnishi, K. Imai, E. Kawano, and S. Igarashi. Estimation of the total saliva volume produced per day in five-year-old children. *Archives of Oral Biology*, 40(8):781–782, 1995.

Nathan Wetzel. McDonald's french fries: would you like small or large fries? *STATS*, 43:12–14, 2005.

WHO. WHO child growth standards: Length/height-for-age, weight-for-age, weight-for-length, weight-for-height and body mass index-for-age. Methods and development, 2006.

J. P. Willems, J. T. Saunders, D. E. Hunt, and J. B. Schorling. Prevalence of coronary heart disease risk factors among rural blacks: a community-based study. *Southern Medical Journal*, 90(8):814–820, 1997. Data obtained from https://hbiostat.org/data/.

Brett Williams and Mal Boyle. Estimation of external blood loss by paramedics: is there any point? *Prehospital and Disaster Medicine*, 22(6):502–506, 2007.

Jessica L. Williams, Dan Harley, Darcy Watchorn, Lachlan McBurney, and David B. Lindenmayer. Relationship between body weight and elevation in Leadbeater's possum (*Gymnobelideus leadbeateri*). *Australian Journal of Zoology*, 69(5):167–174, 2022.

E. B. Wilson Jr. *An Introduction to Scientific Research.* McGraw-Hill Book Company, New York, 1952.

Gary W. Witmer and Michael J. Pipas. A preliminary field evaluation of candidate repellents to reduce black bear damage to western larch trees. In *Proceedings of the Vertebrate Pest Conference*, volume 29, 2020.

Laura A. Wojcik, Darryl G. Thelen, Albert B. Schultz, James A. Ashton-Miller, and Neil B. Alexander. Age and gender differences in single-step recovery from a forward fall. *Journal of Gerentology*, 54A(1): M44–M50, 1999.

Gyeong Won Lee, Rock Bum Kim, Se Il Go, Hyun Seop Cho, Seung Jun Lee, David Hui, Eduardo Bruera, and Jung Hun Kang. Gender differences in hiccup patients: analysis of published case reports and case-control studies. *Journal of Pain and Symptom Management*, 51(2):278–283, 2016.

M. Woodward and A. R. P. Walker. Sugar consumption and dental caries: evidence from 90 countries. *British Dental Journal*, 176:297–302, 1994.

Kathleen Woolf, Megan M. St. Thomas, Nicole Hahn, Linda A. Vaughan, Amanda G. Carlson, and Pamela Hinton. Iron status in highly active and sedentary young women. *International Journal of Sport Nutrition and Exercise Metabolism*, 19:519–535, 2009.

Edward Wright, Sven Grawunder, Eric Ndayishimiye, Jordi Galbany, Shannon C. McFarlin, Tara S. Stoinski, and Martha M. Robbins. Chest beats as an honest signal of body size in male mountain gorillas (*Gorilla beringei beringei*). *Scientific Reports*, 11(1):6879, 2021.

Kuan-Sheng Wu, Susan Shin-Jung Lee, Jui-Kuang Chen, Yao-Shen Chen, Hung-Chin Tsai, Yueh-Ju Chen, Yu-Hsiu Huang, and Huey-Shyan Lin. Identifying heterogeneity in the Hawthorne effect on hand hygiene observation: a cohort study of overtly and covertly observed results. *BMC Infectious Diseases*, 18(1):369, 2018.

C. Wunderlich. *Das Verhalten der Eiaenwarme in Krankenheitem.* Otto Wigard, Leipzig, Germany, 1868.

Yihui Xie. *Dynamic Documents with R and knitr.* Chapman and Hall/CRC, Boca Raton, Florida, second edition edition, 2015. URL https://yihui.name/knitr/. ISBN 978-1498716963.

Yihui Xie. *bookdown: authoring Books and Technical Documents with R Markdown*. Chapman and Hall/CRC, Boca Raton, Florida, 2016. URL https://github.com/rstudio/bookdown. ISBN 978-1138700109.

Tianyu Xu, Zonglei Li, Changjiang Lin, and Qiuyue Yu. Design and hydraulic performance of bionic leaf vein-type drip irrigation emitters. *Irrigation and Drainage*, 2023.

Chloe X. Yap, Anjali K. Henders, Gail A. Alvares, David L. A. Wood, Lutz Krause, Gene W. Tyson, Restuadi Restuadi, Leanne Wallace, Tiana McLaren, Narelle K. Hansell, et al. Autism-related dietary preferences mediate autism-gut microbiome associations. *Cell*, 2021.

Robert W. Yeh, Linda R. Valsdottir, Michael W. Yeh, Changyu Shen, Daniel B. Kramer, Jordan B. Strom, Eric A. Secemsky, Joanne L. Healy, Robert M. Domeier, Dhruv S. Kazi, and Brahmajee K. Nallamothu. Parachute use to prevent death and major trauma when jumping from aircraft: randomized controlled trial. *BMJ*, 363:k5094, 2018.

Yuka Yonekura, Masakazu Terauchi, Asuka Hirose, Tamami Odai, Kiyoko Kato, and Naoyuki Miyasaka. Daily coffee and green tea consumption is inversely associated with body mass index, body fat percentage, and cardio-ankle vascular index in middle-aged Japanese women: a cross-sectional study. *Nutrients*, 12(5): 1370, 2020.

Jay L. Zagorsky. Are blondes really dumb? *Economics Bulletin*, 36(1):401–410, 2016.

Mark Ziemann, Yotam Eren, and Assam El-Osta. Gene name errors are widespread in the scientific literature. *Genome Biology*, 17(1):1–3, 2016.

Marketa Zimova, Lindsey S. Barnard, Brandon M. Davis, Alexander V. Kumar, Diana J. R. Lafferty, and L. Scott Mills. Using remote cameras to measure seasonal molts. *Ecosphere*, 11(4):e03084, 2020.

Kelly H. Zou, Kemal Tuncali, and Stuart G. Silverman. Correlation and simple linear regression. *Radiology*, 227:617–628, 2003.

Index

Page numbers in **bold** refer to the Glossary entry.

Kolibri 360 GmbH
Ferdinand-Jühlke-Straße 7
99095 Erfurt, Deutschland
produktsicherheit@kolibri360.de